5G and Beyond

Xingqin Lin • Namyoon Lee
Editors

5G and Beyond

Fundamentals and Standards

Springer

Editors
Xingqin Lin
Ericsson
Santa Clara, CA, USA

Namyoon Lee
Pohang University of Science and Technology (POSTECH)
Pohang, Korea (Republic of)

ISBN 978-3-030-58196-1 ISBN 978-3-030-58197-8 (eBook)
https://doi.org/10.1007/978-3-030-58197-8

This Springer imprint is published by the registered company Springer Nature Switzerland AG
The registered company address is: Gewerbestrasse 11, 6330 Cham, Switzerland

Preface

5G is the next generation of mobile communications systems beyond 4G LTE. While mobile voice telephony and mobile broadband data services remain the primary applications of mobile communications systems, new applications for the Internet of Things (IoT) and the Fourth Industrial Revolution start to help drive the future growth of mobile communications systems. In the 5G era, there is a need for a new generation of mobile communications systems that incorporate more advanced technology solutions to achieve higher data rates, lower latency, greater capacity, and more efficient spectrum utilization.

The next generation wireless access technology New Radio (NR) is the key enabling technology for supporting the diverse usage scenarios and applications envisioned for the 5G era. The heart of 5G NR is a set of fundamental technologies, which make 5G capable of providing much more efficient networks and enable new services, new ecosystems, and new revenues. This book provides a comprehensive treatment of the 5G mobile communications systems. It covers both the fundamentals and the state-of-the-art of the 5G NR standards.

Chapter 1 provides an overview of the evolution of mobile communications systems from 1G to 4G, what 5G is, and a brief look into 6G. This chapter presents an executive summary of the key ingredient technology components that enable the advanced 5G capabilities. This chapter also describes the 5G standardization process and the key organizations that are essential for defining 5G.

The remaining chapters of this book are organized into three parts. Part I consists of six tutorial chapters that describe the fundamental technology components for 5G and beyond. Chapter 2 presents an accessible treatment of advanced channel coding theory with a focus on rate-compatible polar codes. Chapter 3 provides a comprehensive overview of the state-of-the-art multiple access techniques. Chapter 4 gives a tutorial on massive multiple-input multiple-output (MIMO) technology. Chapter 5 focuses on network densification and introduces theoretical models for densified network analysis and design. Chapter 6 discusses the integration of unmanned aerial vehicles in cellular communication networks. Chapter 7 presents a comprehensive forward-looking vision that defines the main principles that will guide the design and development of a 6G system.

Part II of this book, consisting of Chaps. 8, 9, 10, 11, and 12, introduces the basics of 5G NR standards. Chapter 8 offers a guide to the new generation radio access network (NG-RAN) architecture. Chapter 9 outlines the NR physical layer design. Chapter 10 provides an accessible description of NR channel coding design aspects including Polar codes and low-density parity-check (LDPC) codes. Chapter 11 describes cell search and random access procedures in NR. Chapter 12 presents a primer on the NR bandwidth part concept.

Part III, consisting of Chaps. 13, 14, 15, 16, 17, and 18, describes the key 5G NR evolution directions. Chapter 13 provides an overview of NR ultra-reliable low-latency communication (URLLC). Chapter 14 introduces NR operation in unlicensed spectrum. Chapter 15 gives a tutorial on NR positioning. Chapter 16 offers a primer on NR-integrated access and backhaul. Chapter 17 describes NR-based air-to-ground communications. Chapter 18 discusses how to adapt the NR air interface for non-terrestrial networks with a focus on satellite communications.

With 5G systems being switched on, 5G will bring enormous socio-economic benefits. The technologies continue to be evolved to further expand the 5G ecosystem and transform vertical industries. Our hope is that this book offers some assistance to the interested readers who are making 5G and beyond a reality.

Santa Clara, USA Xingqin Lin

Pohang, South Korea Namyoon Lee

Contents

Chapter 1
Introduction to 5G and Beyond

Xingqin Lin and Namyoon Lee

1.1 Book Objective

Over the last 40 years, mobile communications services have witnessed explosive growth. The first commercial cellular telephone system in the United States, known as advanced mobile telephony system (AMPS), was placed into operation in late 1983. Today, mobile communications services are ubiquitous and used by a large percentage of the world's population. According to Ericsson Mobility Report [1], as of Q3 2019 we have 5.9 billion subscribers and 8 billion subscriptions in the world. The total number of mobile subscriptions now exceeds the world's population.

Appreciating the evolution of mobile communications systems is important for understanding how modern mobile communications systems are now able to deliver the services at the remarkable scale. A new generation of mobile communications standards has appeared about every tenth year, with the first generation (1G) being introduced in the 1980s. From 1G to the third generation (3G), mobile voice telephony was the main application driving the growth of mobile communications systems. Since 3G, mobile broadband data applications have become the main force for further evolution of mobile communications systems. Today, the *Long-Term Evolution* (LTE), representing the fourth generation (4G) of mobile communications systems, has been widely deployed to deliver mobile broadband data services. While the mobile voice telephony and mobile broadband data services remain the primary applications of mobile communications systems, new applications for the Internet of Things (IoT) and the Fourth Industrial Revolution start to help further drive the

X. Lin (✉)
Ericsson, Santa Clara, CA, USA
e-mail: xingqin.lin@ericsson.com

N. Lee
Electrical Engineering, Pohang University of Science and Technology (POSTECH), Pohang, Gyeongbuk, Korea
e-mail: nylee@postech.ac.kr

X. Lin, N. Lee (eds.), *5G and Beyond*, https://doi.org/10.1007/978-3-030-58197-8_1

future growth of mobile communications systems, including the fifth-generation (5G) wireless access systems.

The transition from 4G to 5G will support more diverse usage scenarios and applications. The International Telecommunication Union's Radiocommunication Sector (ITU-R) has defined three areas of usage and applications in the 5G era: enhanced mobile broadband (eMBB), ultra-reliable low-latency communications (URLLC), and massive machine type communications (mMTC) [2]. While the evolution of LTE will have the capability to support a wide range of usage scenarios and applications in the 5G era, there is a need for a new generation of mobile communications systems that incorporate more advanced technology solutions to achieve higher data rates, lower latency, greater capacity, and more efficient spectrum utilization [3]. Equipped with these more advanced capabilities, the next-generation wireless access technology *New Radio* (NR), the main subject of this book, is the key enabling technology for supporting the diverse usage scenarios and applications envisioned for the 5G era.

5G will bring enormous socioeconomic benefits. The heart of 5G NR is a set of fundamental technologies, making 5G capable of providing much more efficient networks, enabling new services, new ecosystems, and new revenues. The technologies continue to be evolved to further expand the 5G ecosystem and transform vertical industries. The main objective of this book is to provide a comprehensive treatment of the 5G mobile communications systems. This book will cover both the fundamentals and the state-of-the-art of the 5G NR standards. Now 5G systems are being switched on. A substantial portion of the book will be devoted to the next-generation wireless access system (6G) to discuss the technologies that will come next.

The next sections give more sense of the intended scope of this book. We first provide an overview of the evolution of mobile communications systems from 1G to 4G. We then elaborate more on what 5G is through discussing the main technical requirements behind 5G, the main 5G use cases, and the key ingredient technology components that enable the advanced 5G capabilities. Next, we describe the 5G standardization process and key organizations that are essential for defining what 5G is. The following section provides a brief look into 6G – the next-generation wireless access system. We close the chapter with outlining the contents of the remaining chapters of this book.

1.2 Evolution of Mobile Communications Systems Before 5G

Mobile communication systems have significantly changed our lives. They eliminate the limitations in time and space to transfer information from one place to another. This allows people to exchange or access information whenever, whatever, and wherever they want. Mobile communication systems now have become an essential part of our lives.

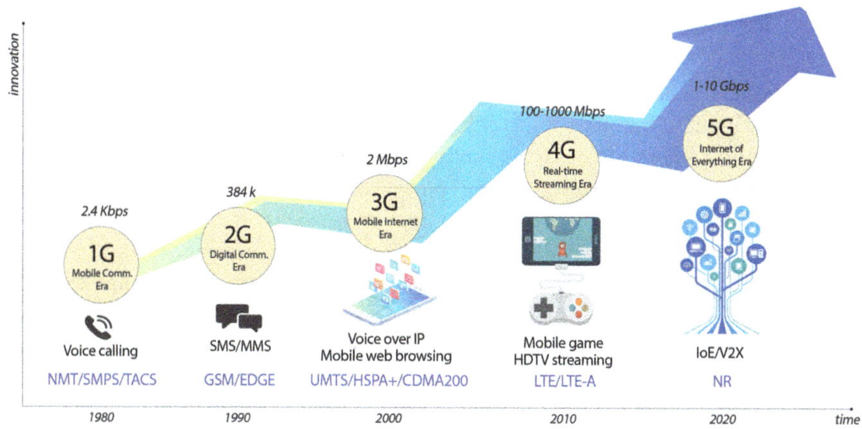

Fig. 1.1 Evolution of mobile communication systems

Mobile communication systems have evolved over several decades. To fully understand where we are in 5G, we need to retrospect what we have developed from the revolutionary 1G to 4G and reflect what has fundamentally changed from one generation to the next (Fig. 1.1).

1.2.1 1G: Analog Mobile Communication Era

1G opened a new mobile communication era. Historically, the initial idea of a cellular system was proposed by Bell Labs in 1947. This idea was commercialized by Nippon Telegraph and Telephone (NTT) in 1979. In early 1980s, the Motorola also released one of the first mobile phones called "DynaTAC" to provide the 1G mobile communication service. Different countries developed their own 1G standards, including Nordic Mobile Telephone (NMT) used in Eastern Europe and Russia, AMPS used in North America, TACS (Total Access Communications System) in the United Kingdom, and TZ-801/2/3 used in Japan.

1G adopted analog communication technologies using about 150 MHz carrier frequencies, and it only offered voice calls services. Although it paved a new way to eliminate the constraint in space for person-to-person communications, 1G had several shortcomings. The communication quality and security level were very low because of the inherent limitation of analog communication technologies. In addition, there was no compatibility across different mobile communication systems because each country developed its own system.

1.2.2 2G: Digital Mobile Communication Era

2G commenced a *digital* mobile communication era. The primary change from 1G was to use digital communication technologies instead of the analog ones. Digital communication technologies significantly improved the quality of communications and the security level, thanks to the power of digital systems. 2G was more than just person-to-person voice communication services. It supported transfer of digitally encrypted messages, including not only voice messages but also send text messages (SMS), picture messages, and multimedia messages (MMS). The numerous message types enriched business opportunities; thereby, the popularity of mobile communication systems exploded in this era.

The first 2G standard was the Global System for Mobile Communications (GSM), which was initially launched in 1991. The GSM supported the data rate of 9.6 kbit/s, and it evolved to General Packet Radio Service (GPRS) and Enhanced Data Rates for GSM Evolution (EDGE), which provided the maximum data rates of 40 and 384 kbit/s, respectively. GSM-based 2G standards adopted time division multiple access (TDMA).

1.2.3 3G: Mobile Internet Era

3G opened the *mobile Internet era*. A big innovation in 3G was the use of *data packet*-based communication technologies. The packet-switched communication technologies offered connectivity to the world wide web, i.e., the mobile Internet, from any location in the world using mobile communication systems. The maximum data rate of 3G was about 2 Mbps, which is approximately four times faster than 2G. This data rate enhancement made possible new services, including voice over IP (e.g., Skype), fast web browsing, and video streaming/conferencing. In this era, an innovative mobile device called *smartphone* was launched, such as the Blackberry in 2002 and the iPhone in 2007. This device expanded the service capabilities of the mobile communication systems.

3G was designed to meet the requirements of International Mobile Telecommunications 2000 (IMT-2000) standards. The first 3G service was launched by NTT DoCoMo in 2001 using the Universal Mobile Telecommunications System (UMTS) system standardized by the 3rd Generation Partnership Project (3GPP). This system evolved to HSPA+, which can yield the uplink and downlink peak data rates up to 28 and 56 Mbit/s, respectively. The Code Division Multiple Access (CDMA) 2000 system was the commercially successful 3G standard in North America and South Korea, standardized by 3GPP2. The follow-up standard, EVDO Rev B standard, enhanced the peak downlink data rates up to 14.7 Mbit/s.

1.2.4 4G: Real-Time Streaming Era

4G started off the *real-time streaming era*. The primary change from 3G to 4G was to provide extremely enhanced data rates up to hundreds of megabit per second. This data rate enhancement stimulated new mobile services, including real-time mobile gaming and high-definition mobile TV. The core technology in improving the data rates was a new multiple access technique, referred to as orthogonal frequency-division multiple access (OFDMA). In addition, multiple-input multiple-output (MIMO) communications technologies further improved the data rates up to an order of magnitude. These two core technologies fundamentally changed the design principles of cellular systems to overcome the channel fading and interference obstacles in wireless environments.

The commercial 4G LTE service was initially offered in Sweden 2009 and spread later in most countries of the world. For example, South Korea and the United Kingdom launched the LTE service in 2011 and 2012. The maximum downlink and uplink data rates of LTE systems are 100 and 50 Mbit/s when using a 20 MHz channel, respectively. This LTE system was enhanced by the follow-up standard, LTE Advanced (LTE-A). LTE-A was standardized in 2010 as part of Release 10 of the 3GPP specification. It used more spectrums and antennas to increase data rates further. LTE-A offered the data rates up to 1000 and 500 Mbit/s in downlink and uplink, respectively. Coordinated multipoint transmission and carrier aggregation technologies were the key ingredients to boost the system capacity.

1.3 What is 5G?

Previous generations of mobile communications systems (1G to 4G) predominantly addressed consumer demands for mobile voice telephony and mobile broadband data services. 5G is the next generation of mobile communications systems [4]. It builds on the successes of the previous generations of mobile communications systems. 5G promises to deliver improved end-user experience and enable new services, new ecosystems, and new revenues.

1.3.1 5G Use Cases

5G is expected to deliver much higher data rates, lower latency, greater capacity, and more efficient spectrum utilization. Equipped with these more advanced capabilities, 5G can support diverse usage scenarios and applications. ITU-R has defined three categories of potential use cases for 5G networks [2], eMBB, URLLC, and mMTC, which are illustrated in Fig. 1.2.

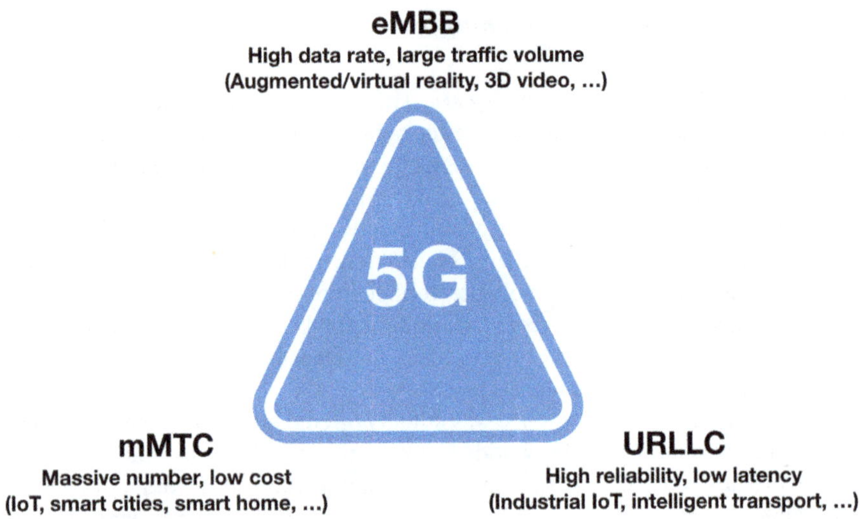

Fig. 1.2 5G usage scenarios

eMBB is a natural evolution of mobile broadband data services that remain the primary applications of mobile communications systems. It is characterized by high data rates and large traffic volumes. Example eMBB services include augmented and virtual reality, high-definition three-dimensional (3D) videos, and 4K streaming. 5G will deliver improved consumer experience by supporting high-speed mobile connectivity for eMBB applications. eMBB may also include applications in enterprise collaboration services.

URLLC corresponds to services require ultra-high reliability and low latency. One exemplary application is wireless control of industrial manufacturing and production processes. The high reliability and low latency characteristics will also be key for traffic safety and control in intelligent transport systems.

mMTC mainly refers to the IoT services that are characterized by a massive number of devices. Example IoT applications include smart cities, smart home, utilities, and remote monitoring. Key requirements of such services include low device complexity, long battery life time, and significant coverage extension, in addition to the support of a massive number of devices. Such services usually have relaxed latency requirements, and each device typically does not require support of high data rates.

There exist more scenarios and applications in the 5G era that may not strictly fall in the three defined categories of eMBB, URLLC, and mMTC. For example, some operators are using 5G as a last-mile access technology to homes. 5G fixed wireless access can help replace the costly deployment of fiber connections to homes. Similarly, mobile voice telephony is still an important application in 5G mobile communications systems, but it requires neither high data rate, nor ultra-high reliability, nor low latency.

1.3.2 5G Technical Requirements

In early 2012, ITU-R started a program, known as IMT-2020, to develop International Mobile Telecommunications for 2020 and beyond [2]. IMT-2020 defines technical performance requirements for 5G, which are summarized in Table 1.1.

Compared to 4G, 5G will increase data rates by 10 times. Specifically, 5G is expected to offer 20 Gbps peak data rate in the downlink (DL) and 10 Gbps peak data rate in the uplink (UL). In dense urban environment, 5G can support user experienced data rate (at the five percentile) of 100 Mbps in the DL and of 50 Mbps in the UL. With the dramatically increased data rates, 5G will deliver much enhanced mobile broadband experience.

5G spectral efficiency is expected to be much increased, with peak spectral efficiency reaching 30 bps/Hz in the DL and 15 bps/Hz in the UL. Different user spectral efficiencies can be supported in a variety of environments for eMBB. For example, in rural areas, 5G can support 0.12 and 0.045 bps/Hz user spectral efficiencies in the DL and UL, respectively, while in indoor hotspot, 0.3 and 0.21 bps/Hz user spectral efficiencies can be reached in the DL and UL, respectively. Accordingly, the DL area traffic capacity in indoor hotspot can reach 10 Mbps/m^2. Overall, 5G spectral efficiency is improved by 3 times compared to 4G.

5G is expected to be able to provide 1 ms user plane latency, which reduces the 4G user plane latency by 10 times. The much improved latency performance can meet the stringent latency requirements of industrial IoT and autonomous transport. In addition, 5G promises much improved reliability performance. In urban macro environment, it is expected that the success probability of transmitting a layer 2 protocol data unit of 32 bytes within 1 ms in channel quality of coverage edge can reach 99.999%.

5G is expected to significantly improve mobility performance. It will provide 0-ms mobility interruption time. 5G can support satisfactory quality of service for mobility speed as high as 500 km/h.

5G is expected to support a connection density of 1 million devices per square kilometer. This greater capacity of 5G is well suited for meeting the requirements of mMTC scenarios.

Energy consumption is also a key consideration. 5G is expected to have high network energy efficiency with support of a high sleep ratio and long sleep duration.

1.3.3 5G Technology Components

While previous generations of mobile communications systems (1G to 4G) were very much radio focused, the entire system of 5G will be transformed to achieve much more efficient networks and enable new services, new ecosystems, and new revenues.

Table 1.1 Summary of 5G key technical requirements

Metric	5G target	Usage scenario
Peak data rate	DL: 20 Gbps; UL: 10 Gbps	eMBB
Five percentile user experienced data rate	DL: 100 Mbps; UL: 50 Mbps	eMBB: dense urban
Peak spectral efficiency	DL: 30 bps/Hz; UL: 15 bps/Hz	eMBB
Five percentile user spectral efficiency	DL: 0.3 bps/Hz; UL: 0.21 bps/Hz	eMBB: indoor hotspot
	DL: 0.225 bps/Hz; UL: 0.15 bps/Hz	eMBB: dense urban
	DL: 0.12 bps/Hz; UL: 0.045 bps/Hz	eMBB: rural
Average spectral efficiency per transmission reception point	DL: 9 bps/Hz; UL: 6.75 bps/Hz	eMBB: indoor hotspot
	DL: 7.8 bps/Hz; UL: 5.4 bps/Hz	eMBB: dense urban
	DL: 3.3 bps/Hz; UL: 1.6 bps/Hz	eMBB: rural
Area traffic capacity	DL: 10 Mbps/m^2	eMBB: indoor hotspot
User plane latency	4 ms	eMBB
	1 ms	URLLC
Control plane latency	20 ms	eMBB/URLLC
Connection density	1,000,000 devices per km^2	mMTC
Success probability	99.999%	URLLC: urban macro
Normalized channel link data rate	UL: 1.5 bps/Hz for mobility speed up to 10 km/h	eMBB: indoor hotspot
	UL: 1.12 bps/Hz for mobility speed up to 30 km/h	eMBB: dense urban
	UL: 0.8 bps/Hz for mobility speed up to 120 km/h	eMBB: rural
	UL: 0.45 bps/Hz for mobility speed up to 500 km/h	eMBB: rural
Mobility interruption time	0 ms	eMBB/URLLC

Radio Access Network

3GPP, a global standard-development organization for mobile communications, has developed a new wireless access technology known as NR. NR is the foundation for 5G radio access networks [5]. Key NR features are summarized as follows:

- Spectrum flexibility: NR supports operation in the spectrum ranging from sub-1 GHz to millimeter wave bands. 5G NR is the first mobile communications

standard that supports operation in millimeter wave bands. Utilizing the large chunks of millimeter wave spectrum enables 5G to deliver multiple gigabit-per-second data rates and also can help mitigate the current spectrum crunch in sub-6 GHz. As in the previous generations of mobile communications systems, operation in sub-6 GHz frequency bands is vital in 5G to provide wide-area coverage. With interworking between high (millimeter wave) and low (sub-6 GHz) frequency bands, 5G NR can enjoy the speed and capacity increases from using high bands and focus on using low bands more for coverage purpose.
- Flexible duplex options: NR supports flexible duplex options including frequency division duplex (FDD), time division duplex (TDD) with semi-statically configured UL/DL configuration, and dynamic TDD. While FDD is often adopted in low-frequency bands where spectra allocations are often paired, TDD becomes increasingly common for higher-frequency bands with unpaired spectra. For large over-the-rooftop cells, semi-static TDD is suitable for handling inter-cell interference issues. Compared to 4G LTE that only supports FDD and semi-static TDD, 5G NR additionally introduces the support of dynamic TDD. In TDD spectrum, for small/isolated cells, dynamic TDD offers the possibility of dynamically allocating radio resources to adapt to UL/DL traffic variations.
- Ultra-lean design: NR embraces ultra-lean design that minimizes always-on transmissions. Take the design of NR reference signals as an example. The NR reference signals are on-demand when possible, and their time and frequency distributions are configurable so that requirements can be met with minimal overhead. In LTE, multiple functions are tied to the always-on cell-specific reference signals (CRS). In contrast, NR reference signal transmission can be extremely sparse at low load. The ultra-lean design leads to higher network energy efficiency and lower interference in 5G NR.
- Forward compatibility: The design of NR encompasses a high degree of forward compatibility that helps to facilitate the introduction of new technologies and applications. First, NR can configure reserved radio resources that are not available for transmission. Through reservation, resources are left blank and thus can be used for future extensions. Second, physical signals and channels are confined in configurable or scheduled radio resources. This yields flexibility for the future while being backward compatible. Third, NR minimizes always-on transmissions. It can be recognized that these forward compatibility designs align with the ultra-lean design principle. The high flexibility of the design and the on-demand principle result in a high degree of forward compatibility.
- Low-latency support: Latency optimization has been an important consideration in NR. Many tools have been introduced in NR to reduce latency. As an example, NR supports "mini-slot" transmission that can start at any OFDM symbol and last only as many symbols as needed for the communication. NR also supports front-loaded reference signals and control signaling that are located at the beginning of the transmission. This can help reduce decoding delay as a device can start to process the received data without buffering. Besides physical layer, certain optimization has also been introduced in higher layer protocols to support low latency.

- Advanced antenna technologies: NR significantly enhances the support of using a large number of antenna elements for both transmission and reception to facilitate beamforming in millimeter wave bands and deploying massive MIMO systems in sub-6 GHz frequency bands. An NR device can support spatial multiplexing of up to eight MIMO layers in the downlink and of up to four MIMO layers in the uplink. Multi-user MIMO capability is much enhanced with the introduction of twelve orthogonal demodulation reference signals (DMRS). NR supports analog beamforming, digital beamforming, or a hybrid combination of both, by carefully designing the physical channels, signals, and procedures. Beam management procedures including beam selection and beam-failure recovery are introduced in NR to support beamforming operations in millimeter wave bands. To facilitate devices to handle the increased phase noise power in millimeter wave bands, transmission of phase tracking reference signals (PTRS) is supported in NR.
- Coexistence with LTE: NR supports functionality to well coexist with LTE. For initial NR deployment, there is an option, known as non-standalone (NSA), that allows NR to focus on user plane functionality by utilizing existing LTE network for control plane functions. NR supports the possibility to have an NR carrier and an LTE carrier overlapping with each other in frequency, thereby enabling dynamic sharing of spectrum between NR and LTE. This facilitates a smooth migration to NR from LTE. Solutions specified to allow this type of operation are the ability for NR physical downlink shared channels to map around LTE CRS, and the possibility of flexible placements of downlink control channels, initial access related reference signals and data channels to minimize collisions with LTE reference signals. NR also supports a so-called supplementary uplink (SUL), which can be used as a low-band complement to the cell's uplink when operating in high frequency bands. A supplementary downlink (SDL), that can be used, for example, in downlink only spectrum, is also supported in NR.

Core Network

In addition to new radio access network, 3GPP has also developed a new 5G core network (5GC) in order to flexibly support a wide range of services with varied performance requirements in the 5G era. 5GC is responsible for functions such as end-to-end connection setup, mobility management, authentication, and charging. 5GC has a service-based architecture, focusing on the services rather than nodes [6]. 5GC features end-to-end flexibility by separating the software functions from the core network hardware. This network softwarization is achieved through software-defined networking (SDN), network functional virtualization (NFV), network slicing, and cloud-based radio access networks (C-RAN).

- SDN separates network control functions from network forwarding functions. Such separation enables network control to become directly programmable and abstracts the physical networking resources such as routers, switches, and gateways. Configuration and management of the physical networking resources can be moved to central data centers. Separating user plane functions from control plane functions is a distinct feature of 5GC. It allows independent scaling, evolution, and flexible deployments of the control plane and user plane, which facilitates the adoption of SDN.

- NFV virtualizes the entire networking functions including network forwarding from the hardware on which it runs. A virtualized network function can run on commercial off-the-shelf hardware, instead of having a custom hardware. With on-demand instantiation of network functions, NFV can facilitate load balancing, upgrades, and scaling and help to reduce the cost of network changes. 5GC has been designed to comprise virtualized, software-based network functions, which enables deployments to use NFV.
- Network slicing allows a shared physical network to be split into multiple virtual networks. A network slice is a dedicated virtual network that has self-contained functionality to support a certain type of service or customer. Network slicing is a key ingredient of 5G to serve the wide range of services with varied performance requirements. The allocated resources to a network slice depend on the service needs. For example, a network slice for eMBB needs to meet the high requirements for bandwidth, while a network slice for URLLC needs to meet the high reliability and low latency requirements. The modularized function design in 5GC can enable flexible and efficient network slicing.
- C-RAN is centralized, cloud computing-based architecture that features centralized processing units and virtualization techniques. Cloudification can facilitate network slice management to enable better support of the wide range of 5G services. Through enabling 5G deployments to use SDN and NFV, 5GC naturally supports cloudification and allows a virtualized network function to be instantiated on a cloud infrastructure.

Backhaul and Fronthaul

Backhaul connects the radio access network to the core network. Compared to the previous generations of mobile communications systems, 5G needs to deliver much higher data rates, lower latency, and greater capacity. Accordingly, 5G backhaul needs to be capable enough to accommodate the 5G technical requirements and should not become the bottleneck of the 5G systems. There are two types of backhaul: wired and wireless. Fiber is a prominent example of wired backhaul. Fiber optics can provide high capacity and high reliability for 5G backhaul. Though the fiber backhaul is often considered as a default option by many operators, its cost may be a concern in some scenarios, such as building the fiber backhaul in suburban and rural areas.

Wireless backhaul is an attractive viable alternative for 5G networks, especially in the scenarios where laying wired backhaul is too costly. Microwave backhaul may use a wide range of frequencies from about 6 to 86 GHz (and even higher frequencies that are under investigation). The range of frequencies enables the microwave backhaul to be used in diverse scenarios from rural areas to dense urban environments. Wireless backhaul may go beyond terrestrial. High-altitude platform systems (HAPS) and satellite technology may also play a role in 5G backhaul. They can complement terrestrial backhaul by offering backhauling to the areas that are difficult to be reached by terrestrial backhaul.

Fronthaul connects the remote radio units (RRU) of a base station to the centralized radio controllers. The logical architecture of 5G NodeB (gNB) is split into two parts called CU (central unit) and DU (distributed unit). The CU and DU are connected by a new interface called F1. The high data rates, low latency, and large capacity requirements of 5G require a fronthaul network to be capable of meeting the stringent demands. The more the centralized functions, the higher the requirements of fronthaul latency and bandwidth. The conventional common public radio interface (CPRI)-based fronthaul cannot meet the 5G technical requirements. The evolved CPRI (eCPRI) will improve the 5G fronthaul capabilities with lower latency and increased bandwidth efficiency and capacity. With eCPRI, it becomes possible to move the beamforming processing from the baseband to the radio, which can simplify the deployment of massive MIMO in 5G.

1.3.4 5G Spectrum

5G NR is designed to operate in a wide range of frequencies. Each spectrum band has its unique characteristics. NR can use new frequency bands defined for 5G, as well as the frequency bands refarmed from the spectrum used by the previous generations of mobile communications systems. To maximize the value of spectrum assets, a service provider should balance and combine the use of low-band, mid-band, and high-band spectrum to achieve quality performance.

- *Low-band* spectrum is below 1 GHz. Due to the desirable propagation properties, low-band spectrum is good for providing wide-area and deep indoor coverage. The channel bandwidths in low-band spectrum are however not wide (e.g., 20 MHz).
- *Mid-band* spectrum is in the range of 1–7 GHz. In mid-band spectrum, channel bandwidths of 50 to 100 MHz are possible. The wide bandwidths can enable networks of large capacity, high data rate, and low latency. Compared to high-band spectrum, mid-band spectrum has better wide-area and indoor coverage properties. Hence, the mid-band spectrum provides good compromise between coverage, capacity, data rate, and latency.
- *High-band* spectrum is in the millimeter wave frequencies above 24 GHz. In high-band spectrum, channel bandwidths up to 400 MHz are possible. The wide-spectrum bandwidths can provide very high data rate and ultra-large capacity. High bands are ideal for localized dense deployments to enable high-throughput and low-latency services. It is however difficult to provide wide-area coverage by using high-band spectrum alone due to the propagation characteristics in millimeter wave frequencies.

5G networks require significantly more spectrum resources to meet the demanding technical requirements. ITU-R identifies and coordinates IMT frequency bands for spectrum harmonization. A new set of frequency bands have been identified, such as the 600 MHz band and the 700 MHz band in the low-band spectrum, the 3.5 GHz band in the mid-band spectrum. ITU-R strives to define a minimum set of bands to facilitate global roaming for devices and economy of scale for equipment.

3GPP continuously defines new operating bands for NR, including both paired bands for FDD and unpaired bands for TDD. Note that there are also unpaired SDL and SUL bands, which are intended to be used together with other bands. In Release 15, frequency bands for NR are divided into two frequency ranges (FR):

- FR1: 410 MHz–7.125 GHz [7].
- FR2: 24.25 GHz–52.6 GHz, commonly referred to as millimeter wave [8].

NR operating bands are defined with prefix of "n" and are numbered from n1 to n512. The range of n1 to n256 is used for the NR bands in FR1, and the range of n257 to n512 is used for the NR bands in FR2. For example, band n71 is a paired band in the 600 MHz frequency for FDD with uplink frequency range of 663–698 MHz and downlink frequency range of 617–652 MHz; band n261 is an unpaired band in the 28 GHz frequency for TDD with uplink and downlink frequency ranges of 27.50–28.35 GHz. If an LTE band is refarmed as an NR band, they share the same band number. The full list of NR frequency bands can be found in 3GPP TS 38.101 [7, 8].

1.4 5G Standardization

Mobile communications standardization is based on consensus of all relevant parties. The history of mobile communications systems has proven that global standards are fundamental to the success of mobile technologies. A globally standardized mobile technology enables global roaming and ensures compatibility, worldwide interoperability, and quality, making the technology more affordable due to economies of scale. The standardization processes are however quite complex, involving standards developing organizations, global and regional regulatory bodies, national administrations, and industry forums.

When it comes to 5G standardization, ITU and 3GPP are the two essential organizations that define 5G. ITU is a specialized agency of the United Nations for information and communication technologies. The main roles of ITU in 5G include spectrum regulation on a global level and setting 5G requirements. 3GPP is a global standard-development organization for mobile communications. It has seven regional telecommunication associations from Asia, Europe, and North America as primary members.

The following two sections describe the main 5G standardization activities in ITU and 3GPP, respectively. Figure 1.3 gives a high level overview of the 5G standardization timeline.

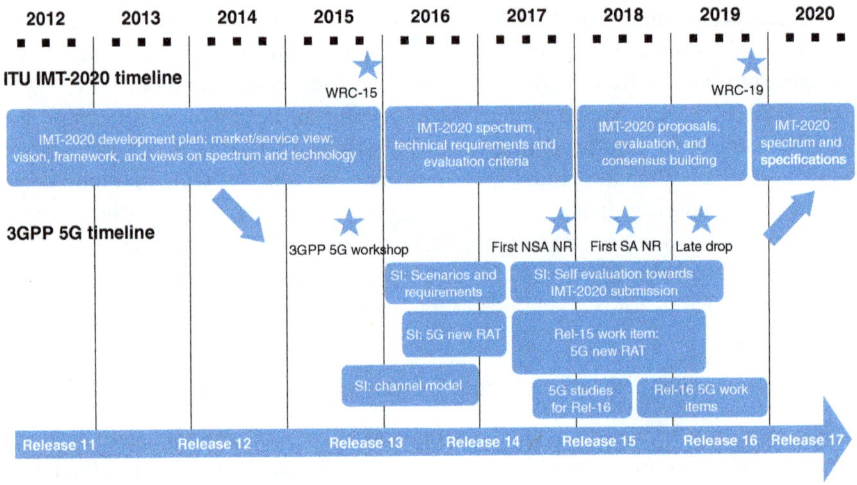

Fig. 1.3 5G standardization timeline

1.4.1 ITU 5G Activities

ITU comprises three sectors: radio communication sector (ITU-R), telecommunication standardization sector (ITU-T), and telecommunication development sector (ITU-D). ITU-R manages radio spectrum and satellite orbits. The mission of ITU-R is to ensure rational, equitable, efficient, and economical use of the radio spectrum by all radio communication services. A key activity of ITU-R is organizing the world radiocommunication conferences (WRC), which are held every 3 to 4 years. At WRC, the *Radio Regulations*, an international treaty that is binding to ITU member states, are reviewed and revised.

ITU-R approves *Recommendations* developed by ITU-R study groups. There are currently six study groups in ITU-R:

- Study group 1: Spectrum management.
- Study group 3: Radiowave propagation.
- Study group 4: Satellite services.
- Study group 5: Terrestrial services.
- Study group 6: Broadcasting services.
- Study group 7: Science services.

Within each study group, subgroups including working parties (WPs) and task groups (TGs) are established to carry out studies. Within the ITU-R study group 5, the working party 5D (WP 5D) is responsible for the overall radio system aspects of IMT systems. The term IMT is the root name that currently encompasses IMT-2000, IMT-Advanced, and IMT-2020, which correspond to 3G, 4G, and 5G mobile communications systems, respectively. WP 5D does not create technical specifications but maintains a set of radio interface specifications (RSPC). For each

IMT generation, there is a set of radio interface technologies (RITs). The RSPC contains overview of each RIT and references to the detailed specifications that are developed and maintained by the corresponding standards developing organizations.

In early 2012, ITU-R WP 5D started a program known as "IMT for 2020 and beyond." This set the stage for 5G research activities worldwide. The Recommendation ITU-R M.2083 [2] depicts the IMT vision on the framework and overall objectives of the future development of IMT for 2020 and beyond. The recommendation examines user and application trends, growth in IMT traffic (detailed in the Report ITU-R M.2370 [9]), technology trends (detailed in the Report ITU-R M.2320 [10]), technical feasibility of IMT between 6 and 100 GHz (detailed in the Report ITU-R M.2376 [11]), and spectrum implications. The recommendation emphasizes that IMT should continue to contribute to connecting the world, new market of information and communications technology, bridging the digital divide, new ways of communication, new forms of education, energy efficiency, social changes, and new art and culture. The recommendation describes diverse usage scenarios including eMBB, URLLC, and mMTC, envisaged for IMT for 2020 and beyond. To support the intended usage scenarios, the recommendation envisions a broad variety of capabilities of IMT-2020.

At WRC 2015 (WRC-15), spectrum for IMT was discussed. WRC-15 decided to make the 700 MHz band (694–790 MHz), a globally harmonized band for providing enhanced mobile broadband capacity. Frequency bands in the L-band (1427–1518 MHz) and in the lower part of the C-band (3.4–3.6 GHz) were also identified. Finding additional spectrum for IMT in the bands above 6 GHz is necessary due to the spectrum crunch in the sub-6 GHz frequency range. Accordingly, WRC-15 decided to have an agenda item (1.13) for the next WRC in 2019 to identify frequency bands in the frequency range between 24.25 and 86 GHz.

After WRC-15, WP 5D continued with spectrum arrangements, setting technical performance requirements, and defining evaluation criteria for IMT-2020. WP 5D published three reports in 2017:

- The Report ITU-R M.2410 [12] describes the minimum technical performance requirements of IMT-2020 candidate radio interface technologies.
- The Report ITU-R M.2411 [13] details the service, spectrum, and technical performance requirements for the IMT-2020 candidate radio interface technologies, as well as the evaluation criteria and submission templates for developing Recommendations and Reports on IMT-2020.
- The Report ITU-R M.2412 [14] elaborates the procedure, the methodology, and the criteria for evaluating the IMT-2020 candidate radio interface technologies.

WP 5D held a workshop on "IMT-2020 Terrestrial Radio Interfaces" in October 2017. The workshop provided information on the process for IMT-2020 standardization and presented potential IMT-2020 technology proponents. Independent evaluation groups also made presentations about their planned actions.

In 2018–2020, IMT-2020 technology proponents submitted their proposals and independent external evaluation groups carried out the evaluation.

WRC 2019 (WRC-19) took a big step in making high frequency millimeter wave spectrum available for 5G including the frequency bands 24.25–27.5 GHz, 37–43.5 GHz, 45.5–47 GHz, 47.2–48.2, and 66–71 GHz. In total, 17.25 GHz of spectrum was identified for IMT after the WRC-19.

The whole IMT-2020 process is planned to be completed in 2020. The key outcome will be the publication of a new ITU-R Recommendation with detailed specifications for the IMT-2020 radio interfaces.

1.4.2 3GPP 5G Standardization

3GPP writes technical specifications for mobile technologies. The standardization work is contribution-driven and consensus based. The project coordination group (PCG) coordinates the overall 3GPP work. The specification work is carried out in three technical specification groups (TSGs): TSG RAN (Radio Access Networks), TSG SA (Services and Systems Aspects), and TSG CT (Core Network and Terminals). The standardization process usually consists of four stages that are overlapping and iterative:

- In stage 1, the service requirements are defined.
- In stage 2, the architecture (including reference points and interfaces) for supporting the service requirements is defined.
- In stage 3, the detailed specifications of the protocols for the defined architecture are produced.
- In stage 4, test specifications are defined.

The 3GPP specifications are divided into releases. Each release consists of a set of features defined in agreed work items.

3GPP started to work on 5G, while ITU began to define IMT-2020. The 3GPP TSG RAN 5G workshop held in September 2015 marked the official start of 5G NR work in 3GPP. At this workshop, there was emerging consensus that a new, non-backward compatible radio access technology would be developed as part of 5G. It was emphasized at the workshop that the new radio access technology should embrace forward compatibility to ensure that the new radio would be capable of meeting new use cases in the future. It was also decided that the work would be split into two phases:

- Phase 1 would be completed by the second half of 2018 (i.e., end of 3GPP Release 15). It was highlighted that in this phase the focus should be on eMBB to facilitate early 5G deployments.
- Phase 2 would be completed by the end of 2019 (i.e., end of 3GPP Release 16) to address all identified use cases and requirements. This would pave the way for the 3GPP IMT-2020 submission to ITU-R.

At the 3GPP TSG RAN 5G workshop, a study item on channel modeling for frequencies up to 100 GHz was also approved. This study prepared channel models for the solution evaluations in developing the new radio access technology that would use new spectrum, particularly millimeter wave. The outcome of this study was documented in the 3GPP TR 38.900 [15] that details the channel models for frequencies from 6 GHz up to 100 GHz. Later, a new 3GPP TR 38.901 [16] was created to capture the channel models not only for frequencies from 6 GHz to 100 GHz but also for frequencies from 0.5 GHz to 6 GHz.

Following the 3GPP TSG RAN 5G workshop, a new study item focusing on developing scenarios and requirements for NR was approved at TSG RAN#70 in December 2015. The outcome of this study was documented in the 3GPP TR 38.913 [17] that details the scenarios, requirements, and key performance indicators for the next-generation access technologies.

The study item on new radio access technology was approved at TSG RAN#71 in March 2016. The scope of this study was to investigate the technology components to be considered for NR. The outcome of this study was documented in the 3GPP TR 38.912 [18] that covers all the RAN aspects of the technology components.

After the series of studies, 3GPP approved a work item at TSG RAN#75 in March 2017 for NR specifications as part of Release 15. At this RAN plenary meeting, 3GPP introduced NSA NR and agreed to accelerate the 5G NR schedule to complete the NSA NR by December 2017, while SA NR was set to be completed by June 2018 as originally scheduled. The reason for accelerating the 5G NR schedule was to enable early large-scale 5G trials and deployments. An intermediate milestone was reached in December 2017 with the approval of the NSA NR specifications. The SA version was completed in June 2018. The last step for Release 15 was a late drop that was completed in March 2019.

The finalization of the Release 15 NR specifications was a major milestone. The focus of NR in Release 15 was eMBB, while URLLC was addressed to some extent. The set of Release 15 specifications can fulfill a subset of the ITU 5G requirements. It is also the basis for further evolution of 5G NR. 3GPP continued NR evolution in Release 16 towards a more complete specifications that can completely fulfill all the ITU 5G requirements. Release 16 is the release that 3GPP uses for IMT-2020 submission. Key Release 16 work items include dual connectivity and carrier aggregation enhancements, URLLC enhancements, industrial IoT, vehicle-to-everything (V2X), MIMO enhancements, integrated access and backhaul, positioning, unlicensed operation, UE power saving, cross-link interference handling, and remote interference management. In addition, 3GPP initiated several studies to broaden the applicability of 5G NR, e.g., exploiting frequencies beyond 52.6 GHz and non-terrestrial radio access (primarily satellites).

During the 5G NR specification work, 3GPP also conducted a study on self-evaluation towards the IMT-2020 submission. The study covers evaluations for two 5G submissions: the first is a set of RITs containing NR and LTE, and the second is a separate RIT for NR only. The outcome of this study was documented in the 3GPP TR 37.910 [19] that includes evaluations against the technical performance requirements, spectrum requirements, and service requirements defined by ITU-R. The study concluded that both the set of RITs and the NR RIT can meet the IMT-2020 requirements.

1.5 What Will 6G Be?

As the 5G NR specification is maturing, research communities have recently started to look at the future of mobile communication systems, a.k.a. 6G systems. Although 6G is still at a premature stage, this section aims to provide an overview of the vision, challenges, and key enabling technologies for 6G. This overview shares common ideas in part with [20–22], yet provides some different angles for 6G.

1.5.1 *Vision for 6G*

What will 6G be? Arguably, 6G systems are expected to provide intelligent and personalized (or task-dependent) services to users at any time and in any place. Billions of wireless devices, including sensors and mobiles, will be placed in homes, cars, buildings, factories, cities, and any environments. These devices will frequently connect to the network whenever they need. This creates meaning information per requested task by interacting with (distributed) data centers, each with high computing capabilities. As a result, the upcoming 6G will make a paradigm shift by not just connecting all wireless devices but also providing optimized information timely for any specific requests of users in a given environment. Providing new intelligent connectivity services can be one vision of 6G.

1.5.2 *Technical Requirements and Applications*

The NR specification for 5G has already made a successful progress towards attaining very high data rates, ultra-high reliability and low latency, and massive connectivity solutions. To provide task-dependent intelligent services, however, 6G systems may need to meet more challenging requirements than 5G. These include (1) extremely high data rates, such as a few terabits per second; (2) super-low latency, less than hundreds of microseconds; and (3) ultra-massive connectivity providing more than 10^7 connections per square kilometer, which cannot be offered by the current 5G standard. The key breakthrough applications that will accelerate the development of such 6G might include duplicated digital twin, pervasive connectivity, and 3D holographic display, among others.

- Digital twin technology [23, 24]: The main idea of the duplicated digital twin is to build an exact digital replica of a complex physical object or system. By feeding a set of data obtained from the real object, this replica mimics the real object behaviors to learn how the real system can evolve under time-varying circumstances and predict the best action for the next to optimize the system. This intelligent control system has already existed. For instance, General Electric developed the digital twins of jet engine components, which are critical to controlling the life-time of the engine. This idea will be explored in many complex systems such as 6G wireless networks, industrial IoT networks, and quantum computers to maximize their performance while providing up-to-date information for a given task request. The extremely high data rates of a few Tbps and super low latency less than several hundred microseconds might be key technical requirements in addition to the availability of cloud computing and the machine learning technologies to enable this application.

- Pervasive connectivity for automation: The number of IoT (or M2M) connections will approximately be 14.7 billons in 2023, according to Cisco Annual Internet Report announced in February 2020. In a particular urban area, a connectivity solution that offers more than 10^7 IoT device connections per square kilometer would be needed. Therefore, 6G networks are required to hold the capability to connect this massive number of sensors in homes, cars, factories, and cities in a seamless manner. This is a realization of pervasive connectivity or "connectivity everywhere." 6G will not just provide such ubiquitous connectivity but also offer intelligent automation services for an enormous variety of wireless devices, thanks to the convergence between communications and intelligent computing capabilities empowered by machine learning technologies. The super-low latency of less than several hundred microseconds will be the key requirement for wireless connection to enable this situational awareness automation services. The energy efficiency will also play a key role in solving scalability obstacles for massive connectivity.
- 3D holographic display [25, 26]: Future wireless devices such as smartphones, tablets, and laptops might have enhanced display systems. 3D holographic display is the most promising technology that can be embedded in future wireless devices. This next-generation display system will be a primary driving force to increase the peak data rates for 6G networks. For instance, the data rate of 4.32 Tbps would be required to send a raw hologram data without any advanced compression technologies [26]. The super-low latency requirement of less than several hundred microseconds will also be critical to synthesize and synchronize the hundreds of different 2D images when building a 3D image in real time.

1.5.3 Key Enabling Technologies

We present five disruptive technologies that can enable the upcoming 6G.

- **Terahertz communications:** An effective way of increasing the data rate by an order of magnitude more than 5G is the use of a larger signal bandwidth. This strategy was already taken in the evolution from 4G to 5G by adopting millimeter wave frequency bands. This trend will continue to the next-generation wireless systems, 6G, to increase data rates more than 5G. In March 2019, the FCC announced a new category of experimental spectrum licenses for frequencies between 95 GHz and 3 THz. This has propelled researchers to have eyes on sub-terahertz bands, specifically in the range of 140 GHz and 300 GHz [27]. The use of THz frequency bands for wireless transmissions will be a critical enabler to provide extremely high-speed data rates for 6G.

Communications using THz frequency bands is very challenging. The signal bandwidth is ultra-wide, and the propagation is highly directional. In addition, the compound channel effects, including blockage and molecular absorption, are not fully understood yet. Pencil beams will also provide two sides of the same coin in managing interference, which will affect medium access control and handover. Terahertz communications will impose another challenge in physical-layer designs. The challenges will include the development of modulation, coding, and beamforming techniques under impairments of RF hardware, low-power AD/DA conversion circuits constraints, and antenna mutual coupling effects.

Obtaining spatial multiplexing gains is also not trivial in sub-terahertz line-of-sight (LOS) MIMO communications. Unlike conventional MIMO channels under 6 GHz frequency bands, in which rich scattering is pronounced, fixed-link LOS MIMO channels in sub-terahertz bands are completely determined by physical parameters such as link distances, antenna array geometries, and wavelengths. Orbital angular momentum (OAM) multiplexing technologies can be suitable for short-range MIMO communications using sub-terahertz bands [28]. The use of reconfigurable uniform linear arrays as a function of system parameters is also a promising technique for sub-terahertz LOS MIMO communications, which was recently proven to be optimal from an information-theoretical viewpoint. This technique can achieve the maximum spectral efficiency under all possible antenna configurations in LOS MIMO channels [29].

- **Pilot-free communications:** Short packet transmissions are essential to enable extremely low-latency communications. By shrinking the packet size, it is possible to dwindle not only air-interface time but also the computation time for decoding; thereby, it possibly enables to meet the stringent delay requirement of less than a hundred microseconds for 6G. The challenge in pilot-based short packet communications is that longer pilots make larger transmission delays. The pilot length can also linearly increase with the number of connected wireless devices to support the massive connectivity. Furthermore, when a channel coherence time is very short in the scenarios such as V2X, the pilot-based communication is not very effective, because the pilots are repeatedly transmitted whenever channel realizations change. Lastly, the pilot-free uplink communications make it possible to eliminate the pilot contamination effects in TDD massive MIMO systems. In theory, this has the potential to achieve an infinite cellular capacity when the number of base station antennas is infinite. Therefore, pilot-free communication technologies will be key enablers for supporting delay-sensitive massive IoT/V2X communications and massive MIMO systems for 6G.

  The pilot free communications will bring a fundamental challenge in reliable message decoding due to the absence of channel knowledge. Several non-coherent communication methods have been developed such as differential modulation and a geometric approach based on the Grassmann manifold [30].

These techniques may not be suitable to meet the stringent requirements of ultra-high reliability and low latency. Joint modulation and coding techniques using compressive sensing and machine learning-aided blind detection technologies are envisaged to enable pilot-free communications for 6G.

- **Convergence of communication, sensing, and computing:** The convergence of communications, sensing, and computing technologies will play an essential role in 6G. Future wireless devices such as self-driving cars and next-generation smartphones are likely to have multi-functionalities offered by communication transceivers, radar, and many sensors. Radar systems measure the reflections of probing signals to detect the presence of objects. In addition, vision sensors can acquire environmental landscapes and scenery surrounding users or cars. They, therefore, provide users with information on complex environments with different angels. Leveraging mobile edge computing technologies with the power of machine learning, this environmental information can be processed to generate user-specific contextual data. This contextual data will contain the multidimensional information of users such as users' activity (e.g., mobile contents) pattern and space-time map of users' geographical information. This multifaceted information can be exploited to predict future behaviors of users to improve the quality of services, which enables context-aware communication systems.
- **Communications using quantum computation:** The availability of quantum computation can bring a breakthrough in the design of communication algorithms. The quantum computation differs from quantum communications that exploit the quantum physical phenomenon called quantum entanglement to transfer information in one place to another. Instead, it uses quantum entanglement to speed up the computation in solving a class of NP-hard optimization problems. The quantum community has recently shown that the exponential speedups in computation is possible for a certain class of NP-hard optimization problems. In this area, the main research challenge is to develop both low-cost quantum processors and advanced algorithms that take advantages of quantum processors. For example, D-Wave is a quantum processor that enables quantum annealing, which is a metaheuristic approach to finding a global optimal solution using the phenomena of quantum fluctuations [31]. The quantum approximate optimization algorithm (QAOA) is another promising quantum-classical hybrid technique that solves combinatorial optimization problems using gate-based noisy quantum devices (e.g., IBM quantum computers) [32]. Harnessing these quantum optimization techniques, one may solve the long-standing computationally challenging optimization problems in communications, including maximum likelihood (ML) detection in massive MIMO, ML decoding of channel codes, and the capacity-maximizing resource allocation algorithms. This new computation capability can propel to improve the data rates and delay performance significantly for 6G.
- **Flexible 4D cellular networks:** Mobile base stations mounted on moving objects such as unmanned aerial vehicles (UAVs), drones, and smart cars can provide an additional degree of freedom in designing a 4D cellular network

[33]. This new degree of freedom can offer a high flexibility that adaptively changes infrastructure configurations in both space and time to optimize network performance. This will be a key feature of 6G cellular networks. By flying the UAV base stations in a densely populated area, one can optimize the network performance by offloading effects. The moving base stations can also provide cellular operators with flexible and scalable space-time coverage maps by eliminating the coverage limitations of the existing cellular networks. When the cellular infrastructures break down due to disastrous events, they are capable of creating wireless connectivity in a cost-effective manner. For example, people may use satellite networks for communications by using UAV base stations as mobile relays/routers under disastrous environments. The mobile base stations will offer new opportunities to provide user-specific contents while guaranteeing high quality of service requirements.

Several critical problems remain open to enable the flexible cellular networks in space-time. Wireless backhaul solutions that can provide ultra-high throughput would be essential. The flying base stations, for example, may use millimeter wave or sub-terahertz LOS MIMO transmission technologies to achieve data rates of a few Tbps. Innovative battery technologies for UAVs and drones are needed for lasting flying base station missions. In addition, low-power mobile edge computing technologies supporting advanced machine learning algorithms will be another essential part of realizing the mobile cellular base stations.

1.6 Book Outline

This book provides an accessible but complete tutorial on the key enabling technologies for 5G and beyond covering both the fundamentals and the state-of-the-art of the 5G standards.

The rest of this book consists of three parts.

- **Part I – Fundamentals of 5G and 6G**: The first part describes the fundamental technology components for 5G and beyond.
  - Chapter 2 presents an accessible treatment of advanced channel coding theory with a focus on rate-compatible polar codes that are capacity-achieving.
  - Chapter 3 provides a comprehensive overview of the state-of-the-art multiple access techniques and discusses rate-splitting multiple access in detail.
  - Chapter 4 gives a tutorial on massive MIMO including a massive MIMO transmission protocol, fundamental aspects, and future research directions.
  - Chapter 5 focuses on network densification and introduces theoretical models based on stochastic geometry for densified network analysis and design.
  - Chapter 6 discusses the integration of unmanned aerial vehicles in cellular communication networks.
  - Chapter 7 presents a comprehensive forward-looking vision that defines the main principles that will guide the design and development of a 6G system.

- **Part II – 5G New Radio Basics**: The second part introduces the basics of 5G NR standards.
  - Chapter 8 offers a guide to the new generation 5G radio access network (NG-RAN) architecture that provides both NR and LTE radio access.
  - Chapter 9 outlines the NR physical layer highlighting aspects around wave-forms and numerologies, bandwidth parts, downlink and uplink control information, downlink and uplink data channels, NR-LTE interworking on the physical layer, power control, and UE capabilities.
  - Chapter 10 provides an accessible description of 5G NR channel coding design aspects. It includes both polar codes for control channels and LDPC codes for data channels
  - Chapter 11 describes cell search and random access procedures in 5G NR.
  - Chapter 12 presents a primer on the bandwidth parts concept in 5G NR, delving into more details beyond the short introduction in Chapter 9.
- **Part III – 5G New Radio Evolution**: The third part describes key 5G NR evolution directions.
  - Chapter 13 provides an overview of NR URLLC by describing the use cases, performance requirements, and standards enhancements.
  - Chapter 14 introduces 5G NR operation in unlicensed spectrum including targeted spectrum and requirements, deployments, and design details.
  - Chapter 15 gives a tutorial on NR positioning. It discusses location services in 5G, fundamentals of positioning, and NR positioning methods and reference signals.
  - Chapter 16 provides an overview of NR integrated access and backhaul (IAB) system architecture, key issues, and designs.
  - Chapter 17 describes how NR can be used for air-to-ground (A2G) communications and discusses potential NR evolution directions for enhanced NR based A2G systems.
  - Chapter 18 discusses how to adapt the NR air interface for non-terrestrial networks with a focus on satellite communications.

Acknowledgement N. Lee is supported by the National Research Foundation of Korea (NRF) grant funded by the Korea government (MSIT) (No. 2020R1C1C1013381).

References

1. Ericsson, Ericsson Mobility Report November 2019, Tech. Rep., Nov 2019. [Online]. Available: https://www.ericsson.com/4acd7e/assets/local/mobility-report/documents/2019/emr-november-2019.pdf
2. ITU-R, IMT vision–framework and overall objectives of the future development of IMT for 2020 and beyond, Recommendation ITU-R M.2083, Sept 2015

3. X. Lin, J. Li, R. Baldemair, T. Cheng, S. Parkvall, D. Larsson, H. Koorapaty, M. Frenne, S. Falahati, A. Grövlen et al., 5G new radio: unveiling the essentials of the next generation wireless access technology. IEEE Commun. Stand. Mag., **3**(3), 30–37 (2019)
4. J.G. Andrews, S. Buzzi, W. Choi, S.V. Hanly, A. Lozano, A.C. Soong, J.C. Zhang, What will 5G be? IEEE J. Sel. Areas Commun. **32**(6), 1065–1082 (2014)
5. X. Lin, Debunking seven myths about 5G new radio, arXiv preprint arXiv:1908.06152, Aug 2019
6. 3GPP, System architecture for the 5G system (5GS), 3GPP TS 23.501, V16.1.0, June 2019
7. 3GPP, NR; user equipment (UE) radio transmission and reception; Part 1: range 1 standalone, 3GPP TS 38.101-1, V16.0.0, July 2019
8. 3GPP, NR; user equipment (UE) radio transmission and reception; Part 2: range 2 standalone, 3GPP TS 38.101-2, V16.0.0, July 2019
9. ITU-R, IMT traffic estimates for the years 2020 to 2030, Report ITU-R M.2370-0, July 2015
10. ITU-R, Future technology trends of terrestrial IMT systems, Report ITU-R M.2320-0, Nov 2014
11. ITU-R, Technical feasibility of IMT in bands above 6 GHz, Report ITU-R M.2376-0, July 2015
12. ITU-R, Minimum requirements related to technical performance for IMT-2020 radio interface(s), Report ITU-R M.2410-0, Nov 2017
13. ITU-R, Requirements, evaluation criteria and submission templates for the development of IMT-2020, Report ITU-R M.2411-0, Nov 2017
14. ITU-R, Guidelines for evaluation of radio interface technologies for IMT-2020, Report ITU-R M.2412-0, Oct 2017
15. 3GPP, Study on channel model for frequency spectrum above 6 GHz, 3GPP TR 38.900, V15.0.0, June 2018
16. 3GPP, Study on channel model for frequencies from 0.5 to 100 GHz, 3GPP TR 38.901, V15.0.0, June 2018
17. 3GPP, Study on scenarios and requirements for next generation access technologies, 3GPP TR 38.913, V15.0.0, July 2018
18. 3GPP, Study on new radio (NR) access technology, 3GPP TR 38.912, V15.0.0, July 2018
19. 3GPP, Study on self evaluation towards IMT-2020 submission, 3GPP TR 37.910, V16.0.0, June 2019
20. Z. Zhang, Y. Xiao, Z. Ma, M. Xiao, Z. Ding, X. Lei, G.K. Karagiannidis, P. Fan, 6G wireless networks: vision, requirements, architecture, and key technologies. IEEE Veh. Technol. Mag. **14**(3), 28–41 (2019)
21. W. Saad, M. Bennis, M. Chen, A vision of 6G wireless systems: applications, trends, technologies, and open research problems. IEEE Netw. **34**(3), 134–142 (2020)
22. E. Calvanese Strinati, S. Barbarossa, J.L. Gonzalez-Jimenez, D. Ktenas, N. Cassiau, L. Maret, C. Dehos, 6G: the next frontier: from holographic messaging to artificial intelligence using subterahertz and visible light communication. IEEE Veh. Technol. Mag. **14**(3), pp. 42–50 (2019)
23. N. Mohammadi, J.E. Taylor, Smart city digital twins, in *2017 IEEE Symposium Series on Computational Intelligence (SSCI)*, Nov 2017, pp. 1–5
24. Z.F.A. Fuller, C. Day, Digital twin: enabling technology, challenges and open research, arXiv preprint arXiv:1911.01276, 2019
25. J. Park, K. Lee, Y. Park, Ultrathin wide-angle large-area digital 3D holographic display using a non-periodic photon sieve. Nat. Commun. 10, 1304 (2019). https://doi.org/10.1038/s41467-019-09126-9
26. X. Xu, Y. Pan, P.P.M.Y. Lwin, X. Liang, 3D holographic display and its data transmission requirement, in *2011 International Conference on Information Photonics and Optical Communications*, Oct 2011, pp. 1–4
27. Y. Xing, T.S. Rappaport, Propagation measurement system and approach at 140 GHz-moving to 6G and above 100 GHz, in *2018 IEEE Global Communications Conference (GLOBECOM)*, Dec 2018, pp. 1–6

28. J. Wang, J.-Y. Yang, I.M. Fazal, N. Ahmed, Y. Yan, H. Huang, Y. Ren, Y. Yue, S. Dolinar, M. Tur, A.E. Willner, Terabit free-space data transmission employing orbital angular momentum multiplexing. Nat. Photonics **6**, 488–496 (2012)
29. H. Do, N. Lee, A. Lozano, Reconfigurable ULAs for line-of-sight MIMO transmission. arXiv preprint arXiv:2004.12039, April (2020)
30. L. Zheng, D.N.C. Tse, Communication on the grassmann manifold: a geometric approach to the noncoherent multiple-antenna channel. IEEE Trans. Inf. Theory **48**(2), 359–383 (2002)
31. C.S.A. Finnila, M. Gomez, J. Doll, Quantum annealing: a new method for minimizing multidimensional functions. Chem Phys. Lett. **5–6**, 343–348 (1994)
32. J.G.E. Farhi, S. Gutmann, A quantum approximate optimization algorithm, arXiv, 2014
33. C. Yan, L. Fu, J. Zhang, J. Wang, A comprehensive survey on UAV communication channel modeling. IEEE Access **7**, 107769–107792 (2019)

Part I
Fundamentals of 5G and 6G

Chapter 2
Advanced Channel Coding

Songnam Hong

2.1 Introduction

In [1], Arikan proposed a new class of channel coding methods, called polar codes, which can achieve the symmetric capacity of the binary-input discrete memoryless channels (BI-DMCs) with a low-complexity successive cancellation (SC) decoding. For finite blocklenghts, the performances of polar codes can be enhanced using list decoding, approaching the optimal maximum-likelihood (ML) performances with a sufficiently large list size [2]. Furthermore, CRC-concatenated polar codes can outperform well-optimized LDPC and turbo codes even for short blocklengths [2]. Recently, due to the attractive performance and low complexity, polar codes have been adopted in 5G NR standard [3].

Mobile communication systems require flexible and adaptive transmission techniques due to time-varying channel environments. In these systems, hybrid automatic repeat request based on incremental redundancy (HARQ-IR) schemes have been widely employed, where parity bits are transmitted in an incremental fashion according to the time-varying channel. For HARQ-IR schemes, *rate-compatible* (RC) codes, consisting of multiple member codes with various code rates, are required. Specifically, the member codes of a RC code satisfy the following requirement: the set of parity bits of a higher-rate member code should be a subset of the set of parity bits of a lower-rate member code. This particular construction enables the receiver which fails to decode at a particular rate, to request only additional parity bits from the transmitter. Thus, there exist various researches on the developments of RC codes based on turbo codes and LDPC codes [4–8]. In addition, RC-LDPC codes were adopted for data channels in 5G NR standard [3].

The successive puncturing has been widely used to construct RC codes. Polar codes can be good candidates for a mother code as they show the capacity-achieving

S. Hong (✉)
Electronic Engineering, Hanyang University, Seoul, South Korea

X. Lin, N. Lee (eds.), *5G and Beyond*, https://doi.org/10.1007/978-3-030-58197-8_2

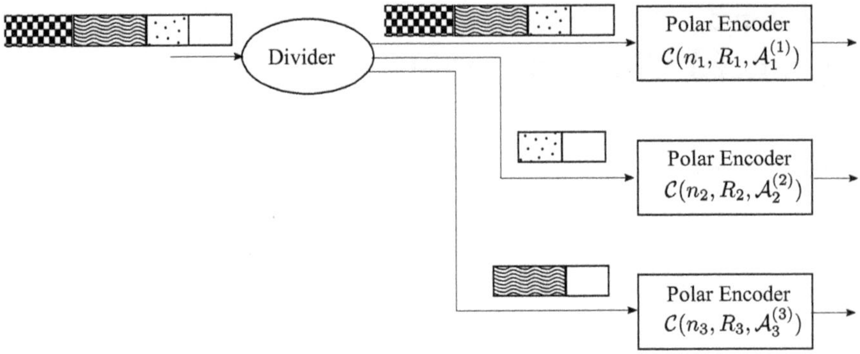

Fig. 2.1 Encoding of RC-Polar codes with $K = 3$

performances for BI-DMCs. However, the RC construction from the successive puncturing is generally not capacity-achieving. In [9–13], good puncturing patterns which can yield good performances have been investigated. More recently, the joint optimization of puncturing patterns and the information sets of polar codes was proposed, being able to outperform LDPC codes [14]. In [9, 11–14], an information set is optimized according to a puncturing pattern. However, they cannot be used to construct a family of RC punctured codes because in HARQ-IR schemes, the identical information set should be used for all the member codes (i.e., punctured polar codes) in the family. In [10], a heuristic algorithm to construct a good puncturing pattern for a given information set was proposed. Despite extensive researches, it is still an open problem to construct an optimal family of RC puncturing patterns and the corresponding common information set.

This chapter introduces a family of RC-Polar codes for which each member code in the family is capacity-achieving. The key idea is to leverage certain common characteristics of polar codes that are optimized for a sequence of successively degraded channels. The proposed code is constructed by the parallel concatenation of multiple (punctured) polar codes. Also, the so-called information-bit divider is used to determine the input of each polar encoder. The encoding structure of the proposed code is depicted in Fig. 2.1. As shown in Figs. 2.3 and 2.4, the proposed code is decoded by a sequence of parallel polar decoders (named sequential decoding). This code is referred to as *parallel concatenated polar* (PCP) codes. We remark that a lower-rate PCP code can be simply obtained by including more constituent polar codes. This particular construction enables incremental retransmissions to yield various transmission rates, and, thus, PCP code can be used for HARQ-IR systems. Remarkable, an optimized polar code is employed at every retransmission, thus being able to achieve the capacity of any class of degraded channels. Since polar codes have the length limitation, PCP code can achieve the capacity only for a sequence of rates satisfying a certain relationship. The corresponding relationship is specified in Theorem 2.5. To address this restriction, capacity-achieving punctured polar codes, which can provide more flexibility on blocklength, are used as constituent codes. In [15], it was proved that a capacity-achieving punctured polar code exists for any given puncturing fraction. Leveraging such punctured polar codes as constituent codes, PCP code can be capacity-achieving for an arbitrary sequence of code rates. However, this does not directly imply that PCP codes yield good performances for practical blocklengths. Especially when an incremental rate is small, PCP code cannot perform well as the length of constituent polar code is too small.

For a RC-Polar code, an efficient approach is to perform puncturing successively from a mother polar code, which is called RC punctured polar (RCPP) code. For the construction of RCPP codes, the main part is to jointly optimize RC puncturing patterns and a common information set which is good for all the member codes in the family. This optimization is not manageable because of its expensive complexity. Thus, numerous heuristic algorithms to generate a puncturing pattern and an associated information set have been proposed (see [9, 11–13] and the references therein). A practical quasi-uniform puncturing (QUP) was developed in [16]. However, its extension to a RC puncturing is not straightforward as we do not know how to construct a good common information set. In [14], an efficient algorithm to optimize a puncturing pattern and an information set simultaneously was developed and shown to outperform LDPC codes. We remark that, in the previous works [12–14], each puncturing pattern uses its own optimized information set. These methods cannot be applied to HARQ-IR schemes since an information set should be maintained during retransmissions, i.e., a common information set should be used for all the member codes in the family.

This chapter describes an efficient method to design a good RCPP code with a practical block length. It is remarkable that there exist good RC puncturing patterns like QUP as long as an optimized information set can be taken for each puncturing pattern. As pointed out before, in HARQ-IR schemes, the member codes in the family should use the same information set, i.e., this information set should be kept during retransmissions. Motivated by this, a novel information-copy technique was proposed in [17], being able to produce *effective* (or virtual) information sets from the common one, each of which is the optimized information set for the corresponding member code in the family. The so-called *hierarchical* puncturing is introduced, which has the special property such that "unknown" frozen bits can be allocated to some frozen-bit channels (carefully chosen according to a puncturing pattern) without affecting the performance. Moreover, we provide a class of hierarchical puncturing patterns to satisfy the rate-compatible constraint. An information-copy technique is described, which repeats some part of information bits to frozen-bit channels. In fact, this method can yield an information-dependent frozen vector, differently from conventional polar codes. Also, the positions of special frozen-bit channels are determined as a function of RC puncturing patterns and the common information sets. From the hierarchical puncturing and information-dependent frozen vector, a systematic method to design a good RCPP code is developed. Specifically, a common information set is optimized for the highest-rate member code in the family, and then it can be updated to generate effective information sets to the other codes in the family, by properly leveraging the common information set and information-dependent frozen vector. Because of the special property of hierarchical puncturing, the impact of unknown (information-dependent) frozen bits can be completely avoided. Via numerical results, it is shown that RCPP code attains a nontrivial performance gain about 2dB over the benchmark code for HARQ-IR schemes. Here, both codes employ the same QUP, while the difference is that the former uses an information-dependent frozen vector and the latter uses the conventional all-zero frozen vector. Therefore, the proposed method, introduced in this chapter, would be crucial for the construction of a practical good RCPP code for HARQ-IR scheme.

2.2 Formal Definition of RC Codes

We provide useful notations and formally define RC codes. Let $[K] \triangleq \{1, 2, \cdots, K\}$ for any positive integer K and let $a^n = (a_1, \ldots, a_n)$ denote a length-n vector. Given k information bits, we define a family of codes or code family:

$$\mathcal{C} = \{\mathcal{C}_1^{\bar{n}_1}, \mathcal{C}_2^{\bar{n}_2}, \cdots, \mathcal{C}_K^{\bar{n}_K}\},$$

with respective lengths $\bar{n}_1 < \bar{n}_2 < \cdots < \bar{n}_K$ and corresponding rates $R_1 > R_2 > \cdots > R_K$, where $R_i = k/\bar{n}_i$. Also, this family of code is said to be rate-compatible if there exists a sequence of encoding functions $\{\bar{e}_i(\cdot)\}_{i\in[K]}$ and a sequence of projection operators $\{\pi_i(\cdot)\}_{i\in[K-1]}$ such that for each $i \in [K-1]$ and for every possible information block $u^k \in \{0, 1\}^k$:

$$\bar{e}_i(u^k) = \pi_i(\bar{e}_{i+1}(u^k)), \tag{2.1}$$

where $\bar{e}_i : \{0, 1\}^k \rightarrow \{0, 1\}^{\bar{n}_i}$ denotes the corresponding encoding function of $\mathcal{C}_i$ for $i \in [K]$ and $\pi_i : \{0, 1\}^{\bar{n}_{i+1}} \rightarrow \{0, 1\}^{\bar{n}_i}$ simply takes $\bar{n}_i$ of the $\bar{n}_{i+1}$ coordinates of its input as output. This sequence of encoding functions is referred to as a sequence of nested encoding functions. Clearly, $\mathcal{C}$ is called rate-compatible if any subfamily of $\mathcal{C}$ denoted by $\mathcal{C}' = \{\mathcal{C}_{i_1}^{\bar{n}_{i_1}}, \mathcal{C}_{i_2}^{\bar{n}_{i_2}}, \cdots, \mathcal{C}_{i_j}^{\bar{n}_{i_j}}\}$, for any $1 \leq i_1 < i_2, \ldots < i_j \leq K$, is rate-compatible. Setting $\bar{n}_0 = 0$ and $n_i = \bar{n}_i - \bar{n}_{i-1}$, for $i \in [K]$, we have that $\bar{n}_i = \sum_{j=1}^{i} n_j$, and we will refer to $\{n_1, n_2, \ldots, n_K\}$ as the set of incremental lengths of $\mathcal{C} = \{\mathcal{C}_1^{\bar{n}_1}, \mathcal{C}_2^{\bar{n}_2}, \cdots, \mathcal{C}_K^{\bar{n}_K}\}$.

Condition 2.1 ensures the rate-compatible constraint such that the set of parity bits of a higher rate code is a subset of the set of parity bits of a lower rate code. Thus, RC code can be employed for HARQ-IR schemes. To be specific, during the i-th transmission, $e_i(u^k) \triangleq \pi_{i-1}^{\perp}(\bar{e}_i(u^k))$ is sent over the channel for a given information block $u^k \in \{0, 1\}^k$. Here, $\pi_{i-1}^{\perp} : \{0, 1\}^{\bar{n}_i} \rightarrow \{0, 1\}^{n_i}$ represents the projection operator orthogonal to $\pi_{i-1}(\cdot)$ that takes the other $n_i = \bar{n}_i - \bar{n}_{i-1}$ coordinates of $\bar{e}_i(u^k)$ not taken by $\pi_{i-1}(\bar{e}_i(u^k))$ as output, where $\bar{n}_0 \triangleq 0$ and $\pi_0^{\perp}(u^k) \triangleq u^k$ denote the identity mapping. Letting $e_i \triangleq \pi_{i-1}^{\perp} \circ \bar{e}_i$, $\{e_i\}_{i\in[K]}$ is called a sequence of *incremental encoding functions*. By definition, RC code has at least one associated sequence of incremental encoding functions. It follows immediately that a family of *linear* codes $\mathcal{C} = \{\mathcal{C}_1^{\bar{n}_1}, \mathcal{C}_2^{\bar{n}_2}, \cdots, \mathcal{C}_K^{\bar{n}_K}\}$ is rate-compatible if and only if there exists a sequence of generator matrices $\{\mathbf{G}_i\}_{i\in[K]}$ such that the columns of $\mathbf{G}_i$ is a subset of those of $\mathbf{G}_{i+1}$ for every $i \in [K-1]$. Accordingly, $\{\mathbf{G}_i\}_{i\in[K]}$ is referred to as a sequence of *nested generating matrices.*

From the definition of RC codes, we obtain the following relationships between the code rates and the incremental blocklengths:

Lemma 2.1 ([15]) For any RC code family of size K, the sets of rates $\{R_i\}_{i\in[K]}$ and incremental block lengths $\{n_i\}_{i\in[K]}$ should meet the following equivalent conditions:

$$\text{(a)}\quad R_i = \frac{R_1}{1 + \sum_{j=2}^{i} \frac{n_j}{n_1}}, \quad \forall i \in \{2, 3, \cdots, K\}, \tag{2.2}$$

$$\text{(b)}\quad n_i = R_1 \left(\frac{1}{R_i} - \frac{1}{R_{i-1}} \right) n_1, \quad \forall i \in \{2, 3, \cdots, K\}. \tag{2.3}$$

Let $W_1 \succeq W_2 \succeq \ldots \succeq W_K$ denote a sequence of successively degraded DMCs with the capacities $I(W_1) > I(W_2) > \ldots > I(W_K)$. Then, the capacity-achievability of RC codes is defined as follows.

Definition 2.2 For $m \in \mathbb{N}$ and $\{k_m\}_{m\in\mathbb{N}}$, a sequence of RC code families:

$$\mathcal{C}_m = \{\mathcal{C}_1^{\bar{n}_{1,m}}, \mathcal{C}_2^{\bar{n}_{2,m}}, \cdots, \mathcal{C}_K^{\bar{n}_{K,m}}\},$$

is said to be *capacity-achieving* with respect to the DMCs $\{W_i\}_{i\in[K]}$ if, for every m, there exists a sequence of decoding functions:

$$\{d_{i,m}(\cdot)\}_{i\in[K]},$$

where $d_{i,m} : \mathcal{Y}^{\bar{n}_{i,m}} \rightarrow \{0, 1\}^{k_m}$ and the corresponding sequence of nested encoding functions $\{e_{i,m}(\cdot)\}_{i\in[K]}$ such that for any $\epsilon > 0$, for every $i \in [K]$, and for all sufficiently large m:

$$R_{i,m} \triangleq k_m/\bar{n}_{i,m} > I(W_i) - \epsilon, \text{ and} \tag{2.4}$$

$$\Pr(u^{k_m} \neq d_{i,m}(y^{\bar{n}_{i,m}})) < \epsilon. \tag{2.5}$$

Here, the joint probability distribution is given by:

$$p(u^{k_m}, y^{\bar{n}_{i,m}}) = 2^{-k_m} W_i^{\bar{n}_{i,m}}(y^{\bar{n}_{i,m}}|e_i(u^{k_m})), \tag{2.6}$$

where $W_i^n(y^n|x^n) \triangleq \prod_{l=1}^{n} W_i(y_l|x_l)$.

In this chapter, we focus on B-DMC W with input alphabet $\mathcal{X} = \{0, 1\}$ and any output alphabet $\mathcal{Y}$, where the transition probabilities are given by $W(y|0)$ and $W(y|1)$ for all $y \in \mathcal{Y}$.

2.3 Capacity-Achieving RC-Polar Code: Asymptotic Results

2.3.1 *Overview of Polar Codes*

Let $2^{\mathbb{N}} = \{2^1, 2^2, \ldots\}$. Given any $n \in 2^{\mathbb{N}}$, $\mathbf{P}_n \triangleq \mathbf{P}_2^{\otimes \log(n)}$ represents the rate-1 generator matrix of all length-n polar codes where $\mathbf{P}_2$ is the 2-by-2 Arikan kernel [1]. In this chapter, zero-frozen bits are assumed. Then, a length-n polar code is fully defined by $\mathbf{P}_n$ and information set $\mathcal{A}$ which specifies the set of good bit-channel locations for carrying information bits. The corresponding code rate is given as:

$$R = |\mathcal{A}|/n.$$

Let $\mathcal{C}(n, R, \mathcal{A})$ be a length-n polar code with rate R and information set $\mathcal{A}$. The corresponding $|\mathcal{A}| \times n$ generator matrix is denoted as $\mathbf{P}_n^{\mathcal{A}}$, which is formed by only taking the rows of $\mathbf{P}_n$ whose indices belong to $\mathcal{A}$. Note that the specific ordering of the rows in $\mathbf{P}_n^{\mathcal{A}}$ is not important in the below.

Given $\{W_i\}_{i\in[K]}$, a sequence of polar codes $\{\mathcal{C}(n, R_i, \mathcal{A}_i)\}_{i\in[K]}$ can be obtained by choosing a different information set $\mathcal{A}_i$ for each channel W_i. This sequence of polar codes is referred to as a sequence of *nested polar codes* if the corresponding information sets are nested, i.e.:

$$\mathcal{A}_1 \supseteq \mathcal{A}_2 \supseteq \ldots \supseteq \mathcal{A}_K. \tag{2.7}$$

For the nested information sets $\mathcal{A}_1 \supseteq \mathcal{A}_2 \supseteq \cdots \supseteq \mathcal{A}_K$, we define an $|\mathcal{A}_1| \times n$ generator matrix $\mathbf{P}_n^{(\mathcal{A}_1,\mathcal{A}_2,\cdots,\mathcal{A}_K)}$ with a partial ordering of rows according to $\{\mathcal{A}_i\}$ as:

$$\mathbf{P}_n^{(\mathcal{A}_1,\mathcal{A}_2,\cdots,\mathcal{A}_K)} = \begin{bmatrix} \mathbf{P}_n^{\mathcal{A}_K} \\ \mathbf{P}_n^{\mathcal{A}_{K-1}\backslash\mathcal{A}_K} \\ \vdots \\ \mathbf{P}_n^{\mathcal{A}_1\backslash\mathcal{A}_2} \end{bmatrix}. \tag{2.8}$$

We remark that the difference between $\mathbf{P}_n^{\mathcal{A}_1}$ and $\mathbf{P}_n^{(\mathcal{A}_1,\mathcal{A}_2,\cdots,\mathcal{A}_K)}$ is that the ordering of rows in the latter is more specifically defined. Also, given any index set $\mathcal{D} \subseteq [n]$, let $\mathbf{P}_n^{(\mathcal{A}_1,\mathcal{A}_2,\cdots,\mathcal{A}_K)}(\mathcal{D})$ denote the submatrix of $\mathbf{P}_n^{(\mathcal{A}_1,\mathcal{A}_2,\cdots,\mathcal{A}_K)}$ which only contains the rows whose indices belong to $\mathcal{D}$. For a channel W_i, $i \in [K]$, and a fixed $\epsilon > 0$, the information set is defined as:

$$\mathcal{A}_{i,n}(\epsilon) = \{j \in [n] : Z(W_{i,n}^{(j)}) \leq \epsilon\}, \tag{2.9}$$

where $W_{i,n}^{(j)}$ denotes the j-th bit channel obtained when a length-n polar code is applied to W_i and $Z(W)$ denotes the Bhattacharyya parameter of bit-channel W, given by:

$$Z(W) \triangleq \sum_{y \in \mathcal{Y}} \sqrt{W(y|0)W(y|1)}.$$

Then, we have:

Lemma 2.1 ([18]) Given $W_1 \succeq W_2 \succeq \ldots \succeq W_K$ and $\epsilon > 0$, $\{\mathcal{C}(n, R_{i,n}\mathcal{A}_{i,n}(\epsilon))\}_{i \in [K]}$ is a sequence of nested polar codes.

It is noticeable that the set of good bit channels can be alternatively defined with respect to the symmetric capacity $I(W)$. Also, the dependency of R_i, $\mathcal{A}_i$ on ϵ will be omitted, if it is clear from the context.

Definition 2.2 A family of RC codes $\mathcal{C} = \{\mathcal{C}_1^{\bar{n}_1}, \cdots, \mathcal{C}_K^{\bar{n}_K}\}$ is referred to as RC-Polar codes if a (punctured) polar code is used in every transmission. Specifically, $\mathcal{C}$ is a RC-Polar code if there exists a sequence of incremental encoding functions $\{e_i(\cdot)\}_{i \in [K]}$ in which each $e_i(\cdot)$ can be implemented by the encoder of a polar code possibly with a puncturing.

The nested property in (2.7) is a key to construct RC-Polar codes in the below. However, since nested polar codes have the same length and various information block sizes, they cannot be straightforwardly used for HARQ-IR schemes, which in contrast assumes a fixed information block size and allows various lengths to yield various rates with a given error probability tolerance. To obtain a family of capacity-achieving RC-Polar codes, we need to construct multiple sequences of nested polar codes as will be described later.

2.3.2 *Capacity-Achieving Punctured Polar Code*

Puncturing is the simplest way to obtain various-length polar codes. A length-n polar code is punctured by removing a set of s columns from the generator matrix. Accordingly, the codeword length is reduced from n to $n - s$. Here, $\alpha = s/n$ is called *puncturing fraction*. Throughout the chapter, it is assumed that the receiver exactly knows the locations of the punctured bits and the decoder estimates both the punctured and transmitted symbols (i.e., the codeword). The punctured polar code can be encoded and decoded similarly as the conventional polar codes. A punctured polar code of post-puncturing length n is characterized by the mother polar code of length $n_u \in 2^{\mathbb{N}}$ and a puncturing pattern $p^{n_u} = (p_1, p_2, \cdots, p_{n_u}) \in \{0, 1\}^{n_u}$ with $p_i = 0$ indicating that the ith coded bit is punctured and thus not transmitted. Given the puncturing pattern p^{n_u}, we let $\pi_{p^{n_u}} : \mathcal{Y}^{n_u} \to \mathcal{Y}^n$ be a projection operator that copies $n = w_H(p^{n_u})$ coordinates of its input as its output based on the p^{n_u}, where $w_H(p^{n_u})$ represents the number of ones in p^{n_u}. That is, $y^n = \pi_{p^{n_u}}(y^{n_u})$ contains the coordinates of y^{n_u} corresponding to the locations of ones in p^{n_u}. The notion of bit channels in polar codes can be directly extended to punctured polar codes as follows. Given a length-n_u polar code and a puncturing pattern p^{n_u}, the transition probability of the i-th bit channel of the resulting punctured polar code is defined as:

$$W^{(i)}(y^n, u^{i-1}, p^{n_u}|u_i)$$
$$= \frac{1}{2^{n_u-1}} \sum_{u_{i+1}^{n_u}} \sum_{y^{n_u} \in \pi_{p^{n_u}}^{-1}(\{y^n\})} W^{n_u}(y^{n_u}|u^{n_u}\mathbf{P}_{n_u}), \quad (2.10)$$

where:

$$W^{n_u}(y^{n_u}|x^{n_u}) = \prod_{j \in [n_u]} W(y_j|x_j), \quad (2.11)$$

and $\pi_{p^{n_u}}^{-1}(S) \triangleq \{y^{n_u} \in \mathcal{Y}^{n_u} : \pi_{p^{n_u}}(y^{n_u}) \in S\}$ denotes the inverse image of $\pi_{p^{n_u}}(\cdot)$. From (2.24), the information set of a punctured polar code can be defined as in (2.9).

Let $\mathcal{C}(n, R, \mathcal{A}, p^{n_u})$ be a punctured polar code of (post-puncturing) length n, code rate R, information set $\mathcal{A}$, and a puncturing pattern p^{n_u}. Also, the corresponding generator matrix $\mathbf{P}_{n,p^{n_u}}$ is defined by removing the columns of $\mathbf{P}_{n_u}$ according to the puncturing pattern p^{n_u}. As in (2.8), $\mathbf{P}_{n,p^{n_u}}^{(\mathcal{A}_1,\mathcal{A}_2,\cdots,\mathcal{A}_K)}$ is defined for any nested information sets $\mathcal{A}_1 \supseteq \mathcal{A}_2 \supseteq \cdots \supseteq \mathcal{A}_K$. In the below, only one puncturing pattern is used for each length n and rate R, and thus, we will omit p^{n_u} without ambiguity. Also, we use the same notations and similar terminologies for both unpunctured and punctured polar codes whenever they are clear from context.

In PCP codes that will be explained in Sect. 2.3.3, any good punctured polar code can be used as constituent code for the length flexibility. The optimization of good punctured polar codes for short lengths is under investigation. In this chapter, focusing on asymptotic performances, the use of capacity-achieving polar codes for any target puncturing fractions are assumed. Note that the existence of such punctured polar codes is shown in the following theorem.

Theorem 2.3 ([15]) *Consider any B-DMC W with $I(W) > 0$. For any fixed $R < I(W)$, $\beta < \frac{1}{2}$, and puncturing fraction $\alpha \in (0, 0.5)$, there exists a sequence of punctured polar codes, each with respective length $n = \lfloor (1-\alpha)2^m \rfloor$ and associated information sets $\mathcal{A}_m \subset [2^m]$, $m \in \mathbb{N}$, such that:*

$$|\mathcal{A}_m| \geq \lfloor 2^m(1-\alpha) \rfloor R = nR, \quad (2.12)$$

$$P_{e,j,m} \leq O(2^{-2^{m\beta}}) = O(2^{-n^\beta}), \quad (2.13)$$

for all $j \in \mathcal{A}_m$ and for all $m \in \mathbb{N}$, where $P_{e,j,m}$ denotes the error probability of j-th bit channel of the m-th punctured polar code.

2.3.3 PCP Codes

This section describes a capacity-achieving RC-Polar code. First of all, the main theorems are provided.

Theorem 2.4 ([15]) *For any sequence of successively degraded channels $W_1 \succeq W_2 \succeq \ldots \succeq W_K$, there exists a sequence of RC-Polar code that is capacity-achieving.*

Theorem 2.5 ([15]) *Suppose that only (non-punctured) polar codes are used as constituent codes. Then, for any sequence of successively degraded channels $W_1 \succeq W_2 \succeq \ldots \succeq W_K$ with corresponding symmetric capacities $I(W_1) > I(W_2) > \ldots > I(W_K) > 0$, there exists a sequence of RC-Polar code that is capacity-achieving if and only if the symmetric capacities satisfy the:*

$$I(W_i) = \frac{I(W_1)}{1 + \sum_{j=2}^{i} 2^{\ell_j}}, \tag{2.14}$$

for each $i \in \{2, \ldots, K\}$ and for some $\ell_j \in \mathbb{Z}$.

We remark that when puncturing for length adaption is not used, the length constraint of a polar code reduces the set of supportable code rates of RC-Polar codes, which are described in Theorem 2.5.

In the rest of this section, the main idea to construct the capacity-achieving RC-Polar code will be explained (see Fig. 2.2). Also, a simple example for $K = 3$ will be provided for better understanding. See [15] for a general code construction of RC-Polar code.

Consider transmission of k information bits over K channels $W_1 \succeq W_2 \succeq \ldots \succeq W_K$ using the RC family of the codes $\{\mathcal{C}_1, \mathcal{C}_2, \ldots, \mathcal{C}_K\}$ with rates $\{R_1, R_2, \ldots, R_K\}$ such that $R_i < I(W_i)$, for $i \in [K]$. From Lemma 2.1, the incremental lengths $\{n_1, n_2, \ldots, n_K\}$ are chosen as:

$$n_i = k\left(\frac{1}{R_i} - \frac{1}{R_{i-1}}\right), \tag{2.15}$$

for $i \in [K]$, where $R_0 = \infty$. Then, the effective length and the effective code rate after i transmissions are, respectively, given as:

$$\bar{n}_i \triangleq \sum_{j=1}^{i} n_j \text{ and } R_i = \frac{k}{\sum_{j=1}^{i} n_j} = \frac{k}{\bar{n}_i}. \tag{2.16}$$

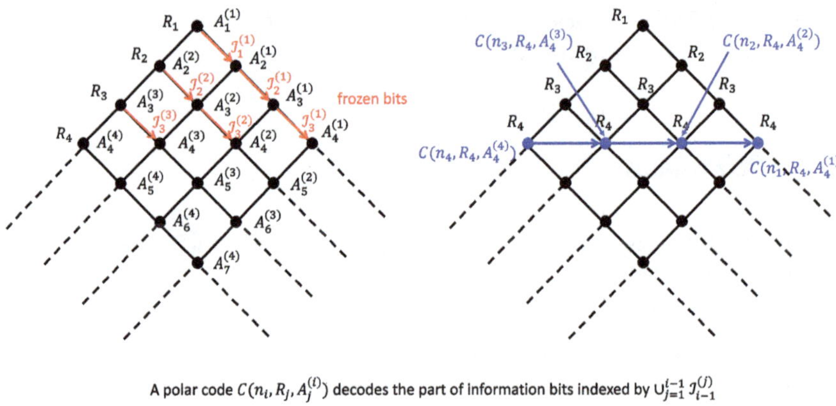

(a) Message Trellis (b) Decoding Trellis

Fig. 2.2 The fundamental ideas of the proposed RC-Polar codes

Given $i \in [K]$, $\mathcal{C}_i$ is constructed from the concatenation of the i number of polar codes with lengths $\{n_1, n_2, \ldots, n_i\}$. The detailed construction will be explained below.

Code construction For each n_i, $i \in [K]$, a sequence of nested (punctured) polar codes $\{\mathcal{C}(n_i, R_j, \mathcal{A}_j^{(i)})\}_{j=i}^K$ with rates $\{R_i, \ldots, R_K\}$ are constructed, such that:

$$|\mathcal{A}_j^{(i)}| = n_i R_j, \tag{2.17}$$

for $j \in [K]$. The exact choice of information sets $\mathcal{A}_j^{(i)}$ are explained in [15]. Then, $\mathcal{C}_i$ is constructed by concatenating the i number of polar codes $\{\mathcal{C}(n_j, R_j, \mathcal{A}_j^{(j)})\}_{j=1}^i$. Remarkable, each polar code $\mathcal{C}(n_i, R_j, \mathcal{A}_j^{(i)})$ encodes the some part of information bits. The set of corresponding indices is defined as follows. Let:

$$\mathcal{I}^{(i)} \triangleq \bigcup_{j=1}^{i-1} \mathcal{I}_{i-1}^{(j)}, \tag{2.18}$$

where $\mathcal{I}^{(1)} = [k]$ and $\mathcal{I}_{i-1}^{(j)}$ contains the indices of information bits corresponding to $\mathcal{A}_{i-1}^{(j)} \setminus \mathcal{A}_i^{(j)}$. We notice that $\mathcal{I}_{i-1}^{(j)}$ contains the indices of information bits which are used as frozen bits in order to transform the rate-R_{i-1} polar code into the rate-R_i polar code.

Encoding As shown in Fig. 2.1, the polar code $\mathcal{C}(n_i, R_i, \mathcal{A}_i^{(i)})$ is used for the i-th transmission, which sends some part of information bits indexed by $\mathcal{I}^{(i)}$. This is available as $\mathcal{I}^{(i)}$ and $\mathcal{A}_i^{(i)}$ are of the same size such as:

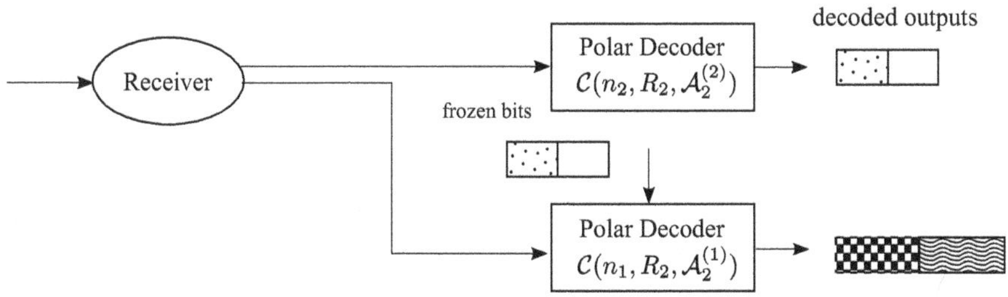

Fig. 2.3 Sequential decoding structure of RC-Polar codes with code rate R_2

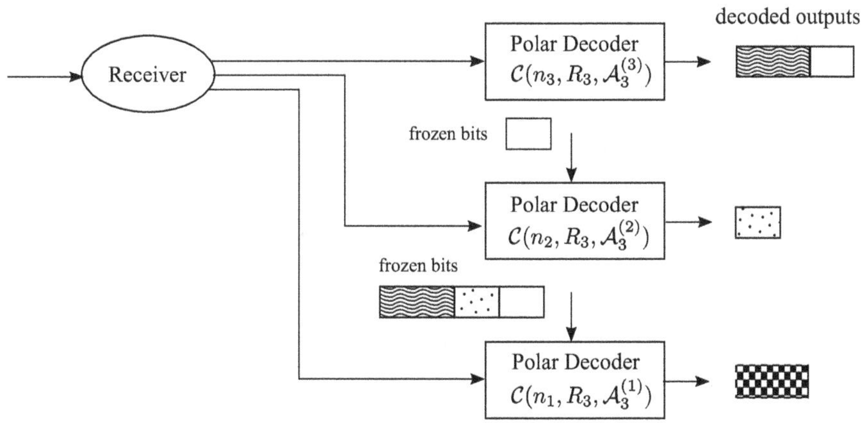

Fig. 2.4 Sequential decoding structure of RC-Polar codes with code rate R_3

Decoding Recall that $\mathcal{C}(n_i, R_i, \mathcal{A}_i^{(i)})$ is the polar code with the highest rate in the sequence of common length n_i. Also, the nested property in (2.7) is satisfied for each such sequence, which will be used in decoding. Consider m retransmissions with $m \in [K]$. Accordingly, rate $I(W_m)$ is the highest rate that can be supported by the channel. The sequential decoder starts by decoding the information bits of the polar code $\mathcal{C}(n_m, R_m, \mathcal{A}_m^{(m)})$. Then, some of the decoded bits are used as the frozen bits of the polar code $\mathcal{C}(n_{m-1}, R_{m-1}, \mathcal{A}_{m-1}^{(m-1)})$. Because of the property in (2.7), this polar code is transformed into the polar code $\mathcal{C}(n_{m-1}, R_m, \mathcal{A}_m^{(m-1)})$ with the lower-rate R_m. Since it is supported by the channel, the information bits of the resulting polar code can be decoded successfully. This process, called sequential decoding, is repeated over m stages (see Figs. 2.3 and 2.4), where, in each stage, the additional information bits are decoded via a polar code of rate R_m. Since the chosen transmit rates R_i defined in (2.16) will approach the corresponding channel capacity $I(W_i)$ as n_i increases for all $i \in [K]$, the RC-Polar code is capacity-achieving.

The main idea will be explained in more detail with the simple example of $K = 3$. Our goal is to design a RC-Polar code which can support code rates $R_1 > R_2 > R_3$. Start with the construction of a sequence of nested (punctured) polar code of length n_1. From Theorem 2.3, one can choose a sufficiently large n_1 and associated information sets $\mathcal{A}_j^{(1)}$, $j = 1, 2, 3$ of size $|\mathcal{A}_j^{(1)}| \geq n_1 R_j$ such that a tolerable error-probability for each bit channel is guaranteed, when the information sets are applied to the channels W_1, W_2, and W_3, respectively. Also, by Lemma 2.1, these information sets meet the nested property $\mathcal{A}_1^{(1)} \supseteq \mathcal{A}_2^{(1)} \supseteq \mathcal{A}_3^{(1)}$. For the simplicity of explanation, we assume that these information sets satisfy $|\mathcal{A}_j^{(1)}| = n_1 R_j$ for all $j = 1, 2, 3$. Note that the nested subset property is preserved by first determining $\mathcal{A}_3^{(1)}$ with $|\mathcal{A}_3^{(1)}| = n_1 R_3$, then choosing additional bit channels from $\mathcal{A}_2^{(1)} \setminus \mathcal{A}_3^{(1)}$ to form a new $\mathcal{A}_2^{(1)}$ with $|\mathcal{A}_2^{(1)}| = n_1 R_2$, and so forth to form a new $\mathcal{A}_1^{(1)}$ with $|\mathcal{A}_1^{(1)}| = n_1 R_1$. As a result, each code with the information set $\mathcal{A}_j^{(1)}$ enables to decode the $n_1 R_j$ information bits. These three information sets are used to support rates $R_1 > R_2 > R_3$ for the chosen length n_1.

In the first transmission, the (punctured) polar code $\mathcal{C}(n_1, R_1, \mathcal{A}_1^{(1)})$ is used to send $k = |\mathcal{A}_1^{(1)}|$ information bits at rate $R_1 = k/n_1$. Recall that $\mathcal{I}^{(1)} = [k]$. We next explain how to identify the information bits for the subsequent transmissions. The index set $\mathcal{I}^{(1)}$ of size k is partitioned into:

$$\mathcal{I}^{(1)} = \mathcal{I}_1^{(1)} \cup \mathcal{I}_2^{(1)} \cup \mathcal{I}_3^{(1)} \tag{2.19}$$

such that $\mathcal{I}_1^{(1)}$, $\mathcal{I}_2^{(1)}$, and $\mathcal{I}_3^{(1)}$ contain the indices of information bits that will be transmitted via the bit channels in $\mathcal{A}_1^{(1)} \setminus \mathcal{A}_2^{(1)}$, $\mathcal{A}_2^{(1)} \setminus \mathcal{A}_3^{(1)}$, and $\mathcal{A}_3^{(1)}$, respectively (see Fig. 2.5). Also, we have:

$$\begin{aligned}
|\mathcal{I}_1^{(1)}| &= |\mathcal{A}_1^{(1)}| - |\mathcal{A}_2^{(1)}| = n_1 R_1 - n_1 R_2 \\
|\mathcal{I}_2^{(1)}| &= |\mathcal{A}_2^{(1)}| - |\mathcal{A}_3^{(1)}| = n_1 R_2 - n_1 R_3 \\
|\mathcal{I}_3^{(1)}| &= |\mathcal{A}_3^{(1)}| = n_1 R_3.
\end{aligned}$$

It is remarkable that $\mathcal{I}_1^{(1)}$ and $\mathcal{I}_2^{(1)}$ contains the information bits to be frozen in the respective codes $\mathcal{C}(n_1, R_1, \mathcal{A}_1^{(1)})$ and $\mathcal{C}(n_1, R_2, \mathcal{A}_2^{(1)})$ so that their rates are reduced to R_2 and R_3, respectively, if subsequent transmissions are needed. Intuitively, it is expected that the information bits in $\mathcal{I}_j^{(1)}$ are assigned to better polarized bit channels as j increases.

Next focus on the second transmission. A new (punctured) polar code of length n_2 is used to transmit the information bits indexed by $\mathcal{I}_1^{(1)}$. Here, n_2 is determined such that $\bar{n}_2 \triangleq n_1 + n_2 = k/R_2$ to guarantee that, after the second transmission, the effective rate equals R_2. From Theorem 2.3, a sequence of nested polar codes can be determined where they have the information sets $\mathcal{A}_j^{(2)}$ of size $|\mathcal{A}_j^{(2)}| = n_2 R_j$ for $j = 2, 3$, and $\mathcal{A}_2^{(2)} \supseteq \mathcal{A}_3^{(2)}$. Then, we have $\mathcal{I}^{(2)} = \mathcal{I}_1^{(1)}$, which contains the information bits for the second transmission. Also, such information bits are transmitted via the (punctured) polar code $\mathcal{C}(n_2, R_2, \mathcal{A}_2^{(2)})$, which is possible as:

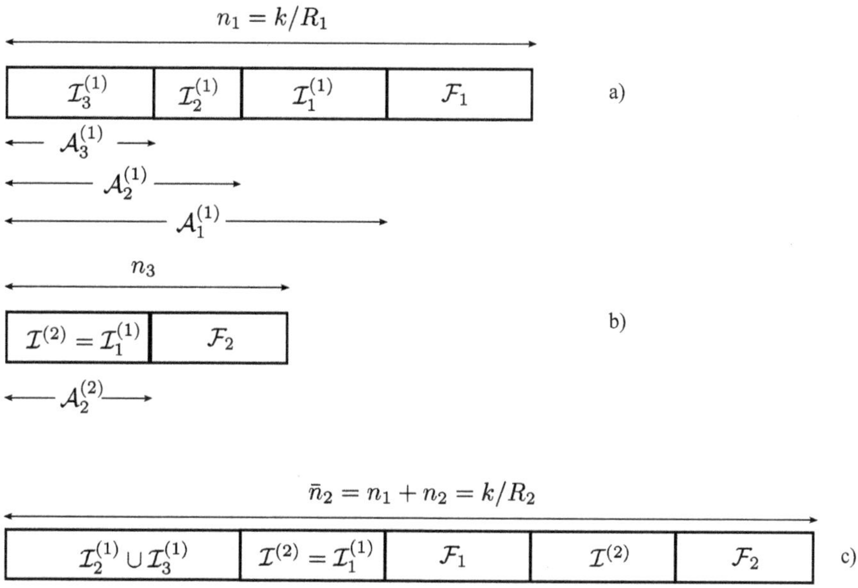

Fig. 2.5 Code construction for R_2. (**a**) Rate-R_1 polar code sent in the first transmission; (**b**) Rate-R_2 polar code sent in the second transmission; (**c**) resulting concatenated codeword sent over two transmissions

$$\begin{aligned}|\mathcal{I}^{(2)}| = |\mathcal{I}_1^{(1)}| &= n_1 R_1 - n_1 R_2 \\ &= k - \frac{n_1 k}{n_1 + n_2} = n_2 R_2 = |\mathcal{A}_2^{(2)}|.\end{aligned}$$

Likewise, for the subsequent transmission, $\mathcal{I}^{(2)}$ is further partitioned into $\mathcal{I}^{(2)} = \mathcal{I}_2^{(2)} \cup \mathcal{I}_3^{(2)}$ such that $\mathcal{I}_2^{(2)}$ and $\mathcal{I}_3^{(2)}$ include the respective indices of information bits which are sent via the bit-channels $\mathcal{A}_2^{(2)} \setminus \mathcal{A}_3^{(2)}$ and $\mathcal{A}_3^{(2)}$. Then, we have:

$$\begin{aligned}|\mathcal{I}_2^{(2)}| &= |\mathcal{A}_2^{(2)}| - |\mathcal{A}_2^{(1)}| = n_2 R_2 - n_2 R_3 \\ |\mathcal{I}_3^{(2)}| &= |\mathcal{A}_3^{(2)}| = n_2 R_3.\end{aligned}$$

Lastly, for the third transmission, the length n_3 is chosen such that $\bar{n}_3 = n_1 + n_2 + n_3 = k/R_3$. This choice ensures that an effective overall rate is equal t R_3 after the third transmission. The information bits indexed by $\mathcal{I}^{(3)} = \mathcal{I}_2^{(1)} \cup \mathcal{I}_2^{(2)}$ are encoded using a (punctured) polar code with the information set $\mathcal{A}_3^{(3)}$ of size $n_3 R_3$. Similarly as before, it is possible as:

$$|\mathcal{I}^{(3)}| = |\mathcal{I}_2^{(1)}| + |\mathcal{I}_2^{(2)}| = (n_1 + n_2) R_2 - (n_1 + n_2) R_3$$

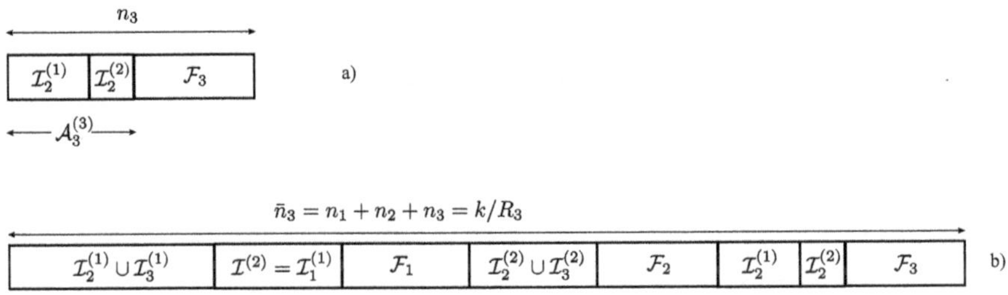

Fig. 2.6 Code construction for R_3. (**a**) Rate-R_3 polar code sent in the third transmission; (**b**) resulting concatenated codeword sent over three transmissions

$$= k - \frac{(n_1 + n_2)k}{n_1 + n_2 + n_3} = n_3 R_3 = |\mathcal{A}_3^{(3)}|.$$

For the code rates R_2 and R_3, the encoding structure and the resulting codeword are, respectively, depicted in Figs. 2.5 and 2.6. We notice that the resulting codeword of length k/R_3 is not a codeword of a polar code. Nevertheless, the decoding procedure will ensure that we can decode a codeword using a polar code at each rate $R_i, i = 1, \ldots, 3$ thereby achieving the capacity of the corresponding channel W_i. In particular, $\mathcal{C}(n_3, R_3, \mathcal{A}_3^{(3)})$ is first used to decode $R_3 n_3$ information bits indexed by $\mathcal{I}^{(3)}$. Then, the decoded information bits indexed by $\mathcal{I}_2^{(2)}$ are used as the frozen bits of $\mathcal{C}(n_2, R_2, \mathcal{A}_2^{(2)})$, which can produce the polar code $\mathcal{C}(n_2, R_3, \mathcal{A}_3^{(2)})$ thereby enabling decoding of $R_3 n_2$ information bits indexed by $\mathcal{I}_3^{(2)}$. Until now, the information bits indexed by $\mathcal{I}^{(3)} \cup \mathcal{I}_3^{(2)}$ are decoded. The rest of information bits are decoded, by converting the code of length n_1 and rate R_1 that was used for the first transmission, into the code of rate R_3. This can be done by taking the decoded information bits indexed by $\mathcal{I}^{(3)} \cup \mathcal{I}_2^{(1)}$ as the frozen bits in the polar code $\mathcal{C}(n_1, R_2, \mathcal{A}_1^{(1)})$ to generate the polar code $\mathcal{C}(n_1, R_3, \mathcal{A}_3^{(1)})$ with $R_3 n_1$ information bits indexed by $\mathcal{I}_3^{(1)}$. Thus, all the information bits are now decoded.

2.4 RCPP Codes for Finite Lengths

In Sect. 2.3, it was shown that PCP code is optimal when the number of information bits is sufficiently large. However, this does not imply that PCP code provides a good performance for practical lengths. This section introduces an efficient method to design a good RCPP code for HARQ-IR schemes. RCPP code can outperform the existing RC codes for practical lengths. First of all, some useful notations will be provided. Given the index subsets $\mathcal{B}, \mathcal{D} \subseteq \{0, \ldots, N-1\}$, $\mathbf{P}_\mathrm{N}(\mathcal{B}, \mathcal{D})$ represents the submatrix of $\mathbf{P}_\mathrm{N}$ obtained by taking the rows and columns whose indices belong to $\mathcal{B}$ and $\mathcal{D}$, respectively. A function $g(\ell) : \{0, \ldots, N-1\} \to \{0, 1\}^n$ is defined,

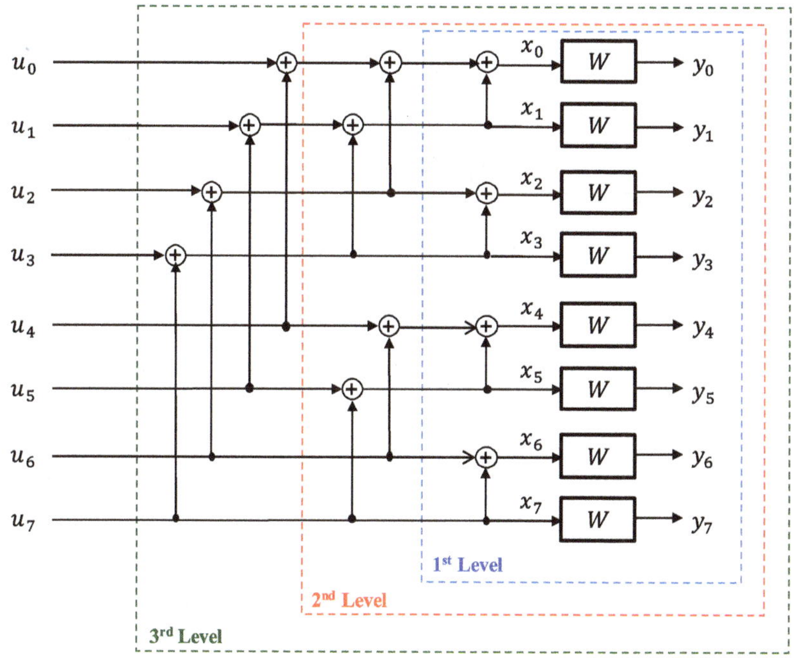

Fig. 2.7 The length-8 polar code with input vector $\mathbf{u}_8 = (u_0, \ldots, u_7)$ and output vector $\mathbf{x}_8 = (x_0, \ldots, x_7)$

which maps ℓ onto a *binary* expansion as:

$$g(\ell) = (b_n^\ell, \ldots, b_1^\ell), \tag{2.20}$$

such that $\ell = \sum_{i=1}^{n} b_i^\ell 2^{i-1}$. Also, let $w_{\mathrm{H}}(\mathbf{b})$ be the number of non-zero elements in a vector $\mathbf{b}$ (called the Hamming weight). Given $\mathbf{u}_{\mathrm{N}} = (u_0, \ldots, u_{N-1})$ and $\mathcal{A} \subset \{0, \ldots, N-1\}$, $\mathbf{u}_{\mathcal{A}}$ denotes the subvector $(u_i : i \in \mathcal{A})$.

Let $\mathbf{u}_{\mathrm{N}} = (u_0, \ldots, u_{\mathrm{N}-1})$ and $\mathbf{x}_{\mathrm{N}} = (x_0, \ldots, x_{\mathrm{N}-1})$ denote the input and output vectors of a length-N polar code, respectively. As seen in Fig. 2.7, the encoding of the polar code is represented as:

$$\mathbf{u}_{\mathrm{N}}\mathbf{P}_{\mathrm{N}} = \mathbf{x}_{\mathrm{N}}. \tag{2.21}$$

In this chapter, the bit-reverse permutation, denoted by $\psi(\cdot)$, is applied to the decoding part, instead of the encoding part as in [1]. Accordingly, SC decoding successively recovers the $\hat{u}_{\psi(i)}$ for $i = 0, \ldots, N-1$ in that order. When $N = 8$, SC decoding order is determined as:

$$\hat{u}_0 \rightarrow \hat{u}_4 \rightarrow \hat{u}_2 \rightarrow \hat{u}_6 \rightarrow \hat{u}_1 \rightarrow \hat{u}_5 \rightarrow \hat{u}_3 \rightarrow \hat{u}_7. \tag{2.22}$$

A punctured polar code of a length $N_p < N$ is constructed by eliminating $N - N_p$ coded bits, which is described by the length-N mother polar code and a puncturing pattern $\mathbf{p}_\mathrm{N} = (p_0, \ldots, p_{N-1}) \in \{0, 1\}^N$ such that $w_\mathrm{H}(\mathbf{p}_\mathrm{N}) = N_p$. Here, $p_i = 0$ implies that the i-th coded bit (e.g., x_i) is punctured and thus not transmitted. The index N will be dropped in $\mathbf{p}_\mathrm{N}$ if it is identified from the context. Given $\mathbf{p}$, the zero-location set which contains the locations of punctured bits is defined as:

$$\mathcal{B}_\mathbf{p} \triangleq \{i \in \{0, \ldots, N-1\} : p_i = 0\}. \tag{2.23}$$

In this section, a puncturing pattern can be specified by either a binary vector $\mathbf{p}$ or a zero-location set $\mathcal{B}_\mathbf{p}$. The corresponding unpunctured coded bits are denoted by $\mathbf{x}_{\mathrm{N_P}} = (x_i : i \in \mathcal{B}_\mathbf{p}^c)$, and, accordingly, the N_p channel observations are denoted by $\mathbf{y}_{\mathrm{N_P}} = (y_i : i \in \mathcal{B}_\mathbf{p}^c)$. Also, the notion of polarized channels in polar codes can be extended into punctured polar codes straightforwardly as follows. Given a BI-DMC $W : \mathcal{X} \to \mathcal{Y}$, where $\mathcal{X} = \{0, 1\}$ and $\mathcal{Y}$ denote the, respectively, input and output alphabets, a length-N polar code, and a puncturing pattern $\mathbf{p}$, the transition probability of the i-th polarized channel of the resulting punctured polar code is defined as:

$$\begin{aligned} &W^{(i)}(\mathbf{y}_{\mathrm{N_p}}, \mathbf{u}_{i-1}, \mathbf{p}|u_i) \\ &= \frac{1}{2^{N-1}} \sum_{u_{i+1},\ldots,u_N} \sum_{\mathbf{y}_\mathrm{N} \in \pi_\mathbf{p}\left(\{\mathbf{y}_{\mathrm{N_p}}\}\right)} W^N(\mathbf{y}_\mathrm{N}|\mathbf{u}_\mathrm{N}\mathbf{P}_\mathrm{N}), \end{aligned} \tag{2.24}$$

where $\pi_\mathbf{p}(S) \triangleq \{\mathbf{y}_\mathrm{N} \in \mathcal{Y}^N : (y_i : i \in \mathcal{B}_\mathbf{p}^c) \in S\}$. Throughout the section, $W_\mathbf{p}^{(i)}$ represents the i-th polarized channel with the transition probability in (2.24). Also, let $I(W_\mathbf{p}^{(i)})$ be the corresponding symmetric capacity. Then, the information set $\mathcal{A}$ of a punctured polar code is obtained by taking the indices of the $|\mathcal{A}|$ highest capacities as:

$$\mathcal{A} = \{\ell_1, \ldots \ell_{\mathcal{A}}\}, \tag{2.25}$$

where $I(W_\mathbf{p}^{(\ell_1)}) \geq I(W_\mathbf{p}^{(\ell_2)}) \geq \cdots I(W_\mathbf{p}^{(\ell_{|\mathcal{A}|})})$. Accordingly, a punctured polar code is denoted as $\mathcal{C}(\mathbf{P}_N, \mathbf{u}_{\mathcal{A}^c}, \mathcal{A}, \mathbf{p})$ with a frozen-bit vector $\mathbf{u}_{\mathcal{A}^c}$. The encoding structure of a punctured polar code is illustrated in Fig. 2.8.

For a given $\mathbf{p}$, the *zero-capacity set*, which contains the zero-capacity polarized channels, is defined as:

$$\mathcal{D}_\mathbf{p}^W \triangleq \{i \in \{0, \ldots, N-1\} : I(W_\mathbf{p}^{(i)}) = 0\}. \tag{2.26}$$

In particular, when W is a perfect channel (i.e., noiseless deterministic channel), the above set is denoted by $\mathcal{D}_\mathbf{p}$. Due to the nesting property of polarized channels, we have:

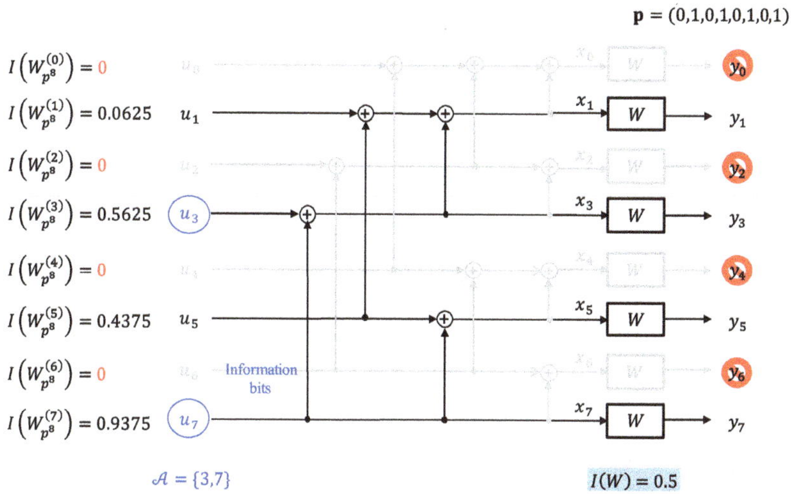

Fig. 2.8 The punctured polar code with $N_p = 4$, $\mathcal{A} = \{3, 7\}$, and $\mathbf{p} = (0, 1, 0, 1, 0, 1, 0, 1)$

$$\mathcal{D}_{\mathbf{p}} \subseteq \mathcal{D}_{\mathbf{p}}^{W}. \tag{2.27}$$

Moreover, it was shown in [17] that:

$$|\mathcal{B}_{\mathbf{p}}| = |\mathcal{D}_{\mathbf{p}}| = N - w_{\mathrm{H}}(\mathbf{p}). \tag{2.28}$$

This implies that the number of zero-capacity synthesized channels is equal to that of punctured coded bits. Thus, all the polarized channels whose indices belong to $\mathcal{D}_{\mathbf{p}}$ should be frozen-bit channels. Given $\{I(W_{\mathbf{p}}^{(i)})) : i = 0, \ldots, N-1\}$, let:

$$\mathcal{L} = \text{max-ind}^{(t)}\{I(W_{\mathbf{p}}^{(i)}) : i = 0, 1 \ldots, N-1\} \tag{2.29}$$

$$\mathcal{S} = \text{min-ind}^{(t)}\{I(W_{\mathbf{p}}^{(i)}) : i = 0, 1, \ldots, N-1\}, \tag{2.30}$$

where $\mathcal{L}$ and $\mathcal{S}$ contain the indices corresponding to the t largest and smallest values in $\{I(W_{\mathbf{p}}^{(i)})\}$, respectively.

2.4.1 A Reciprocal Puncturing

In this section, the so-called *reciprocal* puncturing is introduced, which will be used as the key technology to construct RCPP codes. As noticed before, all the polarized channels according to the zero-capacity set $\mathcal{D}_{\mathbf{p}}$ should be frozen-bit channels, i.e.:

$$u_i = 0 \text{ for } i \in \mathcal{D}_{\mathbf{p}} \Rightarrow \mathcal{A} \cap \mathcal{D}_{\mathbf{p}} = \phi. \tag{2.31}$$

Otherwise, the corresponding punctured polar code surely leads to a frame (or block) error. Thus, we need to identify $\mathcal{D}_{\mathbf{p}}$ for the construction of a good information set $\mathcal{A}$. In [17], it was shown that there exists a class of puncturing patterns to satisfy $\mathcal{D}_{\mathbf{p}} = \mathcal{B}_{\mathbf{p}}$, which are referred to as *reciprocal puncturing*. Furthermore, their sufficient and necessary conditions are obtained as follows.

Theorem 2.1 ([17]) *A puncturing pattern* **p** *is* reciprocal *if and only if the following properties hold:*

$$\textbf{zero}-\textbf{inclusion}: 0 \in \mathcal{B}_{\mathbf{p}}$$

$$\textbf{one}-\textbf{covering}: \textit{For any } i \in \mathcal{B}_{\mathbf{p}}, \textit{ we have } j \in \mathcal{B}_{\mathbf{p}} \textit{ for all } j \textit{ such that } i \succeq_1 j,$$

where $i \succeq_1 j$ *implies that, if* $b_k^{(j)} = 1$, *then* $b_k^{(i)} = 1$ *for all* $k = 1, \ldots, n$ *and* $i \succeq_1 0$ *for every* $i > 0$, *and where* $g(i) = (b_n^{(i)}, \ldots, b_1^{(i)})$ *and* $g(j) = (b_n^{(j)}, \ldots, b_1^{(j)})$.

We remark that the zero-capacity set $\mathcal{D}_{\mathbf{p}}$ of a reciprocal puncturing pattern is directly identified from the puncturing pattern **p** (equivalently, $\mathcal{B}_{\mathbf{p}}$), which makes it much easier to form a good information set $\mathcal{A}$ of the punctured polar code. From Theorem 2.1, the one-covering property implies that if $p_7 = 0$ in a reciprocal puncturing pattern $\mathbf{p} = (p_0, \ldots, p_{15})$, then the following positions should be punctured:

- weight-2 locations: $g^{-1}((0, 0, 1, 1))$, $g^{-1}((0, 1, 0, 1))$, and $g^{-1}((0, 1, 1, 0))$;
- weight-1 locations: $g^{-1}((0, 0, 0, 1))$, $g^{-1}((0, 0, 1, 0))$, and $g^{-1}((0, 1, 0, 0))$.

Next, some useful properties of reciprocal puncturing are derived. Let Π_n be the set of all permutations of $(1, 2, \ldots, n)$ with $|\Pi_n| = n!$. When $n = 3$, we have:

$$\Pi_3 = \{(1, 2, 3), (1, 3, 2), (2, 1, 3), (2, 3, 1), (3, 1, 2), (3, 2, 1)\}.$$

Definition 2.2 Given **p** and $\sigma \in \Pi_n$, a *permuted* puncturing pattern $\mathbf{p}^\sigma$ is defined with its zero-location set:

$$\mathcal{B}_{\mathbf{p}^\sigma} = \{g^{-1}((b^i_{\sigma(n)}, \ldots, b^i_{\sigma(1)})) : i \in \mathcal{B}_{\mathbf{p}}\}, \tag{2.32}$$

where $g(i) = (b^i_n, \ldots, b^i_1)$.

For example, if $\mathbf{p} = (0, 0, 0, 0, 1, 1, 1, 1)$ and $\sigma = (3, 2, 1)$, then we have $\mathbf{p}^{(3,2,1)} = (0, 1, 0, 1, 0, 1, 0, 1)$ (see Fig. 2.9). With this definition, we can get:

Proposition 2.3 ([19]) *If* **p** *is reciprocal, then* $\mathbf{p}^\sigma$ *is also reciprocal for any permutation* $\sigma \in \Pi_n$.

Proposition 2.4 ([19]) *For any reciprocal puncturing pattern* **p**, *we have:*

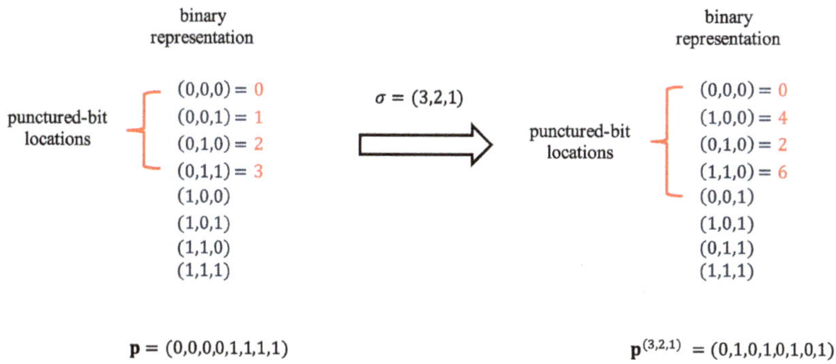

Fig. 2.9 A permuted puncturing pattern

$$\mathbf{P}_{\mathrm{N}}(\mathcal{B}_{\mathbf{p}}, \mathcal{B}_{\mathbf{p}}^{c}) = \mathbf{0}. \tag{2.33}$$

From Proposition 2.4, we know that, when a reciprocal puncturing pattern **p** is used, assigning "unknown" values to the polarized channels belong to $\mathcal{B}_{\mathbf{p}}$ does not impact on the performance of the punctured polar code. This property plays a crucial role in developing an information-copy technique to yield effective information sets of RCPP codes. For the rest of this section, a class of reciprocal puncturing patterns are provided, which are obtained from the so-called successive puncturing and a bit-wise permutation. First we define:

Definition 2.5 Let $\dot{\mathbf{p}}_{N_p}$ denote the *successive* puncturing pattern which punctures the first $N - N_p$ coded bits, i.e., its zero-location set is given by:

$$\mathcal{B}_{\dot{\mathbf{p}}_{N_p}} = \{0, 1, \ldots, N - N_p - 1\}, \tag{2.34}$$

where N_p represents the number of unpunctured coded bits. Then, we have:

- It is reciprocal;
- Its permuted puncturing pattern $\dot{\mathbf{p}}_{N_p}^{\sigma}$ is also reciprocal for any $\sigma \in \Pi_n$.

In [16], $\dot{\mathbf{p}}_{N_p}^{(n,n-1,\ldots,1)}$ is known as *quasi-uniform puncturing* (QUP) in [16], where the QUP algorithm proceeds as:

- Step (1) Initialize the **p** as all ones, and then set the first $N - N_p$ bits as zeros;
- Step (2) Perform bit-reversal permutation on the **p**, and then the resulting puncturing pattern is referred to as QUP.

Consider the example of $N = 8$ and $N_p = 5$. Following the above steps, $\mathbf{p} = (0, 0, 0, 1, 1, 1, 1, 1)$ is chosen as initialization, and then by performing the bit-reversal permutation, $\mathbf{p} = (0, 1, 0, 1, 0, 1, 1, 1)$ is obtained. This is exactly same with $\dot{\mathbf{p}}_5^{(3,2,1)}$ from Definition 2.5.

2.4.2 A Hierarchical Puncturing

In this section, a hierarchical puncturing is defined, and its useful properties for the construction of a RCPP code are derived.

Definition 2.6 A reciprocal puncturing $\mathbf{p}$ with $w_{\mathrm{H}}(\mathbf{p}) = 2^{\bar{n}} = \bar{N}$ for some $\bar{n} < n$ is said to be *hierarchical* if:

$$\mathbf{u}_{\mathrm{N}}\mathbf{P}_{\mathrm{N}}(\{0,\ldots,N-1\},\mathcal{B}^c_{\mathbf{p}}) = (u_i : i \in \mathcal{B}^c_{\mathbf{p}})\mathbf{P}_{\bar{\mathrm{N}}}. \tag{2.35}$$

Since $\mathbf{G}_N(\mathcal{B}_{\mathbf{p}},\mathcal{B}^c_{\mathbf{p}}) = \mathbf{0}$ from Proposition 2.4, the condition (2.35) is equivalent to:

$$\mathbf{P}_{\mathrm{N}}(\mathcal{B}^c_{\mathbf{p}},\mathcal{B}^c_{\mathbf{p}}) = \mathbf{P}_{\bar{\mathrm{N}}}, \tag{2.36}$$

where note that $|\mathcal{B}^c_{\mathbf{p}}| = w_{\mathrm{H}}(\mathbf{p}) = \bar{N}$.

Now, we will explain two key properties of hierarchical puncturing with a simple example. Consider the $N = 8$, $\bar{N} = 4$, and the reciprocal puncturing $\dot{\mathbf{p}}_4^{(3,2,1)}$ is used. Then, the input-output relationship of the punctured polar code is represented as:

$$(x_i : i \in \mathcal{B}^c_{\dot{\mathbf{p}}_4^{(3,2,1)}}) = \mathbf{u}_{\mathrm{N}}\mathbf{P}_{\mathrm{N}}(\{0,\ldots,7\},\mathcal{B}^c_{\dot{\mathbf{p}}_4^{(3,2,1)}}) \tag{2.37}$$

where (a) is due to the fact that:

$$\mathbf{P}_{\mathrm{N}}(\mathcal{B}^c_{\dot{\mathbf{p}}_4^{(3,2,1)}},\mathcal{B}^c_{\dot{\mathbf{p}}_4^{(3,2,1)}}) = \mathbf{P}_{\bar{\mathrm{N}}}$$

$$\mathbf{P}_{\mathrm{N}}(\mathcal{B}_{\dot{\mathbf{p}}_4^{(3,2,1)}},\mathcal{B}^c_{\dot{\mathbf{p}}_4^{(3,2,1)}}) = \mathbf{0}.$$

Clearly, $\dot{\mathbf{p}}_4^{(3,2,1)}$ is hierarchical as it satisfies the condition (2.36) in Definition 2.6. From (2.31), we know that $(u_i : i \in \mathcal{B}^c_{\dot{\mathbf{p}}_4^{(3,2,1)}})$ only carry the information bits. Let $\mathbf{u}_{\bar{\mathrm{N}}} = (u_i : i \in \mathcal{B}^c_{\dot{\mathbf{p}}_4^{(3,2,1)}})$ and $\mathbf{x}_{\bar{\mathrm{N}}} = (x_i : i \in \mathcal{B}^c_{\dot{\mathbf{p}}_4^{(3,2,1)}})$ (i.e., unpunctured coded bits). From (2.37), a length-$\bar{N}$ polar code can be created as:

$$\mathbf{x}_{\bar{\mathrm{N}}} = \mathbf{u}_{\bar{\mathrm{N}}}\mathbf{P}_{\bar{\mathrm{N}}}, \tag{2.38}$$

with $\bar{N} = 4 < N = 8$. From this example, the following two important properties of hierarchical puncturing are identified.

- **Property 1**: As shown in Fig. 2.10, information bits can be decoded using the length-$\bar{N}$ polar decoder with the (unpunctured) observation $\mathbf{y}_{\bar{\mathrm{N}}}$, instead of using the original polar decoder;

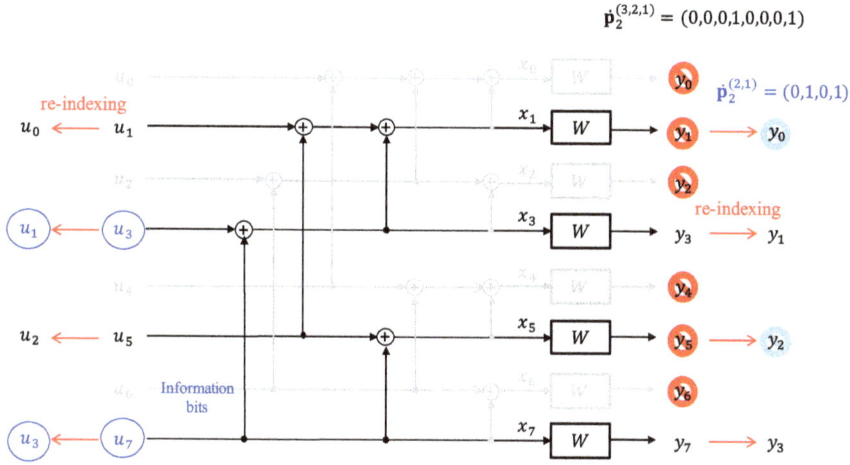

Fig. 2.10 The punctured polar code with $N_p = 2$ and $\dot{\mathbf{p}}_2^{(3,2,1)} = (0, 0, 0, 1, 0, 0, 0, 1)$

- **Property 2**: Since it is also reciprocal, assigning unknown values to the polarized channels corresponding to $\mathcal{B}_{\dot{\mathbf{p}}_4^{(3,2,1)}}$ does not impact on the performance of the length-$\bar{N}$ punctured polar code.

These properties will be leveraged to construct a RCPP code in Sect. 2.4.3.

Now, a class of hierarchical puncturing will be provided. Consider the simple case of $N = 8$ and $\bar{N} = 4$, i.e., half of the coded bits are punctured. From Fig. 2.7, the polar encoding consists of $\log N = 3$ levels. Also, it is easily verified that $\dot{\mathbf{p}}_4 = (0, 0, 0, 0, 1, 1, 1, 1)$ (called successive puncturing) is hierarchical because the following holds:

$$\mathbf{P}_8(\mathcal{B}^c_{\dot{\mathbf{p}}_4}, \mathcal{B}^c_{\dot{\mathbf{p}}_4}) = \mathbf{P}_4. \tag{2.39}$$

Figure 2.8 shows that its permuted puncturing pattern $\dot{\mathbf{p}}_4^{(3,2,1)}$ is also hierarchical. From them, we identify that

$$b^i_3 = 0 \text{ for all } i \in \mathcal{B}_{\dot{\mathbf{p}}_4} \text{ and } b^i_1 = 0 \text{ for all } i \in \mathcal{B}_{\dot{\mathbf{p}}_4^{(3,2,1)}}, \tag{2.40}$$

where $g(i) = (b^i_3, b^i_2, b^i_1)$. Specifically, the third and first levels in Fig. 2.7 are completely eliminated. Also, the remaining parts in the both cases yield the length-4 polar encoding structure, namely, they satisfy the condition (2.36). These arguments can be generalized straightforwardly. Consider that half of the coded bits are punctured (e.g., $\bar{N} = N/2$). If $b^i_j = 0$ for all $i \in \mathcal{B}_{\dot{\mathbf{p}}^{\sigma}_{N/2}}$, then the j-th level in the polar encoding structure is completely eliminated, and, thus, the condition (2.36) is satisfied. From this analysis, we can conclude that $\dot{\mathbf{p}}_{N/2}^{\sigma}$ is hierarchical for any $\sigma \in \Pi_n$, which is generalized for any $\bar{N} = 2^{\bar{n}}$ in Theorem 2.7 below.

Theorem 2.7 ([19]) *For any $\bar{N} = 2^{\bar{n}}$ with $1 \leq \bar{n} < n$, a puncturing pattern $\dot{\mathbf{p}}_{\bar{N}}^{\sigma}$ in Definition 2.5 is hierarchical for any $\sigma \in \Pi_n$.*

To construct a RCPP code efficiently, the puncturing patterns in Remark 1 will be used (see Sect. 2.4.3 for the detailed procedures).

Remark 1 (Rate-compatible, reciprocal, and hierarchical) The puncturing patterns $\dot{p}_{N_p}^{\sigma}$ in Definition 2.5 will be used to construct a RCPP code. In this remark, the key properties of $\dot{p}_{N_p}^{\sigma}$ are provided as follows:

- Given $N_1 > N_2 > \cdots > N_m$, $\dot{p}_{N_1}^{\sigma}, \dot{p}_{N_2}^{\sigma}, \ldots, \dot{p}_{N_m}^{\sigma}$ satisfy the rate-compatible constraint;
- They are reciprocal:
- A punctured polar code, obtained by $\dot{p}_{N_i}^{\sigma}$, can be decoded using a length-$2^{\lceil \log N_i \rceil}$ polar code, rather than a mother polar code.

2.4.3 RCPP Codes

A RCPP code consists of a family of (punctured) polar codes for which the corresponding puncturing patterns satisfy the RC constraint in (2.43). In addition, all the member codes in the family should employ a *common* information set $\mathcal{A}$ as information bits should be unchanged during retransmissions in HARQ-IR schemes. Unfortunately, it is not manageable to find an optimal common information set. Usually, it is optimized for a target code in the family (e.g., the mother polar code or the highest-rate code). Thus, the resulting information set cannot be good for the other member codes in the family, which results in poor performances especially when a rate-change is large. This is main bottleneck to construct a good RCPP code for HARQ-IR schemes.

The above problem is addressed by introducing the so-called *information-copy* technique based on hierarchical (or reciprocal) puncturing. The fundamental idea can be outlined as follows. Some part of information bits are repeated to frozen-bit channels as well as information-bit channels. This can yield an information-dependent frozen vector. Also, the positions of such information bits and frozen-bit channels are determined from RC puncturing patterns and the corresponding optimized information sets. We remark that, in the encoding part, the information-dependent frozen vector is used. In the decoding part, *effective* information sets, which are optimized information sets for the other codes in the family, are obtained by leveraging the common information set and the information-dependent frozen vector. Thus, each member code in the family can be decoded using its own optimized information set. One can concern that the information-dependent frozen bits (i.e., unknown non-zero frozen bits) can cause a performance loss. This problem can be completely avoided due to Property 2 of hierarchical (or reciprocal) puncturing in Sect. 2.4.2. The detailed procedures to construct RCPP codes are provided below.

Suppose that we construct a RCPP code to send k information bits with various rates:

$$r_1 = \frac{k}{N_1} < r_2 = \frac{k}{N_2} < \cdots < r_m = \frac{k}{N_m}. \tag{2.41}$$

The construction method is outlined as follows:

- **step (1)** The mother polar code with the length $\bar{N}_1 = 2^{\bar{n}_1}$ is chosen, where $\bar{n}_1 = \lceil \log N_1 \rceil$.
- **step (2)** A family of RC puncturing patterns are determined, where they are denoted by the length-$\bar{N}_1$ binary vectors as:

$$\mathbf{p}^{(1)}, \mathbf{p}^{(2)}, \mathbf{p}^{(3)}, \ldots, \text{and } \mathbf{p}^{(m)}, \tag{2.42}$$

such that $w_{\mathrm{H}}(\mathbf{p}^{(i)}) = N_i$ for $i \in \{1, \ldots, m\}$. Each $\mathbf{p}^{(i)}$ generates the punctured polar code of rate r_i. Due to the RC constraint, they should satisfy:

$$\text{RC-Condition:} \mathcal{B}_{\mathbf{p}^{(1)}} \subset \mathcal{B}_{\mathbf{p}^{(2)}} \subset \mathcal{B}_{\mathbf{p}^{(3)}} \subset \cdots \subset \mathcal{B}_{\mathbf{p}^{(m)}}. \tag{2.43}$$

The reciprocal (or hierarchical) puncturing patterns in Remark 1 are chosen as:

$$\mathbf{p}^{(i)} \triangleq \dot{\mathbf{p}}_{N_i}^{\sigma}, \tag{2.44}$$

for a fixed $\sigma \in \Pi_{\bar{n}_1}$ and for $i \in \{1, \ldots, m\}$.
- **step (3)** A common information set $\mathcal{A}$ is optimized by considering the puncturing pattern $\mathbf{p}^{(m)}$. Since $\mathbf{p}^{(m)}$ is reciprocal, $\mathcal{A}$ should satisfy $\mathcal{A} \cap \mathcal{B}_{\mathbf{p}^{(m)}} = \phi$.
- **step (4)** The m (punctured) polar codes in the family are denoted by $\{\mathcal{C}^{(i)} : i = 1, \ldots, m\}$ with:

$$\mathcal{C}^{(i)} \triangleq \mathcal{C}(\mathbf{P}_{\bar{N}_1}, \mathbf{u}_{\mathcal{A}^c}, \mathcal{A}, \mathbf{p}^{(i)} = \dot{\mathbf{p}}_{N_i}^{\sigma}), \tag{2.45}$$

where an *information-dependent* frozen vector $\mathbf{u}_{\mathcal{A}^c}$ will be defined in Sect. 2.4.3.

From now on, we will explain how to construct an information-dependent frozen vector and the encoding/decoding procedures of RCPP codes. First, some notations for information sets are provided, which will be used in the sequel.

Definition 2.8 Let $\mathcal{T}^{(i)}$ be the optimal information set for each code $\mathcal{C}^{(i)}$ in the family. This information set can be optimized independently by considering an associated puncturing pattern. Because of a certain requirement (will be defined below), the effective information set obtained by the information-copy technique can be different from the optimal one. To clarify this difference, the effective information set is denoted by $\mathcal{A}^{(i)}$. Accordingly, $\mathcal{C}^{(i)}$ in the family is decoded using the effective information set $\mathcal{A}^{(i)}$.

Information-Dependent Frozen Vector

We introduce an information-copy technique which generates an information-dependent frozen vector (equivalently, *effective* information sets $\mathcal{A}^{(i)}$ for the codes $\mathcal{C}^{(i)}, i = 1, \ldots, m-1$). The main ideas will be explained using the simple case of $K = 2$ and:

$$r_1 = \frac{2}{8} < r_2 = \frac{2}{5} < r_3 = \frac{2}{3}. \tag{2.46}$$

From (2.44) in Step (2), the RC puncturing patterns are obtained as:

$$\mathbf{p}^{(1)} = \mathbf{1} = (1, 1, 1, 1, 1, 1, 1, 1) \tag{2.47}$$

$$\mathbf{p}^{(2)} = \dot{\mathbf{p}}_5^{(3,2,1)} = (0, 1, 0, 1, 0, 1, 1, 1) \tag{2.48}$$

$$\mathbf{p}^{(3)} = \dot{\mathbf{p}}_3^{(3,2,1)} = (0, 0, 0, 1, 0, 1, 0, 1). \tag{2.49}$$

Also, the corresponding information sets are optimized as:

$$\mathcal{T}^{(1)} = \{6, 7\}, \mathcal{T}^{(2)} = \{6, 7\}, \text{ and } \mathcal{T}^{(3)} = \{3, 7\}. \tag{2.50}$$

Following Step (3), the common information set $\mathcal{A}$ is determined by:

$$\mathcal{A} = \mathcal{A}^{(3)} = \mathcal{T}^{(3)} = \{3, 7\}. \tag{2.51}$$

Note that $\mathcal{A} = \mathcal{T}^{(3)}$ is not optimal information set for the code $\mathcal{C}^{(2)}$ as $\mathcal{T}^{(2)} = \{6, 7\} \neq \mathcal{T}^{(3)}$, i.e., $I(W_{\mathbf{p}^{(2)}}^{(6)}) > I(W_{\mathbf{p}^{(2)}}^{(5)})$ (see Fig. 2.11). Nevertheless, in conventional approach, all the member codes in the family use $\mathcal{A}$ as the information sets; whereas, in RCPP codes, each code in the family can employ its own optimized information set, i.e., $\mathcal{C}^{(2)}$ can use $\mathcal{T}^{(2)} = \{6, 7\}$ as its information set. In the below, we will explain how it works.

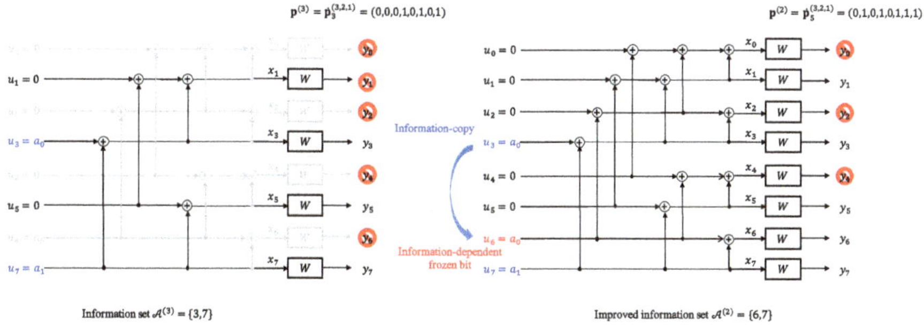

Fig. 2.11 Description of an information-copy technique

From Fig. 2.10, $\mathcal{C}^{(3)}$ can be decoded using the length-4 polar code:

$$\mathcal{C}(\mathbf{P}_4, \{1, 3\}, (0, 0), \mathbf{p} = (0, 1, 1, 1)),$$

rather than the mother polar code $\mathcal{C}(\mathbf{P}_8, \{3, 7\}, \mathbf{u}_{\mathcal{A}^c}, \mathbf{1})$. This is enabled as $\dot{\mathbf{p}}_3^{(3,2,1)}$ is hierarchical (see Property 1 in Sect. 2.4.2). We also remark that the frozen-bit channels belonging to $\mathcal{B}_{\mathbf{p}^{(3)}} = \{0, 2, 4, 6\}$ are not related to this decoding, and, thus, "unknown" frozen bits can be assigned to those frozen-bit channels without degrading the performance of $\mathcal{C}^{(3)}$ (see Property 2 in Sect. 2.4.2). The information-bit $u_3 = a_0$ can be copied to the frozen-bit channel 6 (i.e., $u_6 = u_3 = a_0$). In the decoding of $\mathcal{C}^{(2)}$, u_5 can operate as a frozen bit because the copied one u_6 is decoded earlier. This shows that $\mathcal{C}^{(2)}$ can be decoded using the *effective* information set $\mathcal{A}^{(2)} = \mathcal{T}^{(2)} = \{6, 7\}$, instead of using $\mathcal{A} = \{3, 7\}$. We next focus on the decoding of $\mathcal{C}^{(1)}$. Likewise, the information-bit a_0 can be copied to the frozen-bit channel 4 (i.e., $u_6 = u_3 = u_4 = a_0$). In the decoding of $\mathcal{C}^{(1)}$, both u_3 and u_6 can perform as the frozen bits because the copied one u_4 is decoded earlier. From Property 2 in Sect. 2.4.2, the copied bit u_6 does not impact on the decoding of both $\mathcal{C}^{(2)}$ and $\mathcal{C}^{(3)}$. Therefore, $\mathcal{C}^{(1)}$ can be decoded using the *effective* information set $\mathcal{A}^{(1)} = \mathcal{T}^{(1)} = \{6, 7\}$. From this example, $\mathcal{A}^{(i)} = \mathcal{T}^{(i)}$. But, it is not necessary in general. Consequently, the corresponding information-dependent frozen vector is derived as:

$$\mathbf{u}_{\mathcal{A}^c} = (u_0, u_1, u_2, u_4, u_5, u_6) = (0, 0, 0, u_3 = a_0, 0, u_3 = a_0). \tag{2.52}$$

From the above results, the punctured polar codes in the family are defined as $\mathcal{C}^{(i)} = \mathcal{C}(\mathbf{P}_8, \mathcal{A} = \{3, 7\}, \mathbf{u}_{\mathcal{A}^c} = (0, 0, 0, u_3 = a_0, 0, u_3 = a_0), \mathbf{p}^{(i)})$ for $i = 1, 2, 3$. In this example, an input vector $\mathbf{u}$ (i.e., $\mathbf{u}_{\mathcal{A}}$ and $\mathbf{u}_{\mathcal{A}^c}$) of the mother polar code is determined as:

$$(a_0, a_1) \underbrace{\begin{bmatrix} 0\,0\,0\,1\,1\,0\,1\,0 \\ 0\,0\,0\,0\,0\,0\,0\,1 \end{bmatrix}}_{\triangleq \mathbf{Q}} = (u_0, u_1, \ldots, u_{\bar{N}_1 - 1}), \tag{2.53}$$

where (a_0, a_1) represents the two information bits and the $|\mathcal{A}| \times \bar{N}_1$ matrix $\mathbf{Q}$ is referred to as precoding matrix. Clearly, $\mathbf{Q}$ is formed as a function of $\mathbf{u}_{\mathcal{A}^c}$ and $\mathcal{A}$, where $\mathbf{Q}(0, 3) = \mathbf{Q}(1, 7) = 1$ for $\mathcal{A}^{(3)} = \{3, 7\}$, $\mathbf{Q}(0, 6) = 1$ for $\mathcal{A}^{(2)} \setminus \mathcal{A}^{(3)} = \{6\}$, and $\mathbf{Q}(0, 4) = 1$ for $\mathcal{A}^{(1)} \setminus \mathcal{A}^{(2)} = \{4\}$. In order not to affect the performances of the other codes in the family, information-copy technique should be performed only when the following requirement holds:

- **IC-Requirement:** To construct $\mathcal{A}^{(i)}$ from a given $\mathcal{A}^{(i+1)}$, the indices of the copied frozen-bit channels should belong to $(\mathcal{T}^{(i)} \setminus \mathcal{A}^{(i+1)}) \cap \mathcal{B}_{\dot{\mathbf{p}}_{N_{i+1}}^{(3,2,1)}}$, for $i = 1, 2$. Here, the intersection with $\mathcal{B}_{\dot{\mathbf{p}}_{N_{i+1}}^{(3,2,1)}}$ is required to guarantee that the copied bits do not affect the performance of the code $\mathcal{C}^{(i+1)}$ (see Property 2 in Sect. 2.4.2).

Due to IC-requirement, some information bits in the above example cannot be copied and, accordingly, $\mathcal{T}^{(i)} \neq \mathcal{A}^{(i)}$. Suppose that:

$$\mathbf{p}^{(1)} = \dot{\mathbf{p}}_8^{(3,2,1)} = (1, 1, 1, 1, 1, 1, 1, 1)$$
$$\mathbf{p}^{(2)} = \dot{\mathbf{p}}_5^{(3,2,1)} = (0, 1, 0, 1, 0, 1, 1, 1),$$

and the associated optimal information sets are given as $\mathcal{T}^{(1)} = \{4, 5\}$ and $\mathcal{T}^{(2)} = \mathcal{A}^{(2)} = \{3, 7\}$, respectively. From $\mathbf{p}^{(2)}$, $\mathcal{B}_{\dot{\mathbf{p}}_5^{(3,2,1)}} = \{0, 2, 4\}$. Note that from IC-Requirement, $\mathcal{A}^{(2)}$ should not include the index 5 because of $5 \notin \mathcal{B}_{\dot{\mathbf{p}}_5^{(3,2,1)}}$. Hence, we have that $\mathcal{A}^{(1)} = \{4, 7\} \neq \mathcal{T}^{(2)}$.

Taking the above IC-requirement into account, the systematic algorithms to generate *effective* information sets and a precoding matrix $\mathbf{Q}$ are described in Algorithms 1 and 2, respectively. The precoding matrix $\mathbf{Q}$ is used at the encoding side, and the effective information sets are used at the decoding side. The main ideas of these algorithms are briefly explained as follows. Consider the HARQ-IR scheme to support the m code rates $r_1 = \frac{k}{N_1} < \cdots < r_m = \frac{k}{N_m}$. Then, we have:

- The m information sets $\mathcal{T}^{(1)}, \mathcal{T}^{(2)}, \ldots,$ and $\mathcal{T}^{(m)}$ are optimized by considering the puncturing patterns (i.e., based on $I(W_{\mathbf{p}^{(i)}}^{(j)})$).
- From $\mathcal{T}^{(i)}$'s, Algorithm 1 generates the effective information sets $\mathcal{A}^{(i)}$ for $i = 1, \ldots, m$. Note that $\mathcal{A}^{(i)}$ is not always identical to $\mathcal{T}^{(i)}$.
- From $\mathcal{A}^{(i)}$'s, Algorithm 2 yields a precoding matrix $\mathbf{Q}$ (i.e., information-dependent frozen vector).

Algorithm 1 Effective information sets $\mathcal{A}^{(i)}$'s

Input: Optimized information sets $\mathcal{T}^{(i)}$ for $i \in \{1, \ldots, m\}$ and set $\mathcal{A} = \mathcal{A}^{(m)} = \mathcal{T}^{(m)}$.
Output: Optimized (effective) information sets $\mathcal{A}^{(j)}$ for $j \in \{1, \ldots, m\}$.
Algorithm:

For $j = m - 1, \ldots, 1$

1. Let

$$\mathcal{I}_1 = (\mathcal{T}^{(j)} \setminus \mathcal{A}^{(j+1)}) \cap \mathcal{B}_{\dot{\mathbf{p}}^{\sigma}_{N_{j+1}}} \triangleq \{\ell_1, \ldots, \ell_{|\mathcal{I}_1|}\}$$

$$\mathcal{I}_2 = \text{min-ind}^{(|\mathcal{I}_1|)}\{I(W^{(i)}_{\mathbf{p}^{(j)}}) : i \in \mathcal{A}^{(j+1)}\},$$

where $\psi(\ell_1) < \cdots < \psi(\ell_{|\mathcal{I}|})$ with the bit-reverse permutation $\psi(\cdot)$.
2. Information-copy set $\mathcal{I}_c$:
   - Initialization: $\mathcal{I}_c = \phi$ and $\mathcal{I}_d = \phi$.
   - For $i = 1, \ldots, |\mathcal{I}_1|$
     1. $\mathcal{S} = \{q \in \mathcal{I}_2 \setminus \mathcal{I}_d : \psi(q) > \psi(\ell_i)\}$
     2. If $\mathcal{S} \neq \phi$, then

$$\mathcal{I}_c = \mathcal{I}_c \cup \{\ell_i\} \text{ and } \mathcal{I}_d = \mathcal{I}_d \cup \{q^*\},$$

where $q^* = \min_{q \in \mathcal{T}} \psi(q)$.

3. Effective information set: $\mathcal{A}^{(j)} \triangleq \mathcal{I}_c \cup (\mathcal{A}^{(j+1)} \setminus \mathcal{I}_d)$.

Algorithm 2 Precoding matrix $\mathbf{Q}$

Input: The effective information sets $\mathcal{A}^{(j)}$ for $j \in \{1, \ldots, m\}$ and length $\bar{N}_1$.
Output: The $|\mathcal{A}| \times \bar{N}_1$ precoding matrix $\mathbf{Q}$ where $\mathbf{Q}(i, j)$ denotes the (i, j)-th element of $\mathbf{Q}$ for $i = 0, 1, \ldots, |\mathcal{A}| - 1$ and $j = 0, 1, \ldots, \bar{N}_1 - 1$.
Initialization:

- $\mathcal{A}^{(m)} = \mathcal{A} = \left\{\ell_0, \ldots, \ell_{|\mathcal{A}^{(m)}|-1}\right\}$ and $\mathbf{Q} = \mathbf{0}$.
- Define a mapping $h : \mathcal{A}^{(m)} \rightarrow \{0, 1, \ldots, |\mathcal{A}^{(m)}| - 1\}$, i.e., $h(\ell_j) = j$ for $j = 0, \ldots, |\mathcal{A}^{(m)}| - 1$.
- Let $\mathcal{A}^{(j)} \setminus \mathcal{A}^{(j+1)} \triangleq \{\ell_1^{(j)}, \ldots, \ell_{d_j}^{(j)}\}$ for $j \in \{1, \ldots, J - 1\}$.

Algorithm:

- Assign $\mathbf{Q}(h(\ell_j), \ell_j) = 1$ for $j = 0, \ldots, |\mathcal{A}^{(m)}| - 1$.
- For $j = m - 1, \ldots, 1$
  Assign $\mathbf{Q}(h(i_t^{(j)}), i_t^{(j)}) = 1$ for $t \in [1 : d_j]$.

Encoding and Decoding

The encoding and decoding procedures of RCPP codes are described. First of all, the expression of RC puncturing patterns in (2.44) is simplified by introducing the *seed sequence*.

Definition 2.9 Given $\bar{N}_1$ and σ, a seed sequence is defined by the following length-$\bar{N}_1$ binary vector

$$\mathbf{s}_{\bar{N}_1}^{\sigma} \triangleq (g^{-1}(b_{\sigma(3)}^{0}, b_{\sigma(2)}^{0}, b_{\sigma(1)}^{0}), \ldots, g^{-1}(b_{\sigma(3)}^{\bar{N}_1-1}, b_{\sigma(2)}^{\bar{N}_1-1}, b_{\sigma(1)}^{\bar{N}_1-1})), \tag{2.54}$$

where $(b_n^i, \ldots, b_1^i)$ is the binary representation of "i" for $i = 0, \ldots, \bar{N}_1 - 1$. Also, $\mathbf{p}^{(i)} = \dot{\mathbf{p}}_{N_i}^{\sigma}$ in (2.44) is simply defined with the zero-location set as:

$$\mathcal{B}_{\mathbf{p}^{(i)}} = \{\mathbf{s}_{\bar{N}_1}^{\sigma}(i) : i = 0, 1, \ldots, N_i - 1\}. \tag{2.55}$$

Encoding Consider the RCPP code to support m code rates in (2.41). An input vector $\mathbf{u}$ is first generated using the precoding matrix $\mathbf{Q}$. Then, a polar-encoded output (i.e., codeword) $\mathbf{x}$ is produced. At the i-th (re)transmission, the some part of the encoded output $\mathbf{x} = \mathbf{u}\mathbf{P}_{\bar{N}_1}$ is transmitted:

$$(x_{\mathbf{s}_{\bar{N}_1}^{\sigma}(i)} : i = N_{m-i+2}, \ldots, N_{m-i+1} - 1), \tag{2.56}$$

for $i = 1, \ldots, m$ and with the initial value $N_{m+2} = 0$.

Decoding Focus on the decoding of the RCPP code after the i-th (re)transmission. In the conventional case, the polar decoding is performed using the mother polar code $\mathcal{C}^{(1)} = \mathcal{C}(\mathbf{P}_{\bar{N}_1}, \mathbf{u}_{\mathcal{A}^c}, \mathcal{A}, \mathbf{p}^{(1)})$. Whereas, as shown in Fig. 2.11, RCPP code can be decoded with the length-$\bar{N}_i$ (possibly shorter) polar code, defined as:

$$\mathcal{C}(\mathbf{P}_{\bar{N}_i}, \mathbf{u}_{\bar{\mathcal{A}}^c}, \bar{\mathcal{A}}, \mathbf{p}), \tag{2.57}$$

where $\bar{\mathcal{A}}$ (i.e., *effective* information set), $\mathbf{u}_{\bar{\mathcal{A}}^c}$, and $\mathbf{p}$ will be specified from the given $\mathbf{p}^{(i)}$, $\mathcal{A}^{(i)}$, and $\mathbf{u}_{\mathcal{A}^c}$. Recall that $\mathbf{u}_{\mathcal{A}^c}$ can include the "unknown" values. For the example in (2.53), $\mathbf{u}_{\mathcal{A}^c} = (u_0 = 0, u_1 = 0, u_2 = 0, u_4 = 0, u_5 = 0, u_6 = u_3)$ has the unknown frozen bit $u_6 = u_3$. During the decoding, if one of u_3 and u_6 are decoded, the other bit is immediately copied and operates as the "known" frozen-bit. For simplicity, define:

$$\ell_j^{(i)} \triangleq \text{the } (j+1)\text{-st smallest index in } \mathcal{B}_{\dot{\mathbf{p}}_{\bar{N}_i}^{\sigma}}^c, \tag{2.58}$$

for $j = 0, \ldots, |\mathcal{B}_{\dot{\mathbf{p}}_{\bar{N}_i}^{\sigma}}^c| - 1$. As shown in Fig. 2.10, using the *re-indexing*, we obtain:

$$\begin{aligned}\bar{\mathcal{A}} &= \{j : \ell_j^{(i)} \in \mathcal{A}^{(i)}\}\\ \mathbf{p} &= (p_t^{(i)} : t \in \mathcal{B}_{\dot{\mathbf{p}}_{\bar{N}_i}^{\sigma}}^c)\\ \mathbf{u}_{\bar{\mathcal{A}}^c} &= (u_t : t \in \mathcal{B}_{\dot{\mathbf{p}}_{\bar{N}_i}^{\sigma}}^c \setminus \mathcal{A}^{(i)}),\end{aligned} \tag{2.59}$$

where $\mathbf{p}^{(i)} = \dot{\mathbf{p}}_{N_i}^{\sigma} \triangleq (p_0^{(i)}, \ldots, p_{\bar{N}_1-1}^{(i)})$.

We revisit the simple case in Sect. 2.4.3 and focus on the polar decoding after third transmission. In this case, we have that $\mathcal{A}^{(3)} = \{3, 7\}$, $\mathbf{p}^{(3)} = \dot{\mathbf{p}}_3^{(3,2,1)}(0, 0, 0, 1, 0, 1, 0, 1)$, and $\mathbf{u}_{\mathcal{A}^c} = (u_0 = 0, u_1 = 0, u_2 = 0, u_4 = 0, u_5 = 0, u_6 = u_3)$. When $\bar{N}_3 = 4$, $\mathcal{B}^c_{\dot{\mathbf{p}}_4^{(3,2,1)}} = \{1, 3, 5, 7\}$ is obtained. From (2.58), we have

$$\ell_0^{(3)} = 1, \ell_1^{(3)} = 3, \ell_2^{(3)} = 5, \text{ and } \ell_3^{(3)} = 7,$$

and from (2.59), we can get:

$$\bar{\mathcal{A}} = \{1, 3\}, \mathbf{p} = (0, 1, 1, 1), \text{ and } \mathbf{u}_{\bar{\mathcal{A}}^c} = (u_1 = 0, u_5 = 0).$$

In this case, the decoding operates with:

$$\mathcal{C}(\mathbf{P}_4, \mathbf{u}_{\bar{\mathcal{A}}^c} = (0, 0), \bar{\mathcal{A}} = \{1, 3\}, \mathbf{p} = (0, 1, 1, 1)). \tag{2.60}$$

2.5 Numerical Results

Consider HARQ-IR scheme based on the RCPP code to send $k = 52$ information bits using four different code rates as:

$$r_1 = \frac{52}{256} < r_2 = \frac{52}{192} < r_3 = \frac{52}{128} < r_4 = \frac{52}{64}.$$

In the decoding side, the list decoding with list size 8 and 8-bit CRC is assumed. Thus, the actual length of the information bits, in terms of (punctured) polar codes, is equal to 60. The RCPP code is developed as follows:

- RC puncturing patterns (e.g., QUP) are determined as:

$$\mathbf{p}^{(1)} = \mathbf{1}, \mathbf{p}^{(2)} = \dot{\mathbf{p}}_{192}^{(8,7,\ldots 1)},$$
$$\mathbf{p}^{(3)} = \dot{\mathbf{p}}_{128}^{(8,7,\ldots,1)}, \text{ and } \mathbf{p}^{(4)} = \dot{\mathbf{p}}_{64}^{(8,7,\ldots,1)}. \tag{2.61}$$

- The optimal information set $\mathcal{T}^{(i)}$ of each code $\mathcal{C}^{(i)}$ is optimized independently from the others. Here, each $\mathcal{T}^{(i)}$ is associated with the puncturing pattern $\mathbf{p}^{(i)}$ for $i = 1, 2, 3, 4$. From them, the common information set is determined as $\mathcal{A} = \mathcal{A}^{(4)} = \mathcal{T}^{(4)}$.
- Using Algorithm 1, the effective information sets $\mathcal{A}^{(i)}, i = 1, 2, 3$ are obtained. Also, using Algorithm 2, the precoding matrix $\mathbf{Q}$ is constructed.
- At the decoding side, the (punctured) polar code $\mathcal{C}^{(i)}$ is decoded using the *effective* information set $\mathcal{A}^{(i)}$ for $i = 1, 2, 3, 4$. Moreover, both $\mathcal{C}^{(1)}$ and $\mathcal{C}^{(2)}$ are decoded via the length-256 polar code, $\mathcal{C}^{(3)}$ via the length-128 polar code, and $\mathcal{C}^{(4)}$ via the length-64 polar code.

The performance of RCPP code is compared with the benchmark methods, in order to show its superiority. All the methods use the same RC puncturing patterns in (2.61) (e.g., QUP), while the benchmark methods do not use the information-copy technique, i.e., all the codes in the family are decoded only using the common information set $\mathcal{A}$. Here, the common information sets are optimized from the two different methods:

- **Benchmark method I**: The common information set is optimized for the *highest-rate* code in the family (i.e., $\mathcal{A} = \mathcal{T}^{(4)}$).
- **Benchmark method II**: The common information set is optimized for the *lowest-rate* code in the family (i.e., $\mathcal{A} = \mathcal{T}^{(1)}$).

We remark that, when $\mathcal{A} = \mathcal{T}^{(i)}$, the code $\mathcal{C}^{(j)}$ for some $j > i$ can suffer from a severe error-floor as the common information set cannot guarantee the $\mathcal{B}_{\mathbf{p}^{(j)}} \cap \mathcal{T}^{(i)} \neq \phi$ for some $j > i$. In this example, we observe that if $\mathcal{A} = \mathcal{T}^{(i)}$ is chosen for some $i < 4$, the code $\mathcal{C}^{(4)}$ (in the benchmark method) suffers from a severe error-floor. This can result in a lower throughput for HARQ-IR scheme as shown in Fig. 2.13. Also, from Fig. 2.12, we observe that RCPP code can significantly outperform the benchmark method. As expected, the performance gain is larger as a code rate becomes lower. The corresponding throughput performances for HARQ-IR schemes are provided in Fig. 2.13. Here, *chase combining* (CC) HARQ

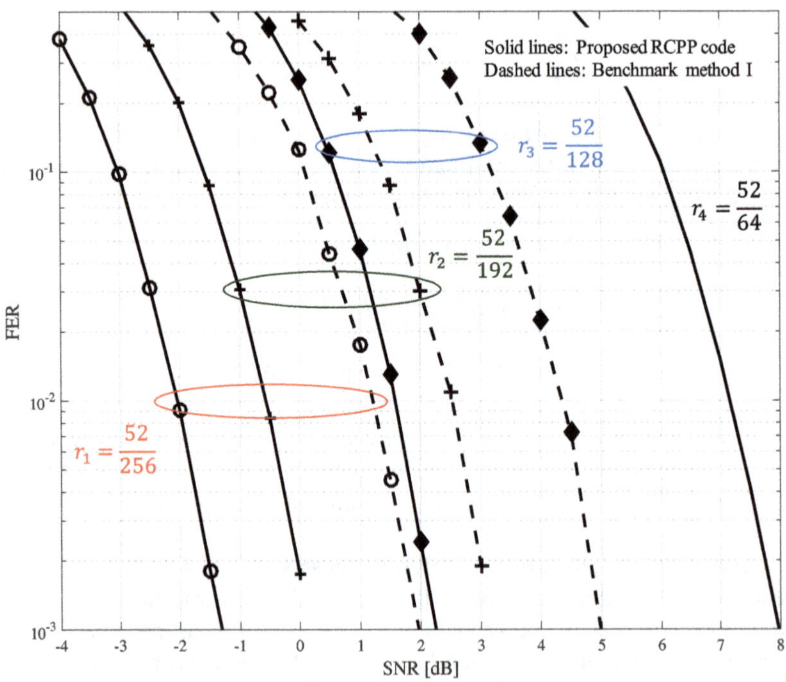

Fig. 2.12 Performance comparisons of the proposed RC-Polar code and benchmark method I

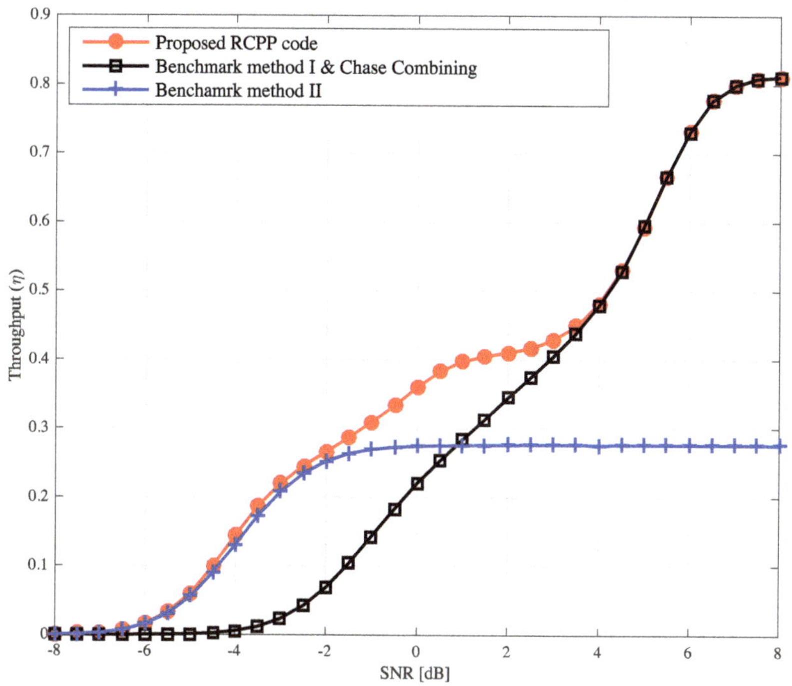

Fig. 2.13 Throughput for various IR-HARQ schemes

scheme is also considered, where the highest-rate polar code of rate r_4 is repeated for every retransmission. From Fig. 2.12, we can see that RCPP code achieves higher throughput than the benchmark schemes. Moreover, we observe that, without leveraging the information-copy technique, HARQ-IR scheme cannot achieve a throughput gain over the simple HARQ-CC scheme. Thus, the information-copy technique would play a key role in constructing a good RCPP code for HARQ-IR schemes. Here, the average throughput (η) is computed as:

$$\eta = \frac{k(1-p_f)}{\sum_{i=1}^{m-1} N_{m-(i-1)}(1-p_i)\prod_{j=1}^{i-1} p_j + N_1 \prod_{j=1}^{m-1} p_j},$$

where p_i represents the conditional probability of a frame error on the i-th transmission, given that all the previous transmissions failed for $i = 1, \ldots, m-1$, and p_f denotes the probability that none of the transmission succeeded. Recall that $N_{m-(i-1)}$ denotes the number of transmitted coded bits until the i-th transmission (see (2.41)). From Fig. 2.13, we obtain that $N_4 = 64$, $N_3 = 128$, $N_2 = 192$, and $N_1 = 256$ for both IR and CC HARQ schemes. In addition, consider another HARQ-IR scheme to send $k = 32$ information bits using five different rates as:

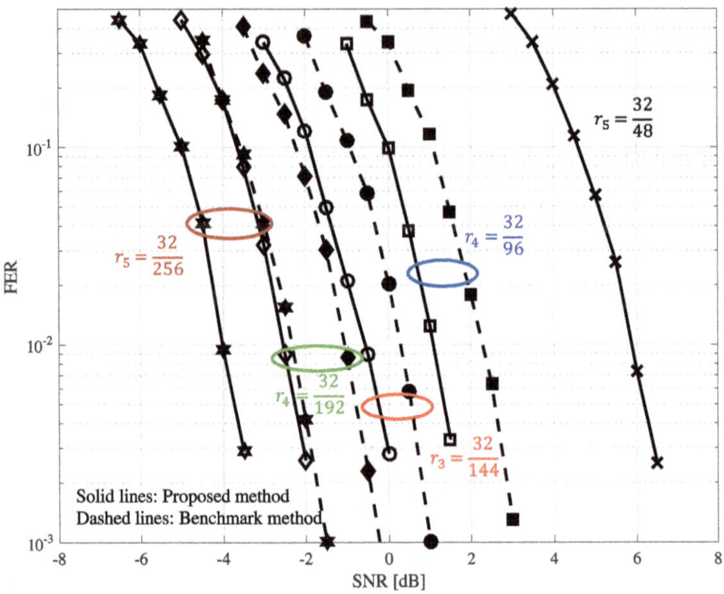

Fig. 2.14 FER comparison of the proposed and benchmark methods for various code rates

$$r_1 = \frac{32}{256} < r_2 = \frac{32}{192} < r_3 = \frac{32}{144} < r_4 = \frac{32}{96} < r_5 = \frac{32}{64}.$$

It is noticeable that the lengths of the (punctured) polar codes in the family are less likely to be the form of power of 2. Likewise the previous example, RCPP code can outperform the benchmark schemes as enhanced information sets are used at the decoding side. Not surprisingly, when RCPP code should support a wide range of code rates, the performance gain of RCPP code becomes much bigger (Fig. 2.14).

2.6 Discussion and Concluding Remarks

This chapter introduced RC-Polar codes which are capacity-achieving with a low-complexity sequential decoding. The main idea is to leverage the common characteristics of polar codes for a sequence of degraded channels. However, this construction cannot support an arbitrary sequence of code rates, as the lengths of polar codes should be the form of a power of 2. This problem was addressed by developing capacity-achieving punctured polar codes. Leveraging such puncturing polar codes, RC-Polar codes, introduced in this chapter, are capacity-achieving for an arbitrary sequence of code rates and for any class of degraded channels. This method is able to use an optimized polar code to generate the proper amount of incremental redundancy at every HARQ (re)transmission, and, thus, it is capacity-achieving.

Focusing on short lengths, reciprocal and hierarchical puncturings were introduced. Using them and the so-called information-copy technique, each code in the family of the RCPP code can be decoded using its own optimized information set. Therefore, this method can alleviate the challenging problem such that in conventional approaches, one common information set is used for all the member codes in the family. This resulted in unbalanced performances. Via simulation results, it was verified that RCPP code can attain a nontrivial performance gain mainly due to the use of improved effective information sets. Therefore, using hierarchical puncturing and information-copy technique would be necessary to construct a good (practical) RCPP code.

References

1. E. Arikan, Channel polarization: a method for constructing capacity-achieving codes for symmetric binary-input memoryless channels. IEEE Trans. Inf. Theory **55**, 3051–3073 (2009)
2. A. Vardy I. Tal, List decoding of polar codes. IEEE Trans. Inf. Theory **61**, 2213–2226 (2015)
3. Y. Blankenship, M. Andersson, L.G. Dennis Hui, S. Sandberg, Channel coding in 5G new radio: a tutorial overview and performance comparison with 4G LTE. IEEE Veh. Technol. Mag. **13**(4), 60–69 (2018)
4. J. Hagenauer, Rate-compatible punctured convolutional codes (rcpc codes) and their applications. IEEE Trans. Commun. **36**, 389–400 (1998)
5. L. Milstein, D. Rowitch, On the performance of hybrid FEC/ARQ systems using rate-compatible punctured turbo (RCPT) codes. IEEE Trans. Commun. **48**, 948–959 (2000)
6. S. McLaughlin, J. Ha, J. Kim, Rate-compatible puncturing of low-density parity-check codes. IEEE Trans. Inf. Theory **50**, 2824–2836 (2004)
7. N. Bhushan, M. El-Khamy, J. Hou, Design of rate-compatible structured LDPC codes for hybrid arq applications. IEEE J. Sel. Areas Commun. **27**, 965–973 (2009)
8. D. Divsalar, T. Chen, K. Vakilinia, R.D. Wesel, Protograph-based raptor-like ldpc codes. IEEE Trans. Commun. **63**, 1522–1532 (2015)
9. A. Eslami, H. Pishro-Nik, Practical approach to polar codes, in *Proceedings of IEEE International Symposium on Information Theory*, 2011, pp. 16–20
10. X. Wang, Q. Yu, L. Zhang, Z. Zhang, Y. Chen, On the puncturing patterns for punctured polar codes, in *Proceedings of IEEE International Symposium on Information Theory*, 2014, pp. 121–125
11. K. Niu, K. Chen, J. Lin, A hybrid arq scheme based on polar codes. IEEE Commun. Lett. **17**, 1996–1999 (2013)
12. R. Wang, R. Liu, A novel puncturing scheme for polar codes. IEEE Commun. Lett. **18**, 2081–2084 (2014)
13. S.-C. Lim, D.-M. Shin, K. Yang, Design of length-compatible polar codes based on the reduction of polarizing matrices. IEEE Trans. Commun. **61**, 2593–2599 (2013)
14. V. Miloslavskaya, Shortened polar codes. IEEE Trans. Inf. Theory **61**(9), 4852–4865 (2015)
15. I. Maric, S.-N. Hong, D. Hui, Capacity-achieving rate-compatible polar codes. IEEE Trans. Inf. Theory **63**(12), 7620–7632 (2017)
16. K. Chen, K. Niu, J.-R. Lin, Beyond turbo codes: rate-compatible punctured polar codes, in *IEEE International Conference on Communications (ICC)*, June 2013

17. S.-N. Hong, D. Hu, On the analysis of puncturing for finite-length polar codes: boolean function approach. [Online]. Available: https://arxiv.org/abs/1801.05095
18. S. Korada, Polar codes for channel and source coding. Ph.D. dissertation, École Polytechnique Fédérale de Lausanne (EPFL), 2009
19. M.-O. Jeong, S.-N. Hong, An efficient construction of rate-compatible punctured polar (RCPP) codes using hierarchical puncturing. IEEE Trans. Commun. **66**(11), 5041–5052 (2018)

Chapter 3
Multiple Access Techniques

Yijie Mao and Bruno Clerckx

Multiple access is an essential physical-layer technique in wireless communication networks that allows multiple mobile users to access the network simultaneously. Driven by the upsurge of devices expected in the fifth generation (5G) and beyond, future wireless communication networks are foreseen to operate in dynamic regimes ranging from underloaded (where the number of scheduled devices is smaller than the number of transmit antennas on each access point) to overloaded (where the number of scheduled devices is larger than the number of transmit antennas on each access point). Besides, each transmitter is required to simultaneously serve devices with heterogeneous capabilities, deployments, as well as qualities of channel state information at the transmitter (CSIT) since the devices for 5G and beyond tend to be more diverse including low-end units such as Internet of Things (IoT) and machine-type communications (MTC)-type devices and high-end equipment such as smartphones with varied user deployments and applications. The resulting requirements for massive connectivity, high throughput, as well as quality of service (QoS) heterogeneity have recently sparked interests in redesigning multiple access techniques for the downlink of communication systems.

This chapter first reviews the state-of-the-art multiple access techniques including their benefits and limitations, followed by introducing the promising multiple access candidate, *rate-splitting multiple access (RSMA)* for 5G and beyond, and a comprehensive comparison among all multiple access techniques. The challenges and future trends of using RSMA will be summarized in the end.

Y. Mao · B. Clerckx (✉)
Communication and Signal Processing Group, Department of Electrical and Electronic Engineering, Imperial College London, London, UK
e-mail: y.mao16@imperial.ac.uk; b.clerckx@imperial.ac.uk

X. Lin, N. Lee (eds.), *5G and Beyond*, https://doi.org/10.1007/978-3-030-58197-8_3

3.1 Evolution of Multiple Access Techniques

The past decades have witnessed the development of multiple access techniques brought by the evolution of cellular networks from the first generation (1G) to 5G. From orthogonal multiple access (OMA) to non-orthogonal multiple access (NOMA) and space-division multiple access (SDMA), multiple access techniques have progressed toward serving more users non-orthogonally in each subcarrier due to the scarcity of spectrum. In this section, those existing multiple access techniques are reviewed.

3.1.1 *Orthogonal Multiple Access (OMA)*

The 1G wireless communication system introduced in the 1980s employs *frequency-division multiple access (FDMA)* where the frequency bandwidth is divided into nonoverlapping frequency sub-channels and each user is allocated with an independent sub-channel. It was used to support the original analog voice services. The second generation (2G) is developed in the 1990s to further enhance the voice service quality as well as to enable short messaging service. The 2G standard system, Global System for Mobile Communications (GSM), adopts *time-division multiple access (TDMA)* where the frequency domain is shared by all users, while the time domain is divided into different time slots and occupied by independent users. The third generation (3G) introduced in the 2000s opens the new dimension of code to design multiple access where *code-division multiple access (CDMA)* is commercially applied to support TV streaming, mobile video calls, and so on. Different from FDMA and TDMA, CDMA enables the simultaneous transmission for multiple users through the same sub-channels by employing the spread spectrum technology to avoid inter-user interference. In 2009, the fourth generation (4G) based on the long-term evolution (LTE) standard is developed to meet the increasing user demand for more sophisticated mobile devices. By employing *orthogonal frequency-division multiple access (OFDMA)* as the standard multiple access technique, the time and frequency resources are further divided into narrow time slots and subcarriers, respectively. The resource blocks formed by the divided time–frequency grids are allocated to the users dynamically. Compared with FDMA, TDMA, and CDMA, OFDMA is more robust and achieves a higher spectral efficiency. The robustness comes from its ability of combatting narrowband co-channel interference and multipath fading by scheduling users over orthogonal subcarriers, while the spectral efficiency comes from its ability of multiplexing users with low data rate into a wider channel with adaptive transmission rate for each user. All of the aforementioned multiple access techniques are categorized into *orthogonal multiple access (OMA)* where users are scheduled in orthogonal dimensions.

3.1.2 Space-Division Multiple Access (SDMA)

Driven by the increasing user demands, access points nowadays are commonly equipped with multiple antennas. The arisen multiple-input multiple-output (MIMO) systems have been widely used in modern wireless standards, including mobile Worldwide Interoperability for Microwave Access (WiMAX) systems, 4G LTE standard, IEEE 802.11n, and so on. The spatial dimension introduced by MIMO systems opens the door to *space-division multiple access (SDMA)*. By utilizing the spatial dimension to separate users, SDMA allows multiple users to be served simultaneously in the same time–frequency resources.

The only strategy that achieves the capacity region of the multiple-input single-output (MISO)/MIMO (Gaussian) broadcast channel (BC) with perfect CSIT is the complex dirty paper coding (DPC) [1], in which the transmitter relies on perfect CSIT to encode the user messages and perform enhanced interference cancellation such that the encoded data stream experiences no interference from previously encoded streams. However, due to the high computational burden of implementing DPC in practice, linear precoding at the transmitter is more practical and attractive since it simplifies the transmitter design [2]. SDMA is therefore commonly implemented using multi-user linear precoding (MU–LP) either in closed-form beamforming or optimized beamforming using optimization tools. Though the beamformer might be suboptimal, SDMA based on MU–LP is shown to be useful especially when users experience semi-orthogonal channels and relatively similar channel strength or long-term signal-to-noise ratio (SNR) [3]. Hence, it is well-acknowledged and becomes the fundamental multiple access of various 4G and 5G techniques such as multi-user MIMO (MU-MIMO), massive MIMO, network MIMO, millimeter-wave MIMO, and coordinated multipoint (CoMP). Figure 3.1 illustrates the system model of K-user SDMA based on MU–LP for MISO BC. The messages $W_1, \ldots, W_K$ intended for K users are independently encoded into data streams $s_1, \ldots, s_K$ and superimposed at the transmitter after linear precoding. Each

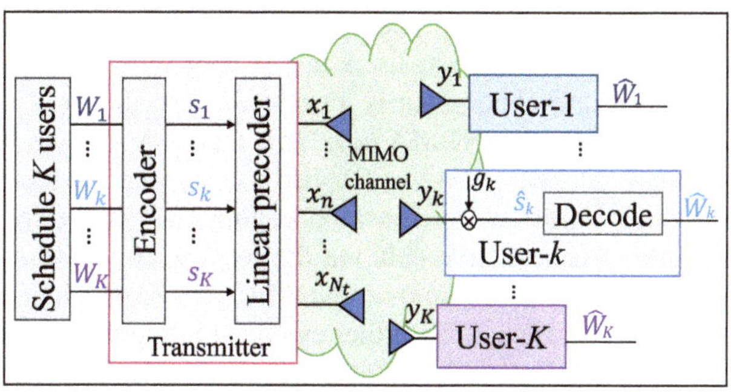

Fig. 3.1 Transmission model of K-user SDMA based on MU–LP

user directly decodes its intended stream by treating interference from streams for all other users as noise.

The main benefit of SDMA is its capability of achieving all spatial multiplexing gains of MISO BC with perfect CSIT. With the use of MU–LP, the precoder and receiver complexity remains low. However, there are three major limitations of SDMA based on MU–LP, which are summarized as follows:

- It is only suited to the underloaded regime, and its performance drops dramatically in the overloaded regime since MU–LP requires more transmit antennas than the number of users in MISO BC so as to generate orthogonal beams to manage multiuser interference efficiently. The current approach to deal with overloaded scenarios at the transmitter is to divide users into groups and schedule user groups over orthogonal resources (e.g., time/frequency). Users in the same group are served by MU–LP. However, such approach may increase latency and decrease QoS.
- The performance of SDMA based on MU–LP is sensitive to the user deployments. It is only suited when users have semi-orthogonal channels with similar channel strengths. Though there exists low-complex scheduling and user pairing algorithms to pair users with semi-orthogonal channels [2], the complexity of the scheduler increases rapidly when considering the optimal scheduling with an exhaustive search.
- Though SDMA based on MU–LP achieves the optimal degrees of freedom[1] (DoF) in MISO BC with perfect CSIT [4], there is a significant DoF and performance loss when CSIT becomes imperfect [5]. As SDMA design is motivated by perfect CSIT, the direct application of SDMA in the presence of imperfect CSIT results in additional interference coming from the imperfect linear precoder design [4].

3.1.3 Non-orthogonal Multiple Access (NOMA)

With the aim of further boosting the system spectral efficiency, *non-orthogonal multiple access (NOMA)* that superposes users in the same time–frequency resources via the power domain or code domain is introduced. Specifically, NOMA can be categorized into power-domain NOMA (e.g., [6]) and code-domain NOMA (e.g., sparse code multiple access (SCMA) [7]). In this chapter, we focus on power-domain NOMA[2] that relies on superposition coding (SC) at the transmitter and successive interference cancellation (SIC) at the receiver [6, 8–10] (which is also denoted in short as SC–SIC). The study of NOMA starts from single-input single-output (SISO) (Gaussian) BC and is further extended to multi-antenna BC. In this

[1]The DoF, also known as spatial multiplexing gain, characterizes the number of interference-free streams that can be transmitted or equivalently the pre-log factor of the rate at high SNR.

[2]In the sequel, power-domain NOMA will be referred to simply by NOMA.

chapter, we denote NOMA in SISO BC as single-antenna NOMA while NOMA in MISO/MIMO BC as multi-antenna NOMA.

Single-Antenna NOMA

The study of single-antenna NOMA is inspired by the well-known result in the literature of information theory that SC–SIC is the capacity-achieving technique for SISO BC [11, 12]. Comparing NOMA and OMA, it is well-known that when there are certain channel strength disparities among users, the capacity region of SISO BC is achieved by NOMA, and it is larger than the rate region achieved by OMA (e.g., TDMA) [12]. However, when users experience the same channel strengths, the advantage of NOMA vanishes, and OMA is sufficient to achieve the capacity region [12].

The major benefit of single-antenna NOMA is its ability to improve the spectral efficiency in an overloaded regime by allowing multiple users (that experience different channel strengths or path losses) to be served by one transmitter with single transmit antenna on the same time–frequency resource. However, its limitation is non-negligible. For a K-user SISO BC, $K - 1$ layers of SIC are required at the user with strongest channel strength to sequentially decode the $K - 1$ streams of all other co-scheduled users before decoding its intended stream. As the number of user increases, the receiver complexity and likelihood of error propagation increase significantly. A practical system requires the number of SIC layers to be small. One method is to divide the users into small groups, apply SC–SIC in each group, and schedule groups over orthogonal resources by using OMA, which, however, would lead to some performance loss and latency issue.

Multi-antenna NOMA

Motivated by the benefits of SC–SIC in SISO BC, NOMA has been further applied to multi-antenna BC. There are two main strategies of multi-antenna NOMA, both of which rely on linearly precoded SC–SIC.

The first strategy, which is simply denoted as "SC–SIC," is a direct application of SC–SIC to MISO/MIMO BC [13–16]. However, contrary to SISO BC, multi-antenna BC is non-degraded, i.e., users cannot be ordered according to their channel strengths in general settings. SC–SIC degrades multi-antenna BC by ordering users based on their effective scalar channels obtained at the transmitter after linear precoding. Users with stronger effective channel strengths are required to decode and remove the streams of users with weaker effective channel strengths in a successive manner. Such strategy forces the multi-antenna non-degraded channel into an effective single-antenna degraded channel since the user with the strongest channel strength is required to decode the messages of all other users. SC–SIC wastes all spatial multiplexing gains in MISO/MIMO BC and is only able to cope with the scenarios when user channels are aligned with certain channel strength

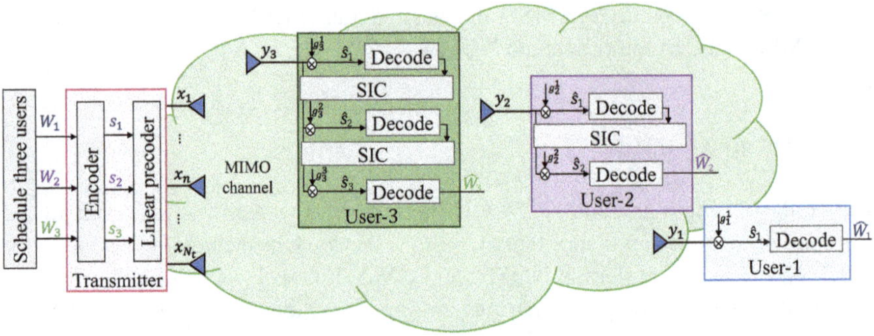

Fig. 3.2 Transmission model of three-user SC–SIC

disparities among them. From the DoF perspective, the sum-DoF achieved by SC–SIC is 1 since one receiver has to decode all streams [17]. It is equal to the DoF achieved by OMA or single-user beamforming. Compared with the sum-DoF $\min\{N_t, K\}$ (where N_t is the number of transmit antennas and K is the number of users in MISO BC) achieved by DPC and MU–LP in a MISO BC with perfect CSIT, SC–SIC in multi-antenna BC results in a significant DoF loss, and such loss comes with a dramatic burden to receivers due to the use of SIC layers. In contrast, MU–LP does not require any SIC at receiver sides, and it achieves a higher spatial multiplexing gain which drives the use of MU–MIMO in 4G [18]. To compensate the DoF loss of SC–SIC, one natural method is to consider dynamic switching between NOMA and SDMA based on the channel states [19]. Figure 3.2 illustrates the transmission model of a three-user SC–SIC with decoding order from user-1 to user-3. Hence, user-3 is required to decode all the three streams.

The second strategy, denoted as "SC–SIC per group," divides users into disparate groups with users in the same group being served by SC–SIC and users across the groups being served by SDMA in order to coordinate inter-group interference [6, 20–24]. By combining SDMA and NOMA in SC–SIC per group, multi-antenna BC is decomposed into non-interfering single-antenna NOMA channels, and the DoF loss of SC–SIC can be recovered. However, it is only suited to an overloaded regime, and users within the same group require almost aligned channels, while users in different groups require (semi-)orthogonal channels. Figure 3.3 illustrates the transmission model of four-user SC–SIC per group with user-1 and user-2 in group 1 while user-3 and user-4 in group 2. The inner-group interference is decoded based on SC–SIC, while the inter-group interference is treated as interference based on MU–LP. By assuming the decoding order in group 1 is from the message of user-1 to that of user-2, user-2 is required to decode the messages of both user-1 and user-2 while fully treating the inter-group interference from user-3 and user-4 as noise. Similarly in group-2, the decoding order from user-3 to user-4 is assumed.

Multi-antenna NOMA also relies on perfect CSIT as SDMA. When CSIT becomes imperfect, extra multiuser interference is introduced for both SC–SIC and SC–SIC per group strategies [16]. Similarly to single-antenna NOMA, the major

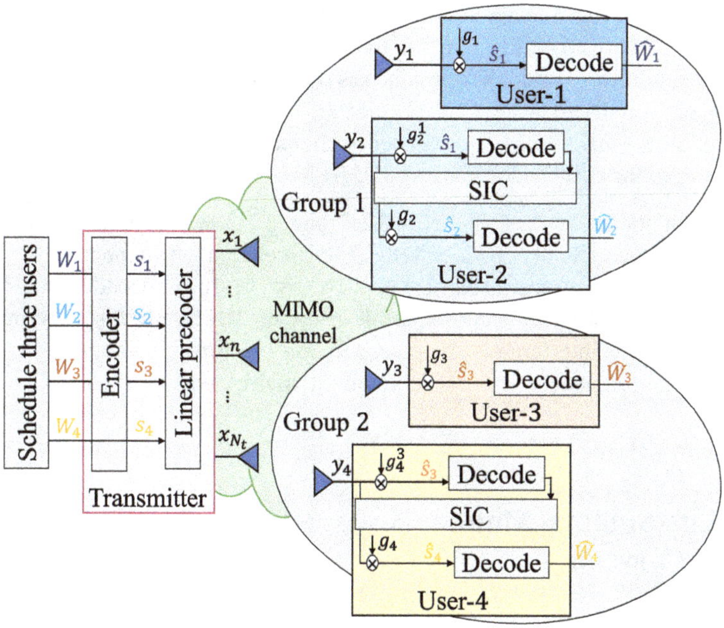

Fig. 3.3 Transmission model of four-user SC–SIC per group

benefit of multi-antenna NOMA is its ability to deal with an overloaded regime with aligned user channels and channel strength disparities. The limitations of multi-antenna NOMA are summarized as follows:

- The DoF loss of multi-antenna NOMA is severe. The fundamental reason that SC–SIC achieves the capacity region of SISO BC is due to the fact that users can be ordered based on channel strengths in such a degraded BC. However, striving to using SC–SIC in non-degraded MISO/MIMO BC degrades multi-antenna BC and results in a waste of spatial resources. Hence, there is an unavoidable DoF loss of SC–SIC in multi-antenna NOMA.
- Multi-antenna NOMA is only suited for specific user deployments when user channels are aligned with a disparity of channel strengths. It is not suited for general settings.
- There is a complexity increase at both the transmitter and the receivers in multi-antenna NOMA. At each receiver, multiple layers of SIC are required to decode and remove the interference from other users. At the transmitter equipped with multiple antennas, the optimization of user grouping, decoding orders, and precoders are coupled since the effective user channels are influenced by the precoders. For example, considering a three-user MISO BC, SC–SIC requires to jointly optimize the precoding vectors of three users and six possible decoding orders, while SC–SIC per group requires the user ordering and grouping to be jointly optimized with precoders. One commonly adopted method to reduce the

complexity at the transmitter in multi-antenna NOMA is to assign the same precoding vector to all users within the same group [6], which, however, would further deteriorate system performance as the overall searching space for optimal precoders is reduced.

- As multi-antenna NOMA is motivated in the presence of perfect CSIT, it is also sensitive to the CSIT inaccuracy as SDMA.

Based on the discussion in Sects. 3.1.2 and 3.1.3, we conclude that SDMA and NOMA are actually two extreme interference management strategies in MISO/MIMO BC where users in NOMA try to fully decode and remove interference created by other users, while users in SDMA always fully treat any residual multiuser interference as noise. Moving toward imperfect CSIT, residual inter-user interference is introduced for both SDMA and NOMA (SC–SIC per group).

3.2 Rate-Splitting Multiple Access (RSMA) for 5G and Beyond

Rate-splitting multiple access (RSMA), based on linearly precoded rate-splitting (RS) at the transmitter and SIC at the receivers, is a more general and powerful multiple access for downlink multi-antenna systems that contains SDMA, NOMA, and OMA as special cases. Apart from SDMA that fully treats interference as noise and NOMA that fully decodes interference, RSMA achieves a more dynamic interference management where the interference is partially decoded and partially treated as noise at each user [4]. At the transmitter that supports RSMA, user messages are split into common and private parts; the common messages are combined and encoded into common streams to be decoded by multiple users, while private messages are independently encoded into private streams to be decoded by the corresponding users. All streams are superimposed at the transmitter and broadcast to the users. Each user relies on layers of SIC to decode the common streams before decoding the intended private stream. By adjusting the power allocation for the common and private streams as well as the message split, RSMA automatically bridges SDMA and NOMA that solely rely on the two extreme interference management strategies or a combination thereof.

3.2.1 Literature Review

The fundamental building block of RSMA is RS technique. The previous study of RS can be categorized into communication and information theory categories. Both are summarized comprehensively in this section.

The information theoretic works on RS are summarized in Table 3.1. The idea of RS is not new. It dates back to Carleial's work and the Han and Kobayashi

Table 3.1 Summary of information theoretic literature on rate-splitting

Timeline	Ref.	Scenarios	CSIT accuracy	Network load	Metric	Main discovery
1981	[25]	SISO IC	Perfect CSIT	Underloaded	Rate region	RS based on Han and Kobayashi (HK) scheme achieves the best-known achievable rate region of the two-user interference channel
1996	[27]	SISO MAC	Perfect CSIT	Underloaded	Capacity region	RS based on successive single-user decoding and interference cancellation achieves the capacity region of the K-user Gaussian MAC
2008	[26]	SISO IC	Perfect CSIT	Underloaded	Capacity region	RS based on the HK scheme achieves rate regions within 1 bit/s/Hz of the capacity region
2013, 2016	[29, 30]	MISO BC	Imperfect CSIT	Underloaded	Sum-DoF	RS (with SIC) achieves the optimum sum-DoF of the K-user underloaded MISO BC with imperfect CSIT
2016	[37]	MISO BC	Imperfect CSIT	Underloaded	DoF region	RS achieves the entire DoF region of the K-user underloaded MISO BC with imperfect CSIT
2016	[34]	MISO BC	Imperfect CSIT	Underloaded	Symmetric DoF	RS achieves a higher symmetric (max–min) DoF than SDMA based on MU–LP for the K-user underloaded MISO BC with imperfect CSIT
2016	[37]	MISO BC	Imperfect CSIT	Overloaded	DoF region	RS achieves the entire DoF region of the K-user overloaded MISO BC with imperfect heterogeneous CSIT qualities
2017	[17]	MISO BC	Perfect CSIT	Overloaded	Symmetric DoF	RS achieves the highest symmetric DoF compared with that of SDMA based on MU–LP and NOMA based on SC–SIC for the K-user overloaded MISO BC with perfect CSIT
2017	[32]	MISO IC	Imperfect CSIT	Underloaded	DoF region	RS achieves the best-known DoF region of the K-cell MISO IC with imperfect CSIT
2017	[33]	MIMO IC,MIMO BC	Imperfect CSIT	Underloaded	DoF region	RS achieves the optimum DoF region of the two-user MIMO IC with imperfect CSIT under certain antenna configurations and CSIT qualities
2017	[36]	MISO BC	Imperfect CSIT	Underloaded	GDoF	RS-assisted interference enhancement approach achieves the entire GDoF region of the two-user underloaded MISO BC with imperfect CSIT

(HK) scheme in 1980s for the two-user SISO interference channel (IC) [25]. Such scheme is further proved in [26] to achieve rate regions within 1 bit/s/Hz of the capacity region. The terminology RSMA is first introduced in [27] for the SISO multiple access channel (MAC), where RS based on successive single-user decoding and interference cancellation has been shown to achieve the capacity region of the K-user Gaussian MAC. However, the uplink RSMA has fundamentally different motivations and structures than the downlink RSMA we considered in this chapter. The use of RS as the building block of RSMA framework is motivated by recent progresses on the fundamental limits of a multi-antenna BC and IC characterized by RS. In contrast with the conventional RS used for MAC or two-user SISO IC, the RSMA technique we introduced here is in a different setup, namely, (1) in a BC and (2) with multiple transmit (and receive) antennas. Note that the study of RS in the multi-antenna BC in both information-theoretical and communication perspectives was initiated a few years ago. In comparison, research on NOMA based on SC–SIC in a BC already appeared for several decades [11, 12]. Up to now, the capacity region of the K-user MISO BC with imperfect CSIT remains an open issue. Instead, attention has been switched to characterizing its DoF region. Surprisingly, the information theoretic upperbound on the sum-DoF of the K-user underloaded MISO BC with imperfect CSIT derived in [28] has been shown to coincides with the sum-DoF achieved by linearly precoded RS with SIC [29, 30]. It is further proved in [31] that RS achieves the entire DoF region of the underloaded MISO BC with imperfect CSIT. In comparison, the sum-DoFs achieved by SDMA based on MU–LP and multi-antenna NOMA are suboptimum. The DoF benefits of RS in imperfect CSIT have also been studied in the underloaded MISO IC [32] and underloaded MIMO IC/BC [33]. The optimum symmetric DoF (also known as max–min DoF) of RS has been studied in [34] for the underloaded MISO BC with imperfect CSIT, where RS achieves a higher symmetric DoF over that of SDMA based on MU–LP. Moving toward the overloaded scenario, the power-partitioning approach that superimposes degraded symbols for no-CSIT users on top of linearly precoded RS symbols for partial-CSIT users has been shown to achieve the entire DoF region of the K-user overloaded MISO BC with imperfect CSIT with heterogeneous CSIT qualities. When CSIT is perfect, the symmetric DoF achieved by RS has been shown to outperform that of SDMA based on MU–LP and NOMA based on SC–SIC in [17] for the K-user overloaded MISO BC with perfect CSIT. To further capture the diversity of channel strengths among users, the generalized DoF (GDoF) has been introduced [26]. The GDoF region of a two-user underloaded MISO BC with imperfect CSIT has been studied in [35, 36] where RS is considered as part of the interference enhancement scheme to achieve the entire GDoF region.

The DoF and GDoF superiority of RSMA over SDMA based on MU–LP and NOMA discovered in the information theoretic literature motivates its recent study at the finite SNR regime for rate enhancement in the practical wireless communication systems. The communication literature on RSMA is summarized in Table 3.2. The DoF improvement of RSMA over SDMA in imperfect CSIT

Table 3.2 Summary of communication literature on rate-splitting

Timeline	Ref.	Scenarios	CSIT accuracy	Network load	Metric	Precoding scheme	RSMA scheme
2016	[38]	MISO BC	Quantized feedback	Underloaded	Ergodic sum rate	Random+ZFBF	1-layer RS
2016	[30]	MISO BC	Imperfect CSIT	Underloaded	Ergodic sum rate, Ergodic rate region	Optimized linear precoding	1-layer RS
2016	[34]	MISO BC	Imperfect CSIT	Underloaded	Max–min rate	Optimized linear precoding	1-layer RS
2016	[37]	MISO BC	Imperfect CSIT	Overloaded	Sum rate	Random+ZFBF	1-layer RS with power splitting
2016	[39]	Massive MIMO	Imperfect CSIT	Underloaded	Sum rate	MRT+RZF	1-layer RS, 2-layer HRS
2017	[40]	MISO BC with hardware impairments	Imperfect CSIT	Underloaded	Sum rate	MRT+RZF	1-layer RS
2017	[17]	Multigroup multicast	Imperfect CSIT	Overloaded	Max–min rate	Optimized linear precoding	1-layer RS
2017	[41]	Millimeter-wave MISO BC	Quantized feedback, statistical CSIT	Underloaded	Sum rate	Hybrid (partially optimized) linear precoding	1-layer RS
2018	[42]	MISO BC	Perfect CSIT	Underloaded, Overloaded	Rate region, WSR	Optimized linear precoding	Generalized RS, 2-layer HRS, 1-layer RS
2018	[43]	Multicell multigroup multicast	Perfect CSIT	Underloaded	EE	Optimized linear precoding	1-layer RS
2018	[44]	MISO BC	Imperfect CSIT	Underloaded	Sum power	Optimized linear precoding	1-layer RS
2018	[45]	MISO BC	Perfect CSIT	Underloaded	EE	Optimized linear precoding	1-layer RS
2018	[46]	MISO BC	Imperfect CSIT	Underloaded	Ergodic sum rate	ZF+THP	1-layer THPRS
2018	[47]	Multi-pair massive MIMO relaying	Imperfect CSIT	Underloaded	Sum rate	MRT+RZF	1-layer RS
2018	[48]	Multi-beam satellite networks	Imperfect CSIT	Underloaded	Sum rate	Optimized power allocation within each beam	1-layer RS
2019	[49]	Cooperative multicell MISO BC	Perfect CSIT	Underloaded	Rate region, WSR	Optimized linear precoding	Generalized RS, 2-layer HRS, 1-layer RS

(continued)

Table 3.2 (continued)

2019	[50]	UAV-assisted C-RAN	Perfect CSIT	Underloaded	WSR	Optimized linear precoding	Generalized RS
2019	[51]	mmWave UAV-assisted MISO BC	Perfect CSIT	Underloaded	EE	Optimized linear precoding	1-layer RS
2019	[52]	C-RAN	Perfect CSIT	Underloaded	WSR	Optimized linear precoding	1-layer RS
2019	[53]	C-RAN	Perfect CSIT	Underloaded	Max–min rate	Optimized linear precoding	Generalized RS
2019	[54]	MISO BC with SWIPT	Perfect CSIT	Underloaded	Rate region	Optimized linear precoding	1-layer RS
2019	[55]	MISO IC with SWIPT	Imperfect CSIT	Underloaded	Sum transmit power	Optimized linear precoding	1-layer RS
2019	[56]	MISO BC with cooperative transmission	Perfect CSIT	Underloaded	Rate region, WSR	Optimized linear precoding	1-layer RS
2019	[57]	Non-orthogonal unicast and multicast	Perfect CSIT, Imperfect CSIT	Underloaded, overloaded	Rate region, WSR, EE	Optimized linear precoding	Generalized RS, 2-layer HRS, 1-layer RS
2019	[58]	MISO BC	Perfect CSIT	Underloaded	WSR	Optimal common stream precoder+ZF	1-layer RS
2019	[59]	Multicarrier multigroup multicast	Perfect CSIT	Overloaded	Max–min rate	Optimized linear precoding	1-layer RS
2020	[60]	MIMO BC	Imperfect CSIT	Underloaded	Sum rate	Regularized block diagonalization	1-layer RS

has been reflected in the rate performance at the finite SNR regime according to recent studies [17, 30, 34, 37, 38]. In the presence of quantized feedback, RS reduces CSIT feedback overhead compared to MU–LP when using random beamforming for the common stream and zero-forcing beamforming (ZFBF) for the private streams [38]. It is further shown that with optimized precoders, RS outperforms MU–LP in the underloaded MISO BC with imperfect CSIT for the ergodic sum rate maximization [30] and the worst-case rate optimization (max–min rate) [34]. When considering the overloaded scenario, RS with power-partitioning strategy has been shown to outperform its time-partitioning counterpart at finite SNR in the overloaded MISO BC with heterogeneous CSIT [37] with low-complex maximum ratio transmission (MRT) or matched filtering beamforming scheme for the common stream and regularized zero-forcing (RZF) beamforming for the private streams. The 2-layer hierarchical RS (HRS) that relies on multiple common messages decoded by different groups of users is proposed in [39] for massive MIMO. Furthermore, the generalized RS scheme of RSMA that embraces 1-layer RS and 2-layer HRS as subcases is proposed in [42] for MISO BC with perfect CSIT, where RSMA shows clear rate region and weighted sum rate (WSR) improvement over SDMA and NOMA. The comparison among SDMA, NOMA and RSMA is further analyzed in the two-user case with optimized precoders to maximize energy efficiency [45] or with low-complex precoding but optimal power allocation for common part and private parts of the user messages [58]. Besides the above studies of RSMA in MISO BC, the transceiver design of RS has been studied in other applications of multi-antenna BC, such as MISO BC with hardware impairment [40], multigroup multicast [40], millimeter-wave (mmWave) systems [41], multi-pair relaying [47], cooperative multicell MISO BC [49], cloud-radio access networks (C-RAN) [52, 53], unmanned aerial vehicle (UAV)-assisted networks [50, 51], simultaneous wireless information and power transfer (SWIPT) [54], cooperative user relaying networks [56, 61], non-orthogonal unicast and multicast [57], multi-carrier systems [59], and so on. All of the above works consider linearly precoded RS at the transmitter, and nonlinear precoder design of RS has been studied as Tomlinson–Harashima Precoded RS (THPRS) [46] and dirty paper coded RS (DPCRS) [62]. Moving toward MIMO BC, different linear combining techniques are studied in [60] with minimum mean-square error (MMSE) combiner showing the best performance.

3.2.2 RSMA Framework

RSMA is a generalized multiple access technique for exploring a larger rate region and the room of QoS enhancement. In the framework of RSMA, there are three commonly studied schemes in the literature, namely, 1-layer RS, 2-layer HRS, and the generalized RS, which are all specified in this section.

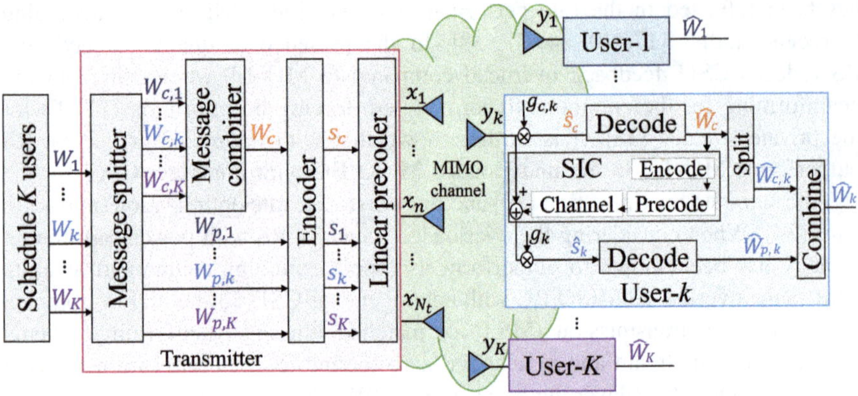

Fig. 3.4 Transmission model of K-user 1-layer RS

1-Layer RS

1-layer RS is the simplest RSMA scheme and it is the building block of the entire RSMA framework. It has been widely studied in the literature of RS in multi-antenna BC and its applications [17, 30, 34, 37, 38, 40–42, 45, 47, 50, 54, 56–59, 61] with both perfect and imperfect CSIT. Figure 3.4 illustrates the transmission model of K-user 1-layer RS with one base station (BS) equipped with N_t transmit antennas simultaneously serving K single-antenna users. The users are indexed by $\mathcal{K} = \{1, \cdots, K\}$.

At the transmitter, the K messages $W_1, \cdots, W_K$ intended for the K users are passed to the message splitter. The message of each user W_k, $k \in \mathcal{K}$ is split into one common part $W_{c,k}$ and one private part $W_{p,k}$.[3] The common parts $W_{c,1}, \cdots, W_{c,K}$ are combined into the common message W_c and encoded into the common stream s_c to be decoded by all users. The private parts $W_{p,1}, \cdots, W_{p,K}$ are independently encoded into K private streams $s_1, \cdots, s_K$ to be decoded by the corresponding users only. The encoded stream vector $\mathbf{s} = [s_c, s_1, \ldots, s_K]^T \in \mathbb{C}^{(K+1)\times 1}$ is linearly precoded via precoding matrix $\mathbf{P} = [\mathbf{p}_c, \mathbf{p}_1, \ldots, \mathbf{p}_K] \in \mathbb{C}^{N_t\times(K+1)}$ with $\mathbf{p}_k \in \mathbb{C}^{N_t\times 1}$, $k \in \{c\} \cup \mathcal{K}$. The resulting transmit signal is

$$\mathbf{x} = \mathbf{P}\mathbf{s} = \mathbf{p}_c s_c + \sum_{k\in\mathcal{K}} \mathbf{p}_k s_k. \tag{3.1}$$

At user sides, the signal received at each user is

[3]Note that it is not necessary to let all users split their messages in some cases. For example, when maximizing the sum rate without QoS rate constraint [30], one user splits its message into common and private parts which is sufficient. However, splitting the messages of all users is more general, and it becomes necessary when user fairness is considered in the design. for instance, when maximizing WSR or max–min fairness or with QoS rate constraint [17, 30, 34].

$$
\begin{aligned}
y_k &= \mathbf{h}_k^H \mathbf{x} + n_k \\
&= \mathbf{h}_k^H \mathbf{p}_c s_c + \mathbf{h}_k^H \mathbf{p}_k s_k + \sum_{j \in \mathcal{K}, j \neq k} \mathbf{h}_k^H \mathbf{p}_j s_j + n_k,
\end{aligned}
\tag{3.2}
$$

where $\mathbf{h}_k \in \mathbb{C}^{N_t \times 1}$ is the channel between the BS and user-k. It may be perfectly known at the transmitter [42, 45, 49, 50, 53, 54, 56–59, 61] or partially known at the transmitter [17, 30, 34, 37–41, 46, 47, 57, 62] due to the quantization error or feedback delay. n_k is the additive white Gaussian noise (AWGN) at user-k that follows the distribution $\mathcal{CN}(0, \sigma_{n,k}^2)$.

Each user firstly decodes the data stream s_c by treating the interference from all private streams as noise.[4] The signal-to-interference-pulse-noise ratio (SINR) of decoding the common stream s_c at user-k is

$$
\gamma_{c,k} = \frac{\left|\mathbf{h}_k^H \mathbf{p}_c\right|^2}{\sum_{j \in \mathcal{K}} \left|\mathbf{h}_k^H \mathbf{p}_j\right|^2 + \sigma_{n,k}^2}. \tag{3.3}
$$

To ensure all users can successfully decode the common stream, its achievable rate should not exceed

$$
R_c = \min\left\{\log_2\left(1 + \gamma_{c,k}\right), \cdots, \log_2\left(1 + \gamma_{c,K}\right)\right\}. \tag{3.4}
$$

Note that R_c is shared by all K users. Denote C_k as the part of rate allocated to user-k for the transmission of $W_{c,k}$, we have

$$
\sum_{k \in \mathcal{K}} C_k = R_c. \tag{3.5}
$$

Once s_c is successfully decoded, it is re-encoded, precoded, and subtracted from y_k. Each user then decodes its intended private stream s_k by treating the interference from the private streams of other users as noise.[5] The SINR of decoding the private stream s_k at user-k is

[4]Please notice that the role of the common stream here is fundamentally different from a multicast stream, though both of them are decoded by all users. The common stream in RS encapsulates parts of private messages of different users. It is not entirely required by all users. In contrast, a multicast stream is encoded by a message originally intended for all users. Each user requires the full message [4].

[5]The decoding order of s_c and s_k can be further optimized. We here follow the rule that the data stream intended for more users has a higher decoding priority [63, 64] for the entire RSMA framework.

$$\gamma_k = \frac{\left|\mathbf{h}_k^H \mathbf{p}_k\right|^2}{\sum_{j\in\mathcal{K}, j\neq k}\left|\mathbf{h}_k^H \mathbf{p}_j\right|^2 + \sigma_{n,k}^2}. \tag{3.6}$$

Its corresponding private rate is $R_k = \log_2(1+\gamma_k)$. Hence, the total achievable rate of user-k, $k \in \mathcal{K}$ is

$$R_{k,tot} = C_k + R_k. \tag{3.7}$$

Following the above-described structure of 1-layer RS, we can design the precoders $\mathbf{p}_c, \mathbf{p}_1, \cdots, \mathbf{p}_K$ with different objectives, such as maximizing the WSR (or sum rate) [30, 42], maximizing the worst-case user rate [34], maximizing EE [45], minimizing transmit power [44], etc.

2-Layer HRS

2-layer HRS is originally introduced for massive MIMO networks [39] with the aim of enhancing the achievable rate of all users and reducing the CSI feedback. In a K-user 2-layer HRS network, the K users are divided into G separated groups indexed by $\mathcal{G} = \{1, \ldots, G\}$ with $\mathcal{K}_g, g \in \mathcal{G}$ users in each group such that $\bigcup_{g\in\mathcal{G}} \mathcal{K}_g = \mathcal{K}$. Different from 1-layer RS where the message of each user is only split into two parts, each user in 2-layer HRS splits its message $W_k, k \in \mathcal{K}_g$ into three different parts $W_k^{\mathcal{K}}, W_k^{\mathcal{K}_g}, W_k^k$ in order to form outer-group common message and inner-group common message. The outer-group common messages $\{W_k^{\mathcal{K}} | k \in \mathcal{K}\}$ of all users are jointly combined into one common message $W_{\mathcal{K}}$ and encoded into the outer-group common stream $s_{\mathcal{K}}$ to be decoded by all users. The inner-group common messages $\{W_k^{\mathcal{K}_g} | k \in \mathcal{K}_g\}$ of users in group-g are jointly combined into the common message $W_{\mathcal{K}_g}$ and encoded into the inner-group common stream $s_{\mathcal{K}_g}$ to be decoded by all users in $\mathcal{K}_g$. The private messages $\{W_k^k | k \in \mathcal{K}\}$ are independently encoded into the private streams $s_1, \cdots, s_K$ for the corresponding users only. The encoded streams $\mathbf{s} = [s_{\mathcal{K}}, s_{\mathcal{K}_1}, \cdots, s_{\mathcal{K}_G}, s_1, \cdots, s_K]^T \in \mathbb{C}^{(K+G+1)\times 1}$ are linearly precoded via precoding matrix $\mathbf{P} = [\mathbf{p}_{\mathcal{K}}, \mathbf{p}_{\mathcal{K}_1}, \ldots, \mathbf{p}_{\mathcal{K}_G}, \mathbf{p}_1, \ldots, \mathbf{p}_K] \in \mathbb{C}^{N_t\times(K+G+1)}$, the shaped transmit signal is

$$\mathbf{x} = \mathbf{P}\mathbf{s} = \mathbf{p}_{\mathcal{K}} s_{\mathcal{K}} + \sum_{g\in\mathcal{G}} \mathbf{p}_{\mathcal{K}_g} s_{\mathcal{K}_g} + \sum_{k\in\mathcal{K}} \mathbf{p}_k s_k. \tag{3.8}$$

At user sides, once each user receives the signal as $y_k = \mathbf{h}_k^H \mathbf{x} + n_k$, it employs two layers of SIC to successfully decode $s_{\mathcal{K}}, s_{\mathcal{K}_g}$ and $s_k, k \in \mathcal{K}_g$. The outer-group common stream $s_{\mathcal{K}}$ is decoded first at all users by treating the interference from all other streams as noise. The corresponding SINR of decoding $s_{\mathcal{K}}$ at user-k is

$$\gamma_k^{\mathcal{K}} = \frac{\left|\mathbf{h}_k^H \mathbf{p}_{\mathcal{K}}\right|^2}{\sum_{g\in\mathcal{G}} \left|\mathbf{h}_k^H \mathbf{p}_{\mathcal{K}_g}\right|^2 + \sum_{j\in\mathcal{K}} \left|\mathbf{h}_k^H \mathbf{p}_j\right|^2 + \sigma_{n,k}^2}. \tag{3.9}$$

Once $s_{\mathcal{K}}$ is successfully decoded with its contributed part removed from the received signal, each user then decodes the inner-group common stream $s_{\mathcal{K}_g}$ by treating interference from other inner-group common streams and private streams as noise. The SINR of decoding $s_{\mathcal{K}_g}$ at user-k is

$$\gamma_k^{\mathcal{K}_g} = \frac{\left|\mathbf{h}_k^H \mathbf{p}_{\mathcal{K}_g}\right|^2}{\sum_{g'\in\mathcal{G}, g'\neq g} \left|\mathbf{h}_k^H \mathbf{p}_{\mathcal{K}_{g'}}\right|^2 + \sum_{j\in\mathcal{K}} \left|\mathbf{h}_k^H \mathbf{p}_j\right|^2 + \sigma_{n,k}^2}. \tag{3.10}$$

After removing $s_{\mathcal{K}_g}$ from the received signal, user-k decodes its private stream s_k. The SINR of decoding the private stream s_k at user-k is

$$\gamma_k = \frac{\left|\mathbf{h}_k^H \mathbf{p}_k\right|^2}{\sum_{g'\in\mathcal{G}, g'\neq g} \left|\mathbf{h}_k^H \mathbf{p}_{\mathcal{K}_{g'}}\right|^2 + \sum_{j\in\mathcal{K}, j\neq k} \left|\mathbf{h}_k^H \mathbf{p}_j\right|^2 + \sigma_{n,k}^2}. \tag{3.11}$$

Following (3.4) and (3.5), we obtain the respective achievable rate of $s_{\mathcal{K}}$, $s_{\mathcal{K}_g}$, and s_k, which are given by

$$\begin{aligned}
\sum_{k\in\mathcal{K}} C_k^{\mathcal{K}} &= \min\left\{\log_2\left(1+\gamma_{k'}^{\mathcal{K}}\right) | k' \in \mathcal{K}\right\}, \\
\sum_{k\in\mathcal{K}_g} C_k^{\mathcal{K}_g} &= \min\left\{\log_2\left(1+\gamma_{k'}^{\mathcal{K}_g}\right) | k' \in \mathcal{K}_g\right\}, \forall g \in \mathcal{G} \\
R_k &= \log_2\left(1+\gamma_k\right), \forall k \in \mathcal{K}.
\end{aligned} \tag{3.12}$$

where $C_k^{\mathcal{K}}$ and $C_k^{\mathcal{K}_g}$ are the parts of the rate allocated to user-k for the transmission of messages $W_k^{\mathcal{K}}$ and $W_k^{\mathcal{K}_g}$, respectively. The total achievable rate of user-k, $k \in \mathcal{K}_g$ is

$$R_{k,tot} = C_k^{\mathcal{K}} + C_k^{\mathcal{K}_g} + R_k. \tag{3.13}$$

Figure 3.5 illustrates the transmission model of four-user 2-layer HRS with one BS equipped with N_t transmit antennas simultaneously serving four single-antenna users. There are two user groups with user-1 and user-2 in group 1 and user-3 and user-4 in group 2. s_{1234} is an outer-group common stream to be decoded by all the four users, while s_{12} and s_{34} are the two inner-group common streams to be decoded by the users within the corresponding groups only. The receiver structures of user-2 and user-4 follow that of user-1 and user-2, respectively.

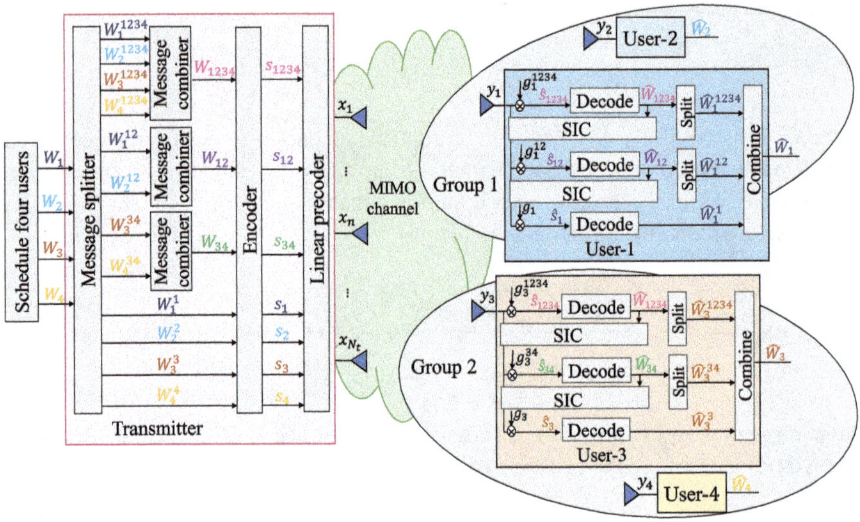

Fig. 3.5 Transmission model of four-user 2-layer HRS

Generalized RS

The generalized RS framework is proposed in [42] with the aim of identifying the largest room for rate and QoS enhancement at the expense of more layers of SIC at each user. In the K-user generalized RS framework, the number of message splits of each user increases with K so as to form common streams intended for different user subsets of $\mathcal{K}$. For any user subset $\mathcal{A} \subseteq \mathcal{K}$, the BS transmits a data stream $s_{\mathcal{A}}$ by loading messages of all users in the subset $\mathcal{A}$, and $s_{\mathcal{A}}$ needs to be decoded by all users in the subset $\mathcal{A}$ while treated as noise by other users. The message of user-k is split into 2^{K-1} parts as $\{W_k^{\mathcal{A}'} | \mathcal{A}' \subseteq \mathcal{K}, k \in \mathcal{A}'\}$. User messages $\{W_{k'}^{\mathcal{A}} | k' \in \mathcal{A}\}$ with the same superscript $\mathcal{A}$ are encoded together into the stream $s_{\mathcal{A}}$.

The concept of *stream order* is introduced here to simplify the explanation. We define the streams to be decoded by l users as *l-order streams*. Hence, the common stream $s_{\mathcal{K}}$ intended for all users is a K-order stream, while the private stream s_k is a 1-order stream since it is only decoded by a single user. In the K-user case, all l-order streams form the stream set $\{s_{\mathcal{A}'} | \mathcal{A}' \subseteq \mathcal{K}, |\mathcal{A}'| = l\}$, and there are in total $\binom{K}{l}$ elements within the set. Specifically, there is one K-order stream $s_{\mathcal{K}}$ and K 1-order streams $s_1, \cdots, s_K$. We further introduce *l-order data stream vector* formed by all l-order streams as $\mathbf{s}_l \in \mathbb{C}^{\binom{K}{l} \times 1}$. Note that when $l = K$, there is one element within the set, and therefore, $\mathbf{s}_K$ reduces to $s_{\mathcal{K}}$. $\mathbf{s}_l$ is then linearly precoded via the precoding matrix $\mathbf{P}_l$ formed by $\{\mathbf{p}_{\mathcal{A}'} | \mathcal{A}' \subseteq \mathcal{K}, |\mathcal{A}'| = l\}$, and the resulting transmit signal is

$$\mathbf{x} = \sum_{l=1}^{K} \mathbf{P}_l \mathbf{s}_l = \sum_{l=1}^{K} \sum_{\mathcal{A}' \subseteq \mathcal{K}, |\mathcal{A}'|=l} \mathbf{p}_{\mathcal{A}'} s_{\mathcal{A}'}. \tag{3.14}$$

At user sides, each user requires $2^{K-1}-1$ layers of SIC to sequentially decode all the intended common streams. The decoding process starts from the K-order stream and then goes down to the 1-order private stream. Note that each user is involved in multiple l-order streams except the 1-order and K-order streams, and the set of l-order streams to be decoded at user-k is $\mathcal{S}_{l,k} = \{s_{\mathcal{A}'} | \mathcal{A}' \subseteq \mathcal{K}, |\mathcal{A}'| = l, k \in \mathcal{A}'\}$. We denote the decoding order of the l-order streams $\mathbf{s}_l$ at all users as π_l. Based on one certain decoding order π_l, we obtain the l-order stream vector to be decoded at user-k as $\mathbf{s}_{\pi_{l,k}} = [s_{\pi_{l,k}(1)}, \cdots, s_{\pi_{l,k}(|\mathcal{S}_{l,k}|)}]^H$, where we assume $s_{\pi_{l,k}(i)}$ is decoded before $s_{\pi_{l,k}(j)}$ if $i < j$. The SINR of user-k to decode the l-order stream $s_{\pi_{l,k}(i)}$ is

$$\gamma_k^{\pi_{l,k}(i)} = \frac{|\mathbf{h}_k^H \mathbf{p}_{\pi_{l,k}(i)}|^2}{I_{\pi_{l,k}(i)} + \sigma_{n,k}^2}, \tag{3.15}$$

where

$$I_{\pi_{l,k}(i)} = \sum_{j>i} |\mathbf{h}_k^H \mathbf{p}_{\pi_{l,k}(j)}|^2 + \sum_{l'=1}^{l-1} \sum_{j=1}^{|\mathcal{S}_{l',k}|} |\mathbf{h}_k^H \mathbf{p}_{\pi_{l',k}(j)}|^2 + \sum_{\mathcal{A}' \subseteq \mathcal{K}, k \notin \mathcal{A}'} |\mathbf{h}_k^H \mathbf{p}_{\mathcal{A}'}|^2$$

is the interference received at user-k when decoding $s_{\pi_{l,k}(i)}$. The first term $\sum_{j>i} |\mathbf{h}_k^H \mathbf{p}_{\pi_{l,k}(j)}|^2$ is the interference from the remaining non-decoded l-order streams in $\mathbf{s}_{\pi_{l,k}}$. The second term $\sum_{l'=1}^{l-1} \sum_{j=1}^{|\mathcal{S}_{l',k}|} |\mathbf{h}_k^H \mathbf{p}_{\pi_{l',k}(j)}|^2$ is the interference from lower-order streams $\{\mathbf{s}_{\pi_{l',k}} | l' < l\}$ to be decoded at user-k, while the third term $\sum_{\mathcal{A}' \subseteq \mathcal{K}, k \notin \mathcal{A}'} |\mathbf{h}_k^H \mathbf{p}_{\mathcal{A}'}|^2$ is the interference received from the streams that are not intended for user-k. The corresponding achievable rate of user-k for the data stream $s_{\pi_{l,k}(i)}$ is $R_k^{\pi_{l,k}(i)} = \log_2(1 + \gamma_k^{\pi_{l,k}(i)})$. Following (3.4), (3.5), and (3.12), the achievable rate of the $|\mathcal{A}|$-order stream $s_{\mathcal{A}}$ ($\mathcal{A} \in \mathcal{K}, 2 \leq |\mathcal{A}| \leq K$) shall not exceed

$$\sum_{k \in \mathcal{A}} C_k^{\mathcal{A}} = \min_{k'} \left\{ R_{k'}^{\mathcal{A}} \mid k' \in \mathcal{A} \right\}. \tag{3.16}$$

where $C_k^{\mathcal{A}}$ is the part of the common rate allocated to user-k ($k \in \mathcal{A}$) for the transmission of $W_k^{\mathcal{A}}$ via $s_{\mathcal{A}}$. Hence, the total achievable rate of user-k is

$$R_{k,tot} = \sum_{\mathcal{A}' \subseteq \mathcal{K}, k \in \mathcal{A}'} C_k^{\mathcal{A}'} + R_k. \tag{3.17}$$

Figure 3.6 illustrates a three-user example of the generalized RS. The message of each user is split into four parts, i.e., the message of user-1 is split into $\{W_1^{123}$,

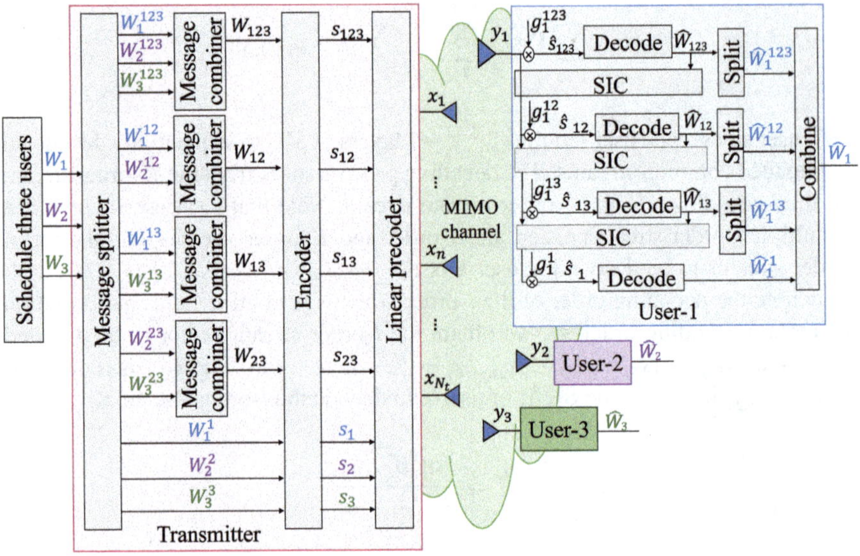

Fig. 3.6 Transmission model of three-user generalized RS

W_1^{12}, W_1^{13}, $W_1^1\}$. There is one 3-order stream, three 2-order streams, and three 1-order streams. The corresponding stream vectors are denoted as $\mathbf{s}_1 = [s_1, s_2, s_3]^T$, $\mathbf{s}_2 = [s_{12}, s_{13}, s_{23}]^T$, and s_{123}, respectively. Each user requires three layers of SIC to sequentially decode the intended streams. The instance of decoding order π_2 for the 2-order streams illustrated in Fig. 3.6 is $12 \rightarrow 13 \rightarrow 23$. All users follow the rule that s_{12} is decoded before s_{13} and s_{23} is decoded lastly. At user-1, only the 2-order streams s_{12} and s_{13} are decoded. The decoding order based on π_2 at user-1 is $\pi_{2,1} = 12 \rightarrow 13$. We have $s_{\pi_{2,1}(1)} = s_{12}$ and $s_{\pi_{2,1}(2)} = s_{13}$.

1-Layer RS vs. 2-Layer HRS vs. Generalized RS

The inclusive relation of the above three RSMA schemes is illustrated in Fig. 3.7. The generalized RS is the most general scheme that embraces 2-layer HRS and 1-layer RS as two sub-schemes. 2-layer HRS is a sub-scheme when only the K-order stream, $|\mathcal{K}_g|$-order streams, and 1-order streams are active (with a nonzero power allocation) in the generalized RS, while 1-layer RS is the sub-scheme of 2-layer HRS when only the K-order stream and 1-order streams are active. All other inactive streams are allocated with zero transmit power.

In terms of the computational complexity and hardware complexity at the BS and users, 1-layer RS achieves the lowest complexity, and the generalized RS has the opposite highest complexity in the RSMA framework. In the K-user case, each user in the 1-layer RS system only requires one layer of SIC without any scheduling requirement at the transmitter, while each user requires two layers of SIC and the

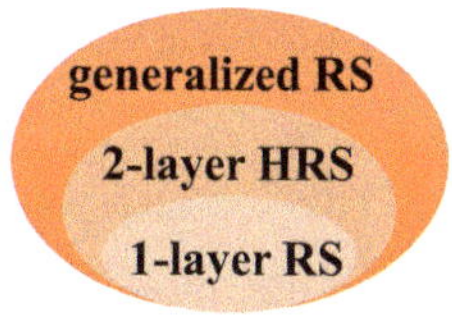

Fig. 3.7 RSMA framework and the schemes included

BS requires to consider the issue of user grouping in 2-layer HRS. Both 1-layer RS and 2-layer HRS maintain relative low transceiver complexities and are practical for implementation since the number of SIC layers deployed at each user is independent from the number of users. In comparison, the generalized RS is more complex to be implemented since the number of SIC layers increases rapidly with the number of users. In the K-user case, $2^{K-1} - 1$ layers of SIC are required at each user, and the decoding order of the common streams needs to be optimized at the transmitter. However, readers are reminded that the motivation of introducing the generalized RS contrasts with the previous two low-complex schemes, that is, to identify the best possible performance of the network at the scarifies of more SIC layers at the receivers.

3.2.3 *RSMA vs. NOMA/SDMA/OMA*

Framework Comparison

Table 3.3 summarizes the comparison of different multiple access techniques. Compared with the existing multiple access techniques, the major and unique characteristic of RSMA is its ability of partially decoding interference and partially treating interference as noise. Gaining benefits from its dynamic interference management capability, RSMA framework generalizes and encompasses multi-antenna NOMA (including SC–SIC and SC–SIC per group), SDMA (based on MU–LP), OMA (TDMA/FDMA), and multicasting as sub-schemes. Their relation is further illustrated in Fig. 3.8.

SDMA is a sub-scheme of RSMA when all common streams are turned off and the transmit power is fully allocated to the private streams. RSMA boils down to NOMA (based on SC–SIC) when each common stream is fully encoded by the entire message of a single user. OMA is a sub-scheme of SDMA, NOMA, and RSMA when the transmit power is fully allocated to a single user. Physical-layer multicasting is achieved by encoding the messages of all users into the K-order stream. Hence, multicasting is a special instance of RSMA with full transmit power being allocated to the K-order common stream (embracing partial messages of all users). Based on the above discussion, we obtain that SDMA based on MU–LP, multi-antenna NOMA, OMA, and multicasting are sub-schemes of RSMA. Most importantly, RSMA smoothly bridges all sub-schemes without hard switching among them.

Table 3.3 Comparison of different multiple access techniques

Multiple access	OMA	SDMA	NOMA		RSMA
Strategy	LP	MU–LP	SC–SIC	SC–SIC per group	All forms of RS
Design principle	Orthogonal resource allocation to get rid of interference	Fully treat interference as noise	Fully decode interference	Fully decode interference in each group and treat interference between groups as noise	Partially decode interference and partially treat interference as noise
Decoder architecture	Treat interference as noise	Treat interference as noise	SIC at receivers	SIC at receivers	SIC at receivers
User deployment scenario	Any angle between channels and any disparity in channel strengths	User channels are (semi-)orthogonal with similar channel strengths	Users experience aligned channel directions and a large disparity in channel strengths	Users in each group experience aligned channel directions and a large disparity in channel strengths. Users in different groups experience orthogonal channels	Any angle between channels and any disparity in channel strengths
Network load	Only one active user (in each resource block)	More suited to overloaded network	More suited to overloaded network	More suited to underloaded network	Suited to any network load

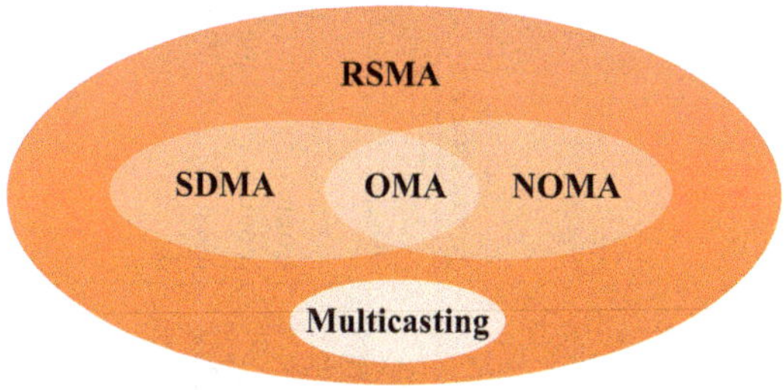

Fig. 3.8 RSMA framework and the schemes included

	s_1	s_2	s_c
SDMA	W_1	W_2	-
NOMA	W_1	-	W_2
OMA	W_1	-	-
Multicasting	-	-	W_1, W_2
RS	$W_{p,1}$	$W_{p,2}$	$W_{c,1}$, $W_{c,2}$

decoded by its intended user and treated as noise by the other user — decoded by both users

Fig. 3.9 Mapping of messages to streams

Figure 3.9 illustrates the different mappings of the messages to the streams in the two-user case for all multiple access techniques. When $K = 2$, the generalized RS boils down to 1-layer RS automatically with one common stream s_c containing one part of message $W_{c,1}$ for user-1 and one part of message $W_{c,2}$ for user-2. Other parts $W_{p,1}$ and $W_{p,2}$ are independently encoded into private streams s_1 and s_2. SDMA is obtained by allocating no power to the common stream ($\|\mathbf{p}_c\|^2 = 0$) such that W_k is encoded into s_k directly. NOMA is obtained by encoding the message of one user, i.e., W_2 entirely into s_c and W_1 into s_1 while s_2 is turned off ($\|\mathbf{p}_2\|^2 = 0$). In this example, user-1 fully decodes the interference from the message of user-2. OMA is obtained when only one user is scheduled ($\|\mathbf{p}_c\|^2 = \|\mathbf{p}_2\|^2 = 0$). Multicasting is obtained when the messages of both users W_1, W_2 are combined into s_c and the private streams are turned off ($\|\mathbf{p}_1\|^2 = \|\mathbf{p}_2\|^2 = 0$).

Complexity Comparison

The qualitative complexity of different strategies is compared in Table 3.4. SDMA based on MU–LP and OMA based on point-to-point linear precoding have the lowest receiver and encoder complexities. However, the scheduling complexity is relatively high due to the subcarrier/time-slot allocation for OMA and user selection for SDMA. As mentioned previously, SDMA based on MU–LP is only suited when the user channels are semi-orthogonal. Accurate CSIT is required to carefully design user scheduling for interference coordination.

Both SC–SIC per group and 2-layer HRS have the highest user grouping complexity. The total number of user grouping methods to be considered in both

Table 3.4 Qualitative comparison of the complexity of different multiple access techniques

Multiple access	OMA	SDMA	NOMA		RSMA		
Strategy	LP	MU–LP	SC–SIC	SC–SIC per group	1-layer RS	2-layer HRS	generalized RS
Encoder complexity	Encode K streams	Encode K streams	Encode K streams	Encode K streams	Encode $K+1$ streams	Encode $K+G+1$ streams	encode 2^K-1 streams
Scheduler complexity	Complex as OMA requires subcarrier/time-slot allocation to all users	Complex as MU–LP requires to pair together semi-orthogonal users with similar channel gains	Very complex as it requires to find aligned users and decide upon $K!$ user ordering	Very complex as it requires to divide users into orthogonal groups, with aligned users in each group. Decide upon $\sum_{k=1}^{K} S(K,k)$ grouping method and at most $K!$ decoding order for each grouping method	Simpler user scheduling as it copes with any user deployment scenario, does not rely on user grouping or user ordering	Decide upon $\sum_{k=1}^{K} S(K,k)$ grouping method without decoding order problem	Complex as it requires to decide upon $\prod_{k=2}^{K-1}\binom{K}{k}!$ decoding orders
Receiver complexity	Does not require any SIC	Does not require any SIC	Requires $K-1$ layers of SIC. Subject to error propagation	Requires $\lvert\mathcal{K}_g\rvert-1$ layers of SIC in each group. Subject to error propagation	Requires one layer of SIC at each user. Less subject to error propagation	Requires two layers of SIC at each user. Less subject to error propagation	Requires $2^{K-1}-1$ layers of SIC. Subject to error propagation

schemes is $\sum_{k=1}^{K} S(K,k)$, where $S(K,k) = \frac{1}{k!}\sum_{i=0}^{k}(-1)^i\binom{k}{i}(k-i)^K$, also known as a Stirling set number [65], is the total number of methods to partition a set of K elements into k nonempty sets. As each user in 2-layer HRS sequentially decodes the outer-group common stream and the inner-group common stream followed by the intended private stream, the decoding order is determined without introducing additional scheduling complexity and only requires two layers of SIC at each user. In comparison, at most $K!$ decoding orders are required to be considered in SC–SIC per group for each grouping method, and each user is required to have $|\mathcal{K}_g|-1$ layers of SIC. For example, for a four-user system with two groups and two users in each group, we have to consider three different user grouping methods and four different decoding orders for each grouping method. Generally, SC–SIC per group has the highest scheduling complexity compared with other schemes since the decoding order and user grouping are required to be jointly decided. Note that $K!$ is the total number of decoding orders when there is one user group. In such scenario, SC–SIC per group reduces to SC–SIC. Different from the single-antenna NOMA in SISO BC where the optimal decoding order of NOMA is determined based on the channel gain, multi-antenna NOMA based on SC–SIC requires the decoding order to be jointly decided with the precoders at the transmitter. As SC–SIC is only suited for aligned user channels with certain channel strength disparities, additional scheduler complexity is introduced for a proper user scheduling algorithm. Hence, the scheduler complexity of SC–SIC is relatively high, and each user requires $K-1$ layers of SIC in the K-user SC–SIC system. Compared with SC–SIC per group, SC–SIC simplifies the scheduling complexity at the transmitter (since there is no requirement of user grouping) but increases the receiver complexity.

Compared with existing multiple access techniques, RSMA is able to achieve a better trade-off between performance and complexity. All RS strategies including 1-layer RS, 2-layer HRS, and generalized RS are suited for users with any channel strength disparity and channel angle in between. Specifically, 1-layer RS has the lowest scheduling complexity compared with all other schemes since it does not have any issue of user scheduling, grouping, and ordering. It also maintains very low receiver complexity since only one layer of SIC is required at each user in the K-user scenario. 1-layer RS is a sub-scheme of 2-layer HRS and the generalized RS. Compared with 1-layer RS, the complexity at the transmitter and receivers for 2-layer HRS is higher due to a higher dimensional message splits. The receiver complexity of 2-layer HRS is still low compared with other schemes since the number of SIC layers required at 1-layer RS and 2-layer HRS is independent from the number of user K. The receiver complexity is much reduced compared with SC–SIC or SC–SIC per group or the generalized RS. Though the generalized RS achieves the highest flexibility of interference management compared with all other schemes, it has a higher transmitter and receiver complexity. The generalized RS requires the decoding order of multiple streams with the same stream order to be jointly decided with the precoders, and each user requires an exponentially increasing number of SIC layers to decode the intended streams sequentially. For

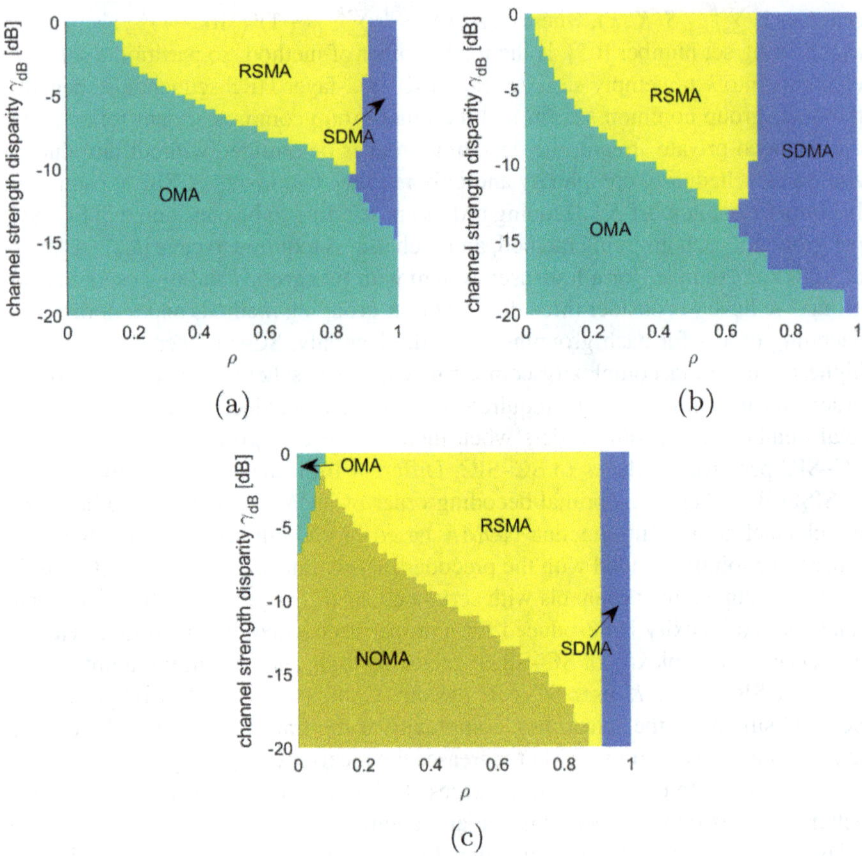

Fig. 3.10 Regions of operation for different multiple access techniques, $K = 2$, SNR = 20 dB, $\epsilon = 0.01$. (**a**) $u_1 = 10^{0.5}$, $u_2 = 1$. (**b**) $u_1 = 1$, $u_2 = 1$. (**c**) $u_1 = 1$, $u_2 = 10^{0.5}$

example, each user requires to decode two 2-order streams in Fig. 3.6, and in total, three layers of SIC are required at each user.

The multi-antenna NOMA and the generalized RS schemes have a number of SIC layers increasing with the number of user K, which not only lead to an increase of the scheduler and receiver complexity but also impel more error propagation in SIC. They are preferred to be applied in the scenarios when K is small so as to achieve a better trade-off between the performance improvement and transmitter/receiver complexity.

Performance Comparison

Figure 3.10 illustrates the preferred regions for the operation of OMA, SDMA, NOMA, and RSMA with perfect CSIT. Following the evaluations in [58], we assume that the BS equipped with $N_t = 2$ transmit antennas is serving two single-antenna users ($K = 2$). The channel vectors are $\mathbf{h}_1 = 1/\sqrt{2}[1, 1]^H$ and $\mathbf{h}_1 = \gamma/\sqrt{2}[1, e^{j\theta}]^H$. As there are only two users, SC–SIC per group boils down to SC–SIC, and the generalized RS boils down to 1-layer RS. The precoders are optimized based on the weighted minimum mean square error (WMMSE) precoding optimization framework developed in [30, 42, 66] with the aim of maximizing the sum rate $\sum_{k=1,2} R_{k,tot}$. The total achievable rate of user-k for RS is given as (3.7). The sum rate formulas of SDMA and NOMA are illustrated in [42]. The colors in Fig. 3.10 illustrate the strategy that achieves the maximized WSR as a function of $\rho = 1 - \frac{|\mathbf{h}_1^H \mathbf{h}_2|^2}{\|\mathbf{h}_1\|^2 \|\mathbf{h}_1\|^2}$ (ranging from 0 to 1) and $\gamma_{\text{dB}} = 20\log_{10}(\gamma)$ (ranging from 0 to -20 dB), i.e., user-1 and user-2 have a long-term SNR of 20dB and 0dB $\leq$ 20dB $+\gamma_{\text{dB}} \leq$ 20dB, respectively. As the WSR of RSMA is always larger than or equal to that of other strategies, we follow the rules below to select the strategy:

(i) if $|\text{WSR}_{\text{RSMA}} - \text{WSR}_{\text{OMA}}| < \epsilon$, the preferred strategy is OMA.
(ii) if $|\text{WSR}_{\text{SDMA}} - \text{WSR}_{\text{OMA}}| > \epsilon$ and $|\text{WSR}_{\text{RSMA}} - \text{WSR}_{\text{SDMA}}| < \epsilon$, the preferred strategy is SDMA.
(iii) if $|\text{WSR}_{\text{NOMA}} - \text{WSR}_{\text{SDMA}}| > \epsilon$ and $|\text{WSR}_{\text{RSMA}} - \text{WSR}_{\text{NOMA}}| < \epsilon$, the preferred strategy is NOMA.
(iv) if $|\text{WSR}_{\text{RSMA}} - \text{WSR}_{\text{SDMA}}| > \epsilon$ and $|\text{WSR}_{\text{RSMA}} - \text{WSR}_{\text{NOMA}}| > \epsilon$, the preferred strategy is RSMA.

Option (iv) is selected when RSMA does not boils down to any other multiple access techniques. We observe from the figure that when equal or higher weight is allocated to the user with a stronger channel, NOMA has no benefit over SDMA at all. Only when the user fairness is taken into consideration with a higher weight allocated to the weaker user, NOMA outperforms SDMA. But NOMA is only preferred for the deployment with small ρ, i.e., users are closely aligned. SDMA is preferred whenever ρ is sufficiently large. In comparison, for all different user weights, RSMA always provides the same or better performance than SDMA, NOMA, and OMA. It unifies and outperforms existing multiple access techniques.

Figure 3.11 further illustrates the ergodic rate region of different multiple access techniques over 100 random channel realizations with imperfect CSIT. The BS is equipped with $N_t = 2$ antennas and serves two single-antenna users. The channel model specified in [30, 34] is adopted, i.e., $\mathbf{h}_k = \widehat{\mathbf{h}}_k + \widetilde{\mathbf{h}}_k$. The estimated channel of each user $\widehat{\mathbf{h}}_k$ and channel error $\widetilde{\mathbf{h}}_k$ have independent and identically distributed (i.i.d.) complex Gaussian entries that follow the distributions $\mathcal{CN}(0, \sigma_k^2)$ and $\mathcal{CN}(0, \sigma_{e,k}^2)$, respectively. The variance of error $\sigma_{e,k}^2$ scales exponentially with SNR as $\sigma_{e,k}^2 \sim \mathcal{O}(P_t^{-\alpha})$, where $\alpha \in [0, \infty)$ is interpreted as the quality of CSIT in the high SNR regime [5, 28–30, 67]. The rate region improvement of RSMA over NOMA and SDMA is significant in all subfigures. Thanks to its flexible interference management capability, RSMA is more robust to CSIT inaccuracy and channel strength disparities between the users. In contrast, NOMA is only suited when there is a certain channel strength disparity between the two users, while SDMA is suited

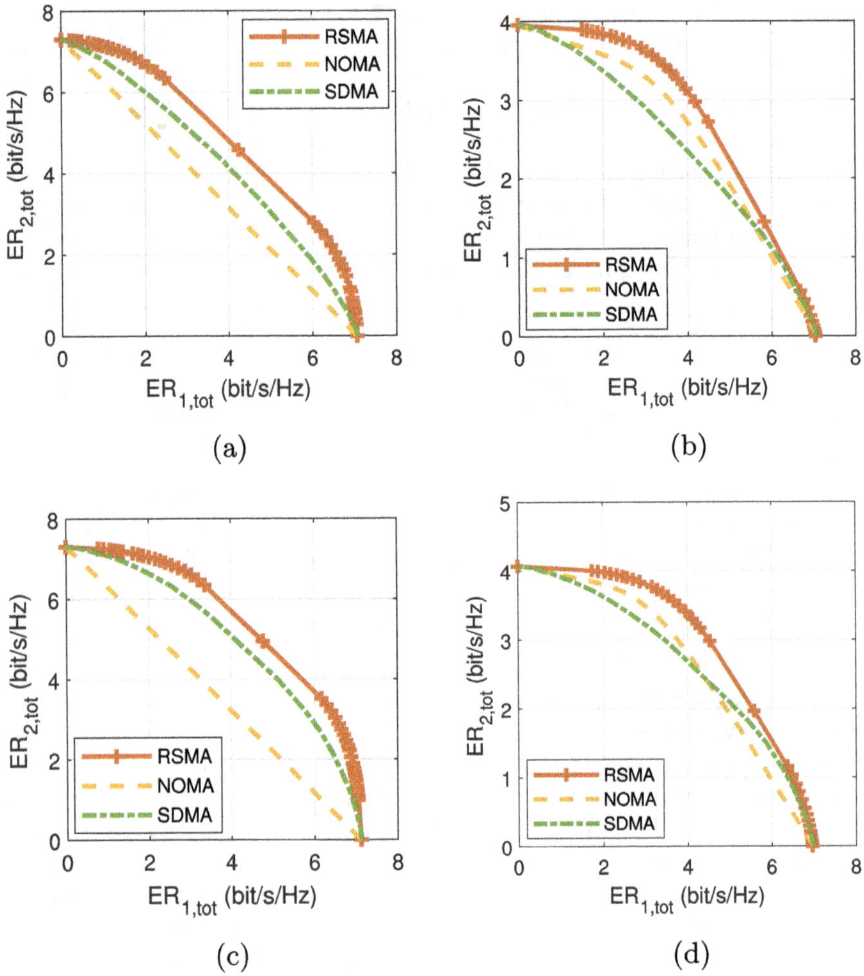

Fig. 3.11 Ergodic rate region comparison of different multiple access techniques with partial CSIT, $K = 2$, SNR = 20 dB. (**a**) $\alpha = 0.6, \sigma_2^2 = 1$. (**b**) $\alpha = 0.6, \sigma_2^2 = 0.09$. (**c**) $\alpha = 0.9, \sigma_2^2 = 1$. (**d**) $\alpha = 0.9, \sigma_2^2 = 0.09$

when users have equal channel strength. Moreover, the performance of SDMA drops as CSIT becomes inaccurate.

We further consider the three-user case. The generalized RS does not reduce to 1-layer RS, and SC–SIC per group does not reduce to SC–SIC. Figure 3.12 illustrates the ergodic sum rate versus CSIT accuracy α of different strategies over 100 random channel realizations with imperfect CSIT. Figure 3.12(a) considers an underloaded regime, while Fig. 3.12b, c shows the results of an overloaded regime but with different channel strength disparities among users. The precoders are designed to maximize the ergodic sum rate (where users have equal weights) subject to a QoS rate constraint of each user. For $\alpha = [0.2, 0.4, 0.6, 0.8, 1]$, the corresponding rate constraint for user-k ($k \in \{1, 2, 3\}$) changes as $\mathbf{r}_k^{th} = [0.1, 0.2, 0.3, 0.4, 0.5]$ bit/s/Hz. In all subfigures, the ergodic sum rate of SC–SIC and MU–LP drops dramatically as α decreases. In contrast, the generalized RS further boosts the system performance and achieves explicit rate gain over all other strategies especially when CSIT is severely inaccurate or in the overloaded regime.

In an extremely overloaded scenario, we further show the WSR improvement of 1-layer RS with a much lower receiver complexity compared with SC–SIC in Fig. 3.13. The BS is equipped with two antennas and serves ten users. The rate of each user is averaged over the ten randomly generated channels. As SNR increases as $[0, 5, \ldots, 30]$ dB, the QoS rate constraint of each user increases as

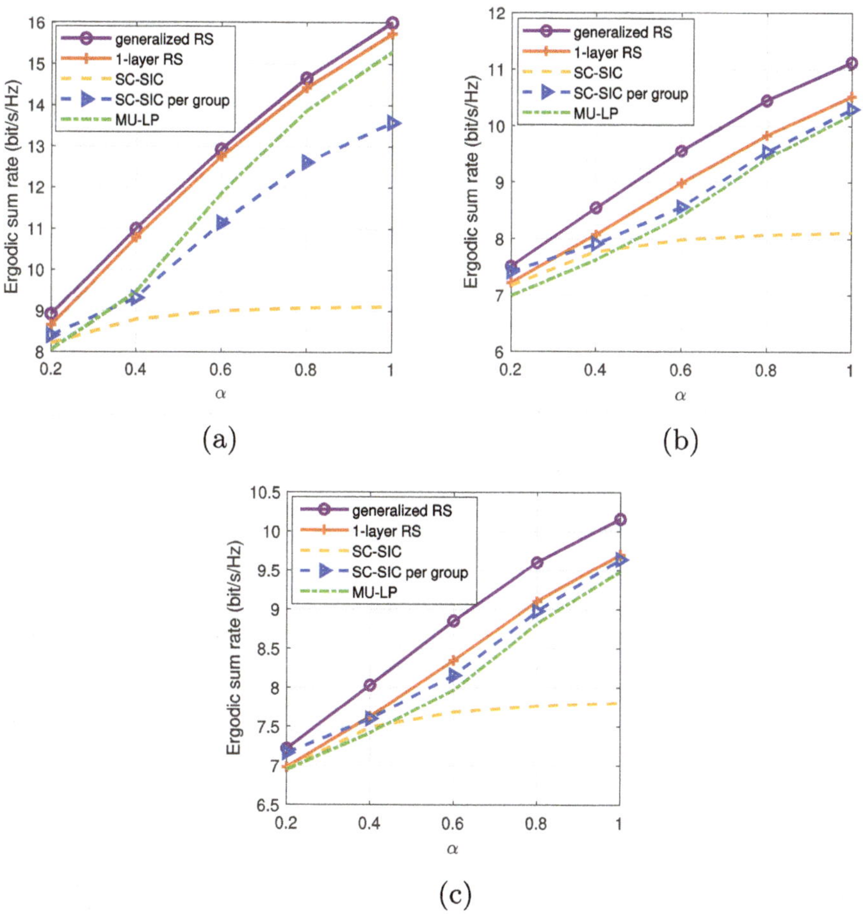

Fig. 3.12 Ergodic sum rate versus CSIT inaccuracy comparison of different multiple access techniques, averaged over 100 random channel realizations, $K = 3$, SNR = 20 dB. (**a**) $\sigma_1^2 = \sigma_2^2 = \sigma_3^2 = 1$, $N_t = 4$. (**b**) $\sigma_1^2 = \sigma_2^2 = \sigma_3^2 = 1$, $N_t = 2$. (**c**) $\sigma_1^2 = \sigma_2^2 = 1$, $\sigma_3^2 = 0.09$, $N_t = 2$

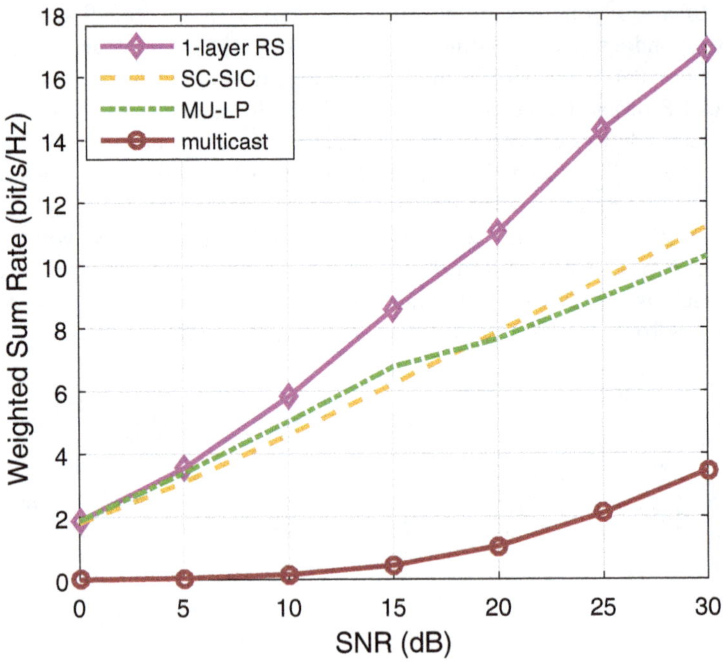

Fig. 3.13 Weighted sum rate versus SNR comparison of different multiple access techniques for overloaded ten-user deployment with perfect CSIT. $\sigma_1^2 = 1, \sigma_2^2 = 0.9, \ldots, \sigma_2^2 = 0.1, N_t = 2$

[0, 0.001, 0.004, 0.01, 0.03, 0.06, 0.1] bit/s/Hz. We observe that 1-layer RS exhibits explicit WSR improvement over all other strategies. It achieves a sum-DoF of 2 with only a single layer of SIC deployed at each user. In contrast, the slopes of the WSRs of SC–SIC and MU–LP are the same and smaller than 1-layer RS. It implies that SC–SIC and MU–LP achieve a sum-DoF of 1. However, SC–SIC requires nine layers of SIC at each user. RS is able to exploit the largest DoF in such overloaded deployment by using the common stream to pack messages from eight users while using the two private streams to serve the remaining two users. In contrast, SC–SIC and MU–LP allocate most of power to a single user, which limits their achievable DoF.

3.2.4 Advantages of RSMA

Based on the above comparison from framework, complexity, and performance aspects, we here summarize the major advantages of RSMA:

- **Universal**: RSMA is a more general multiple access framework that outperforms and unifies OMA, SDMA based on MU–LP, and multi-antenna NOMA as sub-schemes.
- **Flexible**: RSMA is suited to all user deployments (with a diversity of channel directions, channel strengths) and network loads (underloaded and overloaded regimes). It implies that RSMA is capable of managing all different kinds of interference flexibly. RSMA automatically reduces to other multiple access techniques according to the channel conditions, i.e., it reduces to SDMA when user channels are orthogonal in the underloaded MISO BC with perfect CSIT. When the channels are aligned with certain channel strength disparities, it automatically boils down to NOMA. For other channel conditions, RS takes advance to the common streams and achieves a better interference management by partially decoding the interference and partially treating the remaining interference as noise.
- **Robust**: RSMA is robust to CSIT inaccuracy. As RSMA is primarily motivated by multi-antenna deployments with multiuser interference coming from imperfect CSIT, it compensates the DoF loss of other multiple access techniques in imperfect CSIT and is therefore less sensitive to CSIT inaccuracy.
- **Spectrally efficient**: The spectral efficiency of RSMA is always larger than or equal to that of existing multiple access techniques. Considering a MISO BC without QoS constraints, the rate region of RSMA comes much closer to the optimal DPC region than SDMA and NOMA when CSIT is perfect. When CSIT becomes imperfect CSIT, linearly precoded RSMA is able to achieve a larger rate region than complex DPC in multi-antenna BC. As RSMA achieves the optimal DoF in both perfect and imperfect CSIT, it optimally exploits the spatial dimensions and the availability of CSIT. This contrasts with SDMA and NOMA that are suboptimal.
- **Energy efficient**: As RSMA is more general than SDMA and NOMA, its energy efficiency is also larger than or equal to that of existing multiple access techniques in a wide range of user deployments.
- **Enhancing QoS and fairness**: RSMA exhibits a more explicit performance gain over other multiple access techniques when there is a QoS rate constraint for each user or when a higher weight is allocated to the user with a weaker channel condition. Therefore, the ability of a wireless network architecture to partially decode interference and partially treat interference as noise leads to enhanced QoS and user fairness.
- **Reducing complexity**: The performance gain of RSMA can come with a lower transmitter and receiver complexity than multi-antenna NOMA. In contrast to multi-antenna NOMA that requires user grouping, ordering, and switching (between NOMA and SDMA) at the transmit scheduler and multiple layers of SIC at the receivers, 1-layer RS without any user ordering, grouping, or dynamic switching at the transmit scheduler and with only one layer of SIC at each receiver is capable of achieving significant performance gain over NOMA (as illustrated in Fig. 3.13). In contrast to SDMA that requires user pairing to pair users with semi-orthogonal channels, RSMA is suited to all channel conditions,

and it does not require complex user scheduling and pairing. Moreover, RSMA is capable of further reducing CSI feedback overhead [41, 68] in the presence of quantized feedback.

3.3 Emerging Applications of RSMA

RSMA is originally proposed for MIMO BC in cellular communication networks. Recently, the applications of RSMA in other 5G technologies-enabled networks have attracted substantial interests. In massive MIMO system, 2-layer HRS proposed in [39] has been shown to achieve superior sum rate performance over conventional two-tier precoding schemes based on SDMA [69–71], and 1-layer RS has been shown to be a more robust strategy for massive MIMO in the presence of phase and amplified thermal noise since its sum rate does not saturate at high SNR [40]. The application of RSMA in the multigroup multicasting system has been shown to boost the DoF in the high SNR regime as well as to enhance the system performance in the low SNR regime [17]. By using one common stream to encapsulate parts of the multicast messages for different multicast groups, RSMA based on 1-layer RS enables the ability of partially decoding the interference and partially treating the interference as noise. Recent researches have shown that RSMA is more energy efficient in the multicell multigroup systems [43] as well as enhancing the user fairness in the multicarrier multigroup multicast systems [59]. In mmWave MIMO communication systems, the authors in [41] employ 1-layer RS and propose a one-stage feedback scheme which effectively reduces the complexity of the signaling and feedback procedure. The benefits of RSMA have been further discovered in other applications such as non-orthogonal unicast and multicast transmission (NOUM) [57], coordinated multipoint (CoMP) [49], cloud-radio access networks (C-RAN) [52, 53], simultaneous wireless information and power transfer (SWIPT) [54, 55], cooperative relaying [56, 61], wireless caching [37, 72], and unmanned aerial vehicle (UAV)-aided wireless communications [50, 51], which will not be specified here. Motivated by the benefits of RSMA discovered in cellular

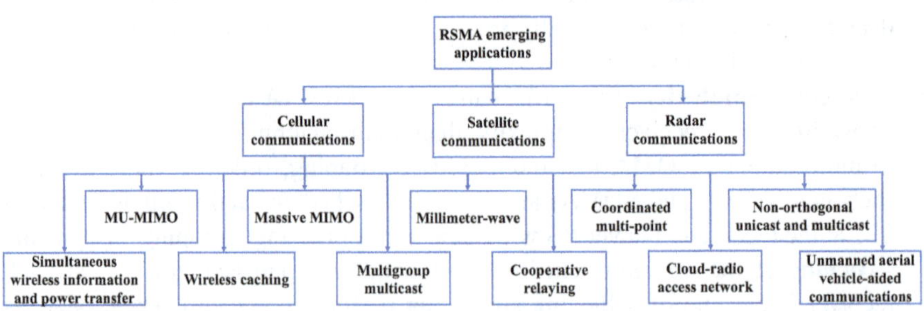

Fig. 3.14 Emerging applications of RSMA

communications, RSMA has been applied to other communication networks such as radar communications [73] and satellite communications [48, 74], which are summarized in Fig. 3.14.

3.4 Challenges and Future Trends of RSMA

The study of RSMA is still in its infancy. Even for the applications specified in Sect. 3.3, there are still many challenges and open issues that remain to be addressed. RSMA is a goldmine of research problems for academia and standard specification issues for industry. The multifarious attractive and potential research directions of RSMA are summarized in Fig. 3.15.

There are various applications of RSMA in other techniques besides those described in Sect. 3.3. Some of the techniques are complementary, and the investigation of RSMA in the combination of those techniques may collide with different sparks. For example, RSMA has shown its performance benefits respectively in cooperative relaying and NOUM. There is also a great potential of applying RSMA in NOUM with cooperative relaying. In such networks, the system performance will be further enhanced since the common stream to be forwarded from the relaying users to other users will help to enhance not only the rate of the multicast message for all users but also the unicast messages for the corresponding users.

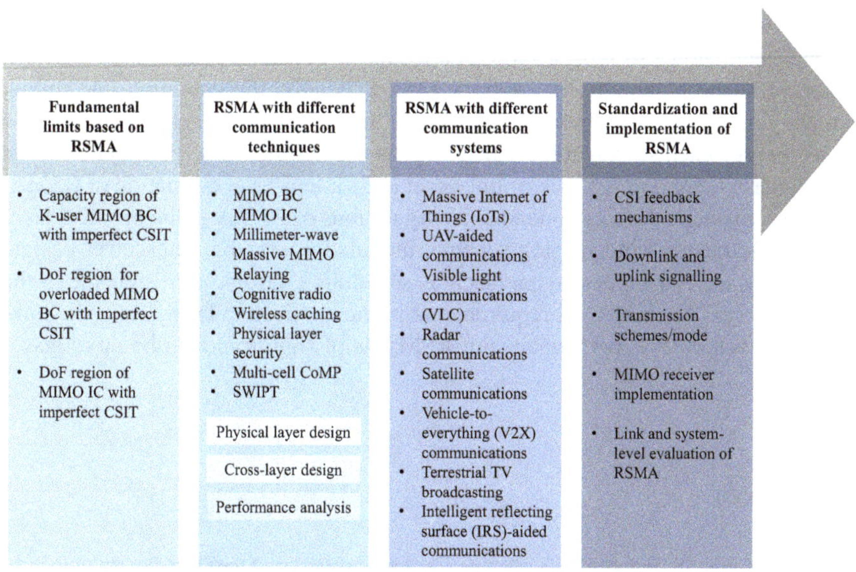

Fig. 3.15 Structure of future research directions

In addition to the promising systems summarized in Fig. 3.15, there are many other combinations of RSMA that are worth to be studied, such as RSMA in UAV-aided, radar, or satellite communication systems. In the UAV-aided communications, one major challenge is the UAV deployment and trajectory optimization. However, perfectly tracking the rapidly changed channels of the entire location map is impossible which would result in strong co-channel interference. As RSMA is superior in robust interference management and it achieves higher performance gain when CSIT is imperfect, the application of RSMA in the UAV-aided multi-antenna broadcast channel has a great potential to overcome that challenge.

The key technologies required to implement RSMA are MU–MIMO/CoMP, superposition coding at the transmitter, SIC at receivers, and non-orthogonal unicast and multicast transmission. Though the standardization of RSMA has not been taken by the 3rd Generation Partnership Project (3GPP) yet, some current work items in 3GPP can be leveraged for the implementation of RSMA. MU–MIMO and CoMP are the key technologies in LTE, which are included in 3GPP Release 8 [75] and 3GPP Release 11 [76], respectively. One major receiver technique used in RSMA is SIC, which has been incorporated in 3GPP Release 12 for network-assisted interference cancellation and suppression (NAICS) [77]. In 3GPP Release 13, superposition coding at the transmitter and successive decoding at each receiver has been further considered for LTE downlink mobile broadband (MBB) services as multiuser superposition transmission (MUST) [78]. The multicast functionality is recently included in 3GPP Release 17 for 5G with the name new radio (NR) multicast/broadcast. Besides the necessary machinery discussed or approved by 3GPP, there are some implementation issues specific to RSMA which require further study. First of all, the CSI feedback mechanisms of RSMA are unclear even though RS has been shown to reduce the CSIT feedback overhead compared to MU–LP in the presence of quantized feedback [38]. Secondly, the downlink and uplink signaling of RS remains obscure. The issue of synchronizing the knowledge of how to split/merge each stream at the transmitter and receivers needs to be tackled. Last but not least, there is still a lack of link-level and system-level evaluation of RSMA. To further evaluate the recommended configurations of RSMA in the physical layer (such as frequency band, coding scheme, modulation scheme, transceiver design, topography, etc.) or higher layers (such as scheduling, error-control scheme in the multiple access layer or QoS requirements in the application layer, etc.), the link-level and system-level performance of RSMA is of significance to be investigated thoroughly.

References

1. H. Weingarten, Y. Steinberg, S.S. Shamai, The capacity region of the Gaussian multiple-input multiple-output broadcast channel. IEEE Trans. Inf. Theory **52**(9), 3936–3964 (2006)
2. B. Clerckx, C. Oestges, *MIMO Wireless Networks: Channels, Techniques and Standards for Multi-antenna, Multi-user and Multi-cell Systems* (Academic Press, New York, NY, USA 2013)

3. T. Yoo, A. Goldsmith, On the optimality of multiantenna broadcast scheduling using zero-forcing beamforming. IEEE J. Sel. Areas Commun. **24**(3), 528–541 (2006)
4. B. Clerckx, H. Joudeh, C. Hao, M. Dai, B. Rassouli, Rate splitting for MIMO wireless networks: a promising PHY-layer strategy for LTE evolution. IEEE Commun. Mag. **54**(5), 98–105 (2016)
5. N. Jindal, MIMO broadcast channels with finite-rate feedback. IEEE Trans. Inf. Theory **52**(11), 5045–5060 (2006)
6. Y. Saito, Y. Kishiyama, A. Benjebbour, T. Nakamura, A. Li, K. Higuchi, Non-orthogonal multiple access (NOMA) for cellular future radio access, in *Proceedings of the IEEE 77th Vehicular Technology Conference (VTC Spring)*, June 2013, pp. 1–5
7. H. Nikopour, H. Baligh, Sparse code multiple access, in *Proceedings of the IEEE Annual International Symposium on Personal Indoor Mobile Radio Communications (PIMRC)*, Sept 2013, pp. 332–336
8. L. Dai, B. Wang, Y. Yuan, S. Han, I. Chih-lin, Z. Wang, Non-orthogonal multiple access for 5G: solutions, challenges, opportunities, and future research trends. IEEE Commun. Mag. **53**(9), 74–81 (2015)
9. Z. Ding, Y. Liu, J. Choi, Q. Sun, M. Elkashlan, I. Chih-lin, H.V. Poor, Application of non-orthogonal multiple access in LTE and 5G networks. IEEE Commun. Mag. **55**(2), 185–191 (2017)
10. W. Shin, M. Vaezi, B. Lee, D.J. Love, J. Lee, H.V. Poor, Non-orthogonal multiple access in multi-cell networks: theory, performance, and practical challenges. IEEE Commun. Mag. **55**(10), 176–183 (2017)
11. T. Cover, Broadcast channels. IEEE Trans. Inf. Theory **18**(1), 2–14 (1972)
12. D. Tse, P. Viswanath, *Fundamentals of Wireless Communication* (Cambridge University Press, Cambridge, U.K. 2005)
13. M.F. Hanif, Z. Ding, T. Ratnarajah, G.K. Karagiannidis, A minorization-maximization method for optimizing sum rate in the downlink of non-orthogonal multiple access systems. IEEE Trans. Signal Process. **64**(1), 76–88 (2016)
14. J. Choi, Minimum power multicast beamforming with superposition coding for multiresolution broadcast and application to NOMA systems. IEEE Trans. Commun. **63**(3), 791–800 (2015)
15. Q. Sun, S. Han, I. Chih-lin, Z. Pan, On the ergodic capacity of MIMO NOMA systems. IEEE Wireless Commun. Lett. **4**(4), 405–408 (2015)
16. Q. Zhang, Q. Li, J. Qin, Robust beamforming for nonorthogonal multiple-access systems in MISO channels. IEEE Trans. Veh. Technol. **65**(12), 10231–10236 (2016)
17. H. Joudeh, B. Clerckx, Rate-splitting for max-min fair multigroup multicast beamforming in overloaded systems. IEEE Trans. Wireless Commun. **16**(11), 7276–7289 (2017)
18. C. Lim, T. Yoo, B. Clerckx, B. Lee, B. Shim, Recent trend of multiuser MIMO in LTE-advanced. IEEE Commun. Mag. **51**(3), 127–135 (2013)
19. Z. Chen, Z. Ding, X. Dai, G.K. Karagiannidis, On the application of quasi-degradation to MISO–NOMA downlink. IEEE Trans. Signal Process. **64**(23), 6174–6189 (2016)
20. Z. Ding, F. Adachi, H.V. Poor, The application of MIMO to non-orthogonal multiple access. IEEE Trans. Wireless Commun. **15**(1), 537–552 (2016)
21. J. Choi, On generalized downlink beamforming with NOMA. J. Commun. Netw. **19**(4), 319–328 (2017)
22. W. Shin, M. Vaezi, B. Lee, D.J. Love, J. Lee, H.V. Poor, Coordinated beamforming for multi-cell MIMO-NOMA. IEEE Commun. Lett. **21**(1), 84–87 (2017)
23. V.D. Nguyen, H.D. Tuan, T.Q. Duong, H.V. Poor, O.S. Shin, Precoder design for signal superposition in MIMO–NOMA multicell networks. IEEE J. Sel. Areas Commun. **35**(12), 2681–2695 (2017)
24. M. Zeng, A. Yadav, O.A. Dobre, G.I. Tsiropoulos, H.V. Poor, Capacity comparison between MIMO-NOMA and MIMO-OMA with multiple users in a cluster. IEEE J. Sel. Areas Commun. **35**(10), 2413–2424 (2017)
25. T. Han, K. Kobayashi, A new achievable rate region for the interference channel. IEEE Trans. Inf. Theory **27**(1), 49–60 (1981)

26. R.H. Etkin, D.N.C. Tse, H. Wang, Gaussian interference channel capacity to within one bit. IEEE Trans. Inf. Theory **54**(12), 5534–5562 (2008)
27. B. Rimoldi, R. Urbanke, A rate-splitting approach to the Gaussian multiple-access channel. IEEE Trans. Inf. Theory **42**(2), 364–375 (1996)
28. A.G. Davoodi, S.A. Jafar, Aligned image sets under channel uncertainty: settling conjectures on the collapse of degrees of freedom under finite precision CSIT. IEEE Trans. Inf. Theory **62**(10), 5603–5618 (2016)
29. S. Yang, M. Kobayashi, D. Gesbert, X. Yi, Degrees of freedom of time correlated MISO broadcast channel with delayed CSIT. IEEE Trans. Inf. Theory **59**(1), 315–328 (2013)
30. H. Joudeh, B. Clerckx, Sum-rate maximization for linearly precoded downlink multiuser MISO systems with partial CSIT: a rate-splitting approach. IEEE Trans. Commun. **64**(11), 4847–4861 (2016)
31. E. Piovano, B. Clerckx, Optimal DoF region of the K-user MISO BC with partial CSIT. IEEE Commun. Lett. **21**(11), 2368–2371 (2017)
32. C. Hao, B. Clerckx, MISO networks with imperfect CSIT: a topological rate-splitting approach. IEEE Trans. Commun. **65**(5), 2164–2179 (2017)
33. C. Hao, B. Rassouli, B. Clerckx, Achievable DoF regions of MIMO networks with imperfect CSIT. IEEE Trans. Inf. Theory **63**(10), 6587–6606 (2017)
34. H. Joudeh, B. Clerckx, Robust transmission in downlink multiuser MISO systems: a rate-splitting approach. IEEE Trans. Signal Process. **64**(23), 6227–6242 (2016)
35. A.G. Davoodi, S.A. Jafar, GDoF of the MISO BC: bridging the gap between finite precision CSIT and perfect CSIT, in *Proceedings of the IEEE International Symposium on Information Theory (ISIT)*, July 2016, pp. 1297–1301
36. A.G. Davoodi, S.A. Jafar, Transmitter cooperation under finite precision CSIT: a GDoF perspective. IEEE Trans. Inf. Theory **63**(9), 6020–6030 (2017)
37. E. Piovano, H. Joudeh, B. Clerckx, Overloaded multiuser MISO transmission with imperfect CSIT, in *Proceedings of the 50th Asilomar Conference on Signals, Systems and Computers*, Nov 2016, pp. 34–38
38. C. Hao, Y. Wu, B. Clerckx, Rate analysis of two-receiver MISO broadcast channel with finite rate feedback: a rate-splitting approach. IEEE Trans. Commun. **63**(9), 3232–3246 (2015)
39. M. Dai, B. Clerckx, D. Gesbert, G. Caire, A rate splitting strategy for massive MIMO with imperfect CSIT. IEEE Trans. Wireless Commun. **15**(7), 4611–4624 (2016)
40. A. Papazafeiropoulos, B. Clerckx, T. Ratnarajah, Rate-splitting to mitigate residual transceiver hardware impairments in massive MIMO systems. IEEE Trans. Veh. Technol. **66**(9), 8196–8211 (2017)
41. M. Dai, B. Clerckx, Multiuser millimeter wave beamforming strategies with quantized and statistical CSIT. IEEE Trans. Wireless Commun. **16**(11), 7025–7038 (2017)
42. Y. Mao, B. Clerckx, V.O.K. Li, Rate-splitting multiple access for downlink communication systems: bridging, generalizing, and outperforming SDMA and NOMA. EURASIP J. Wireless Commun. Netw. **2018**(1), 133 (2018)
43. O. Tervo, L. Trant, S. Chatzinotas, B. Ottersten, M. Juntti, Multigroup multicast beamforming and antenna selection with rate-splitting in multicell systems, in *Proceedings of the IEEE International Workshop on Signal Processing Advances in Wireless Communications (SPAWC)*, June 2018, pp. 1–5
44. M. Medra, T.N. Davidson, Robust downlink transmission: an offset-based single-rate-splitting approach, in *Proceedings of the IEEE International Workshop on Signal Processing Advances in Wireless Communications (SPAWC)*, June 2018, pp. 1–5
45. Y. Mao, B. Clerckx, V.O.K. Li, Energy efficiency of rate-splitting multiple access, and performance benefits over SDMA and NOMA, in *Proceedings of the IEEE International Symposium on Wireless Communication Systems (ISWCS)*, Aug 2018, pp. 1–5
46. A.R. Flores, B. Clerckx, R.C. de Lamare, Tomlinson-harashima precoded rate-splitting for multiuser multiple-antenna systems, in *Proceedings of the IEEE International Symposium on Wireless Communication Systems (ISWCS)*, Aug 2018, pp. 1–6

47. A. Papazafeiropoulos, T. Ratnarajah, Rate-splitting robustness in multi-pair massive MIMO relay systems. IEEE Trans. Wireless Commun. **17**(8), 5623–5636 (2018)
48. M. Caus, A. Pastore, M. Navarro, T. Ramirez, C. Mosquera, N. Noels, N. Alagha, A.I. Perez-Neira, Exploratory analysis of superposition coding and rate splitting for multibeam satellite systems, in *Proceedings of the IEEE International Symposium on Wireless Communication Systems (ISWCS)*, Aug 2018, pp. 1–5
49. Y. Mao, B. Clerckx, V.O.K. Li, Rate-splitting multiple access for coordinated multi-point joint transmission, in *IEEE International Conference on Communications Workshops (ICC Workshops)*, May 2019, pp. 1–6
50. A.A. Ahmad, J. Kakar, R. Reifert, A. Sezgin, UAV-assisted C-RAN with rate splitting under base station breakdown scenarios, in *IEEE International Conference on Communications Workshops (ICC Workshops)*, May 2019, pp. 1–6
51. A. Rahmati, Y. Yapici, N. Rupasinghe, I. Guvenc, H. Dai, A. Bhuyan, Energy efficiency of RSMA and NOMA in cellular-connected mmwave UAV networks, in *IEEE International Conference on Communications Workshops (ICC Workshops)*, May 2019, pp. 1–6
52. A. Alameer Ahmad, H. Dahrouj, A. Chaaban, A. Sezgin, M. Alouini, Interference mitigation via rate-splitting and common message decoding in cloud radio access networks. IEEE Access **7**, 80350–80365 (2019)
53. D. Yu, J. Kim, S. Park, An efficient rate-splitting multiple access scheme for the downlink of C-RAN systems. IEEE Wireless Commun. Lett. **8**(6), 1555–1558 (2019)
54. Y. Mao, B. Clerckx, V.O.K. Li, Rate-splitting for multi-user multi-antenna wireless information and power transfer, in *Proceedings of the IEEE International Workshop on Signal Processing Advances in Wireless Communications (SPAWC)*, July 2019, pp. 1–5
55. X. Su, L. Li, H. Yin, P. Zhang, Robust power- and rate-splitting-based transceiver design in k-user MISO SWIPT interference channel under imperfect CSIT. IEEE Commun. Lett. **23**(3), 514–517 (2019)
56. J. Zhang, B. Clerckx, J. Ge, Y. Mao, Cooperative rate splitting for MISO broadcast channel with user relaying, and performance benefits over cooperative NOMA. IEEE Signal Process. Lett. **26**(11), 1678–1682 (2019)
57. Y. Mao, B. Clerckx, V.O.K. Li, Rate-splitting for multi-antenna non-orthogonal unicast and multicast transmission: spectral and energy efficiency analysis. IEEE Trans. Commun. **67**(12), 8754–8770 (2019)
58. B. Clerckx, Y. Mao, R. Schober, H.V. Poor, Rate-splitting unifying SDMA, OMA, NOMA, and multicasting in MISO broadcast channel: a simple two-user rate analysis. IEEE Wireless Commun. Lett. 1–1 **9**(3), 349–353 (2020)
59. H. Chen, D. Mi, B. Clerckx, Z. Chu, J. Shi, P. Xiao, Joint power and subcarrier allocation optimization for multigroup multicast systems with rate splitting. IEEE Trans. Veh. Technol. 1–1 **69**(2), 2306–2310 (2020)
60. A.R. Flores, R.C. De Lamare, B. Clerckx, Linear precoding and stream combining for rate splitting in multiuser MIMO systems. IEEE Commun. Lett. 1–1 **24**(4), 890–894 (2020)
61. Y. Mao, B. Clerckx, J. Zhang, V.O.K. Li, M. Arafah, Max-min fairness of K-user cooperative rate-splitting in MISO broadcast channel with user relaying. IEEE Trans. on Wireless Commun. **19**(10), 6362–6376 (2020)
62. Y. Mao, B. Clerckx, Beyond dirty paper coding for multi-antenna broadcast channel with partial CSIT: a rate-splitting approach. IEEE Trans. on Commun. (early access, 2020)
63. J. Zhao, D. Gündüz, O. Simeone, D. Gómez-Barquero, Non-orthogonal unicast and broadcast transmission via joint beamforming and LDM in cellular networks. IEEE Trans. Broadcast. 1–13 **66**(2), 216–228 (2019)
64. Y.F. Liu, C. Lu, M. Tao, J. Wu, Joint multicast and unicast beamforming for the MISO downlink interference channel, in *Proceedings of the IEEE International Workshop on Signal Processing Advances in Wireless Communications (SPAWC)*, July 2017, pp. 1–5
65. J. Riordan, *Introduction to combinatorial analysis* (Courier Corporation, Chelmsford, MA, USA 2012)

66. S.S. Christensen, R. Agarwal, E.D. Carvalho, J.M. Cioffi, Weighted sum-rate maximization using weighted MMSE for MIMO-BC beamforming design. IEEE Trans. Wireless Commun. **7**(12), 4792–4799 (2008)
67. G. Caire, N. Jindal, M. Kobayashi, N. Ravindran, Multiuser MIMO achievable rates with downlink training and channel state feedback. IEEE Trans. Inf. Theory **56**(6), 2845–2866 (2010)
68. C. Hao, Y. Wu, B. Clerckx, Rate analysis of two-receiver MISO broadcast channel with finite rate feedback: a rate-splitting approach. IEEE Trans. Commun. **63**(9), 3232–3246 (2015)
69. J. Chen, V.K.N. Lau, Two-tier precoding for FDD multi-cell massive MIMO time-varying interference networks. IEEE J. Sel. Areas Commun. **32**(6), 1230–1238 (2014)
70. J. Park, B. Clerckx, Multi-user linear precoding for multi-polarized massive MIMO system under imperfect CSIT. IEEE Trans. Wireless Commun. **14**(5), 2532–2547 (2015)
71. D. Kim, G. Lee, Y. Sung, Two-stage beamformer design for massive MIMO downlink by trace quotient formulation. IEEE Trans. Commun. **63**(6), 2200–2211 (2015)
72. J. Zhang, P. Elia, Fundamental limits of cache-aided wireless BC: interplay of coded-caching and CSIT feedback. IEEE Trans. Inf. Theory **63**(5), 3142–3160 (2017)
73. C. Xu, B. Clerckx, S. Chen, Y. Mao, J. Zhang, Rate-splitting multiple access for multi-antenna joint communication and radar transmissions. IEEE International Conference on Communications Workshops (ICC Workshops), Dublin, Ireland, 2020, pp. 1–6
74. L. Yin, B. Clerckx, Rate-splitting multiple access for multibeam satellite communications. arXiv preprint arXiv:2002.01731 (2020)
75. Evolved universal terrestrial radio access (E-UTRA); LTE physical layer; General description (Release 8), 3GPP TS 36.201, Tech. Rep., Mar 2009
76. Coordinated multi-point operation for LTE physical layer aspects (Release 11), 3GPP TR 36.819, Tech. Rep., Aug 2016
77. Study on network-assisted interference cancellation and suppression (NAIC) for LTE (Release 12), 3GPP TR 36.866, Tech. Rep., Mar 2014
78. Study on downlink multiuser superposition transmission (MUST) for LTE (Release 13), 3GPP TR 36.859, Tech. Rep., Dec 2015

Chapter 4
Massive MIMO

Hien Quoc Ngo

Massive multiple-input multiple-output (MIMO) technology has become one of the key technologies for fifth generation (5G) wireless systems and beyond due to its potential to offer high spectral efficiency simultaneously to many users, with simple signal processing. These remarkable gains are obtained through the use of many antennas at the base station to spatially multiplex many users on the same time-frequency resource. This chapter provides a comprehensive overview of massive MIMO. A completed massive MIMO transmission protocol under time-division duplex operation is first presented. Then, fundamental aspects including favorable propagation, channel hardening, pilot contamination, and use-and-then-forget capacity bounding technique are discussed in detail. Finally, a range of important topics for future research on massive MIMO is suggested.

4.1 Introduction

Point-to-point multiple-input multiple-output (MIMO) antenna technology can offer high spectral efficiency through multiplexing techniques and high communication reliability through the diversity techniques [1–3]. Thus, since it was first introduced around the 1990s, it has gained significant research attention from both academia and industry. MIMO technology has been incorporated into fourth and fifth generation (4G and 5G) systems. One of the main limitations of MIMO is the aspect of multiplexing gain. More precisely, the multiplexing gain of MIMO technology cannot be fully achieved if we are working in low signal-to-noise ratio (SNR)

H. Quoc Ngo (✉)
School of Electronics, Electrical Engineering and Computer Science, Queen's University Belfast, Belfast, UK
e-mail: hien.ngo@qub.ac.uk

X. Lin, N. Lee (eds.), *5G and Beyond*, https://doi.org/10.1007/978-3-030-58197-8_4

regimes or in the propagation environments having ill-behaved channels such as line-of-sight or double-scattering channels.

By distributing the antenna arrays at the transmitter or receiver side, we have another version of MIMO system called multi-user multiple-input multiple-output (MU-MIMO) system. In MU-MIMO, a base station equipped with several antennas simultaneously serves several users [4, 5]. In this system, the spatial multiplexing gain is exploited to obtain very high spectral efficiency. In addition, different from point-to-point MIMO, MU-MIMO still achieves full spatial multiplexing gain in ill-behaved channels. Therefore, significant efforts have been put into the developments of MU-MIMO during the last decades [6–8]. However, in the conventional MU-MIMO, since the number of base station antennas is small, the system performance in terms of spectral efficiency and communication reliability is limited by mutual interference [9]. To obtain optimal performance, complicated and impractical nonlinear processing and resource allocations need to be used. This makes the conventional MU-MIMO system unscalable in the sense that the system is not feasible when the number of users/base station antennas grows large.

A practical and scalable version of MU-MIMO is massive MIMO where the base station equipped with many antennas simultaneously serves many users in the same frequency bands [10]. With many antennas at the base station, massive MIMO can offer high diversity and array and multiplexing gains, and hence, it can provide very high energy efficiency and spectral efficiency with simple linear processing such as maximum-ratio (MR), zero-forcing (ZF), and minimum mean-square error (MMSE) processing. Therefore, since massive MIMO was first introduced in [11, 12], it has gained significant research attention with many researches from various perspectives ranging from fundamental analysis [13–16], signal processing [17–20], resource allocations [21–23], to practical implementations [24–26]. Nowadays, massive MIMO has become a key technology for 5G wireless networks. The first phase of 5G New Radio (NR) with massive MIMO was standardized by 3GPP. This chapter provides an overview of massive MIMO and some important future research topics.

The remaining of this chapter is organized as follows. Section 4.2 presents a completed time-division duplex (TDD) massive MIMO system including the uplink training, uplink payload data transmission, and downlink pilot data transmission phases. Section 4.3 discusses some fundamentals of massive MIMO including the favorable propagation, channel hardening, pilot contamination, and use-and-then-forget capacity bounding technique. Important topics for future research on massive MIMO are provided in Sect. 4.4. Finally, Sect. 4.5 concludes the chapter.

Notations: The bold lower- and uppercase letters are used for vectors and matrices. The expectation of the random variable is denoted by $\mathbb{E}\{\cdot\}$, while its variance is $\mathrm{Var}\{\cdot\}$. $\mathcal{CN}(\cdot,\cdot)$ is a circularly symmetric Gaussian distribution. The Euclidean norm is $\|\cdot\|$ and the Frobenius norm is $\|\cdot\|_F$. The regular and Hermitian transposes of a matrix are given as $(\cdot)^T$ and $(\cdot)^H$, respectively.

4.2 Massive MIMO Systems

For simplicity, we consider a single-cell massive MIMO system where a base station equipped with M antennas simultaneously serves K single-antenna users distributed within the cell in the same frequency band. We assume that $M > K$. See Fig. 4.1. The channel between the k-th user and the m-th base station antenna includes both large-scale fading and small-scale fading effects. More precisely, it can be modelled as

$$g_{mk} = \sqrt{\beta_{mk}} h_{mk}, \tag{4.1}$$

where h_{mk} represents small-scale fading, while β_{mk} represents large-scale fading. Some specific assumptions regarding the channel model are as follows:

- Large-scale fading coefficients from given user to M base station antennas are the same, i.e.,

$$\beta_{mk} = \beta_k, \quad m = 1 \ldots M. \tag{4.2}$$

  This assumption is reasonable in many practical scenarios where the distance between base station antennas is much smaller than the distance between the base station and the users.

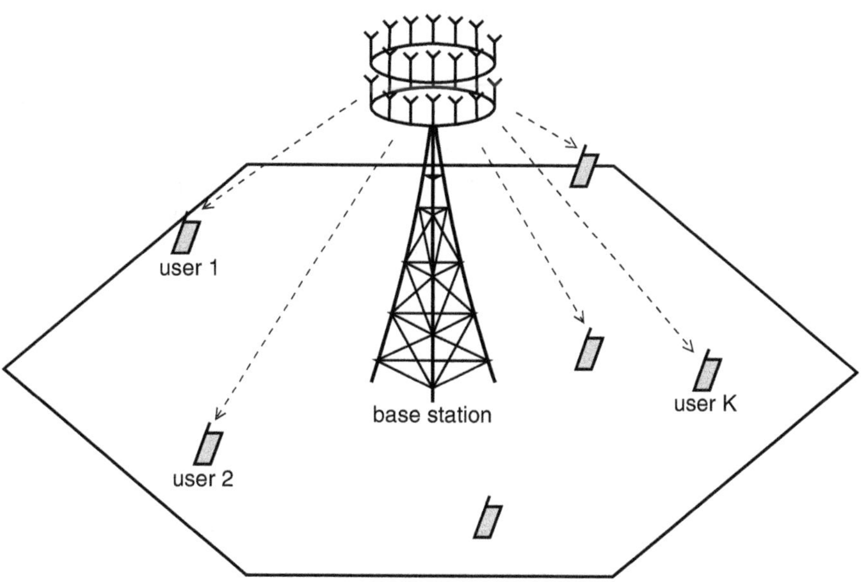

Fig. 4.1 Massive MIMO system model where M-antenna base station serves all K users in the same time-frequency resource

- The small-scale fading channel is Rayleigh fading channel. More precisely, h_{mk} is Gaussian distributed random variable (RV) with zero mean and unit variance, i.e., $h_{mk} \sim \mathcal{CN}(0, 1)$. Furthermore, we assume that $\{h_{mk}\}$, $m = 1, \ldots, M$, $k = 1, \ldots, K$, are independent.
- The small-scale fading is unchanged during each coherence interval and changes independently from one coherence interval to the next.
- The base station and the users have perfect knowledge of the large-scale fading β_k. This assumption is reasonable since the large-scale fading coefficients are independent of base station antenna indices and change very slowly with time. As a result, it is easy to acquire β_k with very high accuracy. One simple method to estimate β_k is discussed in [27].
- The channel is reciprocal, i.e., the uplink channel and downlink channel gains are the same. This reciprocity can be reasonably obtained by a good calibration of the hardware chains [28].

Some further notations which are used throughout this chapter are as follows:

- The channel between the base station and user k is denoted by $\mathbf{g}_k$ which is an $M \times 1$ vector whose m-th element is g_{mk}. The corresponding small-scale fading vector is denoted by $\mathbf{h}_k$ which is equal to $[h_{1k}, \ldots, h_{Mk}]^T$. From (4.1) and (4.2), we have

$$\mathbf{g}_k = \sqrt{\beta_k}\mathbf{h}_k. \tag{4.3}$$

- The $M \times K$ channel matrix between the base station and all K users is denoted by $\mathbf{G}$ whose k-th column is $\mathbf{g}_k$. The corresponding small-scale fading matrix is denoted by $\mathbf{H}$ whose k-th column is $\mathbf{h}_k$.

In general, the uplink and downlink transmission can be performed under frequency-division duplex (FDD) or TDD operation. In FDD operation, the downlink and uplink transmissions perform in the same time, but different frequency bands which are separated by more than a coherence bandwidth. By contrast, in TDD operation, the downlink and uplink transmissions perform in the same frequency bands, but different time. In FDD, the channel estimation overhead depends on the number of base station antennas. While in TDD, the channel estimation overhead depends only on the number of users. Therefore, TDD is preferable in massive MIMO. In this chapter, we will only discuss massive MIMO under TDD operation. With TDD, there are three main phases for each coherence interval (which equals the product of the coherence time and the coherence bandwidth): uplink training, uplink payload data transmission, and downlink payload data transmission. The TDD transmission protocol is shown in Fig. 4.2 where τ_{c}, τ_{p}, τ_{u}, and τ_{d} denote the coherence interval, training duration, uplink payload data transmission duration, and downlink payload data transmission duration (in symbols), respectively, satisfying $\tau_{\mathrm{p}} + \tau_{\mathrm{u}} + \tau_{\mathrm{d}} = \tau_{\mathrm{c}}$. Note that the order of these three phases can be changed. Also, there may be downlink training phase which is

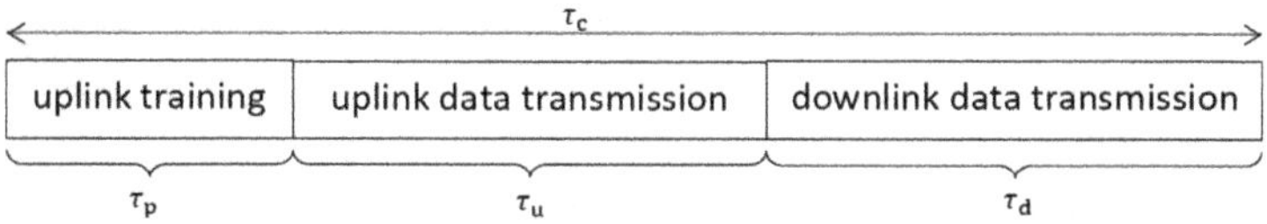

Fig. 4.2 TDD massive MIMO transmission protocol for each coherence interval

used for the channel estimation at the users [29]. In the following sections, we will provide the details of the above three main phases in massive MIMO.

4.2.1 Uplink Training

In this phase, the base station will estimate the channels to all users via the pilots sent from the users. All K users will simultaneously send their pilot sequences to the base station. Let $\sqrt{\tau_p}\boldsymbol{\varphi}_k \in \mathbb{C}^{\tau_p \times 1}$, where $\|\boldsymbol{\varphi}_k\|^2 = 1$, be the pilot sequence assigned to the k-th user, $k = 1, \ldots, K$, and ρ_p be the normalized transmit power of each pilot symbol (normalized by the noise power at the base station). Then, the base station receives an $M \times \tau_p$ pilot matrix:

$$\mathbf{Y}_p = \sqrt{\tau_p \rho_p} \sum_{k=1}^{K} \mathbf{g}_k \boldsymbol{\varphi}_k^H + \mathbf{W}_p, \tag{4.4}$$

where $\mathbf{W}_p$ is an $M \times \tau_p$ noise matrix with (i.i.d.) $\mathcal{CN}(0, 1)$ RVs.

From the received pilot signal (4.5), the base station will estimate the channels to all K users. More precisely, to estimate the channel to user k, the base station first projects $\mathbf{Y}_p$ onto the pilot sequence sent by this user to obtain

$$\begin{aligned} \check{\mathbf{y}}_{p,k} &\triangleq \mathbf{Y}_p \boldsymbol{\varphi}_k \\ &= \sqrt{\tau_p \rho_p}\mathbf{g}_k + \sqrt{\tau_p \rho_p} \sum_{k' \neq k}^{K} \mathbf{g}_{k'} \boldsymbol{\varphi}_{k'}^H \boldsymbol{\varphi}_k + \mathbf{W}_p \boldsymbol{\varphi}_k. \end{aligned} \tag{4.5}$$

Then it uses MMSE estimation scheme to estimate $\mathbf{g}_k$. The MMSE estimate of $\mathbf{g}_k$ given $\check{\mathbf{y}}_{p,k}$ is

$$\begin{aligned} \hat{\mathbf{g}}_k &= \mathbb{E}\left\{\mathbf{g}_k \check{\mathbf{y}}_{p,k}^H\right\} \left(\mathbb{E}\left\{\check{\mathbf{y}}_{p,k} \check{\mathbf{y}}_{p,k}^H\right\}\right)^{-1} \check{\mathbf{y}}_{p,k} \\ &= \frac{\sqrt{\tau_p \rho_p}\beta_k}{\tau_p \rho_p \sum_{k'=1}^{K} \beta_{k'} \left|\boldsymbol{\varphi}_{k'}^H \boldsymbol{\varphi}_k\right|^2 + 1} \check{\mathbf{y}}_{p,k}. \end{aligned} \tag{4.6}$$

Remark 1 The base station can estimate the channels by directly using its received signal $\mathbf{Y}_{\rm p}$ without the projection step. In general, this will yield more accurate channel estimates. However, most of work in massive MIMO literature uses the projection step. This is due to the fact that the projection step simplifies the system performance analysis. More importantly, the difference on the channel estimate quality between the cases without and with projection step is very small. In addition, if all K pilots are chosen from a set of orthogonal sequences, then the MMSE estimate based on $\mathbf{Y}_{\rm p}$ and the MMSE estimate based on the projection $\check{\mathbf{y}}_{{\rm p},k}$ are the same (since $\check{\mathbf{y}}_{{\rm p},k}$ is a sufficient statistic to estimate $\mathbf{g}_k$).

From (4.6), we can see that the channel estimate vector $\hat{\mathbf{g}}_k$ includes M i.i.d. Gaussian RVs with zero mean and variance:

$$\begin{aligned}\gamma_k &\triangleq \mathbb{V}\text{ar}\left\{\left[\hat{\mathbf{g}}_k\right]_m\right\}\\ &= \frac{\tau_{\rm p}\rho_{\rm p}\beta_k^2}{\tau_{\rm p}\rho_{\rm p}\sum_{k'=1}^{K}\beta_{k'}\left|\boldsymbol{\varphi}_{k'}^H\boldsymbol{\varphi}_k\right|^2+1}.\end{aligned} \tag{4.7}$$

The channel estimation error is denoted by

$$\tilde{\mathbf{g}}_k = \mathbf{g}_k - \hat{\mathbf{g}}_k. \tag{4.8}$$

Substituting (4.6) into (4.8), we obtain

$$\begin{aligned}\tilde{\mathbf{g}}_k &= \mathbf{g}_k - \frac{\sqrt{\tau_{\rm p}\rho_{\rm p}}\beta_k}{\tau_{\rm p}\rho_{\rm p}\sum_{k'=1}^{K}\beta_{k'}\left|\boldsymbol{\varphi}_{k'}^H\boldsymbol{\varphi}_k\right|^2+1}\check{\mathbf{y}}_{{\rm p},k}\\ &=\mathbf{g}_k - \frac{\sqrt{\tau_{\rm p}\rho_{\rm p}}\beta_k}{\tau_{\rm p}\rho_{\rm p}\sum_{k'=1}^{K}\beta_{k'}\left|\boldsymbol{\varphi}_{k'}^H\boldsymbol{\varphi}_k\right|^2+1}\left(\sqrt{\tau_{\rm p}\rho_{\rm p}}\mathbf{g}_k+\sqrt{\tau_{\rm p}\rho_{\rm p}}\sum_{k'\neq k}^{K}\mathbf{g}_{k'}\boldsymbol{\varphi}_{k'}^H\boldsymbol{\varphi}_k+\mathbf{W}_{\rm p}\boldsymbol{\varphi}_k\right).\end{aligned} \tag{4.9}$$

Therefore, $\tilde{\mathbf{g}}_k$ includes M i.i.d. Gaussian RVs with zero mean and variance:

$$\epsilon_k \triangleq \mathbb{V}\text{ar}\left\{\left[\tilde{\mathbf{g}}_k\right]_m\right\} = \frac{\tau_{\rm p}\rho_{\rm p}\sum_{k'\neq k}^{K}\beta_k\beta_{k'}\left|\boldsymbol{\varphi}_{k'}^H\boldsymbol{\varphi}_k\right|^2+\beta_k}{\tau_{\rm p}\rho_{\rm p}\sum_{k'=1}^{K}\beta_{k'}\left|\boldsymbol{\varphi}_{k'}^H\boldsymbol{\varphi}_k\right|^2+1}. \tag{4.10}$$

Furthermore, from the MMSE estimation property, $\tilde{\mathbf{g}}_k$ is uncorrelated with $\hat{\mathbf{g}}_k$. Since $\tilde{\mathbf{g}}_k$ and $\hat{\mathbf{g}}_k$ are Gaussian vectors, they are independent. We can see that

$$\gamma_k + \epsilon_k = \beta_k. \tag{4.11}$$

Remark 2 The quantity ϵ_k given in (4.10) represents the mean square error of the channel estimation. If all pilot sequences of all K users are pairwisely orthogonal, i.e., $\boldsymbol{\varphi}_{k'}^H \boldsymbol{\varphi}_k = 0$ for $k \neq k'$, then

$$\epsilon_k \to 0, \text{ as } \rho_\text{p} \to \infty. \tag{4.12}$$

This implies that by increasing the pilot power we can obtain a channel estimate with any level of accuracy we want. However to guarantee all pilot sequences are pairwisely orthogonal, it requires that $\tau_\text{p} \geq K$. In the scenarios where the number of users is large and/or the coherence interval is short, the above condition cannot be satisfied. It means that pilot sequences assigned for K users are not pairwisely orthogonal. In this case, we have

$$\epsilon_k \to \frac{\sum_{k' \neq k}^{K} \beta_k \beta_{k'} \left|\boldsymbol{\varphi}_{k'}^H \boldsymbol{\varphi}_k\right|^2}{\sum_{k'=1}^{K} \beta_{k'} \left|\boldsymbol{\varphi}_{k'}^H \boldsymbol{\varphi}_k\right|^2}, \text{ as } \rho_\text{p} \to \infty. \tag{4.13}$$

We can see that with non-orthogonal pilots, the mean square error of the channel estimation is bounded even when ρ_p goes to infinity. We cannot make the channel estimate as good as we want. The channel estimation error caused by the non-orthogonality of pilots will reduce the system performance significantly, even when the number of base station antennas goes to infinity. This is called pilot contamination effect. The detail of pilot contamination is discussed in Sect. 4.3.4.

4.2.2 Uplink Payload Data Transmission

In this phase, the users send data to the base station. The base station will use the channel state information acquired in the training phase to combine and decode the desired data.

The symbol sent from user k is denoted by $s_{\text{u},k}$ which is assumed to be a zero mean and unit variance variable. We further assume that the symbols from all K users are independent. Since all K users share the same time/frequency resource, the base station receives an $M \times 1$ signal vector:

$$\mathbf{y}_\text{u} = \sqrt{\rho_\text{u}} \sum_{k=1}^{K} \mathbf{g}_k \sqrt{\eta_{\text{u},k}} s_{\text{u},k} + \mathbf{w}_\text{u}, \tag{4.14}$$

where $\rho_\text{u}\eta_k$ is the normalized transmit power of user k with $\eta_{\text{u},k} \leq 1$ is the corresponding power control coefficient, and $\mathbf{w}_\text{u}$ is the noise vector with i.i.d. $\mathcal{CN}(0, 1)$ elements.

The base station wants to detect all $s_k, k = 1, \ldots, K$. To do this, the base station first uses linear processing techniques such as MR, ZF, and MMSE processing to combine the signals received from its M antennas. Then it will detect the desired symbols based on the combined signal. One of the nice things in massive MIMO is that, under many propagation environments such as the isotropic (rich) scattering propagation environments in this chapter, the channel is favorable, and hence, linear processing is nearly optimal. The favorable propagation property is one of the key properties in massive MIMO which will be discussed in detail in Sect. 4.3.1.

The received signal vector at the base station after using a linear processing scheme is given by

$$\mathbf{r}_{\mathrm{u}} = \mathbf{A}^H \mathbf{y}_{\mathrm{u}}, \tag{4.15}$$

where $\mathbf{A}$ is an $M \times K$ combining matrix which is constructed from the channel estimates $\{\hat{\mathbf{g}}_k\}$. Three common linear processing techniques are MR, ZF, and MMSE processing. In this chapter, we consider MR and ZF processing since they are simple for the performance analysis. More importantly, at low SNR, MR perform as well as MMSE, while at high SNR, the performance of ZF is very close to that of MMSE. Therefore, the performance of MR and ZF can cover the performance of MMSE in all SNR regimes. The combining matrix $\mathbf{A}$ corresponding to MR and ZF is given by

$$\mathbf{A} = \begin{cases} \hat{\mathbf{G}} & \text{for MR} \\ \hat{\mathbf{G}}\left(\hat{\mathbf{G}}^H\hat{\mathbf{G}}\right)^{-1} & \text{for ZF.} \end{cases} \tag{4.16}$$

To detect the symbol sent from user k, the base station uses the k-th element of $\mathbf{r}_{\mathrm{u}}$ which is given by

$$r_{\mathrm{u},k} = \mathbf{a}_k^H \mathbf{y}_{\mathrm{u}}, \tag{4.17}$$

where $\mathbf{a}_k$ is the k-th column of $\mathbf{A}$. The substitution of (4.14) into (4.17) yields

$$r_{\mathrm{u},k} = \sqrt{\rho_{\mathrm{u}}}\mathbf{a}_k^H \mathbf{g}_k \sqrt{\eta_{\mathrm{u},k}} s_{\mathrm{u},k} + \sqrt{\rho_{\mathrm{u}}} \sum_{k' \neq k}^{K} \mathbf{a}_k^H \mathbf{g}_{k'} \sqrt{\eta_{\mathrm{u},k'}} s_{\mathrm{u},k'} + \mathbf{a}_k^H \mathbf{w}_{\mathrm{u}}. \tag{4.18}$$

The first term in (4.18) represents the desired signal part, while the second and third terms represent the inter-user interference and noise. In massive MIMO, in the case of orthogonal pilots, thanks to the favorable propagation property, the powers of the inter-user interference and noise are very small relatively compared to the power of the desired signal part, and hence, we can obtain very good performance.

4.2.3 *Downlink Payload Data Transmission*

In this phase, the base station simultaneously transmits signals to all K users. More precisely, the base station first uses the channels estimated during the uplink training phase to precode the symbols intended for all users and then sends the precoded version to them. Note that this is feasible because the channels are reciprocal.

As discussed in the uplink phase, in massive MIMO, linear processing is preferable. Therefore, in this chapter, we consider only linear precoders. With linear precoding techniques, the $M \times 1$ transmitted signal vector at the base station is given by

$$\mathbf{x}_{\mathrm{d}} = \sqrt{\rho_{\mathrm{d}}} \sum_{k=1}^{K} \sqrt{\eta_{\mathrm{d},k}} \mathbf{b}_k s_{\mathrm{d},k}. \tag{4.19}$$

In (4.19), we have the following notation:

- ρ_{d} is the normalized (normalized by noise power) transmit power at the base station.
- $s_{\mathrm{d},k}$ is the symbol intended for the k-th user. We assume that $s_{\mathrm{d},1}, \ldots, s_{\mathrm{d},K}$ are i.i.d. zero mean and unit variance RVs. In addition, we denote $\mathbf{s}_{\mathrm{d}} = [s_{\mathrm{d},1}, \ldots, s_{\mathrm{d},K}]^T$ which represents the vector of K symbols intended for all K users.
- $\mathbf{b}_k$ is the $M \times 1$ precoding vector associated with user k which is constructed based on the channel estimation matrix $\hat{\mathbf{G}}$. Let $\mathbf{B} = [\mathbf{b}_1, \ldots, \mathbf{b}_K]$. Then, the $M \times K$ precoding matrices $\mathbf{B}$ corresponding to maximum-ratio and zero-forcing processing are

$$\mathbf{B} = \begin{cases} \hat{\mathbf{G}}^* & \text{for MR,} \\ \hat{\mathbf{G}}^* \left(\hat{\mathbf{G}}^T \hat{\mathbf{G}}^* \right)^{-1} & \text{for ZF.} \end{cases} \tag{4.20}$$

- $\eta_{\mathrm{d},k}$ is the power control coefficient associated with user k. The choice of $\eta_{\mathrm{d},k}$ must meet the power constraint at the base station:

$$\mathbb{E}\left\{ \|\mathbf{s}_{\mathrm{d}}\|^2 \right\} \leq \rho_{\mathrm{d}}. \tag{4.21}$$

Substituting (4.19) into (4.21), we obtain

$$\mathbb{E}\left\{ \left\| \sqrt{\rho_{\mathrm{d}}} \sum_{k=1}^{K} \sqrt{\eta_{\mathrm{d},k}} \mathbf{b}_k s_{\mathrm{d},k} \right\|^2 \right\} \leq \rho_{\mathrm{d}}. \tag{4.22}$$

Since $\{s_{\mathrm{d},k}\}$ are i.i.d zero mean and unit variance RVs, constraint (4.22) is equivalent to

$$\sum_{k=1}^{K} \eta_{\mathrm{d},k}\, \mathbb{E}\left\{ \|\mathbf{b}_k\|^2 \right\} \leq 1. \tag{4.23}$$

For specific MR and ZF, the constraint (4.23) becomes

$$\begin{cases} \sum_{k=1}^{K} \eta_{\mathrm{d},k} \gamma_k \leq 1/M & \text{for MR,} \\ \sum_{k=1}^{K} \eta_{\mathrm{d},k} / \gamma_k \leq M - K & \text{for ZF,} \end{cases} \tag{4.24}$$

where in (4.24) we have used the following identities

$$\mathbb{E}\left\{ \|\hat{\mathbf{g}}_k\|^2 \right\} = M\gamma_k,$$

and

$$\mathbb{E}\left\{ \left\| \left[\hat{\mathbf{G}}^* \left(\hat{\mathbf{G}}^T \hat{\mathbf{G}}^* \right)^{-1} \right]_k \right\|^2 \right\} = \mathbb{E}\left\{ \left[\left(\hat{\mathbf{G}}^T \hat{\mathbf{G}}^* \right)^{-1} \right]_{k,k} \right\} = \frac{1}{(M-K)\gamma_k}.$$

With the transmit signal vector (4.19) and under the assumption of perfect channel reciprocity, the received signal at user k is given by

$$\begin{aligned} y_{\mathrm{d},k} &= \mathbf{g}_k^T \mathbf{x}_{\mathrm{d}} + w_{\mathrm{d},k} \\ &= \sqrt{\rho_{\mathrm{d}}} \sum_{k'=1}^{K} \sqrt{\eta_{\mathrm{d},k'}} \mathbf{g}_k^T \mathbf{b}_{k'} s_{\mathrm{d},k'} + w_{\mathrm{d},k} \\ &= \sqrt{\rho_{\mathrm{d}}} \sqrt{\eta_{\mathrm{d},k}} \mathbf{g}_k^T \mathbf{b}_k s_{\mathrm{d},k} + \sqrt{\rho_{\mathrm{d}}} \sum_{k' \neq k}^{K} \sqrt{\eta_{\mathrm{d},k}} \mathbf{g}_k^T \mathbf{b}_{k'} s_{\mathrm{d},k'} + w_{\mathrm{d},k}. \end{aligned} \tag{4.25}$$

In (4.25), the first term is the desired signal part, the second term is the inter-user interference which includes the signals transmitted from other users, and the last term is the noise at the user. We assume that $w_{\mathrm{d},k} \sim \mathcal{CN}(0, 1)$. In massive MIMO, owing to the favorable propagation, even with linear processing, the desired signal power dominates the powers of the interference and noise, and hence, we can obtain very good performance.

4.3 Fundamentals of Massive MIMO

This section provides details about key properties of massive MIMO (including favorable propagation and channel hardening), the useful use-and-then-forget capacity technique, and the fundamental limitation of massive MIMO related to pilot contamination effect.

4.3.1 Favorable Propagation

(a) *Definition of Favorable Propagation:* The channel offers favorable propagation if

(i) $\{\mathbf{g}_k\}$ are non-zero vectors; and
(ii) the inner product between two channel vectors $\mathbf{g}_k$ and $\mathbf{g}_{k'}$, for $k \neq k'$, is equal to 0, i.e.,

$$\mathbf{g}_k^H \mathbf{g}_{k'} = 0, \text{ for } k \neq k'. \tag{4.26}$$

We need the first condition because in practice when time (or frequency) division multiple access is used, at a given time (or frequency), there is only one non-zero channel vector (corresponding to the user that is active in this time or frequency).

The second condition will never be exactly satisfied in practice. But as long as

$$\mathbf{g}_k^H \mathbf{g}_{k'} \ll \|\mathbf{g}_k\|^2,$$

we can say the channel is favorable.

(b) *Why Favorable Propagation Is Important?* Let us consider the uplink payload data transmission. Assume that MR processing is used, and the channel estimation is perfect, i.e., $\hat{\mathbf{G}} = \mathbf{G}$. Then the received signal at the base station after using MR processing (4.18) becomes

$$r_{\mathrm{u},k} = \sqrt{\rho_{\mathrm{u}}}\sqrt{\eta_{\mathrm{u},k}}\|\mathbf{g}_k\|^2 s_{\mathrm{u},k} + \sqrt{\rho_{\mathrm{u}}}\sum_{k' \neq k}^{K} \mathbf{g}_k^H \mathbf{g}_{k'}\sqrt{\eta_{\mathrm{u},k'}} s_{\mathrm{u},k'} + \mathbf{g}_k^H \mathbf{w}_{\mathrm{u}}. \tag{4.27}$$

If the channel is favorable (i.e., $\mathbf{g}_k^H \mathbf{g}_{k'} = 0$ for $k \neq k'$), then the inter-user interference disappears, and hence, (4.28) becomes

$$r_{\mathrm{u},k} = \sqrt{\rho_{\mathrm{u}}}\sqrt{\eta_{\mathrm{u},k}}\|\mathbf{g}_k\|^2 s_{\mathrm{u},k} + \mathbf{g}_k^H \mathbf{w}_{\mathrm{u}}. \tag{4.28}$$

As a result, the instantaneous achievable rate of user k is

$$R_{\mathrm{u},k} = \log_2\left(1 + \frac{\rho_{\mathrm{u}}\eta_{\mathrm{u},k}\|\mathbf{g}_k\|^4}{\mathbb{E}\left\{|\mathbf{g}_k^H \mathbf{w}_{\mathrm{u}}|^2\right\}}\right) = \log_2\left(1 + \rho_{\mathrm{u}}\eta_{\mathrm{u},k}\|\mathbf{g}_k\|^2\right), \tag{4.29}$$

and the corresponding instantaneous achievable sum rate is

$$R_{\mathrm{u,sum}} = \sum_{k=1}^{K} R_{\mathrm{u},k} = \sum_{k=1}^{K} \log_2\left(1 + \rho_{\mathrm{u}}\eta_{\mathrm{u},k}\|\mathbf{g}_k\|^2\right). \tag{4.30}$$

It is well-known that the maximum instantaneous achievable sum rate (or the instantaneous sum capacity) of the uplink channel (4.14) is [30]

$$C_{\mathrm{u,sum}} = \log_2 \det\left(\mathbf{I}_K + \mathbf{G}^H \mathbf{P} \mathbf{G}\right), \tag{4.31}$$

where $\mathbf{P} \triangleq \mathrm{diag}\left\{\eta_{\mathrm{u},1}, \ldots, \eta_{\mathrm{u},K}\right\}$. If the channel is favorable, then

$$\begin{aligned} C_{\mathrm{u,sum}} &= \log_2 \det\left(\mathbf{I}_K + \begin{bmatrix} \eta_{\mathrm{u},1}\|\mathbf{g}_1\|^2 & \cdots & 0 \\ \vdots & \ddots & \vdots \\ 0 & \cdots & \eta_{\mathrm{u},K}\|\mathbf{g}_K\|^2 \end{bmatrix}\right) \\ &= \sum_{k=1}^{K} \log_2\left(1 + \rho_{\mathrm{u}}\eta_{\mathrm{u},k}\|\mathbf{g}_k\|^2\right), \end{aligned} \tag{4.32}$$

which is identical to (4.30). The same insights can be obtained for other linear processing techniques (e.g., ZF) and for the downlink payload data transmission. This implies that under favorable propagation, linear processing is optimal in the sense that it maximizes the achievable rate.

(c) *Favorable Propagation in Massive MIMO:*

With the channel model provided in Sect. 4.2, from the law of large numbers, we have

$$\frac{\mathbf{g}_k^H \mathbf{g}_{k'}}{\|\mathbf{g}_k\|^2} = \frac{\mathbf{g}_k^H \mathbf{g}_{k'}/M}{\|\mathbf{g}_k\|^2/M} \overset{a.s.}{\to} 0, \text{ as } M \to \infty, \tag{4.33}$$

where $\overset{a.s.}{\to}$ denotes almost sure convergence. In (4.33), we have used the law of large numbers:

$$\frac{\mathbf{g}_k^H \mathbf{g}_{k'}}{M} \overset{a.s.}{\to} 0, \text{ as } M \to \infty, \tag{4.34}$$

and

$$\frac{\|\mathbf{g}_k\|^2}{M} \overset{a.s.}{\to} \beta_k, \text{ as } M \to \infty. \tag{4.35}$$

Result (4.33) implies that when the number of base station antennas M is large enough, with high probability, $\mathbf{g}_k^H \mathbf{g}_{k'} \ll \|\mathbf{g}_k\|^2$, and hence, channel is favorable. In other words, massive MIMO under the channel model in Sect. 4.2 can offer favorable propagation. This holds true for many other channel models such as Rician channels and double-scattering channels. Favorable propagation is one of the key properties of massive MIMO. This allows us to use linear processing in massive MIMO to obtain good performance since as discussed in

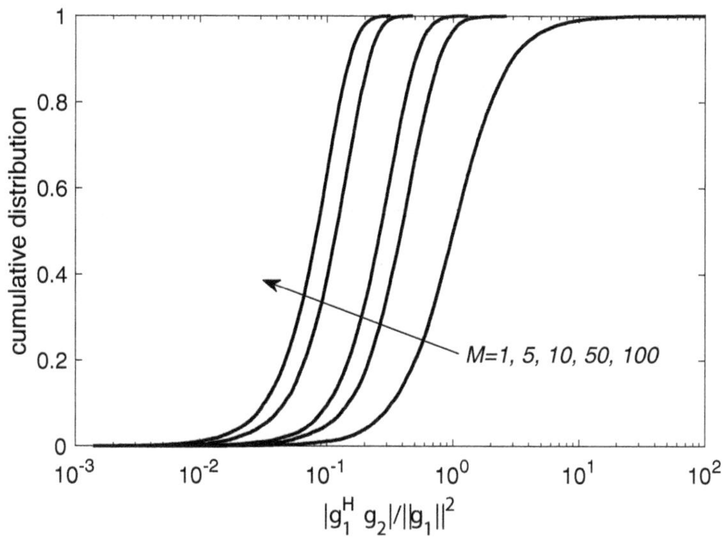

Fig. 4.3 The cumulative distribution of $\frac{|\mathbf{g}_1^H \mathbf{g}_2|}{\|\mathbf{g}_1\|^2}$ for different M. Here we choose $\beta_1 = \beta_2 = 1$

previous section, linear processing is optimal under favorable propagation. Note that in massive MIMO, since the numbers of base station antennas and users are large, we have to deal with high-dimensional matrices and vectors. Thus it is hard and complicated to use nonlinear processing techniques.

Figure 4.3 shows the cumulative distribution of $\frac{|\mathbf{g}_1^H \mathbf{g}_2|}{\|\mathbf{g}_1\|^2}$ for different M. Here we choose $\beta_1 = \beta_2 = 1$. We can see from the figure that, when M increases, $\frac{|\mathbf{g}_1^H \mathbf{g}_2|}{\|\mathbf{g}_1\|^2}$ concentrates around 0. This shows the favorable propagation property of massive MIMO.

4.3.2 *Channel Hardening*

(a) *Definition of Channel Hardening:* The channel offers hardening property if the norms of the channel vectors between the base station and the users are deterministic (do not depend on the small-scale fading coefficients), i.e.,

$$\|\mathbf{g}_k\|^2 = \text{constant (depends only on } \beta_k\text{)}, \text{ for all } k. \tag{4.36}$$

The above condition will never be exactly satisfied in practice. But as long as

$$\|\mathbf{g}_k\|^2 \approx \text{constant (depends on } \beta_k \text{ only)}, \;\; \text{for all } k, \tag{4.37}$$

we can say that we have channel hardening.

(b) *Why Channel Hardening Is Important:*

Let us consider again the uplink payload data transmission with MR processing and perfect channel estimation as in Sect. 4.3.1. We assume further that the channel offers favorable propagation. Then (4.28) becomes

$$r_{\mathrm{u},k} = \sqrt{\rho_{\mathrm{u}}}\sqrt{\eta_{\mathrm{u},k}}\|\mathbf{g}_k\|^2 s_{\mathrm{u},k} + \mathbf{g}_k^H \mathbf{w}_{\mathrm{u}}. \tag{4.38}$$

If the channel is favorable, i.e., $\|\mathbf{g}_k\|^2 = c_k$ which depends only on β_k, then we have

$$r_{\mathrm{u},k} = \sqrt{\rho_{\mathrm{u}}}\sqrt{\eta_{\mathrm{u},k}} c_k s_{\mathrm{u},k} + \mathbf{g}_k^H \mathbf{w}_{\mathrm{u}}. \tag{4.39}$$

Thus, it means that to detect $s_{\mathrm{u},k}$, the base station can use c_k which depends only on large-scale fading. This is more important for the downlink payload data transmission. More precisely, if the channel hardens, then the users do not need to estimate the instantaneous channels for signal detection. It needs to know only large-scale fading coefficients which change very slowly with time. So there will no downlink pilots for the small-scale fading channel estimation which simplifies the system operation a lot. In addition, from (4.39), we can obtain an achievable rate as

$$\begin{aligned} R_{\mathrm{u},k} &= \log_2\left(1 + \frac{\rho_{\mathrm{u}}\eta_{\mathrm{u},k}|c_k|^2}{\mathbb{E}\left\{|\mathbf{g}_k^H \mathbf{w}_{\mathrm{u}}|^2\right\}}\right) \\ &= \log_2\left(1 + \frac{\rho_{\mathrm{u}}\eta_{\mathrm{u},k}|c_k|^2}{\beta_k M}\right), \end{aligned} \tag{4.40}$$

which is very simple and depends only on the large-scale fading coefficients. As a result, the system designs such as power controls and user scheduling are simple and can be done over large-scale fading time scale.

(c) *Channel Hardening in Massive MIMO:* With the channel model provided in Sect. 4.2, from the law of large numbers, we have

$$\frac{\|\mathbf{g}_k\|^2}{\mathbb{E}\left\{\|\mathbf{g}_k\|^2\right\}} = \frac{\|\mathbf{h}_k\|^2}{M} \overset{a.s.}{\to} 1, \text{ as } M \to \infty, \tag{4.41}$$

where again $\overset{a.s.}{\to}$ denotes almost sure convergence. This result implies that when M is large, with high probability, $\|\mathbf{g}_k\|^2$ is very close to its mean value $\mathbb{E}\left\{\|\mathbf{g}_k\|^2\right\} = M\beta_k$. Thus, under our considered channel models, we have channel hardening in massive MIMO. This is the reason why in massive MIMO literature, the mean value of the effective channel gain (after linear processing is

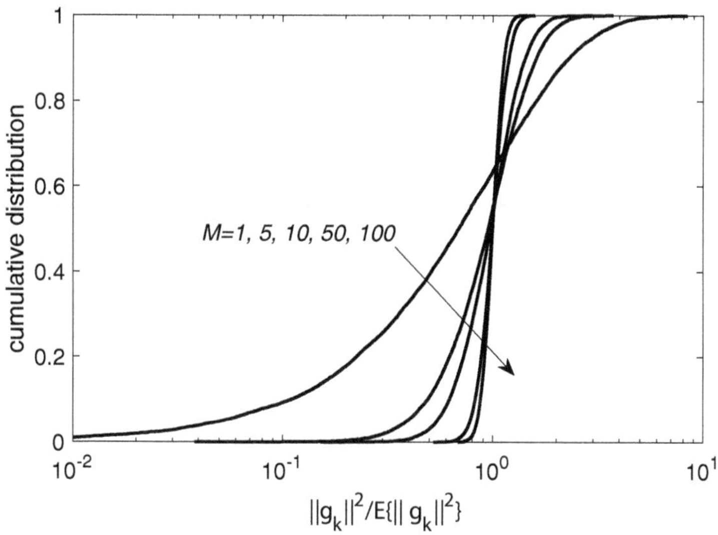

Fig. 4.4 The cumulative distribution of $\frac{\|\mathbf{g}_k\|^2}{\mathbb{E}\{\|\mathbf{g}_k\|^2\}}$ for different M

applied) is used as the true one for signal detection, and the corresponding use-and-then-forget capacity bounding technique is widely used (see Sect. 4.3.3).

Figure 4.4 shows the cumulative distribution of $\frac{\|\mathbf{g}_k\|^2}{\mathbb{E}\{\|\mathbf{g}_k\|^2\}}$ for different number of base station antennas M. We can see that when M is large, $\frac{\|\mathbf{g}_k\|^2}{\mathbb{E}\{\|\mathbf{g}_k\|^2\}}$ is more concentrated around 1. This means when M is large, with high probability, $\|\mathbf{g}_k\|^2$ is very close to $\mathbb{E}\left\{\|\mathbf{g}_k\|^2\right\}$, and hence, we have channel hardening property. Note that, we have a good level of channel hardening even when $M = 50$.

4.3.3 Use-and-Then-Forget Capacity Bounding Technique

In massive MIMO, the use-and-then-forget capacity bounding technique is widely used. There are two main reasons for this: (i) the use-and-then-forget capacity bounding technique yields a very simple analytical rate expression which enables us to get important insights as well as design the systems efficiently and (ii) it performs very well due to the channel hardening property of massive MIMO. In this section, we provide details of the use-and-then-forget bounding technique for both uplink and downlink payload data transmission.

(a) *Uplink Payload Data Transmission:* The processed signal at the base station used to detect signal sent by user k is given by (4.18) which can be rewritten as

$$r_{\mathrm{u},k} = \sqrt{\rho_{\mathrm{u}}\eta_{\mathrm{u},k}}\,\mathbb{E}\left\{\mathbf{a}_k^H\mathbf{g}_k\right\}s_{\mathrm{u},k} + \sqrt{\rho_{\mathrm{u}}\eta_{\mathrm{u},k}}\left(\mathbf{a}_k^H\mathbf{g}_k - \mathbb{E}\left\{\mathbf{a}_k^H\mathbf{g}_k\right\}\right)s_{\mathrm{u},k}$$
$$+ \sqrt{\rho_{\mathrm{u}}}\sum_{k'\neq k}^{K}\mathbf{a}_k^H\mathbf{g}_{k'}\sqrt{\eta_{\mathrm{u},k'}}s_{\mathrm{u},k'} + \mathbf{a}_k^H\mathbf{w}_{\mathrm{u}}. \tag{4.42}$$

The first term of (4.18) is considered as the desired signal part, while the summation of last three terms is treated as the effective noise. Let

$$\tilde{w}_k = \sqrt{\rho_{\mathrm{u}}\eta_{\mathrm{u},k}}\left(\mathbf{a}_k^H\mathbf{g}_k - \mathbb{E}\left\{\mathbf{a}_k^H\mathbf{g}_k\right\}\right)s_{\mathrm{u},k} + \sqrt{\rho_{\mathrm{u}}}\sum_{k'\neq k}^{K}\mathbf{a}_k^H\mathbf{g}_{k'}\sqrt{\eta_{\mathrm{u},k'}}s_{\mathrm{u},k'} + \mathbf{a}_k^H\mathbf{w}_{\mathrm{u}}$$

be the effective noise; then (4.42) becomes

$$r_{\mathrm{u},k} = \sqrt{\rho_{\mathrm{u}}\eta_{\mathrm{u},k}}\,\mathbb{E}\left\{\mathbf{a}_k^H\mathbf{g}_k\right\}s_{\mathrm{u},k} + \tilde{w}_k. \tag{4.43}$$

We can see that the effective noise $\tilde{w}_k$ is uncorrelated with the desired signal $\sqrt{\rho_{\mathrm{u}}\eta_{\mathrm{u},k}}\,\mathbb{E}\left\{\mathbf{a}_k^H\mathbf{g}_k\right\}s_{\mathrm{u},k}$. Thus, (4.43) corresponds to a deterministic single-input single-output channel with uncorrelated noise. For this channel, with Gaussian signalling, Gaussian noise is the worst case. Therefore, we can obtain the following achievable rate (capacity lower bound):

$$R_{\mathrm{u},k} = \log_2\left(1+\frac{\rho_{\mathrm{u}}\eta_{\mathrm{u},k}\left|\mathbb{E}\left\{\mathbf{a}_k^H\mathbf{g}_k\right\}\right|^2}{\mathbb{V}\mathrm{ar}\left\{\tilde{w}_k\right\}}\right)$$
$$= \log_2\left(1+\frac{\rho_{\mathrm{u}}\eta_{\mathrm{u},k}\left|\mathbb{E}\left\{\mathbf{a}_k^H\mathbf{g}_k\right\}\right|^2}{\rho_{\mathrm{u}}\eta_{\mathrm{u},k}\,\mathbb{V}\mathrm{ar}\left\{\mathbf{a}_k^H\mathbf{g}_k\right\} + \rho_{\mathrm{u}}\sum_{k'\neq k}^{K}\eta_{\mathrm{u},k'}\,\mathbb{E}\left\{|\mathbf{a}_k^H\mathbf{g}_{k'}|^2\right\} + \mathbb{E}\left\{\|\mathbf{a}_k\|^2\right\}}\right). \tag{4.44}$$

If the channel estimation overhead is taken into account, we have the following spectral efficiency

$$\mathsf{S}_{\mathrm{u},k} = \frac{\tau_{\mathrm{u}}}{\tau_{\mathrm{c}}}R_{\mathrm{u},k} \tag{4.45}$$

Some remarks regarding (4.44):

- The operational meaning of the use-and-then-forget bounding rate (4.44) is that the base station first uses the channel estimates to combine its received signals (via (4.15)). Then it forgets its channel estimates and just uses the statistical property of the channels, i.e., it treats $\mathbb{E}\left\{\mathbf{a}_k^H\mathbf{g}_k\right\}$ as the true effective channel gain $\mathbf{a}_k^H\mathbf{g}_k$ to detect the desired signal. This is the reason why we call it the "use-and-then-forget" bounding technique.

- By using random matrix theory, we can derive a closed-form expression of (4.44) which is a simple function of large-scale fading coefficients. We skip the derivations here. We can use this closed-form expression to design the systems such as finding the optimal power control coefficients $\{\eta_{\mathrm{u},k}\}$ to maximize the minimum rate of all users. Details can be found at [10].
- We now consider another capacity bound (achievable rate) where we assume that the base station knows perfectly $\mathbf{a}_k^H \mathbf{g}_{k'}$ and $\mathbf{a}_k$. Then from (4.18), we can obtain the following achievable spectral efficiency for user k:

$$S_{\mathrm{u},k}^{\mathrm{perfect}} = \frac{\tau_{\mathrm{u}}}{\tau_{\mathrm{c}}} \mathbb{E}\left\{\log_2\left(1 + \frac{\rho_{\mathrm{u}}\eta_{\mathrm{u},k}\left|\mathbf{a}_k^H \mathbf{g}_k\right|^2}{\rho_{\mathrm{u}}\sum_{k'\neq k}^{K}\eta_{\mathrm{u},k'}|\mathbf{a}_k^H \mathbf{g}_{k'}|^2 + \|\mathbf{a}_k\|^2}\right)\right\}. \tag{4.46}$$

Figures 4.5 and 4.6 compare the spectral efficiency using the use-and-then-forget bounding technique (i.e., expression (4.44)) and the one obtained under the assumption that we have a genie-aided base station (i.e., expression (4.46)) with $\rho_{\mathrm{u}} = \rho_{\mathrm{p}} = 10\,\mathrm{dB}$ and $\rho_{\mathrm{u}} = \rho_{\mathrm{p}} = -10\,\mathrm{dB}$, respectively. Here we choose $\tau_{\mathrm{c}} = 200$, $\tau_{\mathrm{p}} = K = 20$, $\tau_{\mathrm{u}} = \tau_{\mathrm{d}} = (\tau_{\mathrm{c}} - \tau_{\mathrm{p}})/2$, and $\eta_{\mathrm{u},k} = 1$ for all k. Furthermore, we assume that $\beta_k = 1$ for all k, and all pilot sequences of all users are pairwisely orthogonal. We can see from both figures that the use-and-then-forget spectral efficiency is very close to the genie-aided one, even when M is not very large. This means the use-and-then-forget bounding technique works very well and is a good technique to use in massive MIMO.

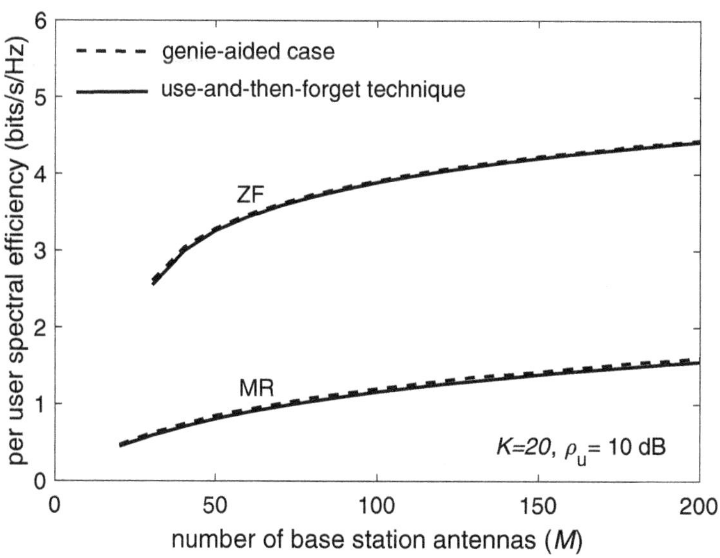

Fig. 4.5 Uplink per user spectral efficiency. Here we choose $\rho_{\mathrm{u}} = \rho_{\mathrm{p}} = 10\,\mathrm{dB}$

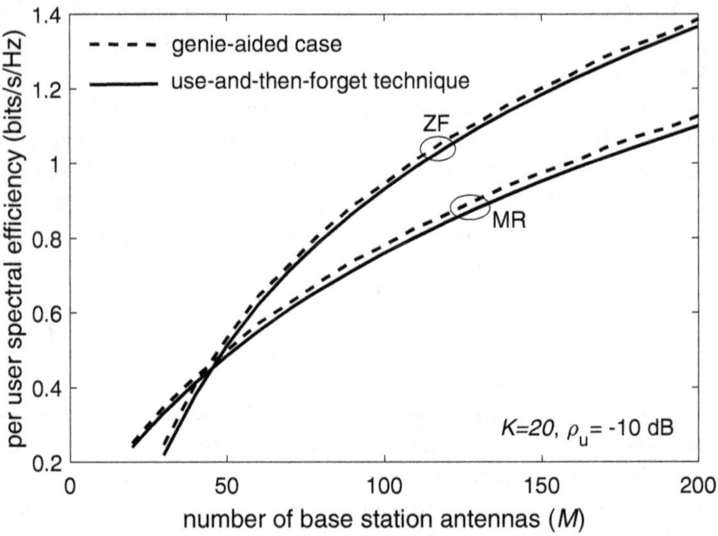

Fig. 4.6 Uplink per user spectral efficiency. Here we choose $\rho_{\rm u} = \rho_{\rm p} = -10\,{\rm dB}$

In addition, ZF outperforms MR in most of the cases, while MR is better than ZF only when the normalized transmit power is low and M is small.

(b) *Downlink Payload Data Transmission:* The signal received at the k-th user is given by (4.25) which can be rewritten as

$$y_{{\rm d},k} = \sqrt{\rho_{\rm d}}\sqrt{\eta_{{\rm d},k}}\,\mathbb{E}\left\{\mathbf{g}_k^T\mathbf{b}_k\right\}s_{{\rm d},k} + \sqrt{\rho_{\rm d}}\sqrt{\eta_{{\rm d},k}}\left(\mathbf{g}_k^T\mathbf{b}_k - \mathbb{E}\left\{\mathbf{g}_k^T\mathbf{b}_k\right\}\right)s_{{\rm d},k}$$
$$+\sqrt{\rho_{\rm d}}\sum_{k'\neq k}^{K}\sqrt{\eta_{{\rm d},k}}\mathbf{g}_k^T\mathbf{b}_{k'}s_{{\rm d},k'} + w_{{\rm d},k}. \tag{4.47}$$

Similar to the approach used in the uplink, we first treat the first term of (4.47) as the desired signal part and the summation of the last as the effective noise. Then by using the worst-case Gaussian noise, we can obtain the following achievable rate

$$R_{{\rm d},k} = \log_2\left(1+\frac{\rho_{\rm d}\eta_{{\rm d},k}\left|\mathbb{E}\left\{\mathbf{g}_k^T\mathbf{b}_k\right\}\right|^2}{\rho_{\rm d}\eta_{{\rm d},k}\,\mathbb{V}{\rm ar}\left\{\mathbf{g}_k^T\mathbf{b}_k\right\}+\rho_{\rm d}\sum_{k'\neq k}^{K}\eta_{{\rm d},k'}\,\mathbb{E}\left\{|\mathbf{g}_k^T\mathbf{b}_{k'}|^2\right\}+1}\right), \tag{4.48}$$

and the corresponding spectral efficiency is

$$\mathsf{S}_{\mathrm{d},k} = \frac{\tau_{\mathrm{d}}}{\tau_{\mathrm{c}}} R_{\mathrm{d},k} \tag{4.49}$$

Remark 3 The operational meaning of the rate (4.48) is that the k-th user treats the $\mathbb{E}\left\{\mathbf{g}_k^T \mathbf{b}_k\right\}$ as the true effective channel gain $\mathbf{g}_k^T \mathbf{b}_k$ for signal detection. The approach works well in massive MIMO since massive MIMO offers channel hardening property, i.e., $\mathbb{E}\left\{\mathbf{g}_k^T \mathbf{b}_k\right\} \approx \mathbf{g}_k^T \mathbf{b}_k$ with high probability.

4.3.4 Pilot Contamination

From (4.6), if all pilot sequences from all K users are pairly orthogonal, i.e., $\boldsymbol{\varphi}_{k'}^H \boldsymbol{\varphi}_k = 0$ for $k' \neq k$, then the channel estimate of $\mathbf{g}_k$ is

$$\hat{\mathbf{g}}_k = \frac{\tau_{\mathrm{p}}\rho_{\mathrm{p}}\beta_k}{\tau_{\mathrm{p}}\rho_{\mathrm{p}}\beta_k + 1}\mathbf{g}_k + \frac{\sqrt{\tau_{\mathrm{p}}\rho_{\mathrm{p}}}\beta_k}{\tau_{\mathrm{p}}\rho_{\mathrm{p}}\beta_k + 1}\mathbf{W}_{\mathrm{p}}\boldsymbol{\varphi}_k, \tag{4.50}$$

which includes the true channel plus noise. By increasing the pilot power and/or the pilot length, we can obtain a channel estimation with very high accuracy. However, to guarantee the mutual orthogonality among K pilot sequences, it requires that $\tau_{\mathrm{p}} \geq K$. This condition cannot be satisfied in many practical scenarios where the coherence interval is not very large and/or the number of users is large. As a consequence, the non-orthogonal pilots have to be used. With non-orthogonal pilots, channel estimate of $\mathbf{g}_k$ includes the non-zero term:

$$\sqrt{\tau_{\mathrm{p}}\rho_{\mathrm{p}}} \sum_{k' \neq k}^{K} \mathbf{g}_{k'} \boldsymbol{\varphi}_{k'}^H \boldsymbol{\varphi}_k$$

which is the interference from other users during the training phase. This yields the concept of *pilot contamination*. In general, pilot contamination comes from the fact that the channel estimated associated with a given user is contaminated by the pilots sent by other users. This pilot contamination effect is more concerning in massive MIMO because it persists even when the number of base station antennas goes to infinity and degrades the system performance significantly. To see this in more details, let us consider the uplink payload data transmission with MR processing. From (4.16) and (4.18), the received signal after deploying MR processing used for detecting $s_{\mathrm{u},k}$ is

$$r_{\mathrm{u},k} = \sqrt{\rho_{\mathrm{u}}}\hat{\mathbf{g}}_k^H \mathbf{g}_k \sqrt{\eta_{\mathrm{u},k}} s_{\mathrm{u},k} + \sqrt{\rho_{\mathrm{u}}} \sum_{k' \neq k}^{K} \hat{\mathbf{g}}_k^H \mathbf{g}_{k'} \sqrt{\eta_{\mathrm{u},k'}} s_{\mathrm{u},k'} + \hat{\mathbf{g}}_k^H \mathbf{w}_{\mathrm{u}}. \tag{4.51}$$

Using (4.6), we have

$$\frac{\hat{\mathbf{g}}_k^H \mathbf{g}_k}{M} = \frac{c_k}{M}\left(\sqrt{\tau_{\rm p}\rho_{\rm p}}\mathbf{g}_k + \sqrt{\tau_{\rm p}\rho_{\rm p}}\sum_{k'\neq k}^{K}\mathbf{g}_{k'}\boldsymbol{\varphi}_{k'}^H\boldsymbol{\varphi}_k + \mathbf{W}_{\rm p}\boldsymbol{\varphi}_k\right)^H \mathbf{g}_k$$
$$= c_k\sqrt{\tau_{\rm p}\rho_{\rm p}}\frac{\|\mathbf{g}_k\|^2}{M} + c_k\sqrt{\tau_{\rm p}\rho_{\rm p}}\sum_{k'\neq k}^{K}\frac{\mathbf{g}_{k'}^H\mathbf{g}_k}{M}\boldsymbol{\varphi}_{k'}^H\boldsymbol{\varphi}_k + c_k\frac{(\mathbf{W}_{\rm p}\boldsymbol{\varphi}_k)^H\mathbf{g}_k}{M}, \tag{4.52}$$

where $c_k = \frac{\sqrt{\tau_{\rm p}\rho_{\rm p}}\beta_k}{\tau_{\rm p}\rho_{\rm p}\sum_{k'=1}^{K}\beta_{k'}\left|\boldsymbol{\varphi}_{k'}^H\boldsymbol{\varphi}_k\right|^2+1}$. Then, by the law of large numbers (4.34) and (4.35), we obtain

$$\frac{\hat{\mathbf{g}}_k^H \mathbf{g}_k}{M} \xrightarrow{a.s.} c_k\sqrt{\tau_{\rm p}\rho_{\rm p}}\beta_k, \text{ as } M \to \infty. \tag{4.53}$$

Similarly, for $k' \neq k$, we have

$$\frac{\hat{\mathbf{g}}_k^H \mathbf{g}_{k'}}{M} = \frac{c_k}{M}\left(\sqrt{\tau_{\rm p}\rho_{\rm p}}\mathbf{g}_{k'}\boldsymbol{\varphi}_{k'}^H\boldsymbol{\varphi}_k + \sqrt{\tau_{\rm p}\rho_{\rm p}}\sum_{k''\neq k'}^{K}\mathbf{g}_{k''}\boldsymbol{\varphi}_{k''}^H\boldsymbol{\varphi}_k + \mathbf{W}_{\rm p}\boldsymbol{\varphi}_k\right)^H \mathbf{g}_{k'}$$
$$= c_k\sqrt{\tau_{\rm p}\rho_{\rm p}}\frac{\|\mathbf{g}_{k'}\|^2}{M}\boldsymbol{\varphi}_{k'}^H\boldsymbol{\varphi}_k + c_k\sqrt{\tau_{\rm p}\rho_{\rm p}}\sum_{k''\neq k'}^{K}\frac{\mathbf{g}_{k''}^H\mathbf{g}_{k'}}{M}\boldsymbol{\varphi}_{k''}^H\boldsymbol{\varphi}_k + c_k\frac{(\mathbf{W}_{\rm p}\boldsymbol{\varphi}_k)^H\mathbf{g}_{k'}}{M}$$
$$\xrightarrow{a.s.} c_k\sqrt{\tau_{\rm p}\rho_{\rm p}}\beta_{k'}\boldsymbol{\varphi}_{k'}^H\boldsymbol{\varphi}_k, \text{ as } M \to \infty, \tag{4.54}$$

and

$$\frac{(\mathbf{W}_{\rm p}\boldsymbol{\varphi}_k)^H\mathbf{g}_{k'}}{M} \xrightarrow{a.s.} 0, \text{ as } M \to \infty. \tag{4.55}$$

Substituting (4.53), (4.54), and (4.55) into (4.51), we obtain

$$\frac{r_{{\rm u},k}}{M} \xrightarrow{a.s.} \sqrt{\tau_{\rm p}\rho_{\rm p}\rho_{\rm u}\eta_{{\rm u},k}}c_k\beta_k s_{{\rm u},k} + \sum_{k'\neq k}^{K}\sqrt{\tau_{\rm p}\rho_{\rm p}\rho_{\rm u}\eta_{{\rm u},k'}}c_k\beta_{k'}\boldsymbol{\varphi}_{k'}^H\boldsymbol{\varphi}_k s_{{\rm u},k'}, \text{ as } M \to \infty. \tag{4.56}$$

From (4.56), we can see that if orthogonal pilots among all K users are used, then when the number of base station antennas goes to infinity,

$$\frac{r_{{\rm u},k}}{M} \xrightarrow{a.s.} \sqrt{\tau_{\rm p}\rho_{\rm p}\rho_{\rm u}\eta_{{\rm u},k}}c_k\beta_k s_{{\rm u},k},$$

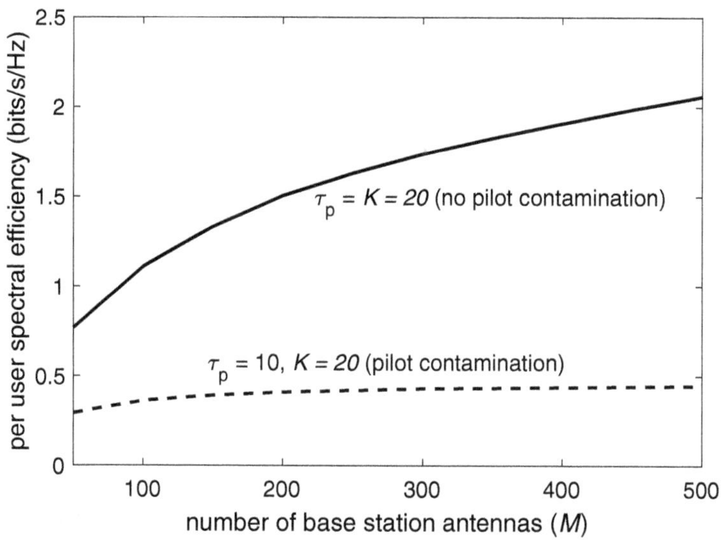

Fig. 4.7 Uplink per user spectral efficiency versus M. Here we choose $\rho_u = \rho_p = 10\,\text{dB}$, $\tau_c = 200$, $K = 20$, and $\eta_{u,k} = 1$

which includes only the desired signal (without interference and noise). Thus the rate increases without bound. However if non-orthogonal pilots are used, the inter-user interference persists. Even when $M \to \infty$, the rate is bounded as

$$R_{u,k} \overset{a.s.}{\to} \log_2 \left(1 + \frac{\eta_{u,k}\beta_k^2}{\sum_{k' \neq k}^{K} \eta_{u,k'}\beta_{k'}^2 |\boldsymbol{\varphi}_{k'}^H \boldsymbol{\varphi}_k|^2} \right).$$

Therefore, the pilot contamination has strong effect on the performance of massive MIMO systems. Figure 4.7 shows the uplink per user spectral efficiency versus the number of base station antennas M with MR processing for different τ_p. Here we use (4.45) to generate the spectral efficiency. Furthermore, we choose $\rho_u = \rho_p = 10\,\text{dB}$, $\tau_c = 200$, $K = 20$, $\tau_u = \tau_d = (\tau_c - \tau_p)/2$, $\beta_k = 1$ and $\eta_{u,k} = 1$ for all k. When $\tau_p = K = 20$, we can assign mutually orthogonal pilot sequences to all K users, and hence, we do not have pilot contamination in this case. But when $\tau_p = 10 < K$, non-orthogonal pilots have to be used. In this case, we assume that K users choose randomly pilots from a predefined set of τ_p orthogonal pilot sequences, and hence, pilot contamination appears. We can see from the figure that, under pilot contamination case, the spectral efficiency is low and is bounded even when M is very large.

4.4 Important Topics for Future Research

During the last decade, after numerous researches ranging from information theory, communication theory, and signal processing to practical experiments, massive MIMO has become mature and has been incorporated in 5G New Radio standards. However, there are still many technical challenges in relation to this technology that are not satisfactorily addressed in previous studies. In this section, we briefly discuss some important topics for future research of massive MIMO.

4.4.1 Massive MIMO with Multiple-Antenna Users

Current massive MIMO research mainly focuses on system with single-antenna users. There are three main reasons for this: (i) single-antenna setup is simple for the analysis; (ii) under the assumption on favorable propagation, the performance of system having one N-antenna user is similar with the one having N autonomous single-antenna users [14]; and (iii) multiple-antenna users are size and cost limited. However, in current wireless systems, users can be equipped with several antennas since they offer additional degrees of freedom. This can improve the spectral efficiency and communication reliability for each communication link. Furthermore, the channel is not very favorable in many scenarios such as the systems with not very large number of base station antennas, or the channels are double-scattering channels with limited number of scatterers. Therefore, research on massive MIMO with multiple-antenna users is of practical importance. There are several works on this topic [31–34]. In [31], the authors showed that adding some antennas to the users can be beneficial in terms of the spectral efficiency when there are few active users in the systems. The authors in [31] further showed that at the users, the performance of the MMSE successive interference cancellation (MMSE-SIC) decoder is the same as the one of the MMSE decoder. This conclusion is drawn from the assumption that no instantaneous channel state information (CSI) is available at the users. However, if some instantaneous CSI is available at the users, and the channel is not very favorable as well as the level of channel hardening is low, then MMSE-SIC is better than MMSE. This was shown in [32]. In [33], the favorable propagation property of massive MIMO with multiple-antenna users was investigated under correlated Rayleigh fading and geometry-based channels. In [34], a novel MR precoder for massive MIMO with multiple-antenna users was proposed. This precoder aimed at increasing the channel hardening level of the systems, and hence improved the spectral efficiency noticeably. There are still many challenges in multiple-antenna user systems which need to be tackled such as how the processing can be done at the user sides or how additional channel information can be beneficial. Therefore, research on this topic is very timely.

4.4.2 Cell-Free Massive MIMO

Cell-free massive MIMO has been proposed in [35] as a practical and scalable version of network MIMO. In cell-free massive MIMO, many access points (APs) distributed in a large area coherently serve all users in the network. There are no cells and cell boundaries in cell-free massive MIMO, and hence, cell-free massive MIMO can overcome the inherent limitation of cellular networks, that is, boundary effect. As a result, cell-free massive MIMO can increase the network connectivity (i.e., it can provide uniformly good services for all users).

Since high network connectivity is one of the main targets of future wireless network, cell-free massive MIMO has attracted a lot of research attention recently. In [36] and [37], the authors investigated the performance of cell-free massive MIMO using zero-forcing processing with max-min rate and total energy efficiency power controls, respectively. The authors in [38] studied the total energy efficiency of cell-free massive MIMO with simple conjugate beamforming schemes. A practical power consumption model including the circuit power consumption, backhaul power consumption, and power amplifier efficiency was taken into account. Paper [38] further proposed some AP selection schemes in which each user is actually served by a small number of APs. This AP selection is similar to the user-centric approaches discussed in [39, 40]. It was shown that user-centric approach is a possible way to implement cell-free massive MIMO in practice since it reduces the backhaul connections among the APs as well between the APs and the central processing units significantly while maintaining a very good performance. Cell-free massive MIMO with local partial zero-forcing scheme was proposed in [41] to improve the spectral efficiency while maintaining the minimum requirement for channel sharing among the APs. The channel acquisition aspects were studied in [42, 43]. In [42], pilot power control was proposed to reduce the pilot contamination effect during the uplink training phase. In [43], the authors proposed to use the downlink pilots so that each user can estimate its effective channel gain. Since under the Rayleigh fading propagation, the level of channel hardening in cell-free massive MIMO is less than that in colocated massive MIMO [44], the proposed downlink pilots in [43] can improve the spectral efficiency significantly. The effects of hardware impairment and limited backhaul constraints were studied in [45–47]. An important conclusion of these works is that cell-free massive MIMO still works well under the limited backhaul and hardware constraints. Recently, papers [48, 49] have addressed an important scalable aspect in cell-free massive MIMO. Both papers prosed some heuristic power controls which are scalable in the sense that they can be implemented when the network size (i.e., the number of APs/users and the coverage area) increases. Since cell-free massive MIMO is a new topic, there are still many research challenges which need to be tackled. Some important research directions of cell-free massive MIMO are (i) scalable signal processing and power controls; (ii) hardware constraints including synchronization, channel reciprocity, signalling, coherent processing, etc.; (iii) channel acquisition aspects;

(iv) practical implementations; and (v) the applications of cell-free massive MIMO to the existing technologies.

4.4.3 Massive MIMO for Massive Access

Future wireless networks have to manage at the time billions of devices with many applications such as the Internet of Things, Internet of Everything, and Smart X. It is very challenging for the current multiple-access technologies to support this massive access. Massive MIMO offers high array gain and multiplexing gain and, hence, is a suitable solution for massive access in the future. More importantly, in massive MIMO, when the number of base station antennas is large, under many practical propagation environments, the channel vectors from the base station to different users are (nearly) pairwisely orthogonal. Thus, if the base station has perfect channel knowledge, it can cancel the inter-user interference with a simple linear processing. However, in practice, the base station needs to estimate the channels from the uplink training. It is very challenging for the base station to obtain high accurate channel estimates in massive access systems. This is because, in massive access systems, the number of users is very large, and hence, non-orthogonal pilot sequences have to be used among the users. With non-orthogonal pilot assignments, channel estimate for a given user is strongly contaminated by the interference from other users, and hence, it is not very accurate. Therefore, channel acquisition is very important in massive MIMO for massive access systems. A joint pilot assignment and data transmission scheme in massive MIMO was proposed in [50] to improve the channel estimation quality. This scheme is based on random access and can serve dense crowds of wireless devices. The channel estimation combined with compressed sensing-based algorithms was investigated in [51, 52]. All above works assumed perfect synchronization which is very challenging in massive access systems. There are also many other problems which are not fully addressed in massive MIMO for massive access systems such as ultra-reliable communications, scalable power controls, high correlated channels, etc. So research on this direction is very important.

4.5 Conclusion

Massive MIMO has become mature and been incorporated into the first version of 5G. It was expected to be the core technology for next versions of 5G as well as 6G. Therefore, it is a good time to review the fundamentals of massive MIMO. In this chapter, we presented basics of massive MIMO. First, we provided a completed TDD system model with simple linear processing (e.g., MR and ZF processing). Then, we summarized the favorable propagation, channel hardening, pilot contamination, and use-and-then-forget capacity bounding techniques. Finally, we suggested some future research directions. These provide a comprehensive theory of massive MIMO for the researchers from both academia and industry to work on this topic.

Acknowledgments This work was supported by the UK Research and Innovation Future Leaders Fellowships under Grant MR/S017666/1.

References

1. A.J. Paulraj, T. Kailath, Increasing capacity in wireless broadcast systems using distributed transmission/directional reception (DTDR), 6 Sep 1994, US Patent 5,345,599
2. G.J. Foschini, M.J. Gans, On limits of wireless communications in a fading environment when using multiple antennas. Wirel. Pers. Commun. **6**(3), 311–335 (1998)
3. İ.E. Telatar, Capacity of multi-antenna Gaussian channels. Eur. Trans. Telecommun. **10**(6), 585–595 (1999)
4. J.H. Winters, Optimum combining in digital mobile radio with cochannel interference. IEEE J. Sel. Areas Commun. **2**(4), 528–539 (1984)
5. J.H. Winters, On the capacity of radio communication systems with diversity in Rayleigh fading environment. IEEE J. Sel. Areas Commun. **5**(5), 871–878 (1987)
6. Q.H. Spencer, C.B. Peel, A.L. Swindlehurst, M. Haardt, An introduction to the multi-user MIMO downlink. IEEE Commun. Mag. **42**(10), 60–67 (2004)
7. L. Liu, R. Chen, S. Geirhofer, K. Sayana, Z. Shi, Y. Zhou, Downlink MIMO in LTE-advanced: SU-MIMO vs. MU-MIMO. IEEE Commun. Mag. **50**(2), 140–147 (2012)
8. M. Vaezi, Z. Ding, H.V. Poor, *Multiple Access Techniques for 5G Wireless Networks and Beyond.* Berlin, Germany: Springer, 2018
9. R. Mochaourab, E.A. Jorswieck, Optimal beamforming in interference networks with perfect local channel information. IEEE Trans. Signal Process. **59**(3), 1128–1141 (2010)
10. T.L. Marzetta, E.G. Larsson, H. Yang, H.Q. Ngo, *Fundamentals of Massive MIMO* (Cambridge University Press, Cambridge, UK, 2016)
11. T.L. Marzetta, How much training is required for multiuser MIMO, in *Fortieth Asilomar Conference on Signals, Systems and Computers (ACSSC '06)*, Pacific Grove, Oct 2006, pp. 359–363
12. T.L. Marzetta, Noncooperative cellular wireless with unlimited numbers of base station antennas. IEEE Trans. Wireless Commun. **9**(11), 3590–3600 (2010)
13. J. Jose, A. Ashikhmin, T.L. Marzetta, S. Vishwanath, Pilot contamination and precoding in multi-cell TDD systems. IEEE Trans. Wireless Commun. **10**(8), 2640–2651 (2011)
14. H.Q. Ngo, E.G. Larsson, T.L. Marzetta, Energy and spectral efficiency of very large multiuser MIMO systems. IEEE Trans. Commun. **61**(4), 1436–1449 (2013)
15. J. Hoydis, S. ten Brink, M. Debbah, Massive MIMO in the UL/DL of cellular networks: how many antennas do we need? IEEE J. Sel. Areas Commun. **31**(2), 160–171 (2013)
16. L. Sanguinetti, E. Björnson, J. Hoydis, Towards massive mimo 2.0: understanding spatial correlation, interference suppression, and pilot contamination. IEEE Trans. Commun. (2019)
17. H. Yin, D. Gesbert, M. Filippou, Y. Liu, A coordinated approach to channel estimation in large-scale multiple-antenna systems. IEEE J. Sel. Areas Commun. **31**(2), 264–273 (2013)
18. Z. Xiang, M. Tao, X. Wang, Massive MIMO multicasting in noncooperative cellular networks. IEEE J. Sel. Areas Commun. **32**(6), 1180–1193 (2014)
19. J. Zhu, R. Schober, V. Bhargava, Secure transmission in multicell massive MIMO systems. IEEE Trans. Wireless Commun. **13**(9), 4766–4781 (2014)
20. R.R. Müller, L. Cottatellucci, M. Vehkaperä, Blind pilot decontamination. IEEE J. Sel. Topics Signal Process. **8**(5), 773–786 (2014)

21. D.W.K. Ng, E.S. Lo, R. Schober, Energy-efficient resource allocation in ofdma systems with large numbers of base station antennas. IEEE Trans. Wireless Commun. **11**(9), 3292–3304 (2012)
22. E. Björnson, L. Sanguinetti, J. Hoydis, M. Debbah, Optimal design of energy-efficient multi-user MIMO systems: is massive MIMO the answer? IEEE Trans. Wireless Commun. **14**(6), 3059–3075 (2015)
23. T. Van Chien, E. Björnson, E.G. Larsson, Joint pilot design and uplink power allocation in multi-cell massive mimo systems. IEEE Trans. Wireless Commun. **17**(3), 2000–2015 (2018)
24. C. Shepard, H. Yu, N. Anand, L.E. Li, T.L. Marzetta, R. Yang, L. Zhong, Argos: practical many-antenna base stations, in *Proceedings of the ACM International Conference on Mobile Computing and Networking (MobiCom)*, Istanbul, Turkey, Aug 2012
25. X. Gao, O. Edfors, F. Rusek, F. Tufvesson, Massive mimo performance evaluation based on measured propagation data. IEEE Trans. Wireless Commun. **14**(7), 3899–3911 (2015)
26. S. Malkowsky, J. Vieira, L. Liu, P. Harris, K. Nieman, N. Kundargi, I.C. Wong, F. Tufvesson, V. Öwall, O. Edfors, The world's first real-time testbed for massive mimo: design, implementation, and validation. IEEE Access **5**, 9073–9088 (2017)
27. A. Ashikhmin, L. Li, T.L. Marzetta, Interference reduction in multi-cell massive mimo systems with large-scale fading precoding. IEEE Trans. Inf. Theory **64**(9), 6340–6361 (2018)
28. F. Kaltenberger, J. Haiyong, M. Guillaud, R. Knopp, Relative channel reciprocity calibration in MIMO/TDD systems, in *Proceedings of the Future Network and Mobile Summit*, Florence, Jun 2010
29. H.Q. Ngo, E.G. Larsson, T.L. Marzetta, Massive MU-MIMO downlink TDD systems with linear precoding and downlink pilots, in *Proceedings of the Allerton*, Urbana-Champaign, Oct 2013
30. P. Viswanath, D.N.C. Tse, Sum capacity of the vector Gaussian broadcast channel and uplink-downlink duality? IEEE Trans. Inf. Theory **49**(8), 1912–1921 (2003)
31. X. Li, E. Björnson, S. Zhou, J. Wang, Massive MIMO with multi-antenna users: when are additional user antennas beneficial? in *Proceedings of the IEEE International Conference on Telecommunications (ICT)*, May 2016, pp. 1–6
32. T.C. Mai, H.Q. Ngo, T.Q. Duong, Cell-free massive mimo systems with multi-antenna users, in *2018 IEEE Global Conference on Signal and Information Processing (GlobalSIP)* (IEEE, 2018), pp. 828–832
33. X. Wu, D. Liu, Novel insight into multi-user channels with multi-antenna users. IEEE Commun. Lett. **21**(9), 1961–1964 (2017)
34. J.A. Sutton, H.Q. Ngo, M. Matthaiou, Performance of a novel maximum-ratio precoder in massive mimo with multiple-antenna users, in *2019 IEEE 30th Annual International Symposium on Personal, Indoor and Mobile Radio Communications (PIMRC)* (IEEE, 2019), pp 1–6
35. H.Q. Ngo, A. Ashikhmin, H. Yang, E.G. Larsson, T.L. Marzetta, Cell-free massive MIMO versus small cells. IEEE Trans. Wireless Commun. **16**(3), 1834–1850 (2017)
36. E. Nayebi, A. Ashikhmin, T.L. Marzetta, H. Yang, B.D. Rao, Precoding and power optimization in cell-free massive MIMO systems. IEEE Trans. Wireless Commun. **16**(7), 4445–4459 (2017)
37. L.D. Nguyen, T.Q. Duong, H.Q. Ngo, K. Tourki, Energy efficiency in cell-free massive mimo with zero-forcing precoding design. IEEE Commun. Lett. **21**(8), 1871–1874 (2017)
38. H.Q. Ngo, L.-N. Tran, T.Q. Duong, M. Matthaiou, E.G. Larsson, On the total energy efficiency of cell-free massive mimo. IEEE Trans. Green Commun. Netw. **2**(1), 25–39 (2017)
39. S. Buzzi, C. D'Andrea, Cell-free massive MIMO: user-centric approach. IEEE Wireless Commun. Lett., **6**(6), 706–709 (2017)
40. F. Riera-Palou, G. Femenias, A.G. Armada, A. Pérez-Neira, Clustered cell-free massive mimo, in *2018 IEEE Globecom Workshops (GC Wkshps)* (IEEE, 2018), pp. 1–6
41. G. Interdonato, M. Karlsson, E. Björnson, E.G. Larsson, Local partial zero-forcing precoding for cell-free massive mimo, arXiv preprint arXiv:1909.01034, 2019

42. T.C. Mai, H.Q. Ngo, M. Egan, T.Q. Duong, Pilot power control for cell-free massive mimo. IEEE Trans. Veh. Tech. **67**(11), 11264–11268 (2018)
43. G. Interdonato, H.Q. Ngo, P. Frenger, E.G. Larsson, Downlink training in cell-free massive mimo: a blessing in disguise. IEEE Trans. Wireless Commun. **18**(11), 5153–5169 (2019)
44. Z. Chen, E. Björnson, Channel hardening and favorable propagation in cell-free Massive MIMO with stochastic geometry. IEEE Trans. Commun. **66**(11), 5205–5219 (2018)
45. J. Zhang, Y. Wei, E. Björnson, Y. Han, S. Jin, Performance analysis and power control of cell-free massive mimo systems with hardware impairments. IEEE Access **6**, 55302–55314 (2018)
46. M. Bashar, K. Cumanan, A.G. Burr, H.Q. Ngo, M. Debbah, P. Xiao, Max–min rate of cell-free massive mimo uplink with optimal uniform quantization. IEEE Trans. Commun. **67**(10), 6796–6815 (2019)
47. M. Bashar, K. Cumanan, A.G. Burr, H.Q. Ngo, E.G. Larsson, P. Xiao, Energy efficiency of the cell-free massive mimo uplink with optimal uniform quantization. IEEE Trans. Green Commun. Netw. **3**(4), 971–987 (2019)
48. G. Interdonato, P. Frenger, E.G. Larsson, Scalability aspects of cell-free massive mimo, in *ICC 2019–2019 IEEE International Conference on Communications (ICC)* (IEEE, 2019), pp. 1–6
49. R. Nikbakht, R. Mosayebi, A. Lozano, Uplink fractional power control and downlink power allocation for cell-free networks. IEEE Wireless Commun. Lett. **9**(6), 774–777 (2020)
50. E. De Carvalho, E. Bjornson, J.H. Sorensen, P. Popovski, E.G. Larsson, Random access protocols for massive mimo. IEEE Commun. Mag. **55**(5), 216–222 (2017)
51. K. Senel, E.G. Larsson, Grant-free massive mtc-enabled massive mimo: a compressive sensing approach. IEEE Trans. Commun. **66**(12), 6164–6175 (2018)
52. L. Liu, W. Yu, Massive connectivity with massive mimo–Part I: device activity detection and channel estimation. IEEE Trans. Signal Process. **66**(11), 2933–2946 (2018)

Chapter 5
Fundamentals of Network Densification

Abhishek K. Gupta, Nithin V. Sabu, and Harpreet S. Dhillon

5.1 Introduction to Densification

With applications in all sectors of human activity, wireless communications is widely regarded as one of the most pervasive technology enablers on the planet. Starting with Marconi's first transatlantic transmission in 1899 to the introduction of worldwide cellular networks in the 1980s and their subsequent evolution from supporting predominantly voice-driven applications to a largely data-driven services, the past 120 years has seen a remarkable transformation of this technology. More recent advancements in Internet-enabled computing and communication devices, primarily smartphones, tablets, wearables, and laptops, have increased mobile data traffic tremendously. According to the well-known predictions by Cisco [1], there has been almost 4000-fold growth in the mobile data traffic over the past 10 years and nearly 400-million-fold growth over the past 15 years. The monthly global mobile data traffic is estimated to be well over 35 exabytes already [2]. Developing efficient techniques to cope up with this data deluge is a key challenge faced by the wireless communications industry today.

A. Gupta gratefully acknowledges the support of the Science and Engineering Research Board (DST, India) under the grant SRG/2019/001459.

H. S. Dhillon gratefully acknowledges the support of the US National Science Foundation (Grant CNS-1617896)

A. K. Gupta · N. V. Sabu
Department of Electrical Engineering, Indian Institute of Technology, Kanpur, Uttar Pradesh, India
e-mail: gkrabhi@iitk.ac.in; nithinvs@iitk.ac.in

H. S. Dhillon (✉)
Wireless@VT, Department of ECE, Virginia Tech, Blacksburg, VA, USA
e-mail: hdhillon@vt.edu

X. Lin, N. Lee (eds.), *5G and Beyond*, https://doi.org/10.1007/978-3-030-58197-8_5

Since history is the best teacher, it will be instructive to understand which technologies have contributed the most to the increase in wireless network capacity in the past. For this, we rely on the well-known observations made by Martin Cooper, Chairman Emeritus of ArrayComm, that for the past 104 years, the number of *conversations* (voice or data) that can be carried out in a given area using all the available radio spectrum has doubled every 2.5 years [3]. This observation is often termed *Cooper's law*. That means there has been a one million fold increase in "capacity" over the past 45 years. Out of this one million fold increase, almost 1600 times increase is attributed to spectrum reuse (equivalently, *denser* deployments), 25 times to more spectrum, 5 times to modulation and coding, and 5 times to frequency division. This fact alone should be sufficient to put the importance of network densification in perspective.

The performance of a wireless system is primarily measured in terms of the achievable data rate, which is further linked to three important metrics, namely, available spectrum, link efficiency, and signal to interference plus noise ratio (SINR), via the famous Shannon-Hartley theorem. Specifically, the throughput of a single user in a cellular network can be expressed as [4]

$$c = m\left(\frac{W}{n}\right)\log_2\left(1+\mathrm{SINR}\right), \tag{5.1}$$

where W denotes the signal bandwidth of the base station (BS), n denotes the number of user equipment (UE) associated with this BS (that are sharing the same bandwidth), and hence W/n is the bandwidth available to each UE. Further, m captures the increase in capacity because of having multiple antennas, e.g., through supporting simultaneous streams of information. Assuming the densities of BSs and UEs (equivalently, average number of nodes per unit area) be λ, and λ_u, respectively, n would be of the order of $\lambda_\mathrm{u}/\lambda$. From the above discussion, it is evident that the UE throughput can be increased by increasing either or all of the four parameters: m, W, SINR, or λ. Here, W can be increased by allocating more spectrum, m can be increased by adding more antennas at the BS and UE, and λ can be increased by adding more BSs, which is also known as *densification* and is the main topic of this chapter. As noted above already, densification alone has contributed almost 1600-fold increase out of the total one million fold increase in capacity over the past 45 years. This is primarily because the addition of more BSs offloads users from the existing BSs, thus providing higher resources to each user, which is often termed the *cell splitting gain* [5].

Early cellular networks were sparse, and hence densification of these networks helped them fill coverage holes by increasing the received serving power at UEs. These BSs providing service to large areas (equivalently, having large coverage footprints) are called macrocell BSs. In the case of third-generation (3G) cellular systems, the primary aim of macro BS densification was to increase the transmission rate in specific areas, for example, macro-BSs deployed in the urban areas [6]. An effect of increased interference due to the densification of BSs was mitigated using frequency reuse and sectorized BS technologies. The density of macro-BSs for 3G cellular systems was not more than 4–5 BSs/km^2. In the fourth-generation (4G) cellular networks, including long-term evolution-advanced (LTE-A), new types of BSs, such as micro-cells, pico-cells, and femto-cells, have been deployed for enabling high-speed data transmission. The targeted density of these new BSs, often collectively called small cells, is about 8–10 BSs/km^2 [6]. The micro-cells and pico-cells are often deployed by service providers to complement the capacity of the existing networks, e.g., to enhance the throughput in specific areas in order to provide in-store services such as in malls and stadiums. On the other hand, femtocells are deployed directly by the users to improve coverage or capacity in small areas, such as in a house or in office. In both 3G and 4G cellular systems, the aim of BS densification was to improve the transmission rate, and the major challenge was interference mitigation. The Third Generation Partnership Project (3GPP) 4G LTE networks included small cell technology in their specifications throughout the second decade of 2000 up to now [7]. Over 14 million small cell BSs have been deployed worldwide till February 2016, and out of this over 12 million BSs were residential.

Now coming to the fifth-generation (5G) cellular networks, the key technologies of 5G are massive multiple-input multiple-output (MIMO) antennas, millimeter wave communications, and small cells. In massive MIMO, hundreds of antennas are used for transmitting gigabit-level data traffic. If we constrain the 5G BS power to about the same as that of 4G BS power, there will be a 10–20-fold reduction in transmission power per antenna compared to 4G BS power. As a result of this, the radius of the cell has to be reduced. Systems will also tend to use millimeter-wave frequencies owing to the availability of hundreds of megahertz bandwidth in these bands. Given the blockage sensitivity of these frequencies, the transmission range in such cases would be limited to about 100 m or so [6]. Therefore, it is expected that 5G networks would consist of small cells deployed at a very high density. Despite that, the interference in these bands remains low and spatially sparse because of highly directional transmission. This opens up the opportunities to deploy various types of services all sharing the same band, thus improving spectrum utilization [8].

With this background, we now revisit (5.1) to express the total throughout per unit area (also termed area spectrum efficiency) as

$$R \approx \lambda_{\mathrm{u}} m \left(\frac{W}{\lambda_{\mathrm{u}}/\lambda} \right) \log_2 (1 + \mathrm{SINR}) \propto \lambda. \tag{5.2}$$

This essentially means that assuming other parameters are not affected with an increase in λ (an assumption that will be scrutinized in the sequel), densification can lead to linear increase in the throughput. This solution works up to the current density of BSs. However, the question is whether this linear relation would remain valid for infinite densification or whether we have already reached the fundamental limit to the gains that can be achieved by densification [5]. Answering this question comprehensively is the main goal of this chapter.

5.2 General System Model and Performance Metrics

We will start the discussion by describing the general cellular network model adopted in this chapter. We will also identify the key performance metrics using which the performance of a cellular network can be quantitatively measured.

Network model We consider a cellular network with multiple BSs and UEs in which the BSs are located in a 2D space with density λ (i.e., the average number of BSs in the unit area is λ), and the UEs are spread in a stationary manner with density λ_u. The set of BSs is denoted as $\mathcal{N}_\mathrm{B}$ and UEs as $\mathcal{N}_\mathrm{U}$. Each UE is associated with one BS, which acts as the serving BS for this user, while the rest of the BSs act as interferers. All the calculations in this chapter will require us to study the impact of the network on the performance of a given UE. Without loss of generality, we will focus on the typical UE that will be placed at the origin.

Channel model Consider an individual BS (let us index it to be the ith BS) located at $\mathbf{x}_i$. The transmit power of this BS is p_i. The signal power attenuates according to a function $\ell(\cdot)$ termed path-loss function. The received signal power from this BS at the typical UE is given as

$$p_{\mathrm{r}i} = p_i G_i \ell(\|\mathbf{x}_i\|)$$

where G_i is a random variable denoting fading caused by various scatterers and obstacles present in the transmitter-receiver path. The path-loss function $\ell(\cdot)$ plays an important role in determining the average received power and depends on the propagation environment. If the propagation environment is free space, the path-loss function is given as the simple power-law relation

$$\ell(r) = \left(\frac{\lambda_s}{4\pi r}\right)^2 \propto r^{-2},$$

where λ_s is the wavelength of the transmitted signal. The above equation is well-known by the name of Friis transmission equation. While this is conceptually simple to work with, it is not valid for environments where the propagation medium consists of blockages, shadowing effects, multiple signal reflections, and scattering. Therefore, it is desirable to use path-loss models that embody the simplicity of the Friis equation but capture the effect of the aforementioned propagation effects reasonably accurately. A widely accepted and used model is the one in which the distance dependence is generalized to $r^{-\alpha}$, where α is the path-loss exponent. The path-loss, thus, is given as

$$\ell(r) = Cr^{-\alpha}$$

where C is a constant termed near-field gain which represents the path-loss at unit distance. The value of α depends upon the transmission frequency and the propagation environment and is generally determined empirically. We would refer to this path-loss model as the *standard path-loss model* throughout the chapter. Since this will be extended further in the chapter to a multi-slope model, we will also refer to this as a *single slope path-loss model* wherever necessary.

SINR model Let us denote the BS that serves the typical UE by index 0. Hence, the received power from this BS at the typical UE is denoted by $S = p_{r0}$. If the UE density is finite, it is possible that some BSs do not have any associated UEs because of which they can suspend their downlink transmission in order to avoid interference to the other UEs. Let $\mathcal{N}_{aB}$ denote the set of all active BSs. Let the active BS density be λ_a, which is essentially equal to the density of the transmitting BSs. If the UE density is infinite, or significantly larger than BS density, or it scales with the BS density as network densifies, all BSs will be considered active, i.e., $\mathcal{N}_{aB} = \mathcal{N}_B$ all the time. The interference power I at the typical UE is given as

$$I = \sum_{i \neq 0,\ i \in \mathcal{N}_{aB}} p_{ri}.$$

The signal to noise power ratio (SNR) is the ratio of serving signal power to the noise power, which is given as

$$\mathrm{SNR} = \frac{p_{r0}}{\sigma^2},$$

where σ^2 is the noise power. Similarly, the SINR at the UE is given as

$$\mathrm{SINR} = \frac{S}{\sigma^2 + I}.$$

In scenarios where thermal noise is negligible compared to the interference power, SIR (signal to interference ratio) is useful to consider, which can be defined as

$$\mathrm{SIR} = \frac{S}{I}.$$

Performance metrics We will consider the following three metrics for evaluating the performance of the cellular network:

1. ***Coverage probability:*** Let γ_s be the SINR threshold required at the typical user for successful transmission. The coverage probability of the typical UE is defined as

$$\mathrm{p_c} = \mathbb{P}\left(\mathrm{SINR} > \gamma_s\right), \tag{5.3}$$

which denotes the probability that the typical UE can achieve the target SINR γ_s. Coverage probability can also be thought as the complementary cumulative distribution function (CCDF) of the SINR. In scenarios where thermal noise is negligible, the coverage probability can also be defined in terms of SIR as

$$\mathrm{p_{cI}} = \mathbb{P}\left(\mathrm{SIR} > \gamma_{\mathrm{s}}\right). \tag{5.4}$$

2. ***Potential throughput (PT):*** The potential throughput of the cellular network is defined as

$$\tau = \lambda_{\mathrm{a}} \mathrm{p_c}(\lambda, \alpha) \times \log(1 + \gamma_{\mathrm{s}}). \tag{5.5}$$

Note that the potential throughput denotes the average number of bits transmitted successfully per unit area. While neglecting thermal noise, the potential throughput is defined as

$$\tau_{\mathrm{I}} = \lambda_{\mathrm{a}} \mathrm{p_{cI}}(\lambda, \alpha) \times \log(1 + \gamma_{\mathrm{s}}). \tag{5.6}$$

3. ***Area spectral efficiency (ASE):*** The ASE of the network is defined as

$$\mathcal{A} = \lambda_{\mathrm{a}} \mathbb{E}\left[\log(1 + \gamma_{\mathrm{s}}) \mathbb{1}\left(\mathrm{SINR} > \gamma_{\mathrm{s}}\right)\right] \tag{5.7}$$

The ASE denotes the theoretical upper limits on the number of bits that can be transmitted successfully per unit area. Its unit is bps/Hz/m^2.

The above metrics naturally depend on the BS density λ. Our goal is to investigate the exact behavior as a function of different environments and system parameters.

5.3 Densification in the Conventional Scenario

We will first discuss the effect of network densification under the conventional assumptions of cellular networks [9] which are as follows:

1. Standard path-loss model is assumed with path-loss exponent $\alpha > 2$

$$\ell(r) = Cr^{-\alpha}.$$

All links are assumed to undergo Rayleigh fading, i.e., $G \sim \exp(1)$.

2. All BSs are assumed to be homogeneous, i.e. they belong to the same class in terms of key parameters, such as the transmit power.
3. UEs are assumed to have significantly larger density than BSs and their density scales with the BS density as the network densifies. We also assume that BSs have full buffer and are always ready to transmit. Therefore, all BSs are active all the time, i.e., $\lambda_{\mathrm{a}} = \lambda$.
4. BSs and UEs are at the same height.

5. All BSs are assumed to be deployed according to a stationary Poisson point process (PPP) in 2D space [10].

Under the above assumptions, the SINR coverage probability of a typical UE is given as [10]

$$\begin{aligned}\mathrm{p_c}(\lambda,\alpha) &= \pi\lambda\int_0^{\infty}\exp\Big(-\pi\lambda v(1+\rho(\gamma_{\mathrm{s}},\alpha))-\gamma_{\mathrm{s}}\sigma^2 v^{\alpha/2}/p\Big)dv\\ &= \pi\int_0^{\infty}\exp\Big(-\pi v(1+\rho(\gamma_{\mathrm{s}},\alpha))-\gamma_{\mathrm{s}}\sigma^2 v^{\alpha/2}\lambda^{-\alpha/2}/p\Big)dv, \qquad (5.8)\end{aligned}$$

where

$$\rho(\gamma,\alpha)=\gamma^{2/\alpha}\int_{\gamma^{-2/\alpha}}^{\infty}\frac{1}{1+u^{\alpha/2}}du. \qquad (5.9)$$

For $\alpha = 4$, $\rho(\gamma,\alpha) = \sqrt{\gamma}\arctan\sqrt{\gamma}$. For $\alpha = 2$, $\rho(\gamma,\alpha) = \infty$. In general, $\rho(\gamma,\alpha)$ is monotonic decreasing function of α. The SIR coverage probability can be obtained from (5.8) as

$$\mathrm{p_{cI}}(\lambda,\alpha)=\frac{1}{1+\rho(\gamma_{\mathrm{s}},\alpha)}. \qquad (5.10)$$

5.3.1 Impact of Densification

It is evident from (5.10) that the SIR distribution is invariant of the BS density λ. Figure 5.1 shows the impact of densification on SIR coverage probability, where the same behavior can be observed. This invariance can be understood with the help of the following example.

Example 1 Consider a cellular network in 2D space with BS density λ. At the typical UE (placed at the origin), the SIR is equal to γ with the serving signal power S and the sum interference I. Suppose the network is densified m times resulting in a BS density of $\lambda' = m\lambda$. As a result, all the BSs will statistically move closer to the origin by a factor of $\sqrt{m}$. Therefore, the new serving power will be equivalent in distribution to

$$S' = p_0\|\mathbf{x}_0'\|^{-\alpha} = p_0 m^{\alpha/2}\|\mathbf{x}_0\|^{-\alpha} = m^{\alpha/2}S.$$

Similarly, the sum interference would be equivalent in distribution to

$$I' = \sum_i p_i\|\mathbf{x}_i'\|^{-\alpha} = \sum_i p_i m^{\alpha/2}\|\mathbf{x}_i\|^{-\alpha} = m^{\alpha/2}I.$$

The new SIR is

$$\mathrm{SIR}' = \frac{S'}{I'} = \frac{m^{\alpha/2}S}{m^{\alpha/2}I} = \gamma$$

which has the same form and hence the same distribution as before.

The SINR distribution follows the same behavior as the SIR distribution, except at low BS density. At a lower value of λ, the serving BS, as well as the interfering BSs, is very far from the typical UE. In this case, the interference is negligible compared to noise, which is termed the *noise-limited scenario*. As the network densifies ($\lambda \to \infty$), the serving BS statistically comes closer to the UE and therefore $\mathrm{p_c}$ increases monotonically as can be seen from (5.8) and Fig. 5.1. At large λ, the SINR coverage probability approaches

$$\lim_{\lambda\to\infty} \mathrm{p_c}(\lambda, \alpha) = \mathrm{p_{cI}}.$$

As the BS density λ reaches the critical BS density λ_1, the noise can be neglected (interference-limited scenario, i.e., $\sigma^2 \approx 0$) and $\mathrm{p_c} \approx \mathrm{p_{cI}}$. This critical density depends on the noise power and the BS transmit power. With further increase in the BS density, the densification no longer improves the SINR of a typical UE as the increase in interference power is counterbalanced by the increase in signal power. This observation that SINR in an interference-limited cellular network does not depend on the BS density is often referred to as the *SINR invariance in cellular networks*.

The network densification reduces the user load on each BS without affecting the SINR characteristics. So the network can achieve an approximately linear increase

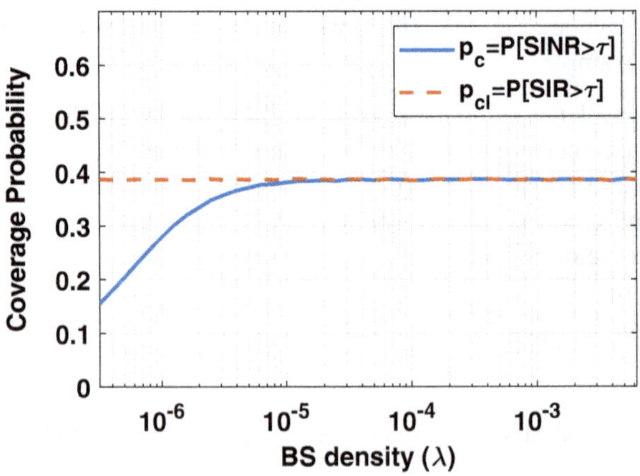

Fig. 5.1 Impact of the BS density on the SINR and SIR coverage probability for a cellular network with the single slope path-loss with $C = 10^{-4}$ and $\alpha = 3$. Here, $\gamma_s = 1$. It can be observed that the SIR distribution is invariant to the BS density

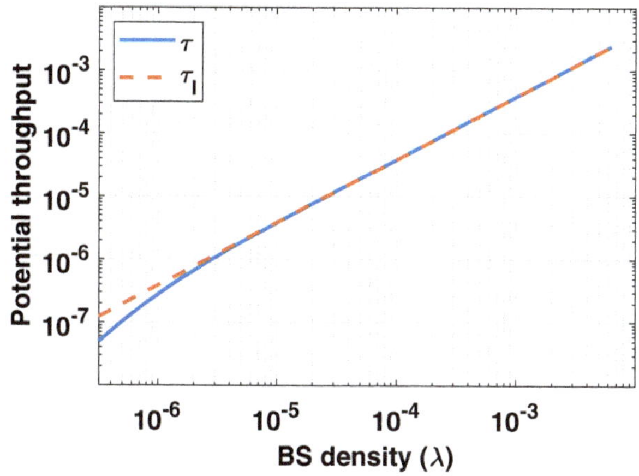

Fig. 5.2 Impact of the BS density on the potential throughput for a cellular network with single slope path-loss with $C = 10^{-4}$ and $\alpha = 3$. Here, $\gamma_s = 1$. The potential throughput grows linearly with the BS densification owing to the SIR invariance observed in Fig. 5.1

in the achievable data rate with the increase in BS density (see Fig. 5.2). As noted before, this gain is termed *cell splitting gain* in the literature. It can be verified using (5.6) and (5.7) that when the SINR invariance property holds, the potential throughput and ASE exhibit linear relation with BS density in the following way:

$$\tau = \lambda \frac{\log(1 + \gamma_s)}{1 + \rho(\gamma_s, \alpha)} \propto \lambda$$

$$\mathcal{A} = \lambda \mathbb{E}\left[\log(1 + \gamma_s) \mathbb{1}\left(\text{SINR} > \gamma_s\right)\right] \propto \lambda.$$

The scaling results in this section are derived based on the assumption that the BSs in the 2D space are distributed according to a homogeneous PPP. However, the real BS deployment is not completely random (nor is it completely regular). Recent studies [11, 12] have shown that a large variety of BS deployment including lattice deployments and hexagonal grid-based deployments have almost similar SIR statistics to that of Poisson deployment, but with a small fixed SIR shift. Moreover, it is known that if all the links in the network undergo significant shadowing that is independent of each other, the network appear Poissonian to the typical UE even if the actual locations are more regular, or even modeled using deterministic grids [13, 14]. Therefore, the results derived in this section are either directly applicable or can be easily modified to apply to more generic scenarios.

Getting back to our main question, the discussion in this section indicates that densifying the network infinitely would keep on increasing the network throughout indefinitely because densification does not impact the coverage probability (equivalently because of the *SINR invariance*). As indicated above already, the SINR invariance property holds for more general setups as well. These include different BS layouts, fading/shadowing assumptions, presence of multiple antennas [15], distribution of antenna azimuths, and the effect of horizontal sectorization [16], to name a few. In fact, this property is even valid for the multi-tier networks [17]. Specifically, for an interference-limited open access network multi-tier network, adding more tiers or BSs does not affect the SINR distribution at the typical UE. That all being said, it is still important to keep in mind that the SINR invariance property may not universally hold because of which jumping to the conclusion that the network throughput will *always* increase linearly with the addition of BSs may be naïve. At the very least, such statements must be qualified with appropriate assumptions, as we demonstrate next.

5.3.2 Effect of the Dual-Slope Path-Loss Model

In this subsection, we revisit our path-loss assumption to understand its impact on SINR invariance and linear scaling of throughput. The standard path-loss model adopted thus far is widely used by researchers as well as in industry; however, it may not be suitable and valid for modern dense networks [5]. When the distance between BS and UE becomes smaller, the relation of the path-loss with distance may change resulting in the change of path-loss exponent itself. As discussed in detail in [5], the region around a BS can be divided into three regions from the perspective of path-loss modeling:

1. **Ground Fresnel region:** This region is located near the ground surface beyond a significant distance from the transmitter. In this region, the direct and ground reflected rays undergo destructive interference resulting in path-loss exponent close to $\alpha = 4$.
2. **Large-scale interference region:** This region lies beyond a certain distance from the transmitter away from the ground surface. In this region, signals coming from various paths can also add constructively resulting in a lower path-loss exponent ($\alpha \approx 2$).
3. **Small-scale interference region:** This region lies around the transmitter up to some finite distance. Due to the absence of obstacles and additional power received from the reflected paths, the path-loss exponent in this region may decrease below free space path-loss exponent of $\alpha = 2$.

When the link distance between a UE and a BS is small, the UE will be inside the path-loss subduction region of the BS where the path-loss exponent becomes smaller than 2. Having different values of the path-loss exponent in different regions will lead to path-loss exhibiting different slopes in these regions. Such path-loss models are termed *multi-slope path-loss models* in the literature [18]. Given their versatility, e.g., in modeling both indoor and outdoor propagation environments, they have been extensively used in 3GPP standards as well. As an aside for now, note that a probabilistic version of such models has also been proposed and validated for blockage sensitive communications including communications at the millimeter-wave frequencies. In these models, links follow different slopes according to a probability distribution that depends on the link distance [19].

A specific example of the multi-slope models is the dual-slope model which is simple yet powerful enough to capture the effects of path-loss exponent reduction. Let the path-loss subduction region be represented by the ball of radius R_c around the receiver where R_c is termed the *corner distance*. The dual-slope (power-law) path-loss function is defined as

$$\ell(r) = \begin{cases} C_0 r^{-\alpha_0}, & \text{if } r \leq R_c \\ C_1 r^{-\alpha_1}, & \text{if } r > R_c \end{cases}. \tag{5.11}$$

Therefore, the space around the receiver is divided into two regions. For BSs lying inside the first region R_1 (i.e., $r \leq R_c$), the path-loss exponent is α_0. The path-loss exponent is α_1 for BSs lying in the second region R_2. We consider that $\alpha_0 \leq \alpha_1$. Here, C_1 is chosen such that the path-loss function is continuous at the boundary between these two regions.

To investigate how the dual-slope path-loss model can affect the scaling laws of densification, we will consider the same assumptions as taken in this section (except of course that we will consider a dual-slope path-loss model instead of the standard path-loss model used thus far). Consider again the typical UE at the origin. BSs will be located in either of two different regions R_1 and R_2. Signals from the BSs inside the ball of radius R_c will undergo lower attenuation compared to the BS outside the ball. The average number of BSs in these two regions will be dependent on the BS density, which will eventually impact the coverage probability.

At low BS density $\lambda \to 0$, all BSs will lie in region R_2. Therefore, signals from each of the BSs will undergo attenuation according to path-loss exponent α_1. Hence, the SIR coverage probability of the network would be equal to $\mathrm{p_{cI}}(\gamma_s, \alpha_1)$ as defined in (5.10). Since the noise would be dominant, the SINR coverage probability will be zero. As BS density increases, initially all BSs would still be located in region R_2. The system will behave exactly like the system with single slope path-loss, and SIR coverage would be invariant of the BS density. However, the SINR coverage would increase with BS density until λ_1 beyond which it would be equal to the SIR coverage probability.

As BS density approaches a critical density λ_3, a few BSs will statistically come closer to the typical UE and would lie in R_1. The serving BS would be one of these BSs as it is the closest BS to the typical UE. At this stage, if the BS density is further increased m times, serving power S will increase by a factor of m^{α_0}, whereas the interfering power I will increase by a factor of m^{α_1} (see Example 1). Here, for the sake of argument we assumed that most of the dominant interferers lie in R_2. Hence, the SIR would decrease by a factor of $m^{\alpha_1 - \alpha_0}$.

As the BS density increases further, another critical density λ_2 approaches where most dominating interfering BSs would lie in the region R_1. At this stage signals from most of the BSs will undergo attenuation according to path-loss exponent α_0. Hence, the SIR coverage probability of the network would be equal to $\mathrm{p_{cI}}(\gamma_s, \alpha_0)$. If

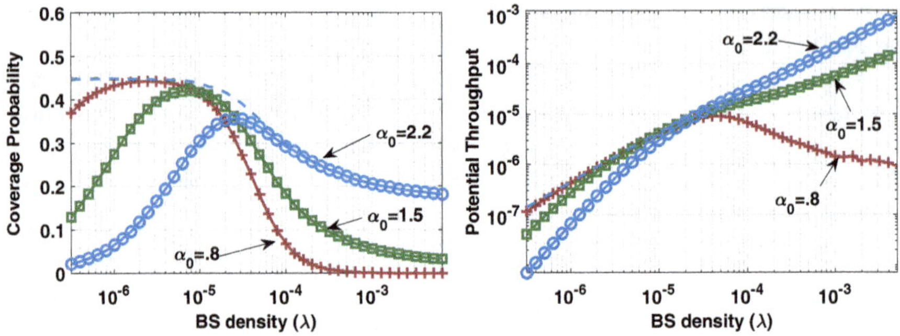

Fig. 5.3 Impact of the BS densification on the SINR and potential throughput under the dual-slope path-loss model with $R_c = 100\,\text{m}$, $C_0 = 10^{-7}$, $\alpha_0 = 0.8, 1.5, 2.2$ and $\alpha_1 = 3.3$. Dashed lines represent respective metrics in the absence of noise

the BS density is further increased m times, serving power S and interfering power I both will increase by a factor of m^{α_0} and SIR would again become invariant of λ. At very high density $\lambda \to \infty$, the SIR and SINR coverage probability of the network would be equal to $\text{p}_{\text{cI}}(\gamma_{\text{s}}, \alpha_0)$.

The critical densities λ_3 and λ_2 at which transition from one region to another region occurs depend on the corner distance R_c. The same behavior of the coverage probability with the BS density is also evident in Fig. 5.3.

Clearly, the SINR invariance property no longer holds under the dual-slope path-loss model due to the aforementioned reasons. Another interesting observation one may have is that at very high density, the SINR coverage probability of the network would be equal to $\text{p}_{\text{cI}}(\gamma_{\text{s}}, \alpha_0)$, which can be zero depending on the value of α_0. It was shown in [18] that under the dual-slope path-loss model, the SIR and SINR coverage probability of a two-dimensional cellular network goes to zero as $\lambda \to \infty$ when $\alpha_0 \leq 2$. This indicates that the ultra-densification of a network can be harmful to the coverage performance. It was also shown that under the dual-slope model, as $\lambda \to \infty$, the potential throughput τ exhibits the following scaling law:

1. τ grows linearly with λ if $\alpha_0 > 2$,
2. τ grows sublinearly with rate $\lambda^{\left(2-\frac{2}{\alpha_0}\right)}$ if $1 < \alpha_0 < 2$,
3. τ decays to zero if $\alpha_0 < 1$.

Contrary to the conclusions drawn in Sect. 5.3.1, a blind densification of the network may not provide gains proportional to the deployment cost and may in fact be even harmful. Apart from the analytical work, it has also been observed by various empirical studies that densification may cause the network throughput to fall and even crash [20]. With these seemingly conflicting conclusions, it is clear that one needs a more careful look at potential factors that may impact the densification gain, which is the topic of the next section.

5.4 Factors Affecting the Densification Gain

In the last section, we saw that adopting a more realistic path-loss model, such as the dual-slope model, changed the conclusions of the densification gain significantly. The naïve assumption that densification can infinitely increase the potential throughput and ASE is clearly not valid. In fact, densification may cause throughput to fall and even crash if done beyond a limit. Apart from path-loss model, there are many other factors that affect the scaling behavior of the performance with densification. In this section, we discuss some of these factors in detail.

5.4.1 *Path-Loss Models*

We have already seen how incorporating two slopes in the path-loss model disrupts the scaling of system performance with densification. There exist other realistic path-loss models suitable for various propagation environment which are discussed below.

Multi-slope Path-Loss Model

The dual-slope model can be extended to a general N-slope path-loss to include propagation environments where more than two path-loss subduction regions exist [21]. The N-slope path-loss is defined as

$$\ell(r) = \begin{cases} \ell_0(r) = C_0 r^{-\alpha_0} & \text{if } r \in [0 = R_0, R_1) \\ \dots \\ \ell_n(r) = C_n r^{-\alpha_n} & \text{if } r \in [R_n, R_{n+1}) \\ \dots \\ \ell_{N-1}(r) = C_{N-1} r^{-\alpha_{N-1}} & \text{if } r \in [R_{N-1}, R_N = \infty) \end{cases} \tag{5.12}$$

Here, $C_0 = 1$ is the near-field gain and $C_n = \prod_{i=1}^{n} R_i^{\alpha_i - \alpha_{i-1}}$ to ensure that path-loss is continuous at the boundaries of the adjacent regions. Also, $0 = R_0 < R_1 < \dots < R_N = \infty$ are corner distances and $0 \le \alpha_0 \le \alpha_1 \le \dots \le \alpha_{N-1}$ are path-loss exponents for N regions. We assume that $\alpha_{N-1} > 2$ to ensure that the sum interference at finite BS density is bounded. We do not require any additional conditions on any other path-loss exponents as the number of BSs lying in all other regions is almost surely finite and, hence, the interference is also almost surely finite. When $N = 2$, this model reduces to the special case of the dual-slope path-loss model (5.11).

As described earlier, the SIR and SINR invariance properties no longer hold for multi-slope models. Initially, SINR improves with densification before the BS density hits the critical density λ_1 after which the network becomes interference limited. After this, SINR and SIR coverage probability become the same. As $\lambda \to \infty$, the coverage probability approach $\mathrm{p_{cI}}(\alpha_0)$ which depends only on the value of α_0 regardless of the number of slopes. Therefore, asymptotic behavior of coverage and potential throughput is exactly the same as the one described above for the dual-slope model.

Probabilistic Two-Regime Model

Given the increasing relevance of millimeter wave communications in cellular networks, it is important to carefully incorporate the blockage sensitivity of these frequencies in the propagation models. In the context of this discussion, it is important to distinguish the line-of-sight (LOS) and non-LOS (NLOS) links as they differ significantly in their propagation characteristics. To model such propagation, a probabilistic two-regime model has been proposed [19, 22] where a link can be LOS or NLOS randomly according to a probability distribution p_{L}. This LOS probability p_{L} depends on the link distance. If the link between a transmitter and a receiver located at a distance of r is LOS, it follows the following path-loss model:

$$\ell_{\mathrm{L}}(r) = C_{\mathrm{L}} r^{-\alpha_{\mathrm{L}}},$$

whereas a NLOS link suffers with the following path-loss

$$\ell_{\mathrm{N}}(r) = C_{\mathrm{N}} r^{-\alpha_{\mathrm{N}}}.$$

The complete model is given as

$$\ell(r) = \begin{cases} \ell_{\mathrm{L}}(r) = C_{\mathrm{L}} r^{-\alpha_{\mathrm{L}}} & \text{with probability } p_{\mathrm{L}}(r) \\ \ell_{\mathrm{N}}(r) = C_{\mathrm{N}} r^{-\alpha_{\mathrm{N}}} & \text{with probability } 1 - p_{\mathrm{L}}(r) \end{cases}.$$

There are two distinguishing properties of these model compared to the two-slope model:

1. The two regimes in this model are probabilistic and can overlap in space, while two regions in the two-slope model are deterministic and mutually exclusive.
2. There is no continuity condition on gains C_{L} and C_{N}. LOS and NLOS links can have different gains event at the unit distance [23].

Throughout this discussion, we assume that each UE is associated with the BS providing the smallest path-loss.

The behavior of SIR and SINR coverage probability with densification under the two-regime model is similar to the two-slope model with some differences. At the low BS density, all BSs will be NLOS. Therefore, signals from each of the BSs will undergo attenuation according to path-loss exponent α_{N}. Hence, the

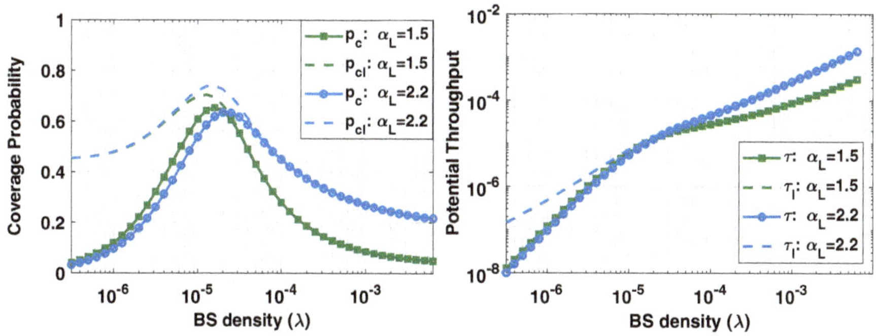

Fig. 5.4 Impact of the BS densification on the potential throughput under the probabilistic two-regime model with $\alpha_{\rm L} = 1.5, 2.2$, $\alpha_{\rm N} = 3.3$, $C_{\rm L} = 10^{-6}$ and $C_{\rm N} = 10^{-7}$. Here, the LOS probability model is assumed to be exponential i.e. $p_{\rm L}(r) = \exp(-\beta r)$ with $\beta = 1/144\,{\rm m}^{-1}$. Dashed lines represent respective metrics in the absence of noise

SIR coverage probability of the network would be constant at $\mathrm{p_{cI}}(\gamma_{\rm s}, \alpha_{\rm N})$. Since the noise would be dominant, SINR coverage probability will increase with BS density. As the BS density increases, a few BSs will become LOS to the typical UE. The serving BS would most likely be one of these BSs. Since the gain of the LOS link is more than the NLOS link, SIR coverage probability would improve. At this stage, if the BS density is further increased, the probability of the serving BS to be LOS increases, and hence the SIR coverage probability improves. After a critical density of BSs, further increase in the BS density causes interfering BSs to become LOS also. This increases the interference severely causing SIR to go down. As the BS density increase further, most dominating interfering BSs would become LOS. At this stage, signals from most of the BSs will undergo attenuation according to path-loss exponent $\alpha_{\rm L}$. Hence, the SIR coverage probability of the network would approach $\mathrm{p_{cI}}(\gamma_{\rm s}, \alpha_{\rm L})$ and becomes constant at this level. The above discussion indicates the existence of an optimal density $\lambda_{\rm opt}$ of that BSs that would maximize the coverage probability (see Fig. 5.4). Densification beyond this optimal density would hurt SIR and even makes it fall to zero if $\alpha_{\rm L} < 2$.

The SIR degradation also impacts the throughput scaling. For some values of $\alpha_{\rm L} < 2$ and $\gamma_{\rm s}$, the densification beyond $\lambda_{\rm opt}$ may reduce the potential throughput (see Fig. 5.4). Asymptotic scaling of potential throughput with densification is the same as scaling under multi-slope model with $\alpha_0 = \alpha_{\rm L}$.

General Multi-regime Multi-slope Probabilistic Path-Loss Model

The multi-slope path-loss model and the probabilistic path-loss model can be combined into a general multi-regime multi-slope path-loss model which is defined as [24, 25]

$$\ell(r) = \begin{cases} \ell_1(r) = \begin{cases} \ell_{1\mathrm{L}}(r), & \text{with probability } p_{1\mathrm{L}}(r) \\ \ell_{1\mathrm{N}}(r), & \text{with probability } (1 - p_{1\mathrm{L}}(r)) \end{cases} & \text{if } 0 \le r \le R_1 \\ \ell_2(r) = \begin{cases} \ell_{2\mathrm{L}}(r), & \text{with probability } p_{2\mathrm{L}}(r) \\ \ell_{2\mathrm{N}}(r), & \text{with probability } (1 - p_{2\mathrm{L}}(r)) \end{cases} & \text{if } R_1 \le r \le R_2 \\ \vdots & \vdots \\ \ell_m(r) = \begin{cases} \ell_{m\mathrm{L}}(r), & \text{with probability } p_{m\mathrm{L}}(r) \\ \ell_{m\mathrm{N}}(r), & \text{with probability } (1 - p_{m\mathrm{L}}(r)) \end{cases} & \text{if } r > R_{m-1} \end{cases} \tag{5.13}$$

where $\ell_{i\mathrm{L}}(r)$ and $\ell_{i\mathrm{N}}(r)$ are the path-loss functions for the LOS and NLOS links, respectively, in ith region. $P_{i\mathrm{L}}(r)$ is the ith piece LOS probability function. $\ell_{i\mathrm{L}}(r)$ and $\ell_{i\mathrm{N}}(r)$ are given as

$$\ell_{i\mathrm{L}}(r) = C_{i\mathrm{L}} r^{-\alpha_{i\mathrm{L}}}, \tag{5.14}$$

$$\ell_{i\mathrm{N}}(r) = C_{i\mathrm{N}} r^{-\alpha_{i\mathrm{N}}}. \tag{5.15}$$

where the parameters can be chosen to match the empirical data. This model is consistent with the ones adopted in 3GPP simulations. We discuss two special cases of this model which are mentioned in 3GPP documents and have been used in 3GPP simulations to evaluate the performance of cellular networks.

3GPP-Model-1

First, we consider a 3GPP model given as [26]

$$\ell_n(r) = \begin{cases} C_{\mathrm{L}} r^{-\alpha_{\mathrm{L}}} & \text{with probability } p_{\mathrm{L}}(r) \\ C_{\mathrm{N}} r^{-\alpha_{\mathrm{N}}} & \text{with probability } (1 - p_{\mathrm{L}}(r)) \end{cases}. \tag{5.16}$$

with linear LOS probability function [27],

$$p_{\mathrm{L}}(r) = \begin{cases} 1 - r/D & \text{when } 0 \le r \le D \\ 0 & \text{when } r > d_1 \end{cases}.$$

Note that the model given in (5.16) is the special case of (5.13) where $m = 2$, $\ell_{1\mathrm{L}}(r) = \ell_{2\mathrm{L}}(r) = C_{\mathrm{L}} r^{-\alpha_{\mathrm{L}}}$, $\ell_{1\mathrm{N}}(r) = \ell_{2\mathrm{N}}(r) = C_{\mathrm{N}} r^{-\alpha_{\mathrm{N}}}$, $p_{1\mathrm{L}}(r) = 1 - \frac{r}{D}$ and $p_{2\mathrm{L}}(r) = 0$. This model was proposed for dense small cell networks.

3GPP-Model-2

The second model considered here is proposed in [26]. This model has the same path-loss function as (5.16) but with an exponential LOS probability function:

$$p_{\mathrm{L}}(r) = \begin{cases} 1 - 5\exp\left(-D_1/r\right) & \text{when } 0 \le r \le D \\ 5\exp\left(-r/D_2\right) & \text{when } r > D \end{cases},$$

where $D = D_1/\ln(10)$. This model is the special case of (5.13) where $m = 2$, $\ell_{1\mathrm{L}}(r) = \ell_{2\mathrm{L}}(r) = C_{\mathrm{L}}r^{-\alpha_{\mathrm{L}}}$, $\ell_{1\mathrm{N}}(r) = \ell_{2\mathrm{N}}(r) = C_{\mathrm{N}}r^{-\alpha_{\mathrm{N}}}$, $p_{1\mathrm{L}}(r) = 1 - 5\exp(-D_1/r)$, and $p_{2\mathrm{L}}(r) = 5\exp(-r/D_2)$.

Figure 5.5 shows the behavior of potential throughput with densification under the two 3GPP path-loss models. The key observations can be summarized as follows. When the network is sparse, the potential throughput quickly increases with BS density. This is due to the fact that the network is noise-limited, and thus adding more BSs immensely benefits the throughput. When the network reaches the practical density (as expected in 4G/5G systems), the scaling trend of the potential throughput is very interesting. Initially throughput exhibits a slowing-down in the rate of growth or even a decrease due to the fast decrease of the coverage probability. As BS density further increases, the growth rate of the throughput picks up. This is because the coverage probability remains almost constant in this region (but at a much lower value than before). The behavior of the throughput depends on the characteristics of the LOS and the NLOS path-loss. The larger the difference between the LOS and the NLOS path-loss exponents, the more the throughput suffers in transition region due to more drastic transition of interference from the NLOS transmission to the LOS transmission. This asymptotic behavior is similar to what is seen in the previous models.

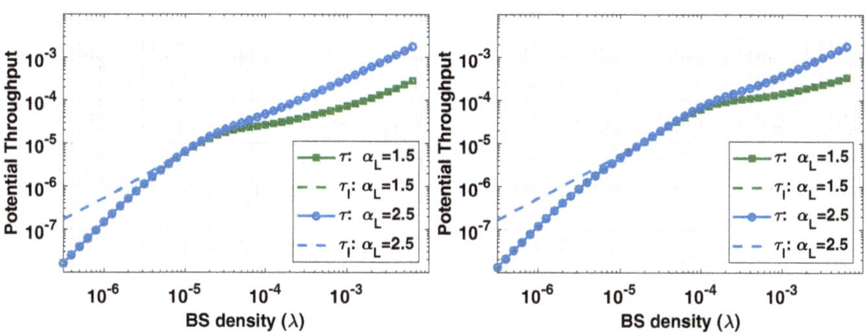

Fig. 5.5 The impact of the BS densification on the potential throughput under the 3GPP-Model-1 and 3GPP-Model-2 with $\alpha_{\mathrm{L}} = 1.5,\ 2.5$, $\alpha_{\mathrm{N}} = 3.75$, $C_{\mathrm{L}} = 10^{-10.38+3\alpha_{\mathrm{L}}}$ and $C_{\mathrm{N}} = 10^{-14.54+3\alpha_{\mathrm{N}}}$. Additionally, for 3GPP-Model-1, $D = 300$ m and for 3GPP-Model-2, $D_1 = 156$ m and $D_2 = 30$ m. Here, $\gamma_{\mathrm{s}} = 1$. Dashed lines represent respective metrics in the absence of noise

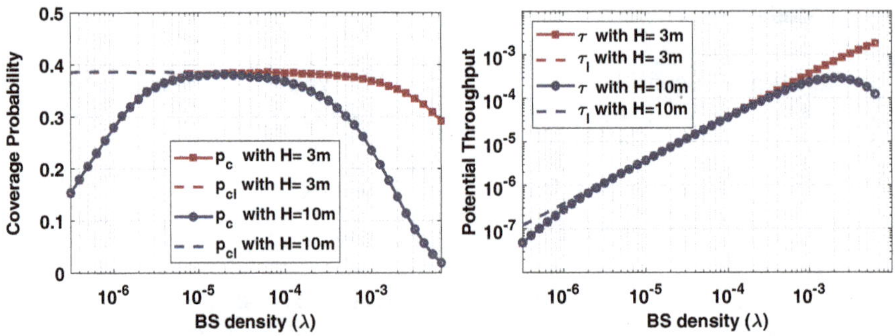

Fig. 5.6 Impact of the BS density on the coverage probability and potential throughput for a cellular network with height difference of H under the single slope path-loss with $C = 10^{-4}$ and $\alpha = 3$. Here, $\gamma_s = 1$. It can be observed that the SIR distribution is invariant to the BS density. Dashed lines represent respective metrics in the absence of noise

5.4.2 Height Difference Between BS and UE Antennas

In deriving the scaling behavior of densification until now, we have assumed that the height of the BSs and UEs are the same. In a practical scenario, there may be some height difference H between the heights of the BSs and UEs. Figure 5.6 shows how SINR and throughput scales with density when H is 3m and 10m. The path-loss model assumed for these plots is the standard single slope path-loss model. We observe that at large density, both the SIR coverage probability and the potential throughput decrease and eventually crash to zero even for path-loss exponent $\alpha > 2$. This is in contrast with results obtained for the scenario with zero height difference (see Figs. 5.1 and 5.2), where throughput was observed to grow linearly with the BS density for the same value of the path-loss exponent. This can be understood by the following example.

Example 2 Consider a 2D cellular network where BSs have height H and UEs are at the ground level. Consider the ith BS at a 2D distance (i.e., distance along the ground) r_i from the typical UE. Hence, the 3D distance between the two is $\sqrt{r_i^2 + H^2}$. The average receiver power at the UE from this BS is $pC(r_i^2 + H^2)^{-\alpha}$. Hence, the SIR is given as

$$\mathrm{SIR} = \frac{\left(r_0^2 + H^2\right)^{-\alpha/2}}{\sum_i \left(r_i^2 + H^2\right)^{-\alpha/2}}.$$

Note that a densification by a factor of m is statistically equivalent to reduction in all 2D distances by a factor $\sqrt{m}$. The new SIR at the typical UE would be equivalent in distribution to

$$\begin{aligned}\mathrm{SIR}' &= \frac{\left((r_0/\sqrt{m})^2+H^2\right)^{-\alpha}}{\sum_i \left((r_i/\sqrt{m})^2+H^2\right)^{-\alpha}}\\ &= \frac{\left((r_0)^2+mH^2\right)^{-\alpha}}{\sum_i \left((r_i)2+mH^2\right)^{-\alpha}}.\end{aligned}$$

If $H = 0$, the above would be equal to the original SIR itself. However, when H is non-zero, for large m,

$$\mathrm{SIR}' \approx \frac{\left(mH^2\right)^{-\alpha}}{\sum_i \left(mH^2\right)^{-\alpha}} = \frac{1}{\sum_i 1} \to 0$$

as there are many BSs around the UE at distance H for large m.

As we increase the BS density, BSs statistically comes closer to the typical UE. If there is a height difference between a BS and the UE, the 3D distance between them cannot go to zero even if the BS density goes to infinity. In fact, the distance between the BS and UE will be lower bounded by the difference in their heights. Therefore the signal power from each BS cannot be larger than $pH^{-\alpha}$. Consequently, at a large density, signal power from each of the serving BSs and each interfering BS would approach this value and result in zero coverage probability. As a result, potential throughput and ASE would also fall to zero. One way to avoid ASE crash is by avoiding the cap on the signal power of the serving BSs, which can be done by deploying BSs at the same height as UE. If that is not possible, reducing the antenna height of BS to that of UE antenna height can delay this crash, but cannot completely avoid it.

5.4.3 *Scaling of the UE Density*

While showing the SINR invariance for single slope path-loss model, we have assumed that the UE density is significantly larger than the BSs or scales with network densification so that all BSs are active all the time. In practice, the UE density is finite. As the BS density reaches the UE density level, some BSs will not have any UEs to serve and can hence be put into idle mode to avoid interfering with the UEs of the other cells and reduce their energy consumption. This will naturally affect the interference distribution and hence the system throughput.

We will continue to use the PPP assumption for the BS deployment here to provide insights into the performance trends under finite UE density. In particular, assume that the BSs are deployed according to a homogeneous PPP with density λ. The typical UE is associated with the closest BS. The rest of the BSs are interfering. It was shown in [28] that the probability that a typical BS is turned-on (which is equal to the probability that a typical BS has at least one UE in its cell) is given as

$$p_{\text{on}} = 1 - 3.5^{3.5}\left(3.5 + \frac{\lambda_u}{\lambda}\right)^{-3.5}.$$

Hence, the active BS density is given as $\lambda_a = \lambda p_{\text{on}}$. It can be quickly verified that

$$p_{\text{on}} \rightarrow 0 \text{ and } \lambda_a \rightarrow \lambda_u \text{ as } \lambda \rightarrow \infty. \tag{5.17}$$

Equation (5.8) can be modified to get the SINR coverage probability for the finite UE density scenario as

$$\begin{aligned} p_c(\lambda, \alpha) &= \pi\lambda \int_0^\infty \exp\left(-\pi\lambda v(1 + p_{\text{on}}\rho(\gamma_s, \alpha)) - \gamma_s\sigma^2 v^{\alpha/2}/p\right)dv \\ &= \pi \int_0^\infty \exp\left(-\pi v(1 + p_{\text{on}}\rho(\gamma_s, \alpha)) - \gamma_s\sigma^2 v^{\alpha/2}\lambda^{-\alpha/2}/p\right)dv, \end{aligned} \tag{5.18}$$

and the SIR coverage probability is given as

$$p_{cI} = \frac{1}{1 + p_{\text{on}}\rho(\gamma_s, \alpha)}. \tag{5.19}$$

The expression (5.18) indicates that the distribution of distance from the serving BS is the same as in the case with infinite UE density. This is because the typical UE still connects to the closest BS from the original PPP. However, compared to the infinite UE density case, the interfering BS density reduces to λ_a when the UE density is finite and therefore the aggregate interference is less.

Note from (5.17) and (5.19) that as $\lambda \rightarrow \infty$, $p_{cI} \rightarrow 1$. This is due to the fact that as the network densifies, the serving BS comes statistically closer to the UE while interfering BS density remains constant at λ_u. This indicates that the SINR invariance does not hold here and densification can in fact improve coverage gains. However, as we will see next, the throughput tells a very different story. For finite UE density case, the potential throughput is given as

$$\tau = \lambda_a \log(1 + \gamma_s)p_c(\gamma_s, \alpha).$$

Note the scaling term λ_a instead of λ owing to the fact that the number of transmissions is equal to the number of active BSs. As the network is densified, the coverage probability increases, but the τ is still upper bounded by $\lambda_u \log(1 + \gamma_s)$. Once the throughput approaches this value, further densification will not give any gains [29]. The same behavior can be observed in the simulation results shown in Fig. 5.7.

Apart from the theoretical bounds on achievable rate, the assumption of finite UE density also raises some practical issues, e.g., loss of multi-user diversity [7]. At large BS density, there would be only one UE in the cell of each active BS.

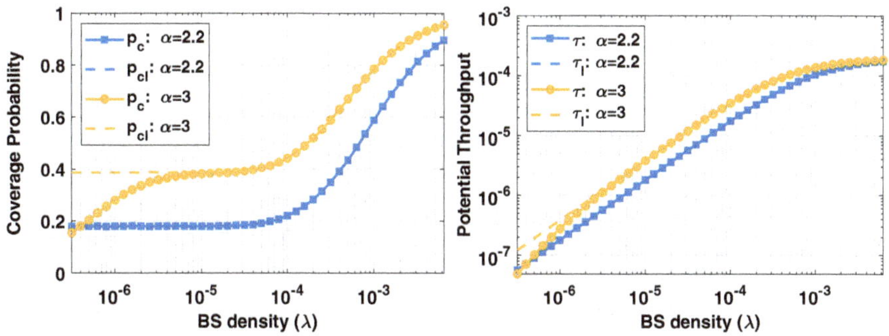

Fig. 5.7 Scaling of the coverage probability and potential throughput with the BS density for a cellular network with the fixed UE density of 200 UE/km^2, under the single slope path-loss. Here, $\gamma_s = 1$, $C = 10^{-4}$ and the path-loss exponent values are taken as $\alpha = 2.2$ and 3. Dashed lines represent respective metrics in the absence of noise

Therefore, BSs lose the opportunity to select the best UE among all connected UEs, which can reduce the practical gains achievable via densification [30].

5.4.4 Traffic Characteristics

The UE density represents the average active traffic load in the network. We have seen in the previous subsection that this average active traffic load has a significant impact on the densification gain. Apart from the average characteristics, instantaneous traffic patterns will naturally have an impact on the densification conclusions as well.

The activity time of a BS depends on the traffic pattern of the UEs served by it. In case of bursty traffic, or when the traffic is sparsely distributed over space, many BSs will not have any traffic requests in their queue. This can reduce the sum interference significantly even when the BS density is high. When there are few active UEs in any BS cell, the downlink and uplink traffic demands become highly dynamic in that cell [7]. As another consequence, the ratio of downlink and uplink traffic becomes highly asymmetric over the space. In such scenario, the same division of resources between uplink and downlink for each cell may lead to inefficient utilization of available resources. To tackle such traffic, dynamic time-division duplex (TDD) has emerged as a promising technology for ultra-dense networks. Dynamic TDD can be seen as a hybrid technology between the conventional half-duplex and the emerging full duplex networks. In dynamic TDD, each BS has the flexibility to choose a custom division of resources between downlink and uplink to match the traffic demands in its cell. However, since the resource division is no longer synchronized among neighboring cells, the communication suffers from cross-link interference. For example, the uplink transmission in a cell may face strong interference from the downlink transmission occurring in the neighboring cell which may degrade the reception significantly.

Since uplink and downlink interference distributions are very different from each other, they can affect the densification gains. As network densifies, downlink interference can severely degrade uplink performance and may render uplink communication unusable. It has also been shown that Dynamic TDD can give significant gains when the mean number of uplink (or downlink) UEs per active BS is less than 1 [31]. This scenario can occur when the uplink/downlink traffic is asymmetric and the network density reaches the UE density. This performance can be improved with the help of interference cancellation and UE power boosting. However, at high network density, the implementation of these schemes may require large overhead, which can eat away all the gains.

5.4.5 Blockages

Blockages can affect both the serving and interfering links, especially at very high transmission frequencies, such as millimeter waves. The impact of blockages can be modeled using the probabilistic two regime (LOS/NLOS) model discussed already in Sect. 5.4.1. Therefore, there exists an optimal BS density at which coverage probability is maximized. This optimal density λ_a ensures that there is a significant probability of having one LOS serving BS while restricting the probability of having a LOS interfering BS. Therefore, λ_a depends on the blockage probability. As blockage probability increases, adequate densification is required to increase the probability of getting at least one BS as LOS which can act as the serving BS [32]. The readers are advised to refer to Sect. 5.4.1 for a more detailed discussion.

5.4.6 Deployment

Most of the existing literature focusing on densification gains considers BS deployment in 2D space. However, in cities (especially, dense downtown areas), BSs are also deployed in the vertical direction, for example, one at each floor of the building. These BSs mainly include user-installed small cells. A user located in the middle floor of a tall building in such an urban environment would see an appreciable number of BSs in every direction. This would seem like a 3D deployment of BSs to these users. Moreover, the increasing maturity of unmanned aerial vehicle (UAV)-assisted communication networks also increases the relevance of 3D networks [33]. Naturally, 3D deployments will have an impact on the densification gains of cellular networks.

The work [34] discussed scaling of densification for general BS deployment in d dimension. For a general d-D network, the required condition for the bounded interference (in almost sure sense) is $\alpha > d$. If the path-loss exponent $\alpha \leq d$, the coverage probability and throughput are both 0. Under the dual-slope path-loss model, the SIR and SINR coverage probability of a general $d-$D system go to zero as $\lambda \to \infty$ for $\alpha_0 \leq d$. As $\lambda \to \infty$, the potential throughput τ exhibit the following scaling behavior:

1. τ grows linearly with λ if $\alpha_0 > d$,
2. τ grows sublinearly with rate $\lambda^{\left(2-\frac{d}{\alpha_0}\right)}$ if $\frac{d}{2} < \alpha_0 < d$,
3. τ decays to zero if $\alpha_0 < \frac{d}{2}$.

For the 3D scenario, the critical value of α_0 is 3. In other words, p_c goes to zero for $\alpha_0 \leq 3$ as BS density goes to infinity. Potential throughput goes to zero if $\alpha_0 < 1.5$. When $1.5 < \alpha_0 < 3$, densification gives sublinear gains to the potential throughput. It is very common for the path-loss exponent of short range systems to be less than these α_0 values, so this is seemingly an important concern for future ultra-dense networks. Figure 5.8 shows the behavior of the coverage probability and potential throughput for a 3D BS deployment with the network density.

5.4.7 Directional Communication

The use of multiple antennas can help improve the performance of wireless systems by providing directionality gains. Directional communication increases the serving power and reduces the aggregate interference. For higher frequencies such as the millimeter waves, directional communication is essential to facilitate reliable communication owing to high propagation losses. As the directionality can improve

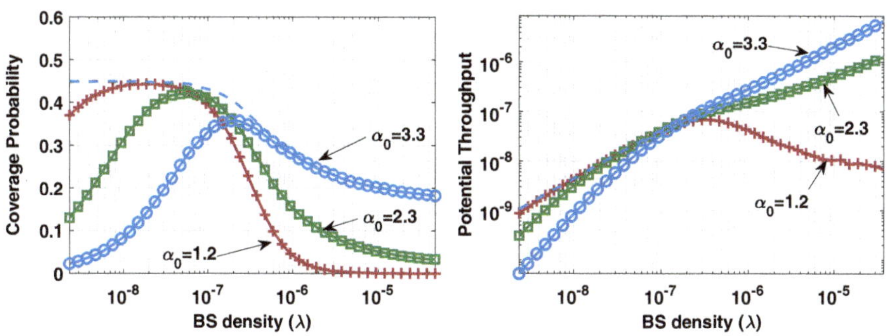

Fig. 5.8 Impact of the BS densification on the SINR and potential throughput for a 3D BS deployment. The path-loss is dual-slope with $R_c = 100\,\text{m}$, $C_0 = 10^{-7}$, $\alpha_0 = 1.2, 2.3, 3.3$, and $\alpha_1 = 4.95$. Dashed lines represent respective metrics in the absence of noise. The behavior is similar to that observed for 2D deployments; however, the critical values of parameters have changed. Dashed lines represent respective metrics in the absence of noise

SINR coverage, network densification gains would increase especially at high densities [35]. Although the introduced directional gain doesn't change the inherent behavior of the scaling laws under densification, it can delay the potential SINR and throughput crashes.

5.4.8 Association Criterion

For systems under single-slope path-loss, we have considered the association criterion that a UE is associated with the BS which provides the highest average received power. There are other association criteria that can be used based on the design objective. One such example is the instantaneous signal power-based association which includes the random fading into consideration while selecting the serving BS. Figure 5.9 shows the comparison of these two association criteria. Instantaneous criterion provides a certain gain to SIR coverage probability. However, at lower density, this gain is not visible in SINR coverage probability. As network densifies, the association criterion needs to be carefully selected.

For millimeter wave systems, an appropriate path-loss model to consider is the probabilistic LOS/NLOS regime model discussed in Sect. 5.4.1. As has been done for the other path-loss models until now, one possible association criterion is to select the closest BS. Figure 5.10 presents scaling results under closest distance-based association compared to the highest average received power-based association. We can observe here that at moderate density of BSs, the distance-based association degrades SINR coverage probability significantly. We also observe that there may not be any optimal BS density which maximizes SIR unlike the case where received power-based association is applied.

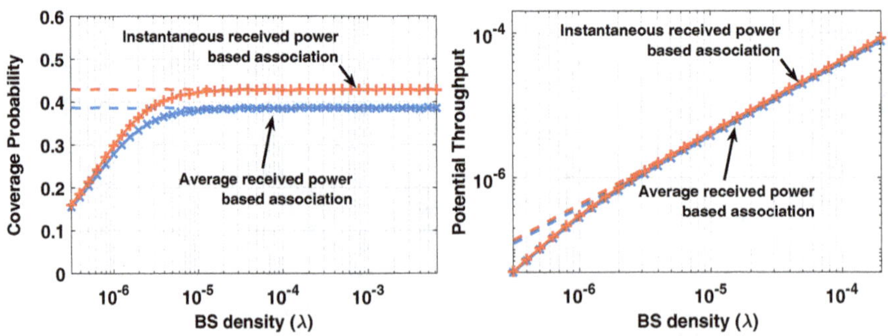

Fig. 5.9 Impact of different association criteria on the densification gain for a cellular network under single slope path-loss model with $C = 10^{-4}$ and $\alpha = 3$. Here, $\gamma_s = 1$. Dashed lines represent respective metrics in the absence of noise

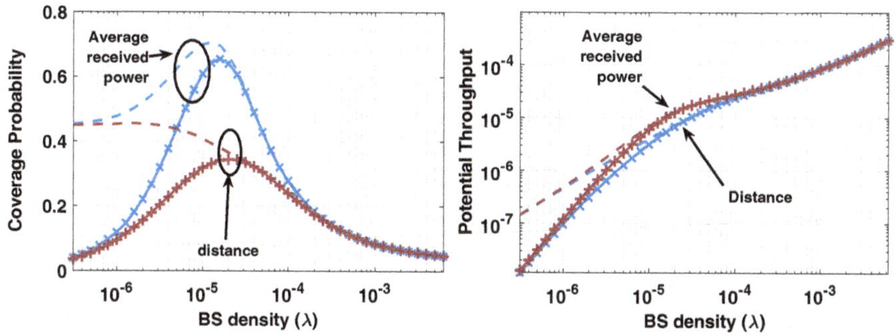

Fig. 5.10 Impact of the implementation of different association criteria on the densification gain for a cellular network under the probabilistic two-regime path-loss model with $\alpha_{\rm L} = 1.5, \alpha_{\rm N} = 3.3$, $C_{\rm L} = 10^{-6}$, and $C_{\rm N} = 10^{-7}$. Here, the LOS probability model is assumed to be exponential, i.e., $p_{\rm L}(r) = \exp(-\beta r)$ with $\beta = 1/144\,{\rm m}^{-1}$. $\gamma_{\rm s} = 1$. Dashed lines represent respective metrics in the absence of noise

5.4.9 Access Restrictions in Multi-tier Networks

As discussed in Sect. 5.1, early deployments of cellular networks involved carefully planned set of large tower-mounted macrocells. As the data demand increased, both service providers and end users have started deploying small BSs, typically in the form of micro-, pico-, and femto-cells, which may share the same spectrum. Each of the tiers is distinguished by its transmit power, BS density, transmission techniques, height, and deployment. Such network consisting of multiple classes (tiers) of BSs is known as a *heterogeneous network (HetNet)* [17, 36]. In the presence of other BS tiers, a cellular network will suffer from additional interference created by their transmissions. Oftentimes, however, UEs of a cellular network may be allowed to connect and use the services of small cells, which can provide additional gain. Whether or not a UE is allowed to connect to small cells (some of which may be owned by the other users) is determined by the *access strategy* [28].

In open access, a UE with subscription to the considered network is allowed to connect to any of the tiers without any restriction. On the other hand, in a closed access strategy, the UE is allowed to connect only to a selected tiers. This access method is mainly used in private infrastructures, e.g., to provide services to its members in a private club. Closed access strategies are often inspired by finite backhaul capacity, security concerns, and the need to reduce the number of hand-offs experienced by UEs as well as the associated signaling overhead.

In both the access schemes, the subscribed tier and all the other tiers will cause interference to the typical UE as they all use the same spectrum. However, in open access, the associated BS is the *best* BS among BSs of all tiers for this UE. On the other hand, in closed access, the associated BS is the *best* BS among all the BSs of the tiers that the UE is subscribed to. It is indeed possible in this case that there is a BS in another tier than can provide better service to the UE, but the UE may be restricted to access it. Therefore, closed access by design leads to a lower coverage probability in this setting. To understand the behavior, we go back to the simple model. Let us assume a K tier network with BSs of each tier deployed according to an independent PPP. For simplicity, we will take identical tiers, however each having different BS density λ_i. Consider a UE of the first network. Denote the set of tiers that allow connection to this UE by $\mathcal{I}$. Suppose the combined density of the tiers that it can connect to is $\mu_1 = \sum_{\mathcal{I}} \lambda_i$ while the combined density of tiers closed to this UE is $\mu_2 = \sum_{[1:K]\setminus\mathcal{I}} \lambda_i$. Let use consider single slope path-loss propagation with α path-loss exponent. Under the above assumptions, the SINR coverage probability of a typical UE is given as [37]

$$\mathrm{p_c}(\lambda, \alpha) = \pi\mu_1 \int_0^\infty \exp\left(-\pi v(\mu_1 + \mu_1\rho(\gamma_\mathrm{s}, \alpha) + \mu_2\beta(\alpha)) - \gamma_\mathrm{s}\sigma^2 v^{\alpha/2}/p\right) dv \tag{5.20}$$

where

$$\rho(\gamma, \alpha) = \gamma^{2/\alpha} \int_{\gamma^{-2/\alpha}}^\infty \frac{1}{1+u^{\alpha/2}} du, \quad \beta(\alpha) = \gamma_\mathrm{s}^{2/\alpha} \int_0^\infty \frac{1}{1+u^{\alpha/2}}. \tag{5.21}$$

For $\alpha = 4$, $\rho(\gamma, \alpha) = \sqrt{\gamma} \arctan \sqrt{\gamma}$ and $\beta(\alpha) = \sqrt{\gamma}\pi/2$. For $\alpha = 2$, $\rho(\gamma, \alpha) = \beta(\alpha) = \infty$. The SIR coverage probability of the typical UE is

$$\mathrm{p_{cI}}(\lambda, \alpha) = \frac{1}{1 + \rho(\gamma_\mathrm{s}, \alpha) + \mu_2/\mu_1\beta(\alpha)} \tag{5.22}$$

In symmetrical networks where each tier densifies equally ($\lambda_i = \lambda\ \forall i$), the closed and open access coverage probability are given as

$$\mathrm{p_{cI,closed}}(\lambda, \alpha) = \frac{1}{1 + \rho(\gamma_\mathrm{s}, \alpha) + (K-1)\beta(\alpha)} \leq \frac{1}{1 + K\rho(\gamma_\mathrm{s}, \alpha)}$$

$$\mathrm{p_{cI,open}}(\lambda, \alpha) = \frac{1}{1 + \rho(\gamma_\mathrm{s}, \alpha)}$$

Closed access can reduce the coverage by factor $\approx K$ as seen in Fig. 5.11. However, in both cases, SIR coverage probability is independent of λ.

In asymmetric networks where one tier has higher density than the other, closed access can severely affect the performance of the latter network. Observe the factor μ_2/μ_1 in (5.22). If μ_1 is fixed, densification of the second network can drastically decrease coverage probability of the first tier under closed access scheme. However, under open access, coverage probability does not depend on this densification. To show this, we consider a two-tier network where the density of the first network is fixed at 30 BS km^2. We densify the second network and the behavior of $\mathrm{p_c}$ and τ is

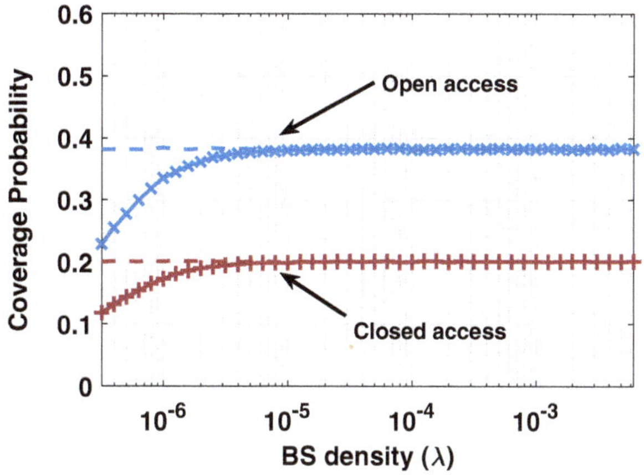

Fig. 5.11 Scaling of the coverage probability with network densification for a two-tier cellular network with different access restrictions. Both networks have the same BS density while densification. The path-loss model is taken as the single slope path-loss with $C = 10^{-4}$ and $\alpha = 3$. Here, $\gamma_s = 1$

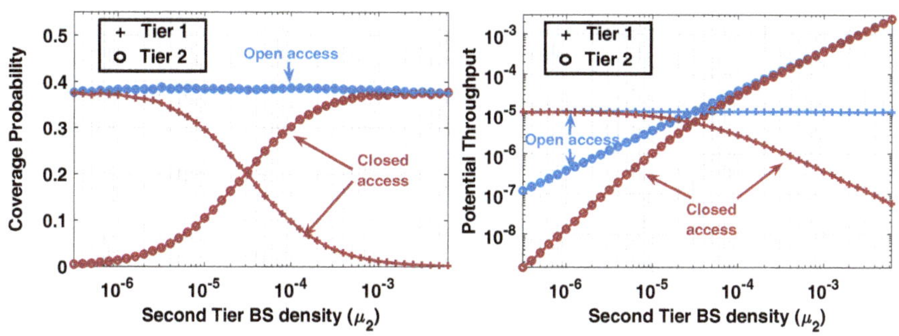

Fig. 5.12 Impact of the BS densification of the second tier on the coverage probability and potential throughput of the first tier in a two-tier asymmetric cellular network with different access restrictions. First tier is fixed density at 30BS/km^2, while the second tier's BS density is varied. The path-loss model is taken as the single slope path-loss with $C = 10^{-4}$ and $\alpha = 3$. Here, $\gamma_s = 1$. Dashed lines represent respective metrics in the absence of noise

shown in Fig. 5.12. In open access, the SIR is invariant to the network density for both the tiers. In closed access, densification of the second tier will cause the first tier's coverage to fall to zero. This is due to the fact that densification of the second tier will increase the interference, while the serving power depends on the first tier's BS density which is fixed. However, the coverage probability of the second tier increases with its densification.

The potential throughput of the first tier remains constant in open access, while it falls to zero in the closed access. However, the throughput of the second tier linearly increases with densification in both access schemes.

This discussion indicates that some tiers are closed for access to some UEs; the densification of one tier can severely affect the other tiers. Therefore, it is important to coordinate the densification of different tiers to ensure reasonable network performance.

5.5 Densification in Modern Networks

In the last section, we described different factors that affect how network performance scales with density and observed their individual impact. In modern networks, many of these factors simultaneously impact the system performance because of which their interplay will decide the asymptotic behavior, which may depart from the scaling performance studied for each factor separately in the previous section. To understand this, let us take a simple example. We observed that a non-zero height difference between the BS and UE could result in zero coverage probability as the network is densified. We also observed that coverage can increase to 1 by densification if the UE density is finite. Clearly, these two factors counter each other, and their cumulative effect depends upon the operational scenarios. This also makes it interesting to investigate the scaling behavior of a network with non-zero height difference and finite UE density, which we do next. For simplicity, we assume the single slope model. Initially, when density is low, the distances along the ground between the typical UE and different BSs will be large compared to the height difference because of which the height can be ignored. When the BS density increases, the SIR remains invariant. As BS density becomes of the order of the UE density, many BSs will start to go in the idle mode (because they do not have any UEs to serve). This reduces the sum interference which increases the SIR coverage probability. At further densification, the number of active BSs becomes constant at λ_u, and hence the potential throughput approaches a constant.

As the BS density approaches ∞, the height difference between the BS and the UE starts showing its effect. In particular, each UE has its serving BS right next to it because of which the serving power approaches a constant value given the lower bound on the path-loss. In addition, since the UE density is finite, the interference statistics would not change with BS densification after BS density gets high enough. This is because each active interfering BS will be located right next to the UE it is serving, and therefore, the point process of the interfering BSs will converge to the point process of the UEs (excluding of course the typical UE). Therefore, the SIR distribution becomes invariant to any further densification. This is contrary to the behavior with finite UE density and zero height difference where SIR coverage probability increases to 1 or under non-zero height difference with infinite UE density where the SIR coverage probability decreases to 0. The potential throughput also remains constant owing to the fact that both the SIR coverage and the active BS density are fixed.

The above discussion necessitates the study involving the interplay of all the factors to understand the exact behavior of a cellular network's performance under densification. In this section, we will consider some case studies where two or more than two factors are considered together.

5.5.1 Finite UE Density Under Multi-slope Path-Loss

We saw that under the single slope path-loss ($\alpha > 2$), with finite UE density, the coverage probability approaches 1. On the other hand, under the multi-slope path-loss with infinite UE density, the coverage probability approaches a constant (may go to zero if $\alpha_0 < 2$). Similarly, for the first case, the potential throughput approaches the constant value, whereas for the second case, the potential throughput increases linearly if $\alpha_0 > 2$, sublinearly if $1 \leq \alpha_0 < 2$ or even go to zero if $\alpha_0 < 1$.

When the UE density is finite, we observe a hybrid behavior under multi-slope path-loss model, as shown in the Fig. 5.13. With densification, the SIR coverage probability first decreases as interference from the BSs inside the corner distance becomes dominant. After the BS density supersedes the UE density, the interference remains bounded, while the serving power increases. This results in coverage probability becoming 1. The potential throughput, therefore, also increases as p_c increases, and then it approaches $\lambda_u \log(1 + \gamma_s)$ asymptotically. This indicates that the densification beyond a point is not beneficial as the throughput gets saturated. Similar behavior is also observed for probabilistic two regime models and the two 3GPP models discussed earlier in the chapter [38].

5.5.2 Height Difference Between BS and UE Under Multi-slope Path-Loss

Figure 5.14 shows the behavior of network's performance under multi-slope path-loss when there is difference between the heights of the BS and UE antennas. We observe that height difference can result in a severe coverage and throughput crash.

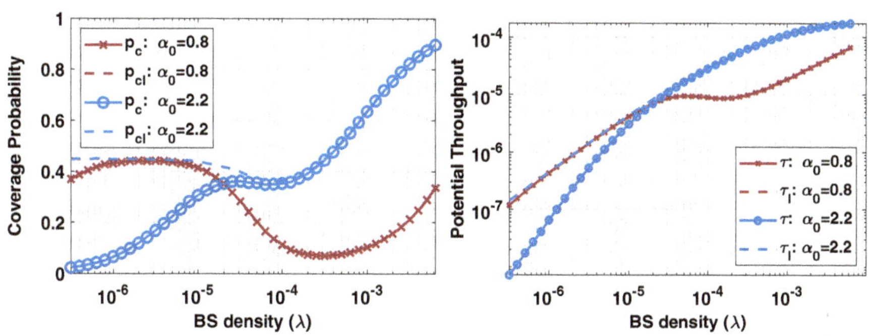

Fig. 5.13 Scaling of the coverage probability and potential throughput under multi-slope path-loss model when the UE density is fixed at 200 UE/km^2. Here, the path-loss model is taken as the dual-slope path-loss with $R_c = 100$ m, $C_0 = 10^{-7}$, $\alpha_0 = 0.8, 2.2$ and $\alpha_1 = 3.3$. $\gamma_s = 1$. Dashed lines represent respective metrics in the absence of noise

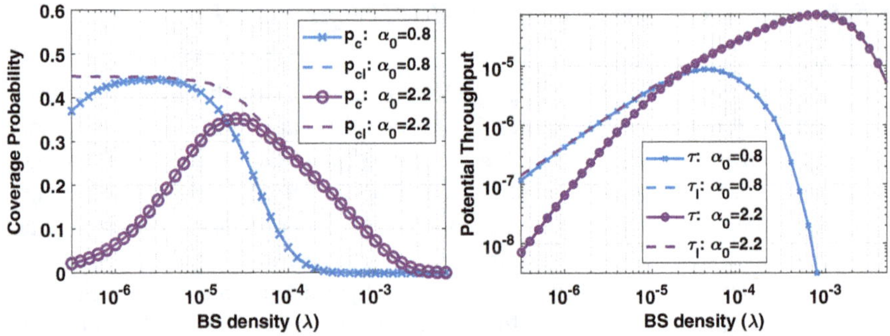

Fig. 5.14 Scaling of the coverage probability and potential throughput under multi-slope path-loss model when there is a height difference of $H = 10\,\text{m}$ between the BS and UE antennas. Here, the path-loss model is taken as the dual-slope path-loss with $R_c = 100\,\text{m}$, $C_0 = 10^{-7}$, $\alpha_0 = 0.8, 2.2$, and $\alpha_1 = 3.3$. $\gamma_s = 1$. Dashed lines represent respective metrics in the absence of noise

Similar behavior is also seen under the probabilistic path-loss model, such as the 3GPP-Model-1 [39, 40].

5.5.3 *Fixed UE Density with Non-zero Height Difference Under Multi-slope and Probabilistic Path-Loss*

A network with fixed UE density and non-zero height difference between BSs and UE was studied in [41] under the probabilistic path-loss model, namely, 3GPP-Model-2. When the BS density is increased, SIR coverage p_{cI} first increases as the serving BS will be LOS with an increasing probability. Then at further densification, p_{cI} decreases as interferers start becoming LOS. As the BS density approaches the UE density, many of the BSs will be inactive, and this puts an upper bound on the interference. The serving power, however, increases with the BS density. Therefore, p_{cI} increases again. When we further densify, after a certain density λ', the serving power gets bounded owing to the fact that the serving BS distance cannot be smaller than the height difference H. The total interference is already bounded. This leads to a constant SIR coverage which may not be 1 and will be invariant of the further densification. The SINR decreases due to the non-zero BS-to-UE antenna height difference H, while it increases due to the BS idleness. This means that the two effects counterbalance each other to some extent. As the active BS density is also bounded, the potential throughput also becomes constant after a certain BS density λ'. Any network densification beyond such a level of BS density is a waste of both money and energy.

Note that in the above discussion, the UE density is finite. If the UE density is infinite (or if it scales with the BS density), the network capacity crashes as discussed before. However, this crash can be avoided if the number of active BSs can be bounded. Instead of serving all users, a fraction of users can be served which will limit the total number of active BSs and hence the interference. This will lead to the similar asymptotic behavior as observed with the finite UE density.

5.5.4 Access Restrictions with Finite UE Density

We now consider a HetNet with two tiers and a finite UE density. Since there are two tiers, we assume that there are equal number of UEs subscribed to each tier. Our aim here is to evaluate the impact of access restriction on coverage and throughput of two tiers. In open access, UEs can connect to any network. Depending on the actual number of UEs associated with each tier, we evaluate the active BS density of that tier. In closed access, the UE can only connect to their own tier; therefore active density of BSs will depend on their UE density only. We first consider the case where both tiers densify equally. The results are shown in Fig. 5.15. For both open and closed access, coverage and throughput follow the same trend as observed in the single tier case with finite UE density. However, as also discussed earlier, closed access can reduce the coverage compared to the open access.

In asymmetric networks where one tier has higher density than the other, closed access can severely affect the performance of the latter network. The results are shown in Fig. 5.16. We consider that the first-tier density is fixed while the second tier densifies. The coverage in open access increases to 1 with densification of the second tier. This is due to the fact that densification of the second tier will cause most UEs to associate with the second tier. The first tier will only get those UEs which have better serving power from the first tier than the potential serving power of the highly dense second tier. The interference on the other hand remains bounded due to the limit on the total number of active BSs.

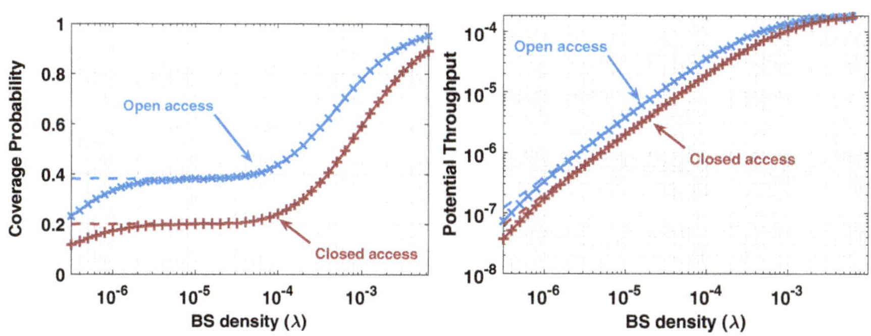

Fig. 5.15 Impact of densification under open and closed access for a two-tier network with the equal scaling of BS density for both tiers ($\mu_1 = \mu_2 = \lambda$) and fixed UE density 200 UE/km^2. The path-loss model is taken as the single slope path-loss with $C = 10^{-4}$ and $\alpha = 3$. Here, $\gamma_s = 1$. Dashed lines represent respective metrics in the absence of noise

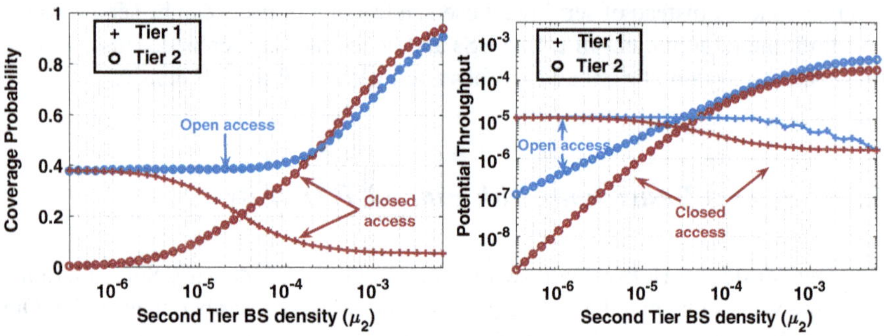

Fig. 5.16 Impact of the BS densification of the second tier on the coverage probability and potential throughput of the first tier in a two-tier asymmetric cellular network with different access restrictions. First tier has the fixed density $\mu_1 = 30\text{BS/km}^2$, while the second tier's BS density (μ_2) is varied. The path-loss model is taken as the single slope path-loss with $C = 10^{-4}$ and $\alpha = 3$. Here, $\gamma_s = 1$

Under closed access, the densification of the second tier drastically decreases coverage probability of the first tier. This is due to the fact that densification of the second tier will increase the interference, while the serving power depends on the first tier's BS density which is fixed. However, the coverage probability of the second tier increases with its densification.

The potential throughput of the first tier goes to zero in open access owing to the fact that the number of UEs associating with the first tier decreases. The potential throughput of the first tier also falls to zero in the closed access, but its fall is earlier than the one observed in open-access scenario, and it saturates afterward. However, the throughput of the second tier linearly increases with densification before saturating in the end, for both access mechanisms.

5.6 Conclusions

Having provided a comprehensive account of the densification gains in a variety of operational scenarios, we get back to the question that we asked early in the chapter: *Is densification the key to the future gains in cellular networks?*. As is the case with many questions in practice, unfortunately, the answer is: *it depends*. In particular, we have seen that while densification helps in many scenarios, it can also cause the SINR and throughput crash in some other scenarios. There are far many important factors that affect the conclusion significantly and may result in drastically different scaling results. Some of these major factors are the path-loss models, the height difference, and scaling of the UE density with the BS density. At the same time, some factors, such as the fading distribution and directionality, may not have as drastic of an impact on the eventual conclusions, especially for the asymptotic results. In practical systems, many of these factors impact system performance simultaneously because of which it is important to understand the interplay between these factors and how they jointly impact the performance. In order to provide key insights about these interplays, we included some important case studies in which two or more such factors were considered jointly. However, the eventual conclusions on the scaling of the network performance still remains highly dependent on the system configuration. This again highlights the importance of using accurate models and operational regimes for the performance analysis of cellular networks.

References

1. Cisco Systems Inc., Cisco visual networking index: global mobile data traffic forecast update, 2015–2020. Growth Lakel. **2011**(4), 2010–2015 (2011)
2. Ericsson, Ericsson mobility report, Nov 2019, accessed: 2020-04-20. [Online]. Available: https://www.ericsson.com/4acd7e/assets/local/mobility-report/documents/2019/emr-november-2019.pdf
3. Arraycomm, Cooper's Law. [Online]. Available: http://www.arraycomm.com/technology/coopers-law
4. N. Bhushan, J. Li, D. Malladi, R. Gilmore, D. Brenner, A. Damnjanovic, R.T. Sukhavasi, C. Patel, S. Geirhofer, Network densification: the dominant theme for wireless evolution into 5G. IEEE Commun. Mag. **52**(2), 82–89 (2014)
5. J.G. Andrews, X. Zhang, G.D. Durgin, A.K. Gupta, Are we approaching the fundamental limits of wireless network densification? IEEE Commun. Mag. **54**(10), 184–190 (2016)
6. X. Ge, S. Tu, G. Mao, C.X. Wang, T. Han, 5G Ultra-Dense Cellular Networks. IEEE Wirel. Commun. **23**(1), 72–79 (2016)
7. M. Ding, D. López-Pérez, H. Claussen, M.A. Kaafar, On the fundamental characteristics of ultra-dense small cell networks. IEEE Netw. **32**(3), 92–100 (2018)
8. A.K. Gupta, A. Alkhateeb, J.G. Andrews, R.W. Heath, Gains of restricted secondary licensing in millimeter wave cellular systems. IEEE J. Sel. Areas Commun. **34**(11), 2935–2950 (2016)
9. J.G. Andrews, F. Baccelli, R.K. Ganti, A tractable approach to coverage and rate in cellular networks. IEEE Trans. Commun. **59**(11), 3122–3134 (2011)
10. J.G. Andrews, A.K. Gupta, H.S. Dhillon, A primer on cellular network analysis using stochastic geometry, arXiv preprint arXiv:1604.03183, 2016
11. A. Guo, M. Haenggi, Asymptotic deployment gain: a simple approach to characterize the sinr distribution in general cellular networks. IEEE Trans. Commun. **63**(3), 962–976 (2015)
12. R.K. Ganti, M. Haenggi, Asymptotics and approximation of the SIR distribution in general cellular networks. IEEE Trans. Wirel. Commun. **15**(3), 2130–2143 (2016)
13. H.P. Keeler, N. Ross, A. Xia, When do wireless network signals appear poisson? arXiv preprint arXiv:1411.3757, 2014
14. N. Ross, D. Schuhmacher, Wireless network signals with moderately correlated shadowing still appear poisson. IEEE Trans. Inf. Theory **63**(2), 1177–1198 (2017)
15. H.S. Dhillon, M. Kountouris, J.G. Andrews, Downlink MIMO HetNets: modeling, ordering results and performance analysis. IEEE Trans. Wireless Commun. **12**(10), 5208–5222, (2013)
16. B. Blaszczyszyn, M.K. Karray, Spatial Distribution of the SINR in Poisson Cellular Networks With Sector Antennas. IEEE Trans. Wirel. Commun. **15**(1), 581–593 (2016)

17. H.S. Dhillon, R.K. Ganti, F. Baccelli, J.G. Andrews, Modeling and analysis of K-tier downlink heterogeneous cellular networks. IEEE J. Sel. Areas Commun. **30**(3), 550–560 (2012)
18. X. Zhang, J.G. Andrews, Downlink cellular network analysis with multi-slope path loss models. IEEE Trans. Commun. **63**(5), 1881–1894 (2015)
19. T. Bai, R.W. Heath Jr., Coverage and rate analysis for millimeter wave cellular networks. IEEE Trans. Wirel. Commun. **14**(2), 1100–1114 (2015)
20. Y. Wu, P. Butovitsch, M. Zhang, Capacity upper bound for adding cells in the super dense cellular deployment scenario, in *Proceedings of the IEEE Vehicular Technology Conference (VTC Spring)*, May 2014, pp. 1–5
21. X. Zhang, J.G. Andrews, Downlink cellular network analysis with a dual-slope path loss model, in *Proceedings of the IEEE International Conference on Communication (ICC)*, vol. 2015, Sept 2015, pp. 3975–3980
22. A.K. Gupta, J.G. Andrews, R.W. Heath, On the feasibility of sharing spectrum licenses in mmWave cellular systems. IEEE Trans. Commun. **64**(9), 3981–3995 (2016)
23. M.N. Kulkarni, E. Visotsky, J.G. Andrews, Correction factor for analysis of mimo wireless networks with highly directional beamforming. IEEE Wireless Commun. Lett. **7**(5), 756–759 (2018)
24. M. Ding, D. Lopez-Perez, G. Mao, P. Wang, Z. Lin, Will the area spectral efficiency monotonically grow as small cells go dense?, in *Proceedings of the IEEE GLOBECOM*, 2015
25. M. Ding, P. Wang, D. López-Pérez, G. Mao, Z. Lin, Performance impact of LoS and NLoS transmissions in dense cellular networks. IEEE Trans. Wirel. Commun. **15**(3), 2365–2380 (2016)
26. 3GPP, 3GPP TR 36.828 (V11. 0.0): further enhancements to LTE Time Division Duplex (TDD) for Downlink-Uplink (DL-UL) interference management and traffic adaptation, Tech. Rep., 2012
27. AHG, Spatial Channel Model, Subsection 3.5. 3, Spatial Channel Model Text Description V6. 0, Tech. Rep., 2003
28. S. Singh, H.S. Dhillon, J.G. Andrews, Offloading in heterogeneous networks: modeling, analysis and design insights. IEEE Trans. Wirel. Commun. **12**(5), 2484–2497 (2013)
29. A.K. Gupta, X. Zhang, J.G. Andrews, Potential throughput in 3D ultradense cellular networks, in *Proceedings of the Asilomar Conference on Signals, System and Computers*, Nov. 2015, pp. 1026–1030
30. D. López-Pérez, M. Ding, H. Claussen, A.H. Jafari, Towards 1 Gbps/UE in cellular systems: understanding ultra-dense small cell deployments. IEEE Commun. Sur. Tutorials **17**(4), 2078–2101 (2015)
31. A.K. Gupta, M.N. Kulkarni, E. Visotsky, F.W. Vook, A. Ghosh, J.G. Andrews, R.W. Heath, Rate analysis and feasibility of dynamic TDD in 5G cellular systems, in *Proceedings of the IEEE International Conference on Communications (ICC)*, May 2016, pp. 1–6
32. J.G. Andrews, T. Bai, M.N. Kulkarni, A. Alkhateeb, A.K. Gupta, R.W. Heath, Modeling and analyzing millimeter wave cellular systems. IEEE Trans. Commun. **65**(1), 403–430 (2017)
33. V.V. Chetlur, H.S. Dhillon, Downlink coverage analysis for a finite 3-D wireless network of unmanned aerial vehicles. IEEE Trans. Commun. **65**(10), 4543–4558 (2017)
34. A.K. Gupta, X. Zhang, J.G. Andrews, SINR and throughput scaling in ultradense urban cellular networks. IEEE Wireless Commun. Lett. **4**(6), 605–608 (2015)
35. A. Merwaday, R. Vannithamby, M.M. Rashid, Z. Yi, C. Clark, X. Wu, On the performance of directional communications in ultra-dense networks, in *Proceedings of the IEEE International Conference on Communication (ICC) Workshop*, 2017, pp. 522–527
36. A.K. Gupta, H.S. Dhillon, S. Vishwanath, J.G. Andrews, Downlink multi-antenna heterogeneous cellular network with load balancing. IEEE Trans. Commun. **62**, 4052–4067 (2014)
37. A.K. Gupta, Asymmetric multi-tier dense networks with access restrictions, 2020, accessed: 2020-04-20. [Online]. Available: http://home.iitk.ac.in/~gkrabhi/asymmtaccess/
38. M. Ding, D. Lopez Perez, G. Mao, Z. Lin, Study on the idle mode capability with LoS and NLoS transmissions, in *Proceedings of the IEEE GLOBECOM*, 2016

39. M. Ding, D.L. Perez, Please lower small cell antenna heights in 5G, in *Proceedings of the IEEE GLOBECOM*, 2016, pp. 1–6
40. M. Ding, D. López-Pérez, Performance impact of base station antenna heights in dense cellular networks. IEEE Trans. Wirel. Commun. **16**(12), 8147–8161 (2017)
41. M. Ding, D. López-Pérez, G. Mao, Z. Lin, Ultra-dense networks: is there a limit to spatial spectrum reuse?, in *Proceedings of the IEEE International Conference on Communication (ICC)*, 2018, pp. 1–6

Chapter 6
UAV-Enabled Cellular Networks

Wonjae Shin and Mojtaba Vaezi

6.1 Introduction

Unmanned aerial vehicles (UAVs) are expected to create more than 100,000 jobs over the 10-year span from 2015 to 2025. UAVs are tomorrow's rules of the sky. UAVs have been used by the military for decades, but there are many other applications for drones: delivery services, infrastructure inspection to search-and-rescue missions, emergency response, disaster relief, health of wildlife, disease control, agriculture, waste management, telecommunication, and Internet access.

UAVs are inherently mobile and, as such, rely on wireless connectivity to support their communication needs for operation (command and control communication) or payload data transmission (e.g., in applications like video streaming for news agencies). For collision avoidance, UAVs may also require to communicate with other nearby UAVs in a distributed manner, i.e., vehicle-to-vehicle (V2V) communication.

This work was supported in part by Basic Science Research Program through the National Research Foundation of Korea (NRF) funded by the Ministry of Science and ICT (NRF-2019R1C1C1006806).

W. Shin (✉)
Department of Electrical and Computer Engineering, Ajou University, Suwon, South Korea
e-mail: wjshin@ajou.ac.kr

M. Vaezi
Electrical and Computer Engineering, Villanova University, Villanova, PA, USA
e-mail: mvaezi@villanova.edu

X. Lin, N. Lee (eds.), *5G and Beyond*, https://doi.org/10.1007/978-3-030-58197-8_6

6.1.1 History of UAV

Drones have been used in the military for over a decade. The first recorded use of a UAV for warfighting occurred in 1849 when the Austrians attacked the Italian city of Venice using unmanned balloons loaded with explosives. In 1915, the British military used aerial photography to their advantage against Germans. The United States began developing UAV technology during the First World War in 1916 and created the first pilotless aircraft. The first massed produced UAV appeared during the Second World War in the United States and was a breakthrough in manufacturing UAVs for the military. In the 1980s, the United States created a program to make inexpensive and unmanned aircraft, which resulted in a medium-sized drone in 1986. In 1990, miniature and micro UAVs were developed.

In 2014, Amazon introduced Amazon Prime Air, a concept which involves using drones for package delivery to customers. A recent Federal Aviation Administration (FAA) report [1] estimates the fleet of drones will reach 2.4 million units by 2022, indicating that it will be more than double in size over 5 years.

6.1.2 UAV Communication Scenarios

In just a few years from now, connected UAVs are expected to perform a large number of tasks in various roles. In tomorrow's wireless networks, UAVs will play various roles such as a user, a base station (BS), or a relay role [2, 3]. These are illustrated in Fig. 6.1 and elaborated on in the following:

- **As users** UAVs need wireless for communication and control. In this role, UAVs introduce a new type of user equipment (UE), referred to as *aerial UEs*. Such a scenario is depicted in Fig. 6.1a in which UAV is communicating with a ground BS. This is the most common use case of UAVs in the current networks.
- **As BSs** UAVs will provide wireless connectivity to the ground or sky. In this role, UAVs introduce a new type of BS called *aerial BS*. As an aerial BS, UAVs are able to adjust their locations swiftly to provide flexible and *on-demand* services to the terrestrial users according to their real-time locations. One application of this scenario in today's network is UAV-assisted disaster management [4]. While such an application results in a sporadic use of UAVs in the current networks, in tomorrow's networks, aerial BS will be ubiquitous and more regular. Figure 6.1b represents such a use case in which two aerial BSs serve terrestrial UEs.
- **As relays** UAVs link aerial or *terrestrial UEs* (users on the ground and inside buildings) to other aerial or terrestrial UEs or vice versa [5, 6]. A use case of UAV as a relay is shown in Fig. 6.1c in which a UAV relay is assisting a terrestrial UE to access a terrestrial BS on the other side of the mountains.

To date, cellular networks have been primarily designed and optimized for terrestrial users, and this is not expected to change radically anytime soon. This

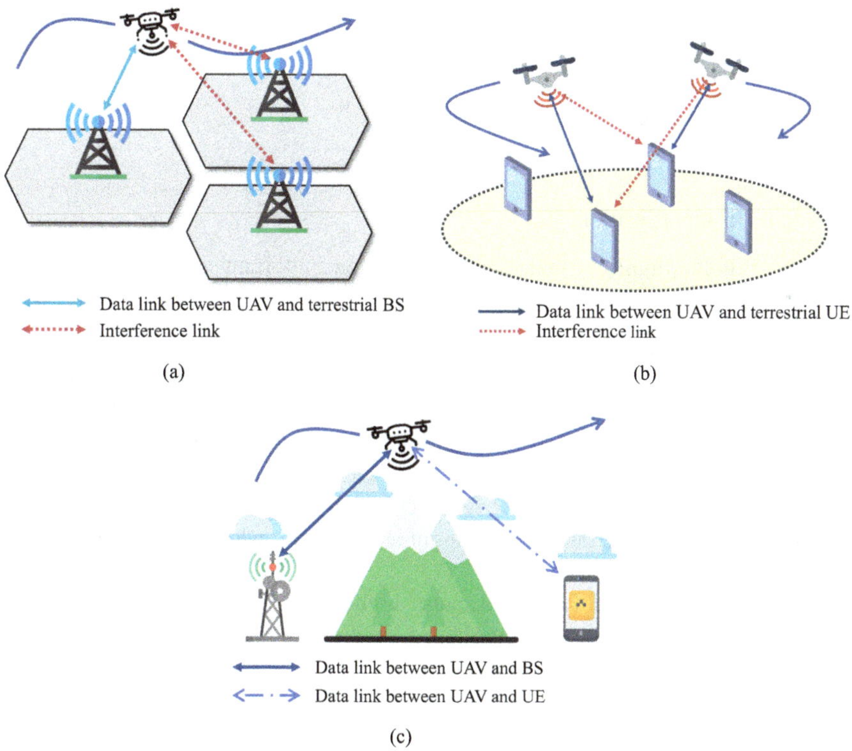

Fig. 6.1 UAV-based cellular communication systems. (**a**) UAV as a user. (**b**) UAV as a BS. (**c**) UAV as a relay

is because a big fraction of the traffic still is expected to be on the ground, even in the future. Given this trend, it is important to understand what is required by the network to fully support UAVs as users, in the short term, and UAVs as BSs and relays, in the long term. The radio environment of aerial UEs in the sky can be very different from that of terrestrial UEs. There are two main factors leading to a different radio environment at higher altitudes: close to free-space propagation and antenna sidelobes [7]. The altitude of UAVs is typically in the range from tens of meters to about 100 m. This makes the favorable *line-of-sight* (LoS) channel without deep fading available for terrestrial users [8]. This is a promising but challenging feature, as we describe it throughout the chapter.

6.1.3 UAV for Beyond 5G: 3GPP Studies

The Third Generation Partnership Project (3GPP) has launched a study item [18] to explore the challenges and opportunities for serving the UAVs as aerial UEs. This study identifies use cases, deployment scenarios, and performance requirements for aerial UEs in the downlink (DL) and uplink (UL). The conclusions of this study item are summarized in four items:

1. DL/UL interference detection
2. DL/UL interference mitigation
3. Mobility
4. Aerial UE identification

The above items highlight that interference is an important issue of UAV-related networks. Some interesting findings are listed below:

- Aerial UEs are more vulnerable than terrestrial UEs to DL interference. This is due to the enhanced LoS between aerial UE and ground BSs, putting them in danger of receiving DL interference from a larger number of cells. As such, compared to terrestrial UEs, a much higher percentage of aerial UEs experience cell-edge-like radio conditions (i.e., poor DL signal-to-interference-plus-noise ratio (SINR)). For example, interference from up to 16 neighboring cells can be experienced by an aerial UE at a height above 50 m, while this number is 8 for terrestrial UEs.
- In the uplink, aerial UEs would cause higher interference to more cells than terrestrial UEs. This is again explained by the fact that aerial UEs experience LoS propagation conditions to more cells with higher probability. Hence, uplink interference increases with the number of aerial UEs since they cause higher uplink interference to multiple cells. This interference degrades the throughput performance of terrestrial UEs too.

A variety of potential solutions for interference detection and mitigation have been proposed too. The main findings are summarized as follows:

- **DL interference detection** can be performed based on measurements reported by the UE, e.g., reference signal received power (RSRP), reference signal received quality (RSRQ), and maximum output power. Other UE-related information, such as mobility history, speed estimation, and timing advance adjustment values, can also be used for interference detection in the network.
- **UL interference detection** may be performed by exchanging information between eNBs.[1] To detect/identify an aerial UE causing interference in UL, eNBs may exchange uplink scheduling information or uplink reference signal (e.g., sounding reference signal (SRS) transmitted by each uplink user).
- **DL interference mitigation** can be achieved in different ways. For example, full-dimension multiple-input and multiple-output (FD-MIMO)[2] [9, 10] is reported to be useful in mitigating DL interference to aerial UEs. Directional, rather than omnidirectional, antenna mitigates DL interference to aerial UEs by decreasing

[1]eNB is an element of an LTE radio access network that communicates with UEs, a BS in LTE.
[2]FD-MIMO is supported since LTE Rel-13.

the interference power coming from a broad range of angles. When aerial UEs have multiple antennas, they can use receive beamforming to mitigate the interference in the downlink. Advanced techniques such as *coordinated multi-point* (CoMP) with *joint transmission* (JT) may also be applied.

- **UL interference mitigation** can also be realized in different ways, including power control-based mechanisms. FD-MIMO with multiple antennas at the eNB is shown to be useful in mitigating interference in the uplink. Directional antennas at the aerial UEs will also reduce uplink interference generated by them.

Simulation and field trial results indicate that mobility-related performance (e.g., handover failure and radio link failure (RLF) handover) of aerial UEs is worse than terrestrial UEs. The above DL and UL interference mitigation techniques are used to improve the mobility performance of aerial UEs. Existing handover procedures can be enhanced to improve the mobility performance.

LTE networks are capable of serving aerial UEs. In 3GPP networks, a permission for a UE to function as an aerial UE can be known from subscription information which is passed to the radio access network (RAN). A flying UE may be identified from the UE-based reporting, e.g., in-flight mode indication, altitude, or location information, by utilizing enhanced measurement reporting mechanism (e.g., the introduction of new events) or by the mobility history information available in the network.

While 3GPP has mostly been concerned with UAVs as aerial UEs, and their connection to cellular networks, industry and academia are exploring the potential of three-dimensional (3D) networks in which UAVs can be users, relays, and BSs. The goal is to boost coverage, spectral efficiency, and user quality of experience for aerial and terrestrial users in the networks.

3GPP has various *study items* and *work items* related to UAVs. The list of study items includes:

1. **UAV traffic requirements:** traffic type can be categorized in one of the following cases: synchronization and radio control, command and control, and application data.
2. **Channel modeling:** based on large measurement campaigns, the 3GPP has proposed rural-macro (RMa), urban-macro (UMa), and urban-micro (UMi) BS deployments models. In academia, however, UAV-related channel models are categorized into air-to-ground (A2G), air-to-air (A2A), and ground-to-air (G2A).
3. **UAV performance analysis:** this study item is about evaluating the performance of cellular networks with both aerial and terrestrial UEs. The conclusion is that UAVs are more likely to experience downlink and uplink interference compared to terrestrial UEs.
4. **Enhancing UAV communications:** this is mostly about addressing interference challenges described previously.

6.2 New Key Features of UAV Communications

6.2.1 Channel Modeling

In this section, we introduce key channel models for UAV communications with an emphasis on new propagation characteristics. A reliable channel modeling is crucial for the design of efficient UAV communication. The common approach in wireless channel modeling is to assume the separability of the large-scale and small-scale channel effects. More specifically, within a coherence time in which the channel can be considered constant, a complex-valued baseband equivalent channel between the transmitter and receiver can be denoted by $g = \sqrt{\beta}h$. Here, the positive real number β represents a large-scale fading arising from the path loss depending on distance and shadowing caused by large obstructions (e.g., building and other large-scale structures in the environment); the complex number h represents a small-scale fading resulting from the constructive and destructive addition of the multiple propagation paths.

In the case of traditional terrestrial networks, the log-distance path loss and Rayleigh fading models are mostly used for the β and h, respectively. For the log-distance path loss model, β can be modeled as

$$\beta\ [\text{dB}] = -\texttt{PL}\ [\text{dB}], \quad \text{where} \quad \texttt{PL}\ [\text{dB}] = 10\alpha \log(d) + X_0 + X_\sigma, \tag{6.1}$$

where d is a distance between the transmitter and receiver in meter; α is the path loss exponent; X_0 is the path loss in dB scale at a reference distance of 1 meter; and X_σ is a normal random variable with zero mean and standard deviation σ that accounts for the shadowing effects.[3]

On the other hand, the UAV channel has the following distinctive characteristics that differ from terrestrial channels: high probability of having an LoS path, high sensitivity to the movement of UAV in 3D space, and communication environment. Therefore, the traditional terrestrial channel model is not suitable for representing UAV channels. Motivated by this, we investigate the dedicated UAV channel models that reflect the characteristics of the UAV channel. UAV channels can be broadly divided into two categories: A2A channels and A2G channels. A2A channel is typically modeled as the simple free-space model due to its high altitude and high LoS opportunity, yet A2G channel is fairly different from A2A channel. This is because the A2G channel is highly dependent on the altitude of the UAV, elevation angle, and type of the propagation environment. In the following, we will discuss several representative models of path loss and small-scale fading for A2A and A2G channels.

[3]Throughout this chapter, $\log(x)$ represents the logarithm of x to the base 10 for ease of exposition.

Path Loss Model

A2A channels mostly experience LoS propagation given that no obstacles exist in the path between the two UAVs in the sky. Hence, a free-space channel model is commonly used for the A2A channel. Although A2G channel is typically different from A2A channel, it can also be modeled as free-space channel model for a special scenario in which LoS path is guaranteed due to high altitude of UAV, or there is almost no blockage and scattering, such as in the rural areas. With this in mind, let us now describe the free-space channel model in more detail.

- **Free-space channel model:** The free-space channel model describes an ideal propagation characteristic under the assumption of no obstructions and reflections between them. Thus, the effect of shadowing is assumed to be negligible due to the dominant LoS link. The free-space path loss is expressed as

$$\mathrm{PL_{FSPL}} = 20\log(d_{3\mathrm{D}}) + 20\log(f) + 20\log\left(\frac{4\pi}{c}\right), \tag{6.2}$$

  where $d_{3\mathrm{D}} = \sqrt{((h_{\mathrm{UAV}} - h_{\mathrm{G}})^2 + {d_{2\mathrm{D}}}^2)}$ and $d_{2\mathrm{D}}$ are a three-dimensional distance and a two-dimensional separation distance between a receiver and a transmitter, respectively (see Fig. 6.2); h_{UAV} and h_{G} denote the heights of the UAV and the ground BS, respectively; f is the signal frequency of interest; and c is the speed of light. The advantage of using this model is that path loss can be easily calculated according to their relative distances.

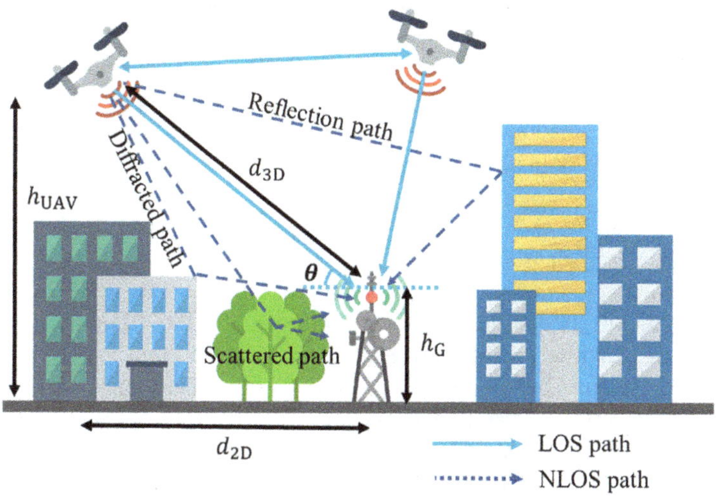

Fig. 6.2 UAV-based wireless communications environment

As for A2G communications, multi-paths may occur in propagation path due to scattering, diffraction, and reflection, especially when the height of UAVs is low or in high-rise buildings environment, as shown in Fig. 6.2. For that reason, free-space channel model is not enough to represent the A2G channel in many cases. The propagation loss in A2G is highly related to the location of UAV in 3D space as well as the surrounding environment (e.g., distribution or height of the building). Also, compared to A2A communication paths experiencing dominant LoS with high probability, A2G channels are more vulnerable to blockage and disconnection by obstacles such as buildings. To reflect these features, more sophisticated channel models based on the altitude of UAV and elevation angle-dependent parameters have been developed as follows:

- **Altitude-dependent path loss model:** When the altitude of UAV is increased, the probability of having a dominant LoS path is increased. Specifically, the effect of signal obstruction and scattering as well as path loss exponent becomes smaller as the altitude of UAV gets higher. Motivated by this, a log-distance path loss model with altitude-dependent parameters was proposed in [11]. The path loss and each parameter are written as

$$\mathtt{PL} = \alpha 10 \log(d_{3\mathrm{D}}) + X_0 + X_\sigma, \tag{6.3}$$

$$\text{where } \alpha = \max(p_{\alpha_1} + p_{\alpha_2} \log(h_{\mathrm{UAV}}), 2), \tag{6.4}$$

$$X_0 = p_{\beta_1} + p_{\beta_2} \log(\min(h_{\mathrm{UAV}}, h_{\mathrm{FSPL}})), \tag{6.5}$$

$$\sigma = p_{\sigma_1} + p_{\sigma_2} \log(\min(h_{\mathrm{UAV}}, h_{\mathrm{FSPL}})), \tag{6.6}$$

  in which $p_{\alpha_1} = 3.9$, $p_{\alpha_2} = -0.9$, $p_{\beta_1} = -8.5$, $p_{\beta_2} = 20.5$, $p_{\sigma_1} = 8.2$, and $p_{\sigma_2} = -2.1$ have been found based on the measurement of real channels, and h_{FSPL} is the height in which free-space propagation is assumed. Recall that X_σ denotes a normal random variable with zero mean and standard deviation σ that accounts for the shadowing effects. It is worth noting that this model suits for the cellular network in rural scenarios with low-altitude UAVs (1.5 m $\leq h_{\mathrm{UAV}} \leq$ 120 m).

  From the above channel model, it can be seen that when the height of the UAV is increased, the standard deviation σ and the path loss exponent α are both reduced. It is noted that the path loss exponent finally converges to 2. As shown in Fig.6.3a, the path loss gets close to free-space path loss $\mathtt{PL}_{\mathrm{FSPL}}$ as the height sufficiently increases regardless of $d_{2\mathrm{D}}$.
- **Probabilistic channel model:** In the A2G channel, there may or may not be an LoS path between UAV and ground node according to the surroundings and infrastructure. In the absence of accurate information on the communication environment, such as the exact locations of communication nodes, number of obstacles or buildings, and geographical environment, the randomness associated with the LoS and non-line-of-sight (NLoS) paths should be taken into account for UAV channel modeling. For that reason, the vast literature on UAV commu-

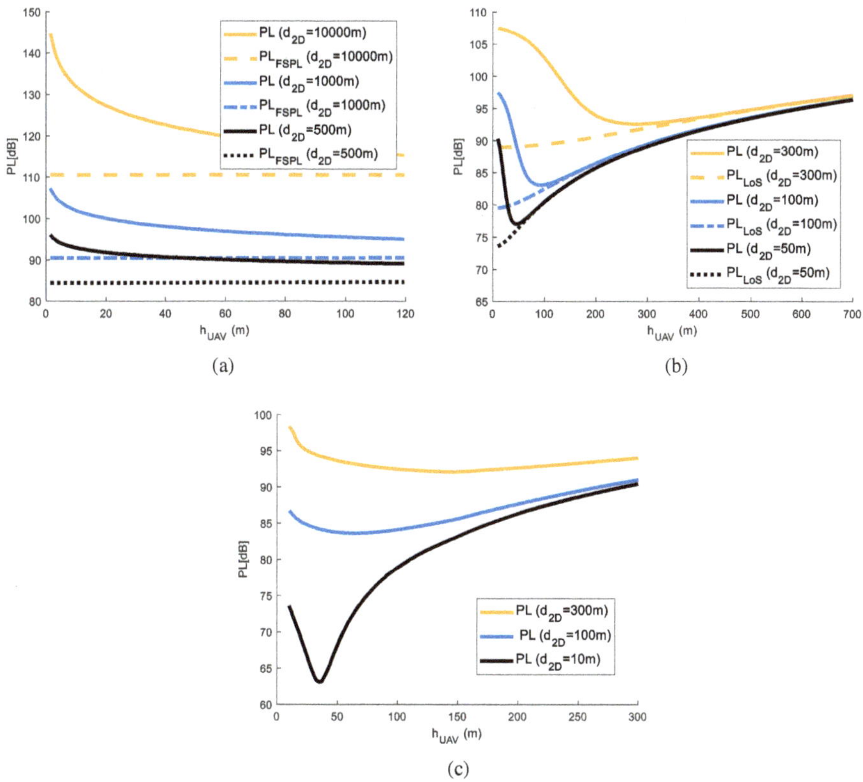

Fig. 6.3 Path loss according to different channel models. (**a**) Altitude-dependent path loss model. (**b**) Probabilistic channel model. (**c**) 3GPP model

nication has adopted the probabilistic path loss model [12–15]. This is a new channel model of integrating two path loss models (LoS and NLoS paths) with consideration of their probability of occurrence. The average path loss for the probabilistic model can be expressed as

$$\mathtt{PL}_{\text{avg}} = P_{\text{LoS}}\mathtt{PL}_{\text{LoS}} + P_{\text{NLoS}}\mathtt{PL}_{\text{NLoS}}, \tag{6.7}$$

where $\mathtt{PL}_{\text{LoS}}$ and $\mathtt{PL}_{\text{NLoS}}$ are path attenuations for the LoS and NLoS paths, respectively; P_{LoS} is the probability of having LoS path; P_{NLoS} is the probability of having NLoS path; and $P_{\text{NLoS}} = 1 - P_{\text{LoS}}$ by definition.

The International Telecommunication Union (ITU) provides a methodology to choose model parameters depending on the characteristics of the environment, such as the density, number, and height of buildings or obstacles [16]. By the use of the parameters provided by the ITU, an elevation angle-dependent probabilistic model is proposed in [17]. In this model, P_{LoS} is given by

$$P_{\text{LoS}} = \frac{1}{1 + a\exp(-b[\theta - a])}, \tag{6.8}$$

where θ is the elevation angle (in degrees) between UAV and ground node as shown in Fig. 6.2. The path loss is modeled as free-space path loss PL_{FSPL} with an excessive path loss component η which is given by

$$\text{PL}_{\text{LoS}} = \text{PL}_{\text{FSPL}} + \eta_{\text{LoS}}, \tag{6.9}$$

$$\text{PL}_{\text{NLoS}} = \text{PL}_{\text{FSPL}} + \eta_{\text{NLoS}}, \tag{6.10}$$

where parameters a, b, η_{LoS}, and η_{NLoS} are determined according to the type of the environment such as suburban, urban, dense urban, or highly dense urban. Since NLoS path tends to undergo more propagation attenuation than LoS path, η_{NLoS} should be set to a larger value than η_{LoS}.

Figure 6.3b shows the path loss according to the height of UAV h_{UAV} in an urban scenario where $a = 9.61$, $b = 0.16$, $\eta_{\text{LoS}} = 1$, and $\eta_{\text{NLoS}} = 20$. When the two-dimensional distance d_{2D} is fixed, an elevation angle θ increases as the height of the UAV h_{UAV} increases. Thus, Fig. 6.3b shows that as the elevation angle θ increases, the path loss converges to the path loss of LoS since the probability of having the LoS path is increased up to 1. Also, it is noted that the height of UAV required for the convergence is somewhat different according to the two-dimensional distance due to its elevation angle.

- **3GPP channel model for aerial vehicles:** The demand for UAV has skyrocketed over recent years, which has led to a 3GPP study on enhanced LTE support for connected UAVs in Release 15. In light of this, 3GPP has defined channel models for aerial vehicles via LTE networks considering three scenarios: rural-macro (RMa), urban-macro (UMa), and urban-micro (UMi) [18].

  Basically, the probabilistic channel model is adopted by the 3GPP channel model. The probability of having LoS path is modeled as a function of d_{2D} and h_{UAV}. When the altitude of UAV is lower than or equal to a specific height threshold h_1, i.e., $h_{\text{UAV}} \leq h_1$, it is reasonable to assume that UAV can be viewed as terrestrial users. Thus, channel modeling for the terrestrial users can be directly used. In contrast to this, when the altitude of UAV is higher than a certain upper threshold h_2, P_{LoS} in RMa and UMa is set to 1 due to the potential dominant LoS path. When $h_1 < h_{\text{UAV}} \leq h_2$, P_{LoS} is expressed as

$$P_{\text{LoS}} = \begin{cases} 1, & d_{\text{2D}} \leq d_1 \\ \frac{d_1}{d_{\text{2D}}} + \exp(\frac{-d_{\text{2D}}}{p_1})(1 - \frac{d_1}{d_{\text{2D}}}), & d_{\text{2D}} > d_1 \end{cases}, \tag{6.11}$$

  where the parameters d_1 and p_1 are functions of h_{UAV}, while h_1 and h_2 are some constant values. These parameters are given according to different scenarios. For example, in RMa scenario, h_1 and h_2 are 10 m and 40 m, respectively; d_1 and p_1 are expressed as

$$d_1 = \max\left(1350.8 \times \log(h_{\text{UAV}}) - 1602, 18\right), \tag{6.12}$$

$$p_1 = \max\left(15021 \times \log(h_{\text{UAV}}) - 16053, 1000\right). \tag{6.13}$$

When $h_1 < h_{\text{UAV}}$, path loss in RMa scenario is modeled as follows:

$$\mathrm{PL}_{\text{LoS}} = \max\left(23.9 - 1.8\log(h_{\text{UAV}}), 20\right)\log(d_{3\text{D}}) + \log(\frac{40\pi f_{\text{GHz}}}{3}),$$

$$\mathrm{PL}_{\text{NLoS}} = \max\left(\mathrm{PL}_{\text{LoS}}, -12 + (35 - 5.3\log(h_{\text{UAV}}))\log(d_{3\text{D}}) + 20\log\left(\frac{40\pi f_{\text{GHz}}}{3}\right)\right),$$

where f_{GHz} is a signal frequency in GHz. In Fig. 6.3c, RMa scenario with the height of BS $= 35$ m, $f_{\text{GHz}} = 3$ GHz is considered, and the average path loss can be obtained by substituting the values of probability and path loss obtained from the above equation into (6.7). This figure shows that as the UAV height increases at a relatively low altitude, the probability of LoS keeps increasing, resulting in a smaller path loss to a certain level. However, when the height is beyond a certain threshold, the path loss rather increases as the UAV gets higher due to the sufficiently long 3D distance $d_{3\text{D}}$.

By combining channel parameters depending on the altitude and probabilistic channel model, the 3GPP model reflects the various features of the ground BS-UAV channel. However, its complicated expressions make it more suitable for numerical simulations rather than theoretical performance analysis.

Antenna Gain

Apart from the path loss, the power of the signal can be affected by the antenna model as well as fading. In the above path loss models, there is an assumption that both the transmitter and receiver are equipped with isotropic antennas (antenna gain $= 0$ dBi) or the antenna gain is negligible. That is, it is assumed that the antenna radiates or receives the same power in all directions for the isotropic antenna model.

However, in practice, uniform radiation cannot be easily implemented in 3D space. Figure 6.4 shows a 3D antenna pattern of cellular BS by using antenna array with eight rows, one column, and 0.8λ vertical antenna element spacing, where λ is the wavelength. It is shown in Fig. 6.4 that there are several sidelobes of the antenna array, the gain of which tends to decrease as theta increases. Indeed, the BS is typically equipped with down-tilted antennas in order to reliably cover its corresponding ground users while mitigating inter-cell interference. When UAV is at high altitudes, it may be served by the sidelobes of the antenna rather than the main lobe of the beam, which results in a lower antenna gain.

Much research has been conducted on modeling the 3D antenna model. Among a variety of models, the two-lobe antenna model [20, 21] is mainly used for simplicity. The two-lobe antenna model considers only two lobes which are the main lobe

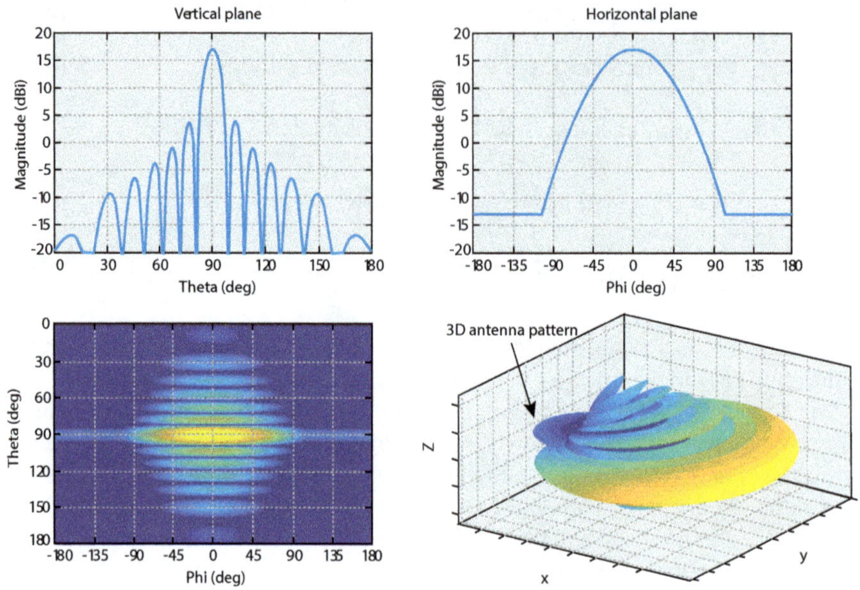

Fig. 6.4 Antenna pattern of the existing cellular BSs [19]

and a single sidelobe. In [22], a path loss model combined with the antenna gain depending on the elevation angle has been developed.

Small-Scale Channel Model

As for small-scale channel models, the existing models for the terrestrial link are commonly used for UAV channels. Rician distribution is usually used to describe the small-scale fading when there is a dominant LoS path, which can be expressed as

$$h = \sqrt{\frac{\kappa}{\kappa+1}} e^{j\theta_s} + \sqrt{\frac{1}{\kappa+1}} h_s, \tag{6.14}$$

where κ denotes the Rician factor. In (6.14), the first term is related to the LoS path arriving with phase θ_s, and the second term is related to a large number of reflected or scattered paths. To be more specific, h_s is assumed to be a complex Gaussian random variable, $h_s \sim \mathrm{CN}(0, 1)$. The Rician K-factor is the ratio between the power in the direct path and the power in scattered paths. A large K-factor means that the LoS path is dominant, and the effect of the scattered path is negligible. If the Rician K-factor, κ is 0 (no dominant path exists), the Rician fading becomes the Rayleigh fading in which $h \sim \mathrm{CN}(0, 1)$.

Due to the high probability of dominant LoS paths, the Rician fading model is usually adopted for UAV communications, unlike conventional terrestrial communications with a rich scattering in which the Rayleigh fading is mainly used. Especially in some studies [23, 24], Rician K-factor is modeled as a function of the height of the UAV for reflecting the effects of ground reflections in A2A and the variation of LoS path strength according to the altitude of UAV in A2G. In addition, other models such as Nakagami-m fading is also taken into account for the small-scale channel model. In Table 6.1, we list several typical small-scale fading models of the literature.

6.2.2 UAV Trajectory Design

In UAV-aided wireless networks, UAVs are expected to move in the 3D space to achieve a better performance metric. A flying trajectory design of UAVs can provide an additional degree of freedom (DoF) for communication performance enhancement by adjusting their locations in 3D locations. As mentioned earlier, the central usage scenarios of UAVs for wireless systems can be classified into three categories: UAV as a relay, UAV as a BS, and UAV as a user. In order to integrate UAVs into existing wireless networks, there has recently been a growing interest in the joint design of UAV trajectory and communication resource allocation for a variety of UAV usage scenarios.

Problem Formulation

To begin with, we now present a general problem formulation for communication system design where the objective is to maximize a utility function such as communication rate over the limited wireless resources. This can be written as

$$\underset{\mathrm{P}_t}{\text{maximize}} \quad u(\mathrm{P}_t) \tag{6.15a}$$

$$\text{Subject to} \quad f_i^{\text{com}}(\mathrm{P}_t) \geq 0, \quad i = 1, \ldots, I_{\text{com}}, \tag{6.15b}$$

where $u(\mathrm{P}_t)$ is the utility function of P_t, a set of communication resources such as power and time/frequency at time instant t to be associated with the problem. The inequalities, $f_i^{\text{com}}(\cdot)$, are expressed as the constraints on wireless networks, and I_{com} denotes the number of associated inequality functions. On the other hand, the optimization problem can also be formulated to minimize a cost function $c(\cdot)$ such as outage probability, i.e., $\min_{\mathrm{P}_t} c(\mathrm{P}_t)$ subject to associated constraints. It is noted that the negative cost function $-c(\mathrm{P}_t)$ can be understood as a utility function $u(\mathrm{P}_t)$, thereby focusing on the utility maximization problem only in the sequel of this chapter.

Table 6.1 Small-scale fading models in the UAV channel

Scenario	Channel	Fading	Parameter	Reference
Suburban	A2G	Rician	Rician K-factor [dB] $= \begin{cases} 3.53 + 0.65h_{\text{UAV}}, 0 < h_{\text{UAV}} \leq 16 \text{ m} \\ 29.6 - 17.4\log(h_{\text{UAV}}), 16 \text{ m} \leq h_{\text{UAV}} \end{cases}$	[23]
Suburban near urban	A2G	Rician	Rician K-factor [dB]: Suburban: $\kappa = 14$ in L-band, $\kappa = 28.5$ in C-band Near urban: $\kappa = 12$ in L-band and $\kappa = 27.4$ in C-band	[25]
Hilly/mountainous	A2G	Rician	Rician K-factor [dB] $= \begin{cases} 12.8 \text{ in the L-band} \\ 29.4 \text{ in the C-band} \end{cases}$	[26]
Outdoor	A2A	Rician	Rician K-factor$= \frac{\rho^2}{2\sigma^2}$ $\rho = 6.469, \sigma = ah_{\text{UAV}}^b + c$ $a = 212.3, b = -2.221, c = 1.289$	[24]
Open field	A2G	Nakagami-m	Nakagami-m factor=4.05	[27]
Suburban	A2G	Nakagami-m	Nakagami-m factor has a log-normal distribution with the mean η and standard deviation ξ $1.12 \leq \eta \leq 1.58, \quad 1.471 \leq \xi \leq 2.705$	[28]

By incorporating UAV trajectory and its associated system parameters into the general communication framework, we can reformulate the optimization problem for UAV-aided wireless communications. The joint optimization problem can be expressed as

$$\underset{\mathrm{P}_t,\mathrm{Q}_t}{\text{maximize}} \quad u(\mathrm{P}_t, \mathrm{Q}_t) \tag{6.16a}$$

$$\text{Subject to} \quad f_i^{\text{com}}(\mathrm{P}_t) \geq 0, \qquad i = 1, \ldots, I_{\text{com}}, \tag{6.16b}$$

$$h_i^{\text{UAV}}(\mathrm{Q}_t) \geq 0, \qquad i = 1, \ldots, I_{\text{UAV}}, \tag{6.16c}$$

$$g_i^{\text{joint}}(\mathrm{P}_t, \mathrm{Q}_t) \geq 0, \quad i = 1, \ldots, I_{\text{joint}}, \tag{6.16d}$$

where at time instant t the trajectory set for M UAVs is denoted by $\mathrm{Q}_t \triangleq \{\mathbf{q}_m(t)| \mathbf{q}_m(t) \triangleq (x(t), y(t), h(t)) \in \mathbb{R}^{3\times 1}, \forall m \in \mathrm{M}\}$, and the location $\mathbf{q}_m(t)$ of the m-th UAV is composed of horizontal locations $(x(t), y(t))$ and a vertical location $h(t)$; we define $\mathrm{M} \triangleq \{1, 2, \ldots, M\}$. The inequalities $h_i^{\text{UAV}}(\mathrm{Q}_t)$ present the constraints on the UAV trajectories, and the inequalities $g_i^{\text{joint}}(\mathrm{P}_t, \mathrm{Q}_t)$ indicate the jointly coupled constraints on both UAV trajectories and communication resources. I_{UAV} and I_{joint} denote the numbers of their corresponding constraints. We assume that the UAV's operating period of interest is T and consider the time instant t where $0 \leq t \leq T$. In some literature, the operating period is discretized into N equal-length time slots, and thus the length of each time slot is $\delta_t = T/N$. For a discrete-time index n, the trajectories for M UAVs can be expressed as $\mathrm{Q}_n \triangleq \left\{\mathbf{q}_m[n]|\mathbf{q}_m[n] \triangleq (x_m[n], y_m[n], h_m[n]) \in \mathbb{R}^{3\times 1}, \forall m \in \mathrm{M}\right\}$.

To better understand the joint optimization approach, let us discuss the related literature on UAV trajectory design, with emphasis on the optimization framework. First of all, we introduce the major utility or cost functions to be optimized in the literature on UAV trajectory designs as follows:

- **Communication data rate**: the achievable rate for the link between UAV and ground user/BS
- **Secrecy rate**: the amount of information that can be sent reliably but also confidentially to legitimate users (UAV or ground user).
- **Outage probability**: the probability of ground user's (or UAV's) signal-to-noise ratio (SNR) being below a predetermined threshold associated with its service requirement.
- **Mission completion time**: the time duration of completing a mission to the UAVs for data transmission to/from ground user/BS.

More details about the object functions for both continuous-time t and discrete-time n in the literature are summarized in Table 6.2.

Next, we introduce several key constraints for the joint optimization of UAV trajectories and communication resource considered in the literature. For ease of exposition, let us focus on the typical constraint on the trajectory of a single UAV, $\mathbf{q}(t)$, unless stated otherwise in the following:

Table 6.2 Objective functions for joint optimization of UAV trajectory and communication resource

	Communication	Secrecy rate	Outage probability	Mission complete time
	throughput [8, 29–36]	[37–39]	probability [40]	complete time [41–46]
Continuous-time	$u(\mathrm{P}_t, \mathrm{Q}_t) =$ $\int_0^T \mathbb{E}[\log_2(1 + \gamma_m(\mathrm{P}_t, \mathrm{Q}_t))]dt$ γ_m : $m -$ th user/UAV's SNR	$u(\mathrm{P}_t, \mathrm{Q}_t)$ $\int_0^T \mathbb{E}\Big[[\log_2(1 + \gamma_{\text{bob}}(\mathrm{P}_t, \mathrm{Q}_t)$ $- \log_2(1 + \gamma_{\text{eve}}(\mathrm{P}_t, \mathrm{Q}_t)]^+\Big]dt$ γ_{bob} : legitimate user's SNR γ_{eve} : eavesdropper's SNR	$c(\mathrm{Q}_t, \mathrm{P}_t)$ $\triangleq \Pr(\gamma_m(\mathrm{P}_t, \mathrm{Q}_t,) < \gamma_{\text{th}})$ γ_{th} : SNR threshold	$c(\mathrm{Q}_t, \mathrm{P}_t) = T$
Discrete-time	$u(\mathrm{P}_n, \mathrm{Q}_n) =$ $\sum_{n=1}^{N} \log_2(1 + \gamma_m(\mathrm{P}_n, \mathrm{Q}_n,)$	$u(\mathrm{P}_n, \mathrm{Q}_n) =$ $\sum_{n=1}^{N}[\log_2(1 + \gamma_{\text{bob}}(\mathrm{P}_n, \mathrm{Q}_n)$ $- \log_2(1 + \gamma_{\text{eve}}(\mathrm{P}_n, \mathrm{Q}_n)]^+$	$c(\mathrm{P}_n, \mathrm{Q}_n)$ $\triangleq \Pr(\gamma_m(\mathrm{P}_n, \mathrm{Q}_n) < \gamma_{\text{th}})$	$c(\mathrm{P}_n, \mathrm{Q}_n) = N$

- **Initial/final UAV's locations**: Specific initial/final locations of UAV can be given as a constraint for the optimization, i.e.,

$$\mathbf{q}(0) = \mathbf{q}_{\text{initial}}, \qquad \mathbf{q}(T) = \mathbf{q}_{\text{final}}, \tag{6.17}$$

  where $\mathbf{q}_{\text{initial}}$ and $\mathbf{q}_{\text{final}} \in \mathbb{R}^{3\times 1}$ are the predetermined initial ($t = 0$) and final ($t = T$) locations, respectively.
- **Maximum speed**: There exist several physical constraints imposed by aircraft's size and weight, e.g., the limited energy, maximum/minimum acceleration, and maximum/minimum speed. Minimum and maximum speed of UAV are two typical constraints for an optimization problem involving UAV trajectory, i.e.,

$$v_{\min} \leq \mathbf{v}(t) \leq v_{\max}, \quad 0 \leq t \leq T, \tag{6.18}$$

  where $\mathbf{v}(t)$ indicates the velocity of UAV at a time t. It is noted that $v_{\min}$ is typically set to 0 in the case of a UAV with rotary-wing, yet $v_{\min} > 0$ in the case of a UAV with fixed-wing due to their different mechanical designs [47].
- **Collision avoidance constraint**: The collision avoidance constraints are largely used to prevent UAVs from crashing with each other. The constraints for M UAVs can be given as follows:

$$d_{\min}^2 \leq \|\mathbf{q}_m(t) - \mathbf{q}_\ell(t)\|^2, \quad \forall m, \ell \in \mathrm{M},\ m \neq \ell \tag{6.19}$$

  where $d_{\min}$ denotes the minimum distance between the m-th UAV and ℓ-th UAV.

Table 6.3 summarizes a broad class of constraints of the joint optimization problems in the literature.

So far, we have studied the general framework for joint trajectory and communication in the context of UAV-aided wireless networks. A specific optimization problem is taken into account as a motivating example in the following:

Example: Trajectory Design for UAV as a Relay

We now take a closer look at a specific example and formulate the trajectory optimization problem for the UAV as a relay, which is one of the most important scenarios among the UAV applications for wireless systems. The motivation behind this scenario is that UAVs can efficiently provide more stable and reliable connectivity between a ground user and a BS due to their high mobility and efficient deployment [48]. As such, UAV is deployed as a relay to achieve seamless coverage for a given geographical area in which a signal blockage by physical objects may occasionally occur.

(i) Problem Description

As shown in Fig. 6.5, we consider a wireless relaying system with a source node (denoted as s), a UAV relay (denoted as r) flying at a fixed altitude h_{UAV} due to some physical limitations, and a destination node (denoted as d). In addition, UAV is assumed to employ the decode-and-forward (DF) relay protocol, where the relay

Table 6.3 Constraint functions for the UAV trajectory and communication resources

UAV trajectory constraints $\left(h_j^{\text{UAV}}\right)$	Communication resources constraints (f_i^{comm})	Joint trajectory and communication constraints $\left(g_p^{\text{joint}}\right)$
▹ Initial/final UAV's locations $\mathbf{q}_m[1] = \mathbf{q}_{\text{initial}},\ \mathbf{q}_m[N] = \mathbf{q}_{\text{final}}$	▹ Average/minimum/maximum power constraints $0 \le p_m[n] \le P_{\max}$ $p_m[n]$: a transmit power of m-th UAV	▹ Information-causality constraints This will be addressed in the example for UAV relay
▹ Maximum speed $v_{\min} \le \lVert\mathbf{v}[n]\rVert \le v_{\max}$	▹ User scheduling and association (UAV as BS) $\sum_{m=1}^{M} \alpha_{k,m}[n] \le 1,\ \forall k, n,$ $\sum_{k=1}^{K} \alpha_{k,m}[n] \le 1,\ \forall m, n,$ $\alpha_{k,m}[n] \in \{0, 1\}$: a user association indicator, i.e., $\alpha_{k,m}[n] = 1$ if user k is associated with UAV m at time n and $\alpha_{k,m}[n] = 0$ otherwise	▹ Minimum SINR constraints $\gamma_m(\mathbf{q}_m[n], p_m[n]) = \dfrac{S(\mathbf{q}_m[n], p_m[n])}{I_{\text{ground}} + I_{\text{aerial}}(\bar{\mathbf{q}}_m[n], \bar{p}_m[n]) + \sigma^2} \ge \Gamma$ $S(\mathbf{q}_m[n], p_m[n])$: the desired signal power I_{ground}: the interference power from exsiting ground transmitters, $I_{\text{aerial}}(\bar{\mathbf{q}}_m[n], \bar{p}_m[n])$: the interference power from other UAVs Γ : SINR threshold associated with quality-of-service (QoS)
▹ Collision avoidance constraints $d_{\min}^2 \le \lVert\mathbf{q}_m[n] - \mathbf{q}_\ell[n]\rVert^2,\ \ \ell \ne m$		
▹ Obstacle avoidance constraints $\lVert\mathbf{q}_m[n] - \mathbf{r}\rVert \ge d_{\min}$ $\mathbf{r}$: a location of the obstacle		
▹ Altitude constraints $H_{\min} \le h_m[n] \le H_{\max}$	▹ Coverage and QoS constraints $\lVert\mathbf{q}_m[n] - \mathbf{g}\rVert \le d$ $\mathbf{g}$: a location of a ground base station d: a coverage radius of a ground base station	

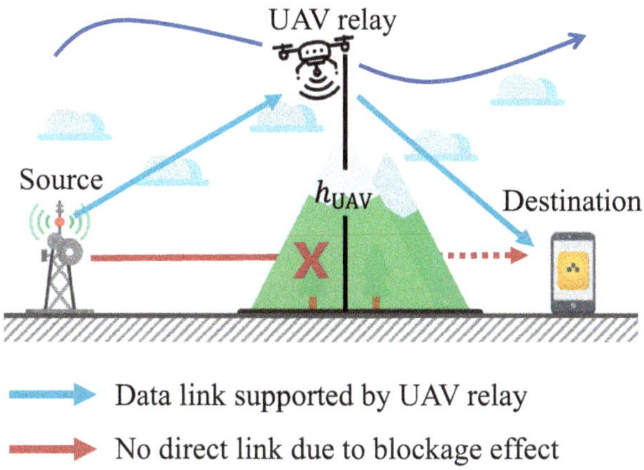

Fig. 6.5 A wireless system assisted by a UAV relay

decodes the signal received from the source and then forwards it to the destination in the subsequent time slot. As mentioned earlier, the free-space channel model can be used for the channel between UAV relay and source/destination nodes due to its dominant LoS links so that the channel power gains can be expressed as

$$\text{s-r link:} \quad h_{\mathsf{r,s}}[n] = \beta_0 d_{\mathsf{r,s}}^{-2}[n] = \frac{\beta_0}{\|\mathbf{q}[n] - \mathbf{w}_{\mathsf{s}}\|^2}, \tag{6.20}$$

$$\text{r-d link:} \quad h_{\mathsf{d,r}}[n] = \beta_0 d_{\mathsf{d,r}}^{-2}[n] = \frac{\beta_0}{\|\mathbf{q}[n] - \mathbf{w}_{\mathsf{d}}\|^2}, \tag{6.21}$$

where β_0 is the channel power for the reference distance d_0; $\mathbf{q}[n] \triangleq (x[n], y[n], h_{\mathrm{UAV}})$, $n = 1, \ldots, N$ denotes the 3D location of the UAV at time slot n; and $h_{m,n}[n]$ and $d_{m,n}$, respectively, present the channel power and the distance between nodes m and n, where $(m, n) \in \{(\mathsf{r}, \mathsf{s}), (\mathsf{d}, \mathsf{r})\}$. In addition, $\mathbf{w}_{\mathsf{s}} = (0, 0, 0)$ and $\mathbf{w}_{\mathsf{d}} = (L, 0, 0)$ are the fixed locations of source and destination nodes, respectively.

The SNR of s-r link (between source and UAV relay) and r-d link (between UAV relay and destination) can be expressed, respectively, as

$$\gamma_{\mathsf{r,s}}[n] = \frac{\gamma_0}{\|\mathbf{q}[n] - \mathbf{w}_{\mathsf{s}}\|^2}, \tag{6.22}$$

$$\gamma_{\mathsf{d,r}}[n] = \frac{\gamma_0}{\|\mathbf{q}[n] - \mathbf{w}_{\mathsf{d}}\|^2}, \tag{6.23}$$

where σ^2 indicates the noise variance at each node and $\gamma_0 \triangleq \beta_0/\sigma^2$ denotes the reference SNR.

Based on the SNR of s-r and r-d links, we can represent the following throughput of them in bits/second/Hz (bps/Hz) for slot n, respectively:

$$R_{\mathsf{s}}[n] = \log_2\left(1 + p_{\mathsf{s}}[n]\gamma_{\mathsf{r,s}}\right), \quad n = 1, \ldots, N, \tag{6.24}$$

$$R_{\mathsf{r}}[n] = \log_2\left(1 + p_{\mathsf{r}}[n]\gamma_{\mathsf{d,r}}\right), \quad n = 1, \ldots, N, \tag{6.25}$$

where $p_m[n]$ denotes the transmission power of node $m \in \{\mathsf{r}, \mathsf{s}\}$ at time n.

In contrast to the conventional static relaying system with instantaneous information forwarding, it is beneficial for relay to buffer the data until the relay arrives at the better position for information forwarding. By considering this, we have *information-causality constraint*, which can be expressed as

$$R_{\mathsf{r}}[1] = 0, \qquad R_{\mathsf{s}}[N] = 0, \tag{6.26}$$

$$\sum_{i=2}^{n} R_{\mathsf{r}}[i] \le \sum_{i=1}^{n-1} R_{\mathsf{s}}[i], \quad n = 2, \ldots, N. \tag{6.27}$$

As previously stated in Table 6.3, a constraint on the movement of UAV should be considered as follows:

$$\|\mathbf{q}[n+1] - \mathbf{q}[n]\|^2 \le v_{\max}^2, \quad n = 1, \ldots, N-1. \tag{6.28}$$

Besides, due to a specific purpose of the UAV operation, the initial and final locations have been given as

$$\mathbf{q}[1] = \mathbf{q}_{\text{initial}}, \qquad \mathbf{q}[N] = \mathbf{q}_{\text{final}}. \tag{6.29}$$

Along with the several constraints, we can formulate the following joint optimization problem of trajectory of UAV and power allocation to maximize the end-to-end throughput from source to destination for a given time duration N:

$$(\mathsf{P1}): \underset{\{Q_n, P_n\}_{n=1}^{n=N}}{\text{maximize}} \quad \sum_{n=2}^{N} \log_2\left(1 + p_{\mathsf{r}}[n]\gamma_{\mathsf{d,r}}[n]\right)$$

$$\text{subject to} \quad \frac{1}{N-1}\sum_{n=1}^{N-1} p_{\mathsf{s}}[n] \le \bar{P}_{\mathsf{s}}, \tag{6.30a}$$

$$\frac{1}{N-1}\sum_{n=2}^{N} p_{\mathsf{r}}[n] \le \bar{P}_{\mathsf{r}}, \tag{6.30b}$$

$$p_{\mathsf{s}}[n] \ge 0, \qquad n = 1, \ldots, N-1, \tag{6.30c}$$

$$p_{\mathsf{r}}[n] \ge 0, \qquad n = 2, \ldots, N, \tag{6.30d}$$

$$(6.27), (6.28), (6.29),$$

where the UAV trajectory is $\left\{Q_n\right\}_{n=1}^{n=N}$ for $Q_n \triangleq \mathbf{q}[n], \forall n$, and the design variables of the transmit power for the source and relay are denoted as $\{P_n\}_{n=1}^{n=N}$ for which $P_n \triangleq \{p_s[n], p_r[n]\}$. (6.30a) and (6.30b) present the average transmit power constraints of the source and relay, respectively. The constraints for the minimum power of the source and relay are denoted as (6.30c) and (6.30d).

Joint optimization problem (P1) is non-convex since it has the non-convex constraint (6.27) with respect to $\left\{Q_n\right\}_{n=1}^{n=N}$. It is difficult for the non-convex problem (P1) to be directly solved in regard to all the coupled variables. To facilitate the optimization, we will introduce an efficient algorithm that enables an optimal trajectory design.

(ii) How to Solve the Problem

The optimization problem of interest is non-convex, and the optimization variables are highly coupled, so a brute-force approach is needed to obtain the globally optimal solution. However, such a method is, in general, computationally infeasible for a large system and does not provide useful system design. In [48], block coordinate descent (BCD) method has been applied to this problem since it is a widely applicable approach in a variety of applications. According to the principle of BCD, all the variables are partitioned into two blocks (subsets), i.e., trajectory block $\{Q_n\}_{n=1}^{n=N}$ and power allocation block $\{P_n\}_{n=1}^{n=N}$. Each iteration of the BCD algorithm [49] consists of successively selecting a single block and updating the block on condition that the remaining blocks are fixed. To this end, we can first express the following subproblem by fixing the trajectory of the relay. Besides, a slack variable $\hat{R}_r[n]$ is introduced to convert the problem into the epigraph form [50]:

$$
\begin{aligned}
(\mathsf{P2}):\quad & \max_{\substack{\{P_n\}_{n=1}^{n=N},\\ \{\hat{R}_r[n]\}_{n=2}^{n=N}}} \sum_{n=2}^{N} \hat{R}_r[n] \\
\text{s.t.}\quad & \sum_{i=2}^{n} \hat{R}_r[i] \le \sum_{i=1}^{n-1} \log_2(1 + p_s[i]\gamma_{s,r}[i]), \quad n = 2, \ldots, N, \qquad (6.31a)\\
& \hat{R}_r[n] \le \log_2(1 + p_r[n]\gamma_{d,r}[n]), \quad n = 2, \ldots, N, \qquad (6.31b)\\
& (6.30a), (6.30b), (6.30c), (6.30d).
\end{aligned}
$$

One can prove that problem (P2) above is convex with respect to $\hat{R}_r[n]$ and $\{P_n\}_{n=1}^{n=N}$, which can be simply solved by the standard convex optimization techniques or existing software tools such as CVX [50].

Let us now turn our attention to optimizing the trajectory design while fixing the power allocation of the resource and relay $\{P_n\}_{n=1}^{n=N}$. By removing the constant terms, the subproblem can be written as

$$(\mathsf{P3}): \quad \max_{\substack{\{\mathrm{Q}_n\}_{n=1}^{n=N},\\ \{\hat{R}_{\mathsf{r}}[n]\}_{n=2}^{n=N}}} \sum_{n=2}^{N} \hat{R}_{\mathsf{r}}[n]$$

$$\text{s.t.} \quad \sum_{i=2}^{n} \hat{R}_{\mathsf{r}}[i] \leq \sum_{i=1}^{n-1} \log_2\left(1 + \frac{\gamma_{\mathsf{s}}[i]}{h_{\mathrm{UAV}}^2 + x^2[n] + y^2[n]}\right), \; n = 2, \ldots, N, \tag{6.32a}$$

$$\hat{R}_{\mathsf{r}}[n] \leq \log_2\left(1 + \frac{\gamma_{\mathsf{r}}[n]}{h_{\mathrm{UAV}}^2 + (x[n] - L)^2 + y^2[n]}\right), \quad n = 2, \ldots, N, \tag{6.32b}$$

$$(6.28), (6.29),$$

where for ease of exposition, $\gamma_{\mathsf{s}}[n] \triangleq p_{\mathsf{s}}[n]\gamma_0$ and $\gamma_{\mathsf{r}}[n] \triangleq p_{\mathsf{r}}[n]\gamma_0$. We can prove that subproblem (P3) is non-convex of $\{\mathrm{Q}_n\}_{n=1}^{n=N}$ due to constraints (6.32a) and (6.32b). To relax the non-convex constraints, we can adopt the successive convex algorithm (SCA) algorithm [51] by which problem (P3) can be efficiently solved. The SCA algorithm is composed of the two basic steps: (1) to approximate a non-concave function as a lower-bounded one and (2) to successively maximize a sequence of the lower bounds.

To begin with, we introduce auxiliary variables to relax the non-convex constraints. Auxiliary variables $\{\delta^{(k)}[n], \zeta^{(k)}[n]\}_{n=1}^{n=N}$ are defined as the trajectory incremental from iteration k to iteration $(k+1)$, i.e., $x^{(k+1)}[n] = x^{(k)}[n] + \delta^{(k)}[n]$ and $y^{(k+1)}[n] = y^{(k)}[n] + \zeta^{(k)}[n]$. The UAV trajectory for iteration k is then expressed as $\{\hat{\mathrm{Q}}_n^{(k)}\}_{n=1}^{n=N}$ where $\hat{\mathrm{Q}}_n^{(k)} \triangleq \mathbf{q}^{(k)}[n], \forall n$.

Next, we exploit the property that function $f(x) = \log_2\left(1 + \frac{\gamma}{a+x}\right)$ for $a, \gamma \geq 0$ is a convex function of x. For a given x_0, inequality $f(x) \geq f(x_0) + f'(x_0)(x - x_0), \forall x$ can be derived by applying the first-order Taylor approximation to function $f(x)$ [50]. When $x_0 = 0$, the inequality is rewritten as

$$\log_2\left(1 + \frac{\gamma}{a+x}\right) \geq \log_2\left(1 + \frac{\gamma}{a}\right) - \frac{\gamma}{a(a+\gamma)\ln 2}x, \quad \forall x. \tag{6.33}$$

For a given $\{x^{(k)}[n], y^{(k)}[n]\}_{n=1}^{n=N}$ in iteration k, the right-hand side (RHS) of (6.32a) can be written as

$$R_{\mathsf{s}}^{(k+1)}[n] \triangleq \log_2\left(1 + \frac{\gamma_{\mathsf{s}}[i]}{h_{\mathrm{UAV}}^2 + x^{(k+1)}[n]^2 + y^{(k+1)}[n]^2}\right) \tag{6.34}$$

$$= \log_2\left(1 + \frac{\gamma_{\mathsf{s}}[i]}{d_{\mathsf{r,s}}^{(k)}[n] + \Delta^{(k)}}\right), \tag{6.35}$$

where $d_{\mathsf{r,s}}^{(k)}[n] \triangleq h_{\mathrm{UAV}}^2 + x^{(k)}[n]^2 + y^{(k)}[n]^2$ and $\Delta^{(k)} \triangleq \delta^{(k)}[n]^2 + \zeta^{(k)}[n]^2 + 2\delta^{(k)}[n]\,x^{(k)}[n] + 2\zeta^{(k)}[n]\,y^{(k)}[n]$. The lower bound of $R_{\mathsf{s}}^{(k+1)}[n]$ is found by the relation between (6.33) and (6.35): $\gamma = \gamma_{\mathsf{s}}[n]$, $a = d_{\mathsf{r,s}}^{(k)}[n]$, and $x = \Delta^{(k)}$. As a result, with the given $x^{(k)}[n]$ and $y^{(k)}[n]$ in iteration $k+1$, the lower bound $R_{\mathsf{s},lb}^{(k+1)}$ can be expressed as

$$\begin{aligned} R_{\mathsf{s}}^{(k+1)}[n] &= \log_2\left(1 + \frac{\gamma_{\mathsf{s}}[i]}{h_{\mathrm{UAV}}^2 + x^{(k+1)}[n]^2 + y^{(k+1)}[n]^2}\right) \\ &\geq -A_{\mathsf{s}}^{(k)}[n]\left(\delta^{(k)}[n]^2 + \zeta^{(k)}[n]^2\right) + B_{\mathsf{s}}^{(k)}[n] \triangleq R_{\mathsf{s},lb}^{(k+1)}[n], \end{aligned} \tag{6.36}$$

$$A_{\mathsf{s}}^{(k)}[n] = \frac{\gamma_{\mathsf{s}}[n]\log_2 e}{d_{\mathsf{r,s}}^{(k)}[n](\gamma_{\mathsf{s}}[n] + d_{\mathsf{r,s}}^{(k)}[n])}, \tag{6.37}$$

$$B_{\mathsf{s}}^{(k)}[n] = R_{\mathsf{s}}^{(k)}[n] - 2x^{(k)}[n]\delta^{(k)}[n]A_{\mathsf{s}}^{(k)}[n] - 2y^{(k)}[n]\zeta^{(k)}[n]A_{\mathsf{s}}^{(k)}[n]. \tag{6.38}$$

In the similar manner as above, we can find a lower bound of the RHS of (6.32b) for iteration $k+1$, denoted as $R_{\mathsf{r},lb}^{(k+1)}[n]$. Due to the space limit, the detail of the lower bound $R_{\mathsf{r},lb}^{(k+1)}[n]$ is omitted [48].

Now, the following relaxed problem can be obtained by replacing the RHS of constraints (6.32a) and (6.32b) with lower-bounded functions $R_{\mathsf{s},lb}^{(k+1)}[n]$ and $R_{\mathsf{r},lb}^{(k+1)}[n]$, respectively:

$$(\mathsf{P4}):\quad \max_{\substack{\{\delta^{(k)}[n],\,\zeta^{(k)}[n]\}_{n=1}^{n=N},\\ \{\hat{R}_{\mathsf{r}}[n]\}_{n=2}^{n=N}}} \sum_{n=2}^{N} \hat{R}_{\mathsf{r}}[n]$$

$$\text{s.t.}\quad \sum_{i=2}^{n} \hat{R}_{\mathsf{r}}[i] \leq \sum_{i=1}^{n-1} R_{\mathsf{s},lb}^{(k+1)}[n], \quad n = 2,\ldots,N, \tag{6.39a}$$

$$\hat{R}_{\mathsf{r}}[n] \leq R_{\mathsf{r},lb}^{(k+1)}[n], \quad n = 2,\ldots,N, \tag{6.39b}$$

$$\|\mathbf{q}^{(k+1)}[n+1] - \mathbf{q}^{(k+1)}[n]\|^2 \leq v_{\max}^2, \quad n = 1,\ldots,N-1, \tag{6.39c}$$

$$\mathbf{q}^{(k+1)}[1] = \mathbf{q}_{\mathrm{initial}},\quad \mathbf{q}^{(k+1)}[N] = \mathbf{q}_{\mathrm{final}}, \tag{6.39d}$$

where $\mathbf{q}^{(k+1)}[n] = \left(x^{(k)}[n] + \delta^{(k)}[n], y^{(k)}[n] + \zeta^{(k)}[n], h_{\text{UAV}}\right)$. We have the properties that lower bounds $R_{\mathsf{s},lb}^{(k+1)}[n]$ and $R_{\mathsf{r},lb}^{(k+1)}[n]$ are concave functions with respect to $\{\delta^{(k)}[n], \ , \zeta^{(k)}[n]\}_{n=1}^{n=N}$, and the super-level set for a concave function is convex [50]. Based on the properties, all the constraints of problem (P4) is convex, and thus problem (P4) is convex. In iteration k, the optimal solution of (P4) can be obtained by the standard convex optimization techniques [50], and then the optimal solution derived in k-th iteration is used for the input of iteration $k+1$. The SCA algorithm for (P3) is summarized in Algorithm 1.

Note that the constraints of the relaxed problem (P4) are tighter than the constraints of (P3) so that the optimal value of (P3) is upper bounded by that of (P4). Also, the value of the optimal solution of (P4) is nondecreasing over each iteration with Algorithm 1. As a result, the numerical stability of Algorithm 1 is ensured.

To relax the non-convex problem to a convex one, the SCA algorithm can be applied to the objective function and constraint, which satisfy the conditions summarized in [51]. Due to the wide application of the SCA algorithm, the algorithm is fairly useful in tackling the non-convex problems of the UAV trajectory [8, 30, 31, 33–36, 38, 39, 46].

Algorithm 1 SCA algorithm for problem (P3)

1: Initialize the UAV trajectory as $\{\hat{\mathrm{Q}}_n^{(k)}\}_{n=1}^{n=N}$, and let $k=0$.
2: **repeat**
3: Given $\{\hat{\mathrm{Q}}_n^{(k)}\}_{n=1}^{n=N}$, find the optimal solution $\{\delta^{\star(k)}[n], \zeta^{\star(k)}[n]\}_{n=1}^{n=N}$ to (P4).
4: Update the UAV trajectory $\{\hat{\mathrm{Q}}_n^{(k+1)}\}_{n=1}^{n=N} \triangleq \left\{x^{(k)}[n] + \delta^{\star(k)}[n], y^{(k)}[n] + \zeta^{\star(k)}[n], h_{\text{UAV}}\right\}_{n=1}^{n=N}$.
5: Update $k = k+1$.
6: **until** convergence of $\{\delta^{\star(k)}[n], \zeta^{\star(k)}[n]\}_{n=1}^{n=N}$ or a maximum number of iterations has been reached

Finally, we can jointly optimize problem (P1) by applying the BCD algorithm. As mention earlier, the whole variables are partitioned into two coordinate blocks, i.e., $\{\mathrm{Q}_n\}_{n=1}^{n=N}$ and $\{\mathrm{P}_n\}_{n=1}^{n=N}$. By solving (P2) and using Algorithm 1, the variables of each block are alternately optimized, and then the obtained outputs are used as inputs in the subsequent iteration.

To be more specific, let us denote the objective function value of (P1) at the beginning of iteration ℓ as $\eta\left(\{\mathrm{Q}^{(\ell)}[n], \mathrm{P}^{(\ell)}[n]\}_{n=1}^{n=N}\right)$. First, for the power allocation block in iteration ℓ, we can obtain $\{\mathrm{P}^{(\ell+1)}[n]\}_{n=1}^{n=N}$ by finding the optimal solution of (P2). We then have

$$\eta\left(\{\mathrm{Q}^{(\ell)}[n], \mathrm{P}^{(\ell)}[n]\}_{n=1}^{n=N}\right) \leq \eta\left(\{\mathrm{Q}^{(\ell)}[n], \mathrm{P}^{(\ell+1)}[n]\}_{n=1}^{n=N}\right). \tag{6.40}$$

Second, for the trajectory block and objective value of (P3), we can obtain $\{Q^{(\ell+1)}[n]\}_{n=1}^{n=N}$ by updating the UAV trajectory and finding the optimal solution of (P4). As a result, it is derived that

$$\eta\left(\{Q^{(\ell)}[n],\ P^{(\ell+1)}[n]\}_{n=1}^{n=N}\right) \leq \eta\left(\{Q^{(\ell+1)}[n],\ P^{(\ell+1)}[n]\}_{n=1}^{n=N}\right). \tag{6.41}$$

The whole process is briefly summarized in Algorithm 2. The convergence of Algorithm 2 is guaranteed by nondecreasing properties of (6.40) and (6.41).

By using the BCD algorithm, we alternately solve the subproblem of each block in iteration ℓ instead of solving a non-convex problem directly; the subproblem is more tractable form as compared to the non-convex problem. Note that the overall framework is of polynomial complexity to be solved. Due to the efficiency of numerical optimization, the BCD algorithm is commonly utilized in the various types of non-convex optimization problems such as UAV trajectory designs [8, 30, 31, 33–36, 38–40, 46].

Algorithm 2 Block coordinate descent algorithm for problem (P1)

1: Initialize a UAV trajectory of (P1) for the constraints and $\ell = 0$.
2: **repeat**
3: Given the UAV trajectory in iteration ℓ, find the optimal power allocation by solving (P2).
4: Given the power allocation in iteration ℓ, update the UAV trajectory by using Algorithm 1.
5: Update $\ell = \ell + 1, \ell \geq 0$.
6: **until** the objective value converges on a stationary point.

6.2.3 Interference-Aware Transmission Design

As depicted in Fig. 6.6, most cellular BSs are equipped with down-tilted antennas in order to focus the power toward the served ground users while mitigating the interference from/to adjacent cells. Due to the down-tilt of the BS, the UAVs with high altitude would be served by the sidelobes rather than the main lobe, which provides a relatively lower antenna gain. As previously stated, the wireless channel between a BS and UAV is dominated by LoS links different from the conventional terrestrial networks. The favorable channel environment is enough to compensate for the reduced antenna gain and further yields a macro-diversity in a variety of UAV usage scenarios, i.e., UAV as a user, UAV as a BS, and UAV as a relay [52].

The LoS-dominant channel is one of the significant opportunities for the UAV-enabled communication scenario. However, there remains a major challenge in exploiting LoS air-to-ground channel, which involves the strong interference imposed by the favorable channel condition in a huge range. To be more specific, in

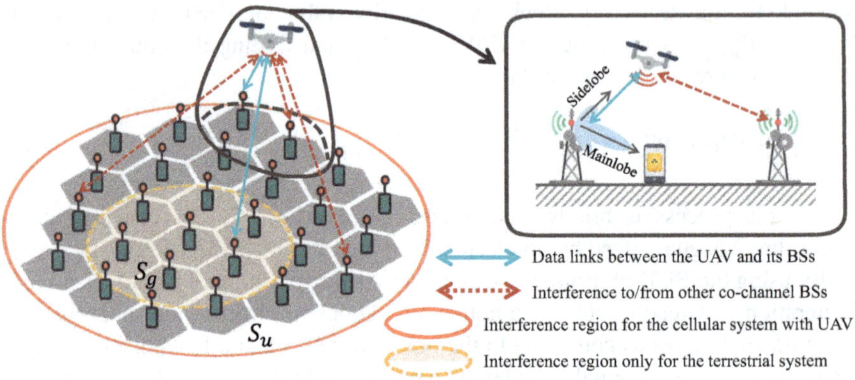

Fig. 6.6 An illustration of interference issues for UAV-based cellular networks

the uplink, the UAV with LoS channel can generate intercell interference (ICI) to an enormous number of co-channel BSs, resulting in a severe loss on data rates of the existing terrestrial users. On the other hand, in the downlink, signals from neighboring BSs in a wide area make the UAV suffer from strong interference, thereby degrading the communication quality of the UAV (see Fig. 6.6). In particular, the interference effect would overwhelm the benefit of a stronger serving link with its associated BS [52]. As a result, interference management is a crucial issue for using UAVs in coexistence with terrestrial cellular systems. Motivated by this, we discuss how to efficiently manage the new type of interference caused by UAVs to ensure the coexistence between the ground and aerial users in the following.

Example: Uplink UAV Communication Scenario

Interference management techniques for ICI have been extensively explored in terrestrial cellular system [53–57]. Much of past research has primarily taken into account the interference caused by the interfering BSs in the first tier owing to the fact that the interference from other tiers is mostly negligible for terrestrial cellular networks (see the interference region S_g in Fig. 6.6). As such, the conventional interference management schemes are insufficient to fully handle the new and more severe interference issues with UAVs at high altitude. Motivated by the above, more sophisticated interference management for a larger region S_u is necessary to achieve the higher spectral efficiency for UAV-enabled wireless communication networks.

We now present an uplink interference management technique for cellular-connected UAV introduced in [58]. As shown in Fig. 6.6, an uplink cellular network is considered, where the terrestrial BSs serve both ground users and a new aerial user with the same spectrum. We assume there are in total J BSs located in the UAV's signal coverage S_u and BS $j \in \mathrm{J} \triangleq \{1, 2, \ldots, J\}$ serves K_j ground users, where $K_j \geq 1$, $\forall j \in \mathrm{J}$ and $K = \sum_{j=1}^{J} K_j$ denotes the number of the entire ground users. For the ease of exposition, it is assumed that the antenna pattern of each BS is fixed, and the UAV centered on the region S_u is with an isotropic antenna pointing downward. Suppose that the total number of orthogonal resource blocks (RBs), the smallest time-frequency resource unit, is N for the uplink scenario, and $N \leq K$ is considered due to frequency reuse.

When assigning an RB to users, it is required to check the availability of an RB $n \in \mathrm{N} \triangleq \{1, 2, \ldots, N\}$ not to generate severe ICI for the uplink communication. To this end, we define $\mathrm{N}_j(q)$, $q \geq 1$ which indicates the set of the first q-tier neighboring BSs of BS j, including itself. To avoid ICI between ground users, BS j may not assign the RB to its serving ground user if the RB is occupied by any other ground users associated with BS ℓ, $\forall \ell \in \mathrm{N}_j(q)$. As a result, BS j does not generate a severe interference toward other occupied BSs in $\mathrm{N}_j(q)$.

The set of occupied BSs for each given RB n is denoted as $\mathrm{J}(n) \subseteq \mathrm{J}$, while the set of unoccupied BSs for that is expressed as $\mathrm{J}^c(n) = \mathrm{J} \backslash \mathrm{J}(n)$. Let $k_j(n)$ be the index of a ground user served by BS j in RB n. The transmit power of ground user $k_j(n)$ is assumed to be $p_{k_j}(n)$, and the channel power gain between the user and BS j in RB n is denoted as $H_j(n)$. $\sigma_j^2(n)$ indicates the total power of background noise for BS j in RB n. With the above notations, we can express the receive SINR of ground user $k_j(n)$ at BS j as

$$\gamma_{k_j}(n) = \frac{p_{k_j}(n) H_{k_j}(n)}{\sigma_j^2(n)}. \tag{6.42}$$

We consider the flying UAV as a new aerial user that accesses the cellular network. The power of channel gain between the UAV and BS j in RB n is denoted as $\tilde{F}_j(n)$. Due to the dominance of LoS propagation for the A2G link, frequency-flat channel is assumed over the spectrum of interest, i.e., $\tilde{F}_j = \tilde{F}_j(n)$, $\forall j, n$. The UAV is assumed to have access to an available RB n to convey the data to one of the unoccupied BSs $j_n \in \mathrm{J}^c(n)$. It is noted that the transmission of UAV with power p_n may cause severe interference to the ground users' signals at the BS $j \in \mathrm{J}(n)$ in the same RBs. Then, the sum rate of all ground users in RB n can be written as

$$R_{g,u}(n) = \sum_{j \in \mathrm{J}(n)} \log_2 \left(1 + \frac{p_{k_j}(n) H_{k_j}(n)}{\sigma_j^2(n) + p_n \tilde{F}_j} \right) \tag{6.43}$$

$$= \sum_{j \in \mathrm{J}(n)} \log_2 \left(1 + \frac{\gamma_{k_j}(n)}{1 + p_n F_j(n)} \right), \tag{6.44}$$

where $F_j(n) \triangleq \tilde{F}_j / \sigma_j^2(n)$, $\forall j, n$. In addition, the achievable rate of the UAV in RB n is denoted as

$$R_u(n) = \log_2(1 + p_n F_{j_n^\star}(n)), \tag{6.45}$$

where $F_{j_n^\star}(n)$ denotes the channel power gain from the UAV to the BS $j_n^\star$ in RB n and $j_n^* = \arg\max_{j \in \mathrm{J}^c(n)} F_j(n)$, $\forall n \in \mathrm{N}$.[4]

To balance between the fairness and the efficiency of the UAV-aided cellular networks, we aim to maximize the weighted sum rate of the UAV and ground user by optimizing the power allocation of UAV across the assigned RBs as follows:

$$
\begin{aligned}
(\mathsf{P5}): \quad & \max_{\{p_n\}_{n\in\mathrm{N}}} \quad \mu_u \sum_{n\in\mathrm{N}} \log_2(1 + p_n F_{j_n^\star}(n)) + \mu_g \sum_{n\in\mathrm{N}} R_{g,u}(n) \\
& \text{s.t.} \quad \sum_{n\in\mathrm{N}} p_n \le P_{\max}, \qquad (6.46\text{a}) \\
& \qquad\quad p_n \ge 0, \ \forall n \in \mathrm{N}, \qquad (6.46\text{b})
\end{aligned}
$$

where $P_{\max}$ is the maximum transmit power at the UAV for the overall RBs. Nonnegative scalars μ_u and μ_g are the weight associated with the UAV and the ground users, respectively.

It is worth noting that problem ($\mathsf{P5}$) is non-concave since the first and the second parts of the objective function are concave and convex, respectively, with respect to the power allocation p_n. By applying the SCA to the non-convex problem, we can effectively obtain a locally optimal solution for the relaxed optimization problem. The basic idea of the SCA is to approximate the non-concave objective function as a concave one given a local point in each iteration. To this end, we note that the second part of the objective function $\sum_{n\in\mathrm{N}} R_{g,u}(n)$ is lower-bounded by the first-order approximation with the given $p_n^{(k)}$ in the k-th iteration due to the property of convex function, which is

$$
\sum_{n\in\mathrm{N}} R_{g,u}(n) \ge A^{(k)} + \sum_{n\in\mathrm{N}} B_n^{(k)}(p_n - p_n^{(k)}), \tag{6.47}
$$

$$
A^{(k)} = \sum_{n\in\mathrm{N}} \sum_{j\in\mathrm{J}(n)} \log_2\left(1 + \frac{\gamma_{k_j}(n)}{1 + p_n^{(k)} F_j(n)}\right), \tag{6.48}
$$

$$
B^{(k)} = -\sum_{j\in\mathrm{J}(n)} \frac{F_j(n)\gamma_{k_j}(n)}{\ln 2(1 + p_n^{(k)} F_j(n) + \gamma_{k_j}(n))(1 + p_n^{(k)} F_j(n))}. \tag{6.49}
$$

Based on the lower bound of (6.47), the problem ($\mathsf{P5}$) can be reformulated as follows:

$$
\begin{aligned}
(\mathsf{P6}): \quad & \max_{\{p_n\}_{n\in\mathrm{N}}} \quad \mu_u \sum_{n\in\mathrm{N}} \log_2(1 + p_n F_{j_n^\star}(n)) + \mu_g \sum_{n\in\mathrm{N}} B_n^{(k)} p_n \\
& \text{s.t.} \quad (6.46\text{a}), (6.46\text{b}),
\end{aligned}
$$

[4]This turns out to be the optimal cell association of the UAV in RB n for the weighted sum rate problem regardless of the UAV's transmit power allocations p_n. This proof is omitted here due to the space limitations. Please refer to [58] for more details.

where the constant terms are omitted in the optimization problem for simplicity.

The problem (P6) is a concave problem which can be efficiently solved by leveraging convex optimization techniques, the Karush-Kuhn-Tucker (KKT) conditions [50]. The near-optimal solution can be obtained by solving the subproblem (P6) of the $\{p_n\}_{n\in N}$ for each iteration step such as in Algorithm 1 mentioned earlier. It is shown in [58] that the SCA-based approximation can achieve a performance close to the primal-dual-based upper bound of the problem optimal value and also significantly improve the performance over existing terrestrial interference management approaches. **Other Approaches**

As can be seen from the aforementioned study, the interference issue is vital for the performance of UAV-based cellular networks. In this context, there are several other approaches that tackle such interference issues arising when integrating UAVs into cellular networks [58–62]. We next briefly introduce two major approaches among them.

- **Cognitive radio approach in the uplink/downlink scenario [61]**
  To overcome the problem of spectrum scarcity in wireless networks, cognitive radio (CR) technique has emerged as a promising solution for efficient use of the spectrum. The integration of CR and UAVs can give rise to a significant benefit, especially for aerial-ground interference management. The UAV in the sky is capable of detecting the terrestrial user's signal over a much larger region than its serving BS on the ground, thanks to the high probability of LoS-dominant A2G connections. In this regard, the RBs assigned to the ground users can be found by using the strong sensing capability of the UAV. To be more specific, in the downlink, the serving BS of the UAV first randomly selects M ($\geq N$) candidate RBs from the available RBs, where N denotes the number of the RBs requested by UAV to satisfy its service requirement. The UAV performs spectrum sensing by measuring its received interference power for each RB n from neighboring BSs that are not associated with it. Then, the N RBs among M candidate RBs are chosen with the lowest interference power measured at the UAV. Similar to the downlink, the CR approach can be adopted in the uplink. M candidate RBs out of all RBs are randomly chosen by the serving BS of the UAV. The UAV next performs spectrum sensing that collects the measured values for its interference power imposed by the UAV in each candidate RB. Finally, the UAV selects the N requested RBs with the lowest measured power among candidate RBs and sends the set of RB index back to its serving BS for uplink transmission. This CR-based approach can be very effective in mitigating the UAV's potential interference that brings a huge benefit for the massive deployment of UAVs.
- **A power-domain NOMA approach in the uplink scenario [59]**
  In order to fulfill massive connectivity, low latency, as well as user fairness for the emerging wireless networks, non-orthogonality-based system designs have been recently developed and gained significant attention of researchers [63–66]. In contrast to orthogonal multiple access (OMA) used in today's wireless networks, non-orthogonal multiple access (NOMA) scheme can share the limited RBs by superposing multiple users' signals with different allocated power

coefficients. Then, the non-orthogonally multiplexed users can be separated by using successive interference cancelation (SIC) at the receiver. This NOMA approach can be applied to UAV-based cellular network in the uplink scenario. This is because the UAV at high altitude typically has much stronger LoS channels with BS than ground users, thereby providing a great opportunity to remove the effect of the strong interference via SIC operation. In [59], a new cooperative NOMA solution is proposed under spectrum sharing between the UAV and existing ground users by exploiting the backhaul links among BSs. By doing so, the proposed framework can handle a much larger region for co-channel BSs as compared to conventional NOMA techniques. Specifically, due to the strong A2G LoS-dominant channels, the UAV can be associated with multiple BSs in its coverage. One (or some) of unoccupied BSs can decode the UAV's signal and forward it to their backhaul-connected BS for interference cancelation. Based on the cooperative NOMA system, the adjacent BSs can cancel the severe interference caused by the UAV, and thus it is expected that the achievable rates for the UAV and ground users are largely improved.

We have addressed the opportunities and challenges generated from new features of UAV communications. However, there still remain many problems awaiting solution. In the following section, we will discuss the interesting problems which are able to be considered as future work.

6.3 Research Challenges and Open Problems

UAV-enabled wireless networks have numerous advantages, including coverage and capacity enhancement, Internet of Things (IoT) support, and fast, flexible, and low-cost deployment particularly for emergency situations and disaster relief. On the other hand, there are various challenges that need further research to be addressed.

- Interference management
- 3D placement optimization
- Channel modeling
- Energy limitation
- UAV antennas
- Simulators and performance analysis
- Mobility management and path planning
- Fronthaul and backhaul connectivity
- Security and privacy issues

In this section, we describe some important future directions.

6.3.1 Interference Management

As discussed earlier, intercell interference in 3D networks is much more severe than that in 2D networks, in both uplink and downlink transmissions. This is caused by the LoS channel between UAVs and ground BSs. As a result of this, not only the interference links will be stronger, but their number will increase rapidly. For example, while the average number of neighbors for a user at the height of 1.5 m is 5, this number increases to 17 at the height of 120 m [11]. To address this important challenge, fundamental and practical research is required. Various interference channel settings such as mixed interference regime [67], in which one of the receivers is subject to strong interference while the other one suffers from weak interference, become more important when one link is LoS while the other one is NLoS. Other information-theoretic models such as symmetric channel [68] may also be relevant to these networks again due to the LoS channel and relatively large distance between the UAV users and ground BSs. In addition to such fundamental models, more practical methods such as CoMP [8] have already been proposed to UAV-based networks. This work proposes joint decoding and optimizes UAV placement and movement designs to improve users' rates. Other cooperative approaches, such as coordinated beamforming [69], may also be useful in mitigating interference. One big challenge is that the number of interfering cells is noticeably high, implying that coordination should be done among a large number of cells.

6.3.2 3D Placement Optimization

The 3D placement of UAVs is one of the main challenges in UAV-based networks and has received significant attention in the literature. This task is challenging because it depends on various factors such as deployment environment (e.g., geographical area), locations of terrestrial and aerial users and A2G channel, and so on. What is more, the simultaneous deployment of multiple UAVs becomes more challenging as it needs intercell interference considerations. While this has been investigated from different perspectives such as maximizing coverage [17], improving energy efficiency of data collection from IoT devices [15], and maximizing the number of terrestrial UEs covered by the UAV [70], more comprehensive studies are needed to better understand 3D deployment and optimize it.

6.3.3 Channel Modeling

As noted earlier, the A2G channel characteristics are remarkably different from those of classical channels in 2D networks. Channel modeling is another critical issue in UAV-based communications which highly affect the performance of UAV-based wireless communications in terms of coverage and capacity. Accurate A2G channel modeling is the first step toward meaningful design and deployment of UAV-based cellular systems. While various independent measurements and works have investigated this problem and proposed channel modeling [11, 22, 71], the accuracy and comprehensiveness of these models need further verification, particularly at new settings such as millimeter-wave frequency operations.

6.3.4 Security and Privacy Issues

UAVs are susceptible to cyber threats such as spoofing and denial of service that face the IoT. They can also be jammed or hijacked for unintended purposes [2]. From a data communication point of view, the security of communication and privacy of users can be compromised in any wireless communication system. However, with the integration of UAVs into wireless networks, the LoS-dominant channels and high mobility of UAVs bring new security challenges in the physical layer where intrinsic characteristics of wireless channels are exploited for security. For example, eavesdroppers may be in positions where their channel is no longer degraded with respect to the main channel (the channel between the legitimate transmitter and receiver). In such cases, information-theoretic secrecy capacity tends to zero if no measure is taken to make legitimate channel better than the other channel [72]. Interestingly, techniques based on *artificial noise* (AN) could be used to degrade the eavesdropper channel [73]. AN-aided techniques are particularly important when the eavesdropper's channel state information (CSI) is not available at the transmitter. Nonetheless, AN injected into directions orthogonal to the main channel can be harmful for other legitimate users in the networks too. Furthermore, due to the mobility of UAV, it will be harder to identify legitimate and malicious nodes in the network. Machine learning-based physical layer security approaches, e.g., [74, 75], are shown to be more agile and could be a promising direction for future research.

References

1. FAA, FAA Aerospace Forecast Forecast Fiscal Years 2017–2038, 2020
2. A. Fotouhi, H. Qiang, M. Ding, M. Hassan, L.G. Giordano, A. Garcia-Rodriguez, J. Yuan, Survey on UAV cellular communications: practical aspects, standardization advancements, regulation, and security challenges. IEEE Commun. Surv. Tutorials **21**(4), 3417–3442 (2019)
3. M. Mozaffari, A.T.Z. Kasgari, W. Saad, M. Bennis, M. Debbah, Beyond 5G with UAVs: foundations of a 3D wireless cellular network. IEEE Trans. Wirel. Commun. **18**(1), 357–372 (2018)
4. M. Erdelj, E. Natalizio, K.R. Chowdhury, I.F. Akyildiz, Help from the sky: leveraging UAVs for disaster management. IEEE Pervasive Comput. **16**(1), 24–32 (2017)
5. M.F. Pinkney, D. Hampel, S. DiPierro, Unmanned aerial vehicle (UAV) communications relay, in *Proceedings of the IEEE Military Communications Conference (MILCOM)*, vol. 1, 1996, pp. 47–51

6. E.P. De Freitas, T. Heimfarth, I.F. Netto, C.E. Lino, C.E. Pereira, A.M. Ferreira, F.R. Wagner, T. Larsson, UAV relay network to support WSN connectivity, in *Proceedings of the IEEE International Congress on Ultra Modern Telecommunications and Control Systems*, 2010, pp. 309–314
7. A. Takacs, X. Lin, S. Hayes, E. Tejedor, Drones and networks: ensuring safe and secure operations, 2018, [Online]. Available: https://www.ericsson.com/en/reports-and-papers/white-papers/drones-and-networks-ensuring-safe-and-secure-operations
8. L. Liu, S. Zhang, R. Zhang, CoMP in the sky: UAV placement and movement optimization for multi-user communications. IEEE Trans. Commun. **67**(8), 5645–5658 (2019)
9. Y.-H. Nam, B.L. Ng, K. Sayana, Y. Li, J. Zhang, Y. Kim, J. Lee, Full-dimension MIMO (FD-MIMO) for next generation cellular technology. IEEE Commun. Mag. **51**(6), 172–179 (2013)
10. B. Mondal, T.A. Thomas, E. Visotsky, F.W. Vook, A. Ghosh, Y.-H. Nam, Y. Li, J. Zhang, M. Zhang, Q. Luo et al., 3D channel model in 3GPP. IEEE Commun. Mag. **53**(3), 16–23 (2015)
11. R. Amorim, H. Nguyen, P. Mogensen, I.Z. Kovács, J. Wigard, T.B. Sørensen, Radio channel modeling for UAV communication over cellular networks. IEEE Wireless Commun. Lett. **6**(4), 514–517 (2017)
12. D. Athukoralage, I. Guvenc, W. Saad, M. Bennis, Regret based learning for UAV assisted LTE-U/WiFi public safety networks, in *Proceedings of the IEEE Global Communications Conference (GLOBECOM)*, 2016, pp. 1–7
13. E. Vinogradov, H. Sallouha, S.D. Bast, M.M. Azari, S. Pollini, Tutorial on UAV: a blue sky view on wireless communication, 2019, [Online]. Available: https://arxiv.org/abs/1901.02306
14. M. Alzenad, A. El-Keyi, H. Yanikomeroglu, 3D placement of an unmanned aerial vehicle base station for maximum coverage of users with different QoS requirements, 2017, [Online]. Available: https://arxiv.org/abs/1709.05235
15. M. Mozaffari, W. Saad, M. Bennis, M. Debbah, Mobile unmanned aerial vehicles (UAVs) for energy-efficient Internet of Things communications. IEEE Trans. Wirel. Commun. **16**(11), 7574–7589 (2017)
16. ITU, Propagation data and prediction methods for the design of terrestrial broadband millimetric radio access systems, Rec. ITU-R. P. 1410-2, 2003
17. A. Al-Hourani, S. Kandeepan, S. Lardner, Optimal LAP altitude for maximum coverage. IEEE Wireless Commun. Lett. **3**(6), 569–572 (2014)
18. 3GPP, Technical specification group radio access network: study on enhanced LTE support for aerial vehicles, 3GPP 36.777 V15.0.0, June 2018
19. X. Lin, V. Yajnanarayana, S.D. Muruganathan, S. Gao, H. Asplund, H.-L. Maattanen, M. Bergstrom, S. Euler, Y.-P.E. Wang, The sky is not the limit: LTE for unmanned aerial vehicles. IEEE Wirel. Commun. Mag. **56**(4), 204–210 (2018)
20. M.M. Azari, F. Rosas, S. Pollin, Reshaping cellular networks for the sky: major factors and feasibility, in *Proceedings of the IEEE International Conference on Communications (ICC)*, 2018, pp. 1–7
21. K. Venugopal, M.C. Valenti, R.W. Heath, Device-to-device millimeter wave communications: interference, coverage, rate, and finite topologies. IEEE Trans. Wirel. Commun. **15**(9), 6175–6188 (2016)
22. A. Al-Hourani, K. Gomez, Modeling cellular-to-UAV path-loss for suburban environments. IEEE Wireless Commun. Lett. **7**(1), 82–85 (2017)
23. X. Ye, X. Cai, X. Yin, J. Rodriguez-Pineiro, L. Tian, J. Dou, Air-to-ground big-data-assisted channel modeling based on passive sounding in LTE networks, in *Proceedings of the 2017 IEEE Globecom Workshops (GC Wkshps)*, 2017, pp. 1–6
24. N. Goddemeier, C. Wietfeld, Investigation of air-to-air channel characteristics and a UAV specific extension to the rice model, in *Proceedings of the 2015 IEEE Globecom Workshops (GC Wkshps)*, 2015, pp. 1–5
25. D.W. Matolak, R. Sun, Air-ground channel characterization for unmanned aircraft systems–part III: the suburban and near-urban environments. IEEE Trans. Veh. Technol. **66**(8), 6607–6618 (2017)

26. R. Sun, D.W. Matolak, Air-ground channel characterization for unmanned aircraft systems part II: hilly and mountainous settings. IEEE Trans. Veh. Technol. **66**(3), 1913–1925 (2017)
27. E. Yanmaz, R. Kuschnig, C. Bettstetter, Achieving air-ground communications in 802.11 networks with three-dimensional aerial mobility, in *Proceedings of the IEEE INFOCOM*, 2013, pp. 120–124
28. W. Khawaja, I. Guvenc, D. Matolak, UWB channel sounding and modeling for UAV air-to-ground propagation channels, in *Proceedings of the IEEE Global Communications Conference (GLOBECOM)*, 2016, pp. 1–7
29. M.M.U. Chowdhury, E. Bulut, I. Guvenc, Trajectory optimization in UAV-assisted cellular networks under mission duration constraint, in *2019 IEEE Radio and Wireless Symposium*, 2019
30. Y. Huang, W. Mei, J. Xu, L. Qiu, R. Zhang, Cognitive UAV communication via joint maneuver and power control. IEEE Trans. Commun. **67**(11), 7872–7888 (2019)
31. Q. Wu, Y. Zeng, R. Zhang, Joint trajectory and communication design for multi-UAV enabled wireless networks. IEEE Trans. Wirel. Commun. **17**(3), 2109–2121 (2018)
32. J. Lyu, Y. Zeng, R. Zhang, Cyclical multiple access in UAV-aided communications: a throughput-delay tradeoff. IEEE Wireless Commun. Lett. **5**(6), 600–603 (2018)
33. H. Wang, G. Ren, J. Chen, G. Ding, Y. Yang, Unmanned aerial vehicle-aided communications: joint transmit power and trajectory optimization. IEEE Wireless Commun. Lett. **7**(4), 522–525 (2018)
34. N. Zhao, X. Pang, Z. Li, Y. Chen, F. Li, Z. Ding, M.-S. Alouini, Joint trajectory and precoding optimization for UAV-assisted NOMA networks. IEEE Trans. Commun. **67**(5), 3723–3735 (2019)
35. X. Jiang, Z. Wu, Z. Yin, Z. Yang, Power and trajectory optimization for UAV-enabled amplify-and-forward relay networks. IEEE Access **6**, 48688–48696 (2018)
36. Y. Chen, N. Zhao, Z. Ding, M.-S. Alouini, Multiple UAVs as relays: multi-hop single link versus multiple dual-hop links. IEEE Trans. Wirel. Commun. **17**(9), 6348–6359 (2018)
37. G. Zhang, Q. Wu, M. Cui, R. Zhang, Securing UAV communications via joint trajectory and power control. IEEE Trans. Wirel. Commun. **18**(2), 1376–1389 (2019)
38. Y. Cai, F. Cui, Q. Shi, M. Zhao, G.Y. Li, Dual-UAV-enabled secure communications: joint trajectory design and user scheduling. IEEE J. Sel. Areas Commun. **36**(9), 1972–1985 (2019)
39. F. Cheng, G. Gui, N. Zhao, Y. Chen, J. Tang, H. Sari, UAV-relaying-assisted secure transmission with caching. IEEE Trans. Commun. **67**(5), 3140–3153 (2019)
40. S. Zhang, H. Zhang, Q. He, K. Bian, L. Song, Joint trajectory and power optimization for UAV relay networks. IEEE Commun. Lett. **22**(1), 161–164 (2017)
41. S. Zhang, Y. Zeng, R. Zhang, Cellular-enabled UAV communication: a connectivity-constrained trajectory optimization perspective. IEEE Trans. Commun. **67**(3), 2580–2604 (2018)
42. E. Bulut, I. Guevenc, Trajectory optimization for cellular-connected UAVs with disconnectivity constraint, in *Proceedings of the IEEE International Conference on Communications (ICC)*, 2018, pp. 1–6
43. S. Zhang, R. Zhang, Trajectory optimization for cellular-connected UAV under outage duration constraint. J. Commun. Inf. Netw. **4**(4), 55–71 (2018)
44. U. Challita, W. Saad, C. Bettstetter, Deep reinforcement learning for interference-aware path planning of cellular-connected UAVs, in *Proceedings of the IEEE International Conference on Communications (ICC)*, 2018, pp. 1–7
45. Y. Zeng, X. Xu, R. Zhang, Trajectory design for completion time minimization in UAV-enabled multicasting. IEEE Trans. Wirel. Commun. **17**(4), 2233–2246 (2018)
46. J. Zhang, Y. Zeng, R. Zhang, UAV-enabled radio access network: multi-mode communication and trajectory design. IEEE Trans. Signal Process. **66**(20), 5269–5284 (2018)
47. Y. Zeng, Q. Wu, R. Zhang, Accessing from the sky: a tutorial on UAV communications for 5G and beyond. Proc. IEEE **107**(12), 2327–2375 (2019)
48. Y. Zeng, R. Zhang, T.J. Lim, Throughput maximization for UAV-enabled mobile relaying systems. IEEE Trans. Commun. **64**(12), 4983–4996 (2016)

49. P. Tseng, Convergence of a block coordinate descent method for nondifferentiable minimization. J. Optim. Theory Appl. **109**(3), 475–494 (2001)
50. S. Boyd, L. Vandenberghe, *Convex Optimization* (Cambridge University Press, Cambridge, UK, 2004)
51. M. Razaviyayn, Successive convex approximation: analysis and applications, Ph.D. dissertation, University of Minnesota, 2014
52. V. Yajnanarayana, Y.-P.E. Wang, S. Gao, S. Muruganathan, X. Lin, Interference mitigation methods for unmanned aerial vehicles served by cellular networks, 2018, [Online]. Available: https://arxiv.org/abs/1802.00223
53. W. Shin, N. Lee, J.-B. Lim, C. Shin, K. Jang, On the design of interference alignment scheme for two-cell MIMO interfering broadcast channels. IEEE Trans. Wirel. Commun. **10**(2), 437–442 (2011)
54. W. Shin, M. Vaezi, B. Lee, D.J. Love, J. Lee, H.V. Poor, Non-orthogonal multiple access in multi-cell networks: theory, performance, and practical challenges. IEEE Commun. Mag. **55**(10), 176–183 (2017)
55. D. Gesbert, S. Hanly, H. Huang, S.S. Shitz, O. Simeone, W. Yu, Multi-cell MIMO cooperative networks: a new look at interference. IEEE J. Sel. Areas Commun. **28**(9), 1380–1408 (2010)
56. C. Suh, M. Ho, D.N.C. Tse, Downlink interference alignment. IEEE Trans. Commun. **59**(9), 2616–2626 (2011)
57. D. Lee, H. Seo, B. Clerckx, E. Hardouin, D. Mazzarese, S. Nagata, K. Sayana, Coordinated multipoint transmission and reception in LTE-advanced: deployment scenarios and operational challenges. IEEE Commun. Mag. **50**(2), 148–155 (2012)
58. W. Mei, Q. Wu, R. Zhang, Cellular-connected UAV: uplink association, power control and interference coordination. IEEE Trans. Wirel. Commun. **18**(11), 5380–5393 (2019)
59. W. Mei, R. Zhang, Uplink cooperative NOMA for cellular-connected UAV. IEEE J. Sel. Topics Signal Process. **13**(3), 644–656 (2019)
60. W. Mei, R. Zhang, Cooperative downlink interference transmission and cancellation for cellular-connected UAV: a divide-and-conquer approach. IEEE Trans. Commun. **68**(2), 1297–1311 (2020)
61. W. Mei, Q. Wu, R. Zhang, UAV-sensing-assisted cellular interference coordination: a cognitive radio approach, 2020, [Online]. Available: https://arxiv.org/abs/2001.01253
62. U. Challita, W. Saad, C. Bettstetter, Interference management for cellular-connected UAVs: a deep reinforcement learning approach. IEEE Trans. Wirel. Commun. **18**(4), 2125–2140 (2019)
63. L. Dai, B. Wang, Y. Yuan, S. Han, C.-l. I, Z. Wang, Non-orthogonal multiple access for 5G: solutions challenges opportunities and future research trends. IEEE Commun. Mag. **53**(9), 74–81 (2015)
64. Y. Saito, Y. Kishiyama, A. Benjebbour, T. Nakamura, A. Li, K. Higuchi, Non-orthogonal multiple access (NOMA) for cellular future radio access, in *Proceedings of the IEEE 77th Vehicular Technology Conference (VTC Spring)*, 2013, pp. 1–5
65. M. Vaezi, Z. Ding, H.V. Poor, *Multiple Access Techniques for 5G Wireless Networks and Beyond* (Cham, Switzerland, Springer, 2019)
66. M. Vaezi, G.A.A. Baduge, Y. Liu, A. Arafa, F. Fang, Z. Ding, Interplay between NOMA and other emerging technologies: a survey. IEEE Trans. Cogn. Commun. Netw. **5**(4), 900–919 (2019)
67. M. Vaezi, H.V. Poor, Simplified Han-Kobayashi region for one-sided and mixed Gaussian interference channels, in *Proceedings of the IEEE International Conference on Communications (ICC)*, 2016, pp. 1–6
68. O. Ordentlich, U. Erez, B. Nazer, The approximate sum capacity of the symmetric Gaussian K-user interference channel. IEEE Trans. Inf. Theory **60**(6), 3450–3482 (2014)
69. W. Shin, M. Vaezi, B. Lee, D.J. Love, J. Lee, H.V. Poor, Coordinated beamforming for multi-cell MIMO-NOMA. IEEE Commun. Lett. **21**(1), 84–87 (2017)
70. I. Bor-Yaliniz, H. Yanikomeroglu, The new frontier in RAN heterogeneity: multi-tier drone-cells. IEEE Commun. Mag. **54**(11), 48–55 (2016)

71. A. Al-Hourani, S. Kandeepan, A. Jamalipour, Modeling air-to-ground path loss for low altitude platforms in urban environments, in *Proceedings of the IEEE Global Communications Conference (GLOBECOM)*, 2014, pp. 2898–2904
72. A.D. Wyner, The wire-tap channel. Bell Syst. Tech. J. **54**(8), 1355–1387 (1975)
73. S. Goel, R. Negi, Guaranteeing secrecy using artificial noise. IEEE Trans. Wirel. Commun. **7**(6) (2008)
74. X. Zhang, M. Vaezi, Deep learning based precoding for the MIMO Gaussian wiretap channel, in *Proceedings of the IEEE Global Communications Conference (GLOBECOM) Workshops*, 2019, pp. 1–6
75. S. Yun, J.-M. Kang, I.-M. Kim, J. Ha, Deep artificial noise: deep learning-based precoding optimization for artificial noise scheme. IEEE Trans. Veh. Technol. **69**(3), 3465–3469 (2020)

Chapter 7
6G Wireless Systems: Challenges and Opportunities

Walid Saad

7.1 Introduction

The 5G wireless cellular system has emerged in response to the need for providing high-speed, reliable wireless connectivity not only to conventional wireless services, such as video streaming, but also to the so-called Internet of Things (IoT) system that interconnects machine-type devices such as sensors and autonomous robots. Indeed, beyond providing high data rates to the so-called enhanced mobile broadband (eMBB) services such as mobile TV, 5G systems are expected to guarantee ultra-reliable, low-latency communication (URLLC) to IoT services. The need for URLLC was hailed as the ultimate game-changer for 5G systems and beyond.

In 5G, URLLC services are essentially viewed as IoT applications that must reliably collect very short packets (few bytes) from small sensors or robots with an uplink over-the-air latency of less than 1 ms. For such IoT services, high reliability is defined by 3GPP as achieving a percentage of successfully delivered packet within the application time deadline constraint within the range 99.9% to 99.999%, depending on the application. Moreover, for such URLLC IoT services, data rate can be completely neglected since the transmitted uplink packets are only a few bytes long.

While such a URLLC design is sound when dealing with IoT sensors networks or factory automation applications, it is questionable whether one can keep restricting URLLC services to uplink short packets in the near future. In particular, the IoT itself is now witnessing an unprecedented revolution that is disrupting its original premise as a mere machine-to-machine communication system and transforming it into a complex Internet of Everything (IoE) system in which machines, people, and complex processes must constantly communicate with one another. As explained

W. Saad (✉)
Electrical and Computer Engineering Department, Virginia Tech, Blacksburg, VA, USA
e-mail: walids@vt.edu

X. Lin, N. Lee (eds.), *5G and Beyond*, https://doi.org/10.1007/978-3-030-58197-8_7

in [1], the IoT can be viewed as "the equivalent of a railroad line, including the tracks and the connections, whereas the IoE is all of that and the trains, ticket machines, staff, customers, weather conditions, etc." and more. Indeed, the IoE will bring forth new wireless services, such as large-scale autonomy (including fully autonomous vehicles, drones, and flying cars), a massive tactile Internet system, wireless brain-computer interfaces, and advanced eXtended reality (XR) applications (encompassing augmented, mixed, and virtual reality (AR/MR/VR)) [2]. These will impose very stringent quality-of-service (QoS) requirements across the rate, reliability, and latency spaces and will blur the boundary between classical 5G URLLC and eMBB services. As such, although 5G may be able to meet the QoS needs of basic XR or autonomous robotics, it will still fall short in meeting the more stringent rate (e.g., above 100 Gbps for some XR applications such as the so-called Ultimate VR class of services), latency (e.g., below 1 ms for wireless brain-computer interfaces), and high reliability (near-zero packet errors at low latency, i.e., extreme reliability [3]) needs of tomorrow's IoE applications.

To overcome the challenges of emerging IoE services, there is a need to develop a novel *sixth generation (6G)* wireless system, whose design is inherently tailored to the need for highly reliable, low-latency, and high-rate services. 6G will be a byproduct of traditional trends in communication technologies (e.g., densification, higher rates, and massive antennas) coupled with recent services and technological advances that include new wireless devices (e.g., body implants, XR apparatus, etc.), emerging artificial intelligence (AI) paradigms [4], and the need for supporting multiple functions that range from imaging to sensing and control. Beyond supporting new IoE services, the road toward 6G must also be able to overcome some of the limitations of 5G that were identified in early rollouts of the system, including:

(1) *High Frequency, High Rate, High Mobility:* At its early stages, 5G was envisioned to be a system that operates almost exclusively at high-frequency millimeter wave (mmWave) bands that can deliver the promised data rates. However, the early deployment of 5G systems is still primarily relying on sub-6 GHz spectrum bands, particularly for highly mobile scenarios. Indeed, thus far, mmWave has been deployed by a handful of operators and only for fixed wireless access. Therefore, the road to beyond 5G systems must revisit the problem of delivering high-speed wireless access at high frequencies for highly mobile environments.

(2) *Elusive Reliability:* Although early deployments of 5G have shown very promising performance in terms of data rate and low latency, reliability targets remain elusive. To date, outside of laboratory trials, there has been no fully fledged deployment that achieved the reliability targets set forth when 5G was conceived, i.e., a five nine (99.999%) reliability at low latency. This shortcoming in terms of reliability can be partially attributed to the lack of URLLC fundamentals, as outlined in [3]. The challenges of reliability will be also further exacerbated when looking at higher frequency such as mmWave bands that will be a hallmark of beyond 5G systems.

(3) *Coverage in Extreme Conditions:* Despite the growing success of 5G communication systems, conservative estimates show that more than half of the global population, mostly in extreme conditions, such as rural areas and disaster-affected areas, will still live in "wireless darkness" post the 5G era. Indeed, providing wireless coverage under extreme conditions has remained a major problem facing wireless systems, ever since the inception of the successful 2G cellular network. Examples of this lack of coverage are abundant. For example, in 2016, the FCC estimated that over 10% of all Americans and 40% of the US rural population did not have access to high-speed wireless connectivity [5]. Meanwhile, in the aftermath of hurricane Harvey, about 95% of cell sites in Houston stopped working, and wireless networks along the Texas coast suffered significant outages. As we enter the era of smart cities, the persistence of such a lack of connectivity will have adverse economic and societal consequences, and, thus, providing connectivity to rural and disaster-affected areas must become a priority for beyond 5G systems.
(4) *Spectral and Energy Efficiency:* Although 5G will provide significant spectrum efficiency gains compared to 4G, as it stands, the originally sought target of a three-times increase in spectrum efficiency has not been met yet. Moreover, early deployments of 5G show that the system is less energy-efficient than 4G which once again motivates more fundamental research in spectrum and energy efficiency whose targets will be even higher for beyond 5G systems.

Motivated by these limitations and the challenges of future IoE services, in this chapter, we provide a holistic overview on how 6G wireless systems will entail. In particular, we develop a bold new vision of 6G systems (detailed in Fig. 7.1) that uncovers the key drivers of 6G across applications, trends, metrics, and technologies. We then present new 6G services and provide a prospective road map to accelerate the leap from 5G toward 6G while going by a "beyond 5G" milestone.

7.2 6G Driving Applications, Metrics, and New Service Classes

As has been the case in every past generation of cellular systems, every new "G" is often motivated by a plethora of emerging wireless applications that bring forth new performance requirements and designs. 6G will be no exception: It will be driven by an unparalleled emergence of exciting new applications, ranging from XR to haptics, robotics and autonomous systems, and brain-computer interface over wireless networks, that will constitute the heart of the IoE and smart cities. These new services will require establishing new target performance metrics for 6G while radically redefining standardized 5G application types such as URLLC, eMBB, and massive machine-type communication (mMTC) as will be evident from Table 7.1. In this section, we first introduce the main driving applications of 6G

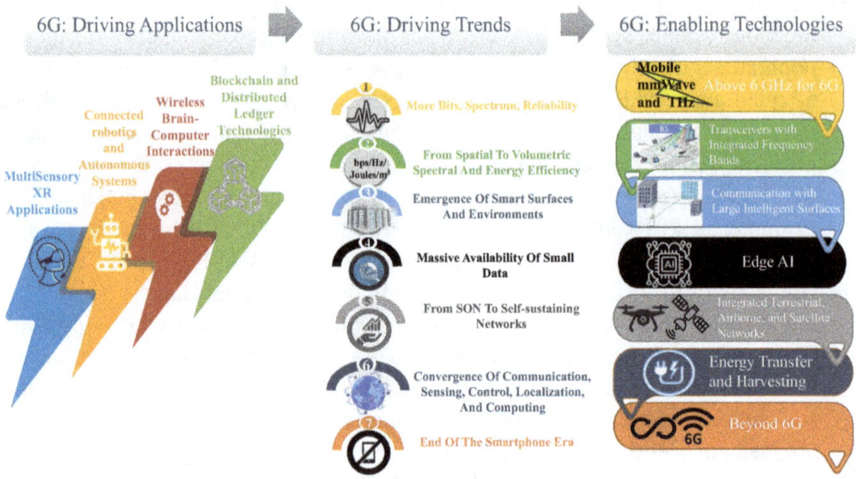

Fig. 7.1 An overview on our broad vision for 6G: Applications, trends, and technologies

cellular systems, and, then, based on those applications, we present some desirable metrics and performance requirements.

7.2.1 6G: Driving Applications and Their Performance Requirements

Even though classical wireless applications, such as video services, live mobile streaming, and voice/messaging services, will still be essential in 6G systems, the performance of 6G will, however, be determined by four new classes of IoE applications that include as follows:

Multi-sensory XR Applications

6G must support several XR applications within the broad AR/MR/VR spectrum. Although 5G will surely support basic VR services, it will not be able to deliver a full immersive XR experience for more advanced AR/MR/VR applications (including holographic teleportation) that require capturing all sensory inputs from the users, thus imposing more stringent latency and reliability constraints than those possible with 5G. Moreover, many XR applications will require highly reliable, ultra-low-latency, and high-rate communications with requirements that cut across traditional eMBB and URLLC services. In addition, providing a realistic and

Table 7.1 Requirements of 5G vs. Beyond 5G vs. 6G

	5G	Beyond 5G	6G
Device types	• Smartphones. • Sensors. • Drones.	• New generation of smartphones. • Sensors. • Drones. • XR equipment.	• Advanced sensors and DLT devices. • Diverse CRAS devices. • XR and BCI devices. • Body implants and intelligent wearables.
Spectral and energy efficiency improvements with respect to today's networks	10× in bps/Hz/m^2/Joules	100× in bps/Hz/m^2/Joules	1000× in bps/Hz/m^3/Joules (volumetric)
Rate requirements	1 Gbps	100 Gbps	1 Tbps
End-to-end delay requirements	5 ms	1 ms	<1 ms
Radio-only delay requirements	100 ns	100 ns	10 ns
Processing delay	100 ns	50 ns	10 ns
End-to-end reliability requirements	99.999%	99.9999%	99.99999%
Frequency bands	• Sub-6 GHz. • MmWave for fixed access.	• Sub-6 GHz. • MmWave for fixed access at 26 GHz and 28GHz.	• Sub-6 GHz. • MmWave for mobile access. • Exploration of THz bands (above 300 GHz). • Non-RF (e.g., optical, VLC, etc.).
Architecture	• Dense deployment of small base stations at sub-6 GHz with assistance from macrocell base stations. • MmWave small cells of about 100 m (for fixed access).	• Denser sub-6 GHz cells. • <100 m dense (and tiny) mmWave cells.	• Cell-free smart surfaces at high frequency supported by mmWave base stations for mobile and fixed access. • Temporary hotspots served by base station carried by drones or implemented on tethered balloons. • Trials of tiny THz cells.

immersive XR user experience requires a *joint system design* that integrates not only engineering (communications, storage, networking, computing) requirements but also *perceptual* requirements related to human cognition, senses, and physiology. Indeed, for future XR services, minimal and maximal human perceptual requirements and limits must be integrated within the communication, computing, and processing functions of the system. In other words, physical and cognitive human perceptions will now be a key determinant of QoS in XR applications. For this purpose, we envision defining a new notion of *quality-of-physical-experience (QoPE)* metric that combines physical/cognitive factors from the human user itself with the more common QoS measures (e.g., delay and rate) and conventional quality-of-experience (QoE) concepts such as the mean-opinion score. This concept of QoPE will be affected by various human-centric factors such as brain cognition/capabilities, body physiology, gestures, and even age. For instance, in our recent work in [7], we have demonstrated that the human brain fails to distinguish between different latency measures, particularly when operating within the URLLC regime. As a result, a network can save its communication resources (particularly energy resources to enhance energy efficiency) by being aware of the human-in-the-loop when performing resource management and network optimization. Meanwhile, in [8], we characterized the impact of visual and haptic perceptions on wireless performance using the just-noticeable difference (JND), an established measure from psychophysics. We have shown how using notions such as JND is necessary to truly understand the performance limits of wireless communication for XR-type services. In summary, the requirements of XR services for 6G will essentially be a combination of classical URLLC and eMBB with the need to integrate perceptual factors.

Connected Robotics and Autonomous Systems (CRAS)

A major driver for 6G is the anticipated proliferation of CRAS applications including drone systems, self-driving cars and platoons, flying vehicles, swarms of drones, and autonomous robotics. Deploying CRAS over cellular systems will not be a simple case of "an additional uplink IoT devices with short packets." On the one hand, next-generation CRAS will require exchange of high-rate data such as high-definition (HD) maps that can serve for navigation purposes and require eMBB-level communication with high reliability. On the other hand, CRAS will have complex control systems that will dictate the latency and reliability requirements for the system. This, in turn, motivates a *cyber-physical* design to the wireless system in which physical constraints, extracted from the control system requirements and the physical world, must be incorporated into traditional, cyber-only QoS metrics. Indeed, here, we can once again redefine the QoPE notion while substituting the human factors with the physical control system factors. In addition, CRAS will undoubtedly open up the door for an unimaginably rich set of applications that we cannot even foresee at this point. Hence, CRAS can be seen as a prime use case for beyond 5G systems that will require not only stringent requirements across the reliability-rate-latency spectrum – a balance not yet available in 5G – but also *a joint cyber-physical system design* that merges control with communications and even navigation/localization.

Wireless Brain-Computer Interactions (BCI)

Traditionally, BCI applications were limited to healthcare scenarios involving electroencephalography (EEG) devices that allow humans to control prosthetic limbs or interact with nearby devices using brain implants. However, the ongoing revolution in the field of wireless brain to computer interfaces will inevitably lead to novel BCI paradigms (e.g., multi-brain-controlled cinema [9]) that will rely on 6G connectivity. Most recently, Neuralink announced that it has successfully built special microchips and flexible fiber electrodes that can be implanted into a human brain enabling it to seamlessly connect to computing systems or neighboring machines. Clearly, these tremendous advances in the field of BCI will soon lead to a broad range of brain implants that can wireless connect to the entire IoE system. In such applications, there is a seamless integration between human cognition and the wireless network, a feature that is not seen in 5G applications. Using BCI, one can envision transmitting brain signals over the wireless channel for various purposes such as communicating with nearby devices or even remotely controlling CRAS devices. As discussed in [6, 10], such wireless cognition can strain the capacity of existing 5G systems and will require very strict rate, latency, and reliability guarantees. Indeed, once implants and BCI devices become more pervasive, we envision a world in which, instead of conventional smartphones, individuals will interact with their environment and other individuals using a range of worn, embedded, and implanted devices. Such interactions will enable control of environments using gestures and communication with others using haptic messages. This, in turn, yields many interesting questions such as what happens when humans communicate through touch instead of words? Meanwhile, the emerging notion of "emotion-driven devices" also called affective computing that helps devices understand our mood and then make suggestions to match it (e.g., rescheduling a meeting due to a mood swing or tiredness) will also leverage BCI devices. Such *affective, emphatic, and haptic communications* will potentially become major use cases for 6G. Compared to the main 5G services, wireless BCI application will need fundamentally different performance indicators. Analogous to XR, wireless BCI applications will require high reliability, high data rates, and extremely low latency. However, compared to XR, BCI applications will be more sensitive to physical perceptions thus requiring performance guarantees in terms of QoPE. In addition, wireless BCI opens the door for new research problems at the intersection of neuroscience and wireless communications.

Blockchain and Distributed Ledger Technologies (DLT)

Many next-generation IoE services will rely on blockchain and DLT ideas in order to provide security and distributed operation. Although the notions of blockchain and DLT are not directly tied to communication, their operations will require a reliable communication infrastructure. In particular, massive machine-type communications (mMTC) services in tomorrow's IoE will largely integrate blockchain and DLT ideas. This, in turn, motivates a need for understanding the synergies between communication and blockchain. For example, it has been recently shown in [11] that wireless network errors can impact the probability of forking in a blockchain. This, in turn, motivates revisiting the fundamentals of URLLC and mMTC services in order to understand how blockchain and DLT applications will impact the requirements. In many ways, blockchain and DLT applications can be seen as massive, distributed IoE sensing applications that need a wireless connection that can provide a synergistic mix of massive machine-type communications (mMTC) and URLLC in order to maintain reliable connectivity, low latency, as well as scalability.

7.2.2 6G: Key Trends and Metrics

Clearly, the driving applications of 6G outlined in Sect. 7.2.1 will yield several novel cross-system trends that will, in turn, determine the objectives that 6G must be able to meet. These trends can be summarized into seven key groups:

- **Trend 1 – More Bits, More Spectrum, More Reliability:** A majority of the applications outlined in Sect. 7.2.1 will require higher data rates and higher spectrum efficiency than what 5G can deliver. For example, we anticipate yet another need for a 1000× increase in data rates, yielding a target of around 1 terabit/second, in order to meet the QoS requirements of next-generation IoE services such as XR or even wireless BCI. This, in turn, once again requires the exploration of more spectrum resources that go beyond sub-6 GHz and mmWave frequency bands. Moreover, much of the aforementioned 6G services will require an even more stringent reliability than basic IoT sensor systems. Therefore, delivering higher reliability is a major challenge for all wireless systems starting from 5G. This challenge is particularly exacerbated by the fact that going higher in frequencies negatively impacts reliability. As such, one key driving trend behind 6G will be the need for higher reliability at higher frequency bands.
- **Trend 2 – From Areal to Volumetric Spectral and Energy Efficiency:** Drones and flying vehicles will become a major component of beyond 5G systems. Indeed, 6G must serve both ground and flying users, encompassing smartphones, XR devices, and BCI implants, along with aerial vehicles and drones. Hence, we are now witnessing a transformation from traditional, two-dimensional wireless systems into fully fledged three-dimensional (3D) wireless systems [12]. This

3D nature of 6G mandates an evolution toward defining volumetric spectrum and energy efficiency, rather than the traditional areal definition. We particularly anticipate that 6G systems must deliver very high spectral and energy efficiency (SEE) requirements quantified in bps/Hz/m^3/Joules. This is aligned with the evolution that we have seen starting from 2G systems (bps) to 3G systems (bps/Hz), then 4G systems (bps/Hz/m^2) to 5G systems (bps/Hz/m^2/Joules).
- **Trend 3 – Emergence of Smart Surfaces and Environments:** All current generations of wireless cellular systems used base stations (BSs), of different sizes, forms, and number of antennas, to transmit data to their users. However, satisfying the very stringent rate, coverage, spectrum efficiency, and delay requirements of the 6G applications discussed in Sect. 7.2.1 through the traditional trends of designing better BS-centric transceivers or using more antennas at BS towers will no longer be possible. One promising solution is to exploit the tremendous recent advances in metamaterials that allows one to transform man-made structures such as buildings, walls, and roads into electromagnetically active metasurfaces with radio frequency (RF) capabilities. This transformation, exemplified by the Berkeley ewallpaper project,[1] will allow future wireless systems to use such smart metasurfaces as large, reconfigurable intelligent surfaces (RISs) and environments to provide pervasive, high-speed wireless communications. This trend that shifts from traditional tower-mounted BSs toward RISs will be a major force behind the 6G architectural evolution.
- **Trend 4 – Massive Availability of Small Data:** AI is witnessing a radical departure from traditional centralized "big data" cloud architectures toward a distributed AI paradigm in which massive, "small" data is dispersed across multiple edge devices and must be processed in a distributed manner using on-device machine learning. This paradigm shift will be further fueled by the emerging IoE discussed in Sect. 7.2.1 in which sending large data volumes to a cloud faces major communication, privacy, and scalability challenges. In consequence, 6G systems are expected to leverage both big and small datasets that are dispersed across the system so as to improve network operation and deliver new services. In consequence, it is necessary to investigate new AI and machine learning techniques that go beyond classical big data analytics so as to address the aforementioned challenges with effective distributed AI techniques that can exploit local, on-device edge data processing tailored to the real-time, private, and mission-critical nature of the IoE services.
- **Trend 5 – From Self-Organizing Networks (SONs) to Self-Sustaining Networks:** Classical or legacy cellular trends, such as SONs, will witness yet another evolution in 6G systems. For instance, SON functions, which have remained elusive or scarcely integrated into 4G/5G systems, become a necessity for 6G given the highly distributed nature of driving applications such as CRAS and DLT technologies. Therefore, 6G must be able to deliver intelligent SON functions that can be used to manage network resources and operations, as well as system

[1] See https://bwrc.eecs.berkeley.edu/projects/5605/ewallpaper.

optimization. In fact, the deployment of CRAS and DLT applications calls for a paradigm shift from traditional SON, using which the cellular system simply adapts its functions to some states of the environment, to a *self-sustaining network (SSN)* that is able to guarantee its key performance indicators (KPIs), *in near-perpetuity*, under largely complex and dynamic environments that will result from the diverse domain of 6G applications and services. SSNs should be capable of not only adapting their network functions but also sustaining their system resource usage and management (e.g., by properly exploiting spectrum resources and potentially harvesting energy) to autonomously maintain stringent, long-term KPIs. Naturally, the deployment of SSN functions will make use of the recent advances in the AI domain. Ultimately, 6G could potentially incorporate AI-powered SSN algorithms.

- **Trend 6 – Convergence of Communications, Computing, Control, Localization, and Sensing (3CLS):** To date, all existing wireless cellular systems were designed with one exclusive purpose: providing wireless connectivity. However, it is expected that 6G will be able to deliver services beyond communications. In particular, 6G will mark a convergence of diverse functions that include communications, control, computing [13], localization, and sensing. We view 6G as a multi-purpose and versatile system that can offer a diverse set of 3CLS applications which are particularly suitable and arguably necessary for services like CRAS, XR, and DLT where tracking, control, localization (e.g., for navigation), and computing are an inherent feature. In addition, by leveraging the use of sensing functions, 6G can build a *3D mapping of the radio environment* across multiple frequency bands. This mapping can then be used to assist in network functions and user operation. In a nutshell, 6G systems must seamlessly integrate and manage a broad range of 3CLS functions.
- **Trend 7 – End of the Smartphone Era:** Smartphones were the driving force behind the wireless revolution from 3G all the way up to 5G. However, as exemplified by the advances in the BCI and XR fields, the next decade will experience an exponential rise in the number of embedded wearable devices and implants whose functionalities will gradually start replacing those of smartphones. For instance, XR and BCI devices that include smart wearables, integrated headsets, and advanced body implants that can take direct sensory inputs from human senses can potentially bring an end to the smartphones era and constitute major 6G use case scenarios. For instance, we may see a shift from traditional BS-to-smartphone communication links, which are integral to all current and previous cellular systems, toward RIS-to-implant communication links in 6G and beyond.

From Table 7.1, we can observe that the aforementioned trends will collectively lead to new desirable performance targets and requirements that will be met in two stages of cellular system evolution: (a) a first (evolutionary) stage that we can call a "beyond 5G" stage and (b) a second, revolutionary 6G stage.

7.2.3 New Service Classes for 6G

In addition to introducing new performance metrics and targets, the aforementioned technological trends will necessitate redefining the different application and service types in 5G by enhancing and potentially combining conventional URLLC, eMBB, and mMTC services while also leading to new types of services (summarized in Table 7.2), as discussed next:

Mobile Broadband Reliable Low-Latency Communication

From the discussion in Sect. 7.2.2, we can easily observe that the existing distinction between eMBB and URLLC is not sustainable to support tomorrow's IoE applications such as advanced XR, wireless BCI, CRAS, or even smart city services. This is due to the fact these applications will need not only very low latency and high reliability but also high data rates (at the level of eMBB services). To cater for these

Table 7.2 Summary of 6G service classes, their performance indicators, and example applications

Service	Performance indicators	Example applications
MBRLLC	• Stringent requirements in the rate-reliability-latency space. • Energy efficiency. • Rate-reliability-latency for mobile scenarios. • Handover failures.	• XR/AR/VR. • Autonomous vehicular systems • Autonomous drones • Legacy eMBB and URLLC
mURLLC	• Ultra-high reliability. • Massive connectivity. • Massive reliability. • Scalable URLLC.	• Conventional IoT. • Device tracking. • Distributed blockchain and DLT. • Massive sensing. • Autonomous robotics.
HCS	• QoPE capturing raw wireless metrics as well as human and physical factors.	• BCI. • Haptics. • Empathic communication. • Affective communication.
MPS	• Control system stability. • Computing delay. • Localization precision and accuracy. • Accuracy of sensing and mapping functions. • Delay and reliability for communications. • Energy.	• CRAS. • Telemedicine. • Environmental mapping and imaging. • Some special cases of XR services.

requirements, we can introduce a new cellular service class that we dub *mobile broadband reliable low-latency communication (MBRLLC)* which will allow 6G to deliver any desirable performance target that lies in the rate-reliability-latency dimensions. In general, MBRLLC can be seen as a generalization of classical URLLC and eMBB services. In addition, energy efficiency will be a major design challenge for MBRLLC not only because of its effect on reliability and data rate but also because devices in 6G will continuously shrink in their size and increase in their functions, thus requiring highly energy-efficient designs.

Massive URLLC

In 5G, URLLC services pertain to guaranteeing reliability and low latency for well-defined uplink IoE applications such as IoT sensors. The fundamentals of URLLC have already been widely explored in the literature (e.g., see [14]). However, in 6G, there is a need to scale traditional URLLC across the device dimension. This, in turn, can yield a new *massive URLLC (mURLLC)* service that combines 5G URLLC with conventional mMTC. mURLLC exhibits a trade-off in reliability-latency-scalability, and, thus, it requires a departure from average-based system designs (e.g., based on average data rate or average latency). Instead, a principled and scalable framework which accounts for latency, reliability, packet size, network architecture, topology (across edge, access, and core), and decision-making under uncertainty is needed [15]. Moreover, in mURLLC one must also deal with very extreme networking conditions as outlined in [3].

Human-Centric Services

6G will have to deal with *human-centric services (HCS)*, a new type of service class that imposes QoPE performance targets (tightly integrated with the human users and their body/physiology, as discussed in Sect. 7.2.1) instead of raw rate-reliability-latency metrics. A prime example of HCS would be wireless BCI applications in which service performance is determined by the cognition, actions, and even physiology of human users. For HCS, a new set of QoPE performance indicators should be defined and quantified as function of traditional (raw) QoE and QoS performance metrics.

Multi-purpose 3CLS and Energy Services

6G systems must jointly deliver 3CLS services and their derivatives. For example, 6G systems can provide navigational and localization inputs to CRAS devices. In addition, using new advances in wireless energy transfer, 6G systems can potentially provide energy to recharge small devices such as IoE sensors. Hence, we anticipate that 6G will have to define a new class of service that goes beyond communication.

These *multi-purpose 3CLS and energy services (MPS)* will be of central importance for CRAS and wireless BCI applications, among others. For MPS, there is a necessity for (a) joint uplink-downlink designs and (b) meeting desirable performance targets for the control (e.g., in terms of control stability), computing (e.g., computing delay), energy (e.g., amount of energy to transfer), localization (e.g., precision of localization), as well as mapping and sensing functions (e.g., accuracy of a mapped environment). MPS services are also suitable to operate cyber-physical systems over the wireless infrastructure.

7.3 6G: Enabling Technologies

To enable all the foreseen 6G services and meet their required QoS and QoPE performance, a broad range of new, disruptive technologies must be integrated into 6G systems. These technologies, their challenges, and their 6G integration are explained next.

7.3.1 6G at Above 6 GHz: From Small Cells Toward Tiny Cells

From Trends 1 and 2, we can see that, in 6G, higher data rates and SEE will be needed anywhere, anytime. This, in turn, motivates the exploration of higher-frequency bands beyond sub-6 GHz which had already started with 5G and mmWave. First, and foremost, as previously mentioned, one of the key limitations of 5G systems is the lack of high-speed wireless access at high frequencies for highly mobile environments. Therefore, a first step in this area requires developing new fundamental science to understand how one can make *mobile mmWave* a reality in early 6G systems or even at the beyond 5G step. The enabling technologies needed to realize the vision of mobile mmWave communications can include a combination of sub-6 GHz and mmWave bands, as well as the use of caching to minimize handover failures as proposed in [16]. As 6G progresses, exploiting frequencies above mmWave, particularly at the terahertz (THz) frequency ranges, will then become critical [10]. To leverage higher THz and mmWave frequencies, the size of some of the 6G cells would have to shrink from small cells to "tiny cells" whose radius is only few tens of meters. As a result, once we go higher in frequency and reach the THz frontier, there will be a need for new network architecture designs that can accommodate much denser deployments of tiny cells. In addition, at higher frequencies, there is a need for developing new mobility management techniques that are tailored to the highly intermittent nature of high-frequency communication links.

7.3.2 Transceivers with Integrated Frequency Bands

Delivering seamless connectivity for mobile 6G services will not be possible by only relying on dense, high-frequency tiny cells. Instead, we must conceive of an integrated system that can exploit multiple frequencies across the microwave/mmWave/THz bands (e.g., using multimode base stations) in order to provide seamless connectivity at both wide and local area network levels. Early works in [17, 18] have shown the potential of exploring such multiband communication. In addition, one can anticipate the integration of RF and non-RF bands. For instance, one can leverage the use of RF, optical, and visible light communication (VLC) to enhance not only communication efficiency but also resilience of the system to surges in traffic (e.g., in hotspot areas or even disaster-affected areas).

7.3.3 Communication with Large Reconfigurable Intelligent Surfaces

In order to deliver higher data rates, wireless research activities have mainly focused on two directions: (a) exploring higher frequencies such as mmWave bands and (b) equipping tower-mounted BSs with a massive number of RF antennas via the so-called massive multiple-input multiple-output (MIMO) communication paradigm. Indeed, both massive MIMO and mmWave bands will be integral to both 5G and 6G because they can help overcome the challenge of spectrum scarcity as well as the wireless channel impediments such as fading and interference, thus delivering better SEE and higher data rates at higher frequencies (Trend 1). However, these two paths toward faster wireless networking have their own limitations. For instance, the number of RF antennas and the type of RF circuits that can be used to create a "truly massive" MIMO system are limited by the hardware capability of BS towers. Moreover, due to the high susceptibility of mmWave frequencies to channel variations (e.g., blockage), reaping their benefits requires maintaining constant line-of-sight (LoS) links to the users, a feat only possible through network densification – deploying a significantly large number of massive MIMO BSs. However, densification is again limited by various geographical and hardware constraints. Hence, ushering in the 6G era will require a major rethinking to the architecture of wireless cellular systems. In particular, for many decades, the focus of wireless research and development efforts has been on designing effective transceivers while assuming the wireless channel and its propagation environment to be uncontrollable. In contrast, owing to the recent advances in metamaterial-based devices, it is now possible to transform man-made structures such as walls, buildings, and roads into electromagnetically active metasurfaces that can be employed as RF transceiver, complementing or even replacing traditional tower-mounted BSs. By doing so, one can build large RISs that can be used to not only design more effective massive MIMO transceivers (with antenna arrays spanning a very large surface and an ability to perform near-field LoS communications) *but to also control the propagation environment* by employing RISs as reflectors of wireless signals. Indeed, for 6G systems, as per Trend 3, we foresee an initial shift from conventional massive MIMO over tower-mounted BSs toward large RISs and smart environments [19–23] that act as both transceivers and reflectors so as to provide massive surfaces for wireless communications and for heterogeneous devices (Trend 7). RIS will hence allow 6G systems to now control the propagation environment, thus yielding many new research challenges and opportunities. They will also enable novel ways for wireless communication such as by using holographic RF radio and holographic MIMO.

7.3.4 Edge AI

AI is experiencing a major interest from the wireless communications community [4] motivated by some of the recent breakthroughs in deep learning, the increase in the data availability (Trend 4), and the emergence of smart devices (Trend 7). We envision at least three 6G use case scenarios for AI: (a) growth in big data analytics through AI, particularly for prediction, caching, and environment mapping tasks; (b) emergence of distributed AI for network optimization and for creating SSNs (Trend 5), through multi-agent reinforcement learning techniques; and (c) rise of on-device, edge AI techniques that exploit advances in federated learning [24] and related areas to enable the 6G system to exploit distributed, small data that is dispersed across its devices. Moreover, AI will allow 6G to automatically provide MPS to its devices and to generate and transmit 3D radio environment maps (Trend 6). Ultimately, we expect that 6G systems will witness a new paradigm of "collective network intelligence" in which network intelligence is further fostered at the edge to provide fully distributed autonomy. This new leap toward edge AI will lead to a 6G system that can support the services of Sect. 7.2, deliver 3CLS, and potentially substitute classical frame structures that were proposed by 3GPP. Indeed, a major open problem here is whether future wireless systems, starting with 6G, will gradually become fully operated and managed by AI functionalities.

7.3.5 Integrated Terrestrial, Airborne, and Satellite Networks

As outlined earlier, providing coverage to rural areas and areas with extreme conditions (e.g., disaster-affected areas) has been a major challenge for wireless systems since the rise of cellular networks a couple of decades ago. One seemingly promising solution for this decades-old problem is through the use of drones that can act as flying wireless BSs. Indeed, drone-BSs can complement terrestrial ground networks by delivering wireless connectivity to hotspots and to rural areas that has little to no infrastructure. Drone-BSs can also be used to provide on-demand wireless access in response to emergency situations in disaster-affected areas. In addition to acting as BSs, drones will also be integral users of 5G infrastructure and beyond. Indeed, drones will require wireless connectivity in order to receive control data, transmit sensing data (e.g., maps or videos), or communicate with ground infrastructure. This dual role of drones in tomorrow's wireless networks means that 6G systems will inherently be 3D wireless systems that must meet volumetric performance targets (Trend 2). In addition, to support communication for drone-BSs as well as terrestrial BSs, there will be a need for guaranteeing satellite connectivity with low orbit satellites (LEO) and CubeSats to provide backhaul links as well as further wide area coverage. Integrating terrestrial, airborne, and satellite networks [12] and [25] into a single wireless system will therefore be an important objective for 6G.

7.3.6 Energy Transfer and Harvesting

One common feature among all the aforementioned IoE services is their need for energy efficiency. In fact, many IoE devices, including wearables, sensors, and implants, have a very small form factor and very limited energy and computing resources. As such, 6G systems must be able to deliver more energy-efficient communications. To do so, one possibility is to exploit emerging energy harvesting and energy transfer technologies. On the one hand, 6G infrastructure may leverage advances in energy harvesting to equip network devices (e.g., BSs or even drones) with solar-powered energy sources that can provide a clean and continuous source of power. On the other hand, RF energy harvesting and transfer can be exploited to provide RF energy to IoE devices. Indeed, 6G can possibly be the first cellular generation that can deliver energy, as well as 3CLS (Trend 6). As wireless energy transfer technologies start to mature, we envision 6G BSs to be capable of providing basic energy transfer for IoE devices, especially implants, wearables, and sensors (Trend 7). Other energy-related ideas such as energy harvesting and backscatter communication will also constitute important enabling technologies for 6G.

Beyond 6G

A few technologies will start to mature along the same timeline as 6G, and, hence, they can potentially play a role toward the final steps of the 6G standardization and research process. One major example is *quantum computing and communications* that can provide security and long-distance networking. While 6G system will likely not leverage much quantum technologies, we foresee that quantum communication will start to become more viable and practical along the same timeline as 6G.

As such, although the specific role of quantum communications and computing in 6G remains rather unclear, the next few years will see more synergies across these two areas. For instance, quantum computing can potentially speed up much of the algorithms that run in a cellular system, thus contributing to reducing latency. Moreover, quantum computing can also be an important enabler for faster AI at the edge of 6G systems and beyond. Last, but not least, emerging areas such as neuro-inspired designs [26] and molecular communications may also have a role in shaping 6G systems.

7.4 6G: Open Research Problems

As is evident from the trends that we have identified in Sect. 7.2 and the enabling technologies that we exposed in Sect. 7.3, 6G will bring forth many interesting open problems and challenges, as summarized in Table 7.3 and discussed next.

7.4.1 3D Rate-Reliability-Latency Fundamentals

Performance analysis is arguably always the first step toward understanding the limits and capabilities of a wireless system. 6G will not be an exception: There is a clear need for characterizing its performance. In particular, for 6G systems, there is a need for new techniques to understand the fundamental 3D performance of the system, in terms of rate-reliability-latency trade-offs as well as volumetric SEE. This analysis must be able to quantify the spectrum, energy, and communication requirements that are needed by 6G in order to support the previously discussed driving applications. Some recent works in [15, 27–29] provide a first step in this direction.

7.4.2 Leveraging Integrated, Heterogeneous High-Frequency Bands

Leveraging high-frequency bands such as mmWave and THz in 6G opens up the door for a diverse set of new research problems. As previously discussed, for mmWave, a central open problem is to develop new mobility management techniques that enable mobile communications at mmWave bands. Here, there is a need for new mobility management protocols that can minimize handover rates (as done in [16]) as well as for fundamental analysis of mobility performance in mmWave networks. It can also be of interest to investigate whether AI techniques can help in improving performance for highly mobile mmWave systems by exploit-

Table 7.3 Summary of key 6G research areas

Research area	Challenges	Open problems
3D rate-reliability-latency fundamentals	• Fundamental communication limits. • 3D nature of 6G systems.	• 3D performance analysis of rate-reliability-latency region. • Analysis of achievable rate-reliability-latency performance targets. • SEE analysis in 3D space. • Quantification of spectrum and energy needs.
Leveraging integrated, heterogeneous high-frequency bands	• Challenges of operation in highly mobile systems. • Susceptibility to molecular absorption and blockages. • Short range. • Lack of propagation models. • Need for high fidelity hardware. • Presence of frequency bands with different characteristics.	• Mobility and handover management for high-frequency THz and mmWave systems. • Cross-band physical, link, and network layer optimization. • Coverage and range improvement. • Design of mmWave and THz tiny cells. • Design of new high fidelity hardware for THz. • Propagation characterization for mmWave and THz bands.
3D networking	• Presence of users and base stations in 3D. • High mobility.	• 3D propagation modeling. • 3D performance metrics. • 3D mobility management and network optimization.
Communication with RISs	• Complexity of metasurfaces and RISs. • Absence of precise models of performance. • Absence of faithful propagation environment models. • Heterogeneity of 6G devices and services. • RIS capability to offer various functions (reflectors, BSs, etc.).	• Optimal deployment and location of RIS surfaces. • RIS reflectors vs. RIS transceiver BSs. • Energy transfer using RISs or other means. • AI-enabled RIS. • RIS across 6G services. • Fundamental analysis of the performance of RIS transmitters and reflectors across frequency bands.

AI for wireless	• Design of low-complexity, edge AI solutions. • Small but massively distributed data.	• SON using reinforcement learning techniques. • Data analytics for both big and small data. • AI-guided network management. • Edge AI operating on wireless networks.
New QoPE metrics	• Incorporate raw metrics with human perceptions. • Accurate modeling of human perceptions and physiology.	• Theoretical development of QoPE metrics. • Empirical QoPE designs. • Practical psychophysics experiments. • Need for precise and realistic QoPE performance targets and measures.
Joint communication and control	• Integration of communication and control performance indicators. • Handling dynamics and multiple time scales.	• Co-design of communication and control systems. • Control-aware wireless communication metrics. • Wireless-enabled control metrics. • Joint system optimization for CRAS.
3CLS	• Integration of multiple functions. • Lack of prior models.	• Design of 3CLS metrics. • Joint 3CLS optimization. • AI-enabled 3CLS. • Energy efficient 3CLS.

(continued)

Table 7.3 (continued)

Research area	Challenges	Open problems
Design of 6G protocols	• 3D network-enabled 6G protocols that can also handle diverse propagation environments. • Need to serve different devices with heterogeneous capabilities and mobility patterns. • Need for adaptive and self-learning protocols across the network stack.	• Design of signaling, scheduling, and network coordination protocols that do not rely on pre-fixed and rigid frame structures. • Development of adaptive multiple access protocols. • Design of adaptive and proactive handover schemes that can handle 3D mobility. • Novel identification and authentication techniques suitable for new 6G devices. • Development of AI-inspired edge protocols for multiple 6G functions.
RF and non-RF link integration	• Different physical nature of RF/non-RF interfaces.	• Hardware for joint RF/non-RF systems. • System-level analysis for systems with joint RF and non-RF capabilities. • Use of RF/non-RF systems for various 6G services.
Holographic radio	• Lack of existing models. • Hardware and physical layer challenges.	• Design of holographic MIMO using RISs. • Performance analysis of holographic RF. • 3CLS over holographic radio. • Network optimization with holographic radio.

ing environmental information (e.g., image modalities [3]) or by predicting the mobility patterns and user behavior [30]. Meanwhile, at the THz frequencies, there is a need for novel transceiver architectures and propagation models [10, 31]. In order to overcome the high THz path loss, new transceivers must exhibit high power, high sensitivity, and low noise figure. Once physical layer challenges are overcome, there is a need to introduce new link-layer, multiple access, and network protocols so as to optimize the use of cross-band resources while factoring in uncertain and dynamically varying mmWave and THz environments. Another important research direction here is to investigate the co-existence of THz, mmWave, and microwave cells across all layers, building on early works such as [18]. In addition, there is a need to understand whether high-frequency bands can indeed provide reliable communication. For example, in [32], we have shown that molecular absorption can significantly affect the distance at which THz communications can provide reliable links to XR users, and in [33], we have shown that blockages will also play an important role in determining whether THz can provide reliable but high-rate links as required by XR services. Building on this work, one envisions a plethora of open problems related to investigating whether MBRLLC is possible at THz frequency or whether it is inevitable to exploit integrated frequency bands for maintaining high reliability.

7.4.3 3D Networking

Because ground and aerial networks are becoming largely integrated, as discussed in Sect. 7.3, 6G will have to support communications in 3D space. This includes providing connectivity to 3D flying users, as well as enabling the deployment of 3D drone-carried BSs (e.g., temporary drone-BSs or tethered balloons). This motivates the need for major research efforts across multiple directions related to 3D networking. First, data-driven and measurement-based modeling of the 3D propagation environment is needed. Such modeling requires both theoretical and experimental efforts that can create realistic propagation models for 3D cellular systems. Second, novel techniques for performing 3D frequency and wireless network planning (e.g., where to place BSs, balloons, or drone-BSs) must be investigated. Our prior results in [12] showed that such 3D planning is substantially different from conventional 2D networks due to the new altitude dimension and the associated degrees of freedom. In particular, we showed that designing a fully fledged 3D cellular system puts forward new challenges from network deployment all the way to optimization and network operation. Last, but not least, new techniques for network optimization, 3D mobility management, routing, and dynamic network resource management are also needed.

7.4.4 Communications with RISs

As a byproduct of Trend 3, 6G will potentially deliver wireless connectivity via intelligent RIS systems that encompass active frequency-selective surfaces, passive metallic reflectors, passive/active reflection arrays, as well as non-reconfigurable and reconfigurable metasurfaces. This naturally leads to many important research problems that range from the optimized deployment of passive reflectors and metasurfaces to the intelligent operation (potentially using edge AI) of reconfigurable metasurfaces. Fundamental performance analysis to understand the performance limitations and benefits of RISs and smart surfaces in terms of data rate, delay, reliability, and achievable coverage is also needed, building on the early works in [19–23]. Other key open problems here include investigating how practical models for metamaterial-based RIS devices and systems can impact the operation of RIS-based RF transceivers or reflectors. In other words, there is a need for metamaterial-informed communication system models that can reflect the real-world constraints of actual RIS devices. Last, but not least, it is of interest to analyze how RISs can leverage high-frequency bands (e.g., THz or mmWave) to provide high-speed connectivity to services such as XR, as studied in our work in [34].

7.4.5 AI for Wireless

AI brings forward many major research directions for 6G. We can distinguish two major areas: (a) AI for wireless communication and (b) wireless communication for AI. In the first area, beyond the need for massive, small data analytics, there is a need for deploying innovative machine learning algorithms that can provide SSN functions to 6G systems. This includes AI-enabled network optimization, resource management, and distributed network control. This area will particularly leverage advances in multi-agent reinforcement learning, artificial neural networks (ANNs), and game theory so as to instill smart, self-sustaining properties into 6G systems. One prominent problem here is the development of ANN-driven reinforcement learning algorithms that enable a network to optimize the usage of its resources while learning from its environment. This is particularly suitable for emerging applications such as XR and drone networking, as done in [35–39]. In the second direction, there is a need to understand how wireless factors, such as fading, mobility, or interference, can impact the performance of edge AI algorithms, such as federated learning. For example, it was shown in [24] and [40] that the convergence of federated learning will be strongly impacted by wireless packet errors and wireless latency. This, in turn, motivates a need for joint design of wireless and learning algorithms, particularly when dealing with edge AI techniques such as federated learning or distributed generative adversarial networks (GANs) [41]. Indeed, to perform critical edge AI application tasks, low-latency, high-reliability, and scalable AI is needed, along with a reliable infrastructure [4] and [8]. This joint design of ML and wireless networks is a very important 6G research area. Another important open problem for AI in 6G systems is the design of training-free machine learning frameworks that can execute different tasks with very limited training data. One step toward this direction is through the idea of experienced deep learning that we introduced in [27] in which a deep learning agent is allowed to gain experience in a virtual environment that can be created using GANs.

7.4.6 QoPE Metrics

The development of QoPE metrics that integrate physical factors from human cognition and physiology (for HCS) or from a control system (e.g., for CRAS) is an important 6G research field, particularly in light of the type of emerging network devices (Trend 7). This requires both realistic psychophysics experiments and new, precise mathematical QoPE expressions that merge QoS, QoE, and human user perceptions. In order to perform theoretical modeling of QoPE metrics, one can explore techniques from other domains that include the field of multi-attribute utility theory in operations research (e.g., see [38]) and the field of machine learning (e.g., see [7]). We anticipate that 6G could be one of the first cellular network generations that can support a whole new range of services (wireless BCI) that can capture the multiple cognitive senses of a human.

7.4.7 Joint Communication and Control

6G will have to provide pervasive connectivity to CRAS services for various purposes such as navigation and control. The performance of services such as CRAS is highly dependent on their practical control systems whose operation requires data input from the wireless communication links of 6G. Hence, effectively integrating CRAS over 6G systems requires a new *communication and control co-design paradigm* in which the wireless communication performance of the 6G links is optimized in a way to satisfy the control system stability. Meanwhile, the control system must be designed in a way to be cognizant of the wireless network state. Due to the traditional radio-centric focus (3GPP and IEEE fora), this joint communication and control co-design aspect has not yet been investigated in depth. Here, we note that prior art on related ideas, such as networked control systems, often abstracts the wireless network specifics and, hence, they cannot be directly applied to real-world cellular communications. As a result, communication and control co-design will be an important research problem for 6G. However, in order to deliver connectivity to cyber-physical systems such as autonomous vehicle, there is a need to couple the performance of the control, communication, and computing systems. For example, in [42, 43], we provided guidelines on how one can design a wireless system that can meet the control system stability requirements (in terms of delay and reliability) of autonomous vehicles. Similar performance analysis and joint communication, control, and computing designs are needed for a broad range of CRAS applications ranging from vehicular platoons to autonomous swarms of drones [44, 45].

7.4.8 3CLS

Beyond joint communications and control, one can also envision a need for joint design across all 3CLS functions. For instance, there is little work that rigorously studies the possible interdependence between computing, communication, control, localization, sensing, energy, and mapping, from an end-to-end perspective. Fundamental problems here range from developing new ways to jointly achieve the target performance of all 3CLS functions to introducing multimodal sensor fusion algorithms for faithfully reconstructing 3D images and allowing autonomous vehicles or robots to navigate in unknown environments. 3CLS will be relevant to several 6G applications that include XR, CRAS, and DLT.

7.4.9 Design of 6G Protocols

From the discussion in Sect. 7.2.2 and the identified challenges, one can see that 6G may require a major redesign of protocols. For instance, conventional 5G protocols may need to be replaced with novel AI-powered protocols for various network functions that range from signaling to scheduling. In contrast to the largely rigid designs of 5G protocols, new 6G protocols must be able to continuously evolve with the dynamic state of the wireless system. Moreover, as the development of 6G progresses, one must investigate the possibility of introducing novel protocols for dynamic multiple access [46]. These multiple access protocols must be able to intelligently switch the type of adopted multiple access (orthogonal or non-orthogonal, random or scheduled) scheme based on the network state and the application requirements. In addition, there is also a need for developing novel protocols for device handover that are cognizant of the 3D nature of 6G and the diverse types of mobile devices that must be served. Authentication and identification protocols will also have to be revisited in order to support a new generation of wireless devices such as vehicles, drones, and implants. Finally, 6G may require all protocols to be distributed in order to exploit the small datasets distributed over the system's edge.

7.4.10 *RF and Non-RF Link Integration*

In 6G, it is expected that multiple forms of RF and non-RF links will co-exist. In particular, a 6G device may potentially be able to leverage optical, visible light communication (VLC), molecular communication, and neuro-communication, among others. Indeed, the design of systems with joint RF/non-RF capabilities is an important research area for 6G and beyond 5G.

7.4.11 *Holographic Radio*

By deploying RISs and similar metastructures, it is conceivable that RF holography (including holographic MIMO) and spatial spectral holography will become possible in 6G systems. In essence, holographic RF provides spatial spectral holography and spatial wave field synthesis capabilities that enable a network to control the entirety of a physical space as well as the full closed-loop of an electromagnetic field. This can significantly enhance spectrum efficiency and system capacity. It also will be an enabler for integrating imaging and wireless communication. Clearly, holographic radio is a widely open research problem for 6G.

Finally, to explore all the aforementioned open areas, there will be a need for a broad range of analytical tools. As a result, in Fig. 7.2, we provide a detailed summary of those analytical tools that will play important roles in 6G systems.

7.5 Conclusions

Although it is too early to assert how 6G systems will look like, in this chapter, we provided a rather comprehensive and holistic view on what the main building blocks of 6G systems will be. In particular, we have identified the main limitations of current 5G systems and outlined some of the driving trends and applications behind the leap toward 6G. We have then provided a holistic view on the trends, technologies, and open problems of 6G wireless systems. Although several topics will emerge as a natural evolution from 5G, new research avenues such as RIS and smart surface communication, 3CLS, wireless networking for BCI, and others will create an exciting research agenda for the next decade that will span multiple disciplines as seen in Fig. 7.2. We have also made a few key observations:

(1) The first near-term step toward 6G systems will be to enable mobile broadband services at high-frequency mmWave bands which will be necessary to sustain high-speed communications at high frequencies.

Fig. 7.2 Necessary foundations as well as associated analytical tools for 6G

(2) The next step toward 6G will be to understand the fundamentals of URLLC, with a focus on the notion of reliability. This will require exploring new tools from economics and statistics to quantify the performance of wireless systems in terms of distributions, rather than averages.
(3) Future wireless services will likely require high reliability, low latency, and high data rates which is a significant departure from traditional short-packet, low-rate URLLC services. This, in turn, requires a new understanding of the fundamental trade-offs governing the rate-reliability-latency spaces.
(4) 6G will experience a transition from the smartphone-BS paradigm into a new era of smart surfaces communicating with implants and human-embedded devices. This transition will lead to many new opportunities, but it will also require new ways to define QoS and to seamlessly integrate human users into the wireless communication loop.
(5) Performance analysis and optimization of 6G requires operating in 3D space and investigating a rich system that integrates drones, satellites, flying vehicles, and traditional wireless infrastructure.

(6) The move toward 6G systems will not be yet another case of leveraging additional, high-frequency spectrum bands to deliver higher network capacity, as has been the case for decades. Instead, it will be driven by a diverse set of applications, technologies, and techniques (see Figs. 7.1 and 7.2) as well as a convergence of 3CLS factors.
(7) AI will play an instrumental role in 6G systems. This role ranges from enabling SSNs to embedding collective network intelligence through new notions of edge AI. A new paradigm of joint learning-communication co-design is particularly necessary to widely deploy emerging edge AI algorithms such as federated learning.

In a nutshell, the next decade presents a rich set of opportunities for wireless research that span multi-disciplinary areas. Indeed, the road toward 6G systems will be an exciting era for wireless technologies in which we will see convergence of technologies ranging from AI to computing, control, and cyber-physical systems within the realm of 6G systems.

References

1. A. Karl, Internet of everything vs. internet of things, Feb 2018. [Online]. Available: http://techgenix.com/internet-of-everything/
2. F. Hu, Y. Deng, W. Saad, M. Bennis, A.H. Avghami, Cellular-connected wireless virtual reality: requirements, challenges, and solutions, Jan 2020. [Online]. Available: https://arxiv.org/abs/2001.06287
3. J. Park, S. Samarakoon, H. Shiri, M.K. Abdel-Aziz, T. Nishio, A. Elgabli, M. Bennis, Extreme urllc: vision, challenges, and key enablers, Jan 2020. [Online]. Available: https://arxiv.org/abs/2001.09683
4. M. Chen, U. Challita, W. Saad, C. Yin, M. Debbah, Artificial neural networks-based machine learning for wireless networks: a tutorial. IEEE Commun. Surv. Tutorials, to appear, 2019
5. F.C. Commission, Broadband progress report, 2016. [Online]. Available: https://www.fcc.gov/reports-research/reports/broadband-progress-reports/2016-broadband-progress-report
6. W. Saad, M. Bennis, M. Chen, Wireless communications and applications above 100 ghz: opportunities and challenges for 6g and beyond. IEEE Netw., to appear 2020
7. A.T.Z. Kasgari, W. Saad, M. Debbah, Human-in-the-loop wireless communications: machine learning and brain-aware resource management. IEEE Trans. Commun., to appear, 2019
8. J. Park, S. Samarakoon, M. Bennis, M. Debbah, Wireless network intelligence at the edge, arXiv preprint arXiv:1812.02858, Dec 2018
9. P. Zioga, F. Pollick, M. Ma, P. Chapman, K. Stefanov, Enheduanna a manifesto of falling live brain computer cinema performance: performer and audience participation, cognition and emotional engagement using multi brain BCI interaction. Front. Neurosci. **12**, 191 (2018)
10. T.S. Rappaport, Y. Xing, O. Kanhere, S. Ju, A. Madanayke, S. Mandal, A. Alkhateeb, G.C. Trichopoulos, Wireless communications and applications above 100 GHz: opportunities and challenges for 6G and beyond. IEEE Access **7**, 78729–78757 (2019)
11. G. Lee, J. Park, W. Saad, M. Bennis, Performance analysis of blockchain systems with wireless mobile miners, June 2019. [Online]. Available: https://arxiv.org/abs/1906.06759
12. M. Mozaffari, A.T.Z. Kasgari, W. Saad, M. Bennis, M. Debbah, Beyond 5G with UAVs: foundations of a 3D wireless cellular network. IEEE Trans. Wirel. Commun. **18**(1), 357–372 (2019)

13. Y. Li, J. Liu, B. Cao, C. Wang, Joint optimization of radio and virtual machine resources with uncertain user demands in mobile cloud computing. IEEE Trans. Multimedia **20**(9), 2427–2438 (2018)
14. G. Durisi, T. Koch, P. Popovski, Toward massive, ultrareliable, and low-latency wireless communication with short packets. Proc. IEEE **104**(9), 1711–1726 (2016)
15. M. Bennis, M. Debbah, H.V. Poor, Ultrareliable and low-latency wireless communication: tail, risk, and scale. Proc. IEEE **106**(10), 1834–1853 (2018)
16. O. Semiari, W. Saad, M. Bennis, B. Maham, Caching meets millimeter wave communications for enhanced mobility management in 5G networks. IEEE Trans. Wireless Commun. **17**(2), 779–793 (2018)
17. O. Semiari, W. Saad, M. Bennis, Joint millimeter wave and microwave resources allocation in cellular networks with dual-mode base stations. IEEE Trans. Wireless Commun. **16**(7), 4802–4816 (2017)
18. O. Semiari, W. Saad, M. Bennis, M. Debbah, Integrated millimeter wave and sub-6 GHz wireless networks: a roadmap for joint mobile broadband and ultra-reliable low-latency. IEEE Wirel. Commun. Mag. **26**(2), 109–115 (2019)
19. S. Hu, F. Rusek, O. Edfors, Beyond massive MIMO: the potential of data transmission with large intelligent surfaces. IEEE Trans. Signal Process. **66**(10), 2746–2758 (2018)
20. M. Jung, W. Saad, M. Debbah, C.S. Hong, On the optimality of reconfigurable intelligent surfaces (RISs): passive beamforming, modulation, and resource allocation. IEEE Trans. Wirel. Commun., submitted 2020. [Online]. Available: https://arxiv.org/pdf/1910.00968.pdf
21. M. Jung, W. Saad, Y. Jang, G. Kong, S. Choi, Reliability analysis of large intelligent surfaces(liss): rate distribution and outage probability. IEEE Wireless Commun. Lett. **8**(6), 1662–1666 (2019)
22. M. Jung, W. Saad, Y. Jang, G. Kong, S. Choi, Performance analysis of large intelligent surfaces (LISs): asymptotic data rate and channel hardening effects. IEEE Trans. Wirel. Commun., to appear 2020
23. M. Jung, W. Saad, G. Kong, Performance analysis of large intelligent surfaces (LISs): uplink spectral efficiency and pilot training. IEEE Trans. Commun. submitted 2020. [Online]. Available: https://arxiv.org/abs/1904.00453
24. M. Chen, Z. Yang, W.Saad, C. Yin, S. Cui, H.V. Poor, A joint learning and communications framework for federated learning over wireless networks, 2019. [Online]. Available: https://arxiv.org/pdf/1909.07972.pdf
25. X. Cao, S. Kim, K. Obraczka, C. Wang, D.O. Wu, H. Yanikomeroglu, Guest editorial airborne communication networks. IEEE J. Sel. Areas Commun. **36**(9), 1903–1906 (2018)
26. R.C. Moioli, P.H.J. Nardelli, M.T. Barros, W. Saad, A. Hekmatmanesh, P. Goria, A.S. de Sena, M. Dzaferagic, H. Siljak, W. van Leekwijck, D. Carrillo, S. Latre, Neurosciences and 6G: lessons from and needs of communicative brains, arXiv:2004.01834, 2020. [Online]. Available: https://arxiv.org/abs/2004.01834
27. A.T.Z. Kasgari, W. Saad, M. Mozaffari, H.V. Poor, Experienced deep reinforcement learning with generative adversarial networks (gans) for model-free ultra reliable low latency communication. IEEE Trans. Wirel. Commun. 2019. [Online]. Available: https://arxiv.org/pdf/1911.03264.pdf
28. A.T.Z. Kasgari, W. Saad, Model-free ultra reliable low latency communication (URLLC): a deep reinforcement learning framework, in *Proceedings of the IEEE International Conference on Communications (ICC)*, Shanghai, China, May 2019
29. R. Amer, W. Saad, N. Marchetti, Mobility in the sky: performance and mobility analysis for cellular-connected UAVs. IEEE Trans. Commun., to appear 2020
30. A. Ferdowsi, U. Challita, W. Saad, Deep learning for reliable mobile edge analytics in intelligent transportation systems. IEEE Vehicular Technology Magazine, Special Issue on Mobile Edge Computing for Vehicular Networks **14**(1), 62–70 (2019)
31. Y. Xing, T.S. Rappaport, Propagation measurement system and approach at 140 GHz-moving to 6G and above 100 GHz, in *Proceedings of the of IEEE Global Communications Conference (GLOBECOM)*, Abu Dhabi, United Arab Emirates, United Arab Emirates, Dec 2018

32. C. Chaccour, R. Amer, B. Zhou, W. Saad, On the reliability of wireless virtual reality at terahertz (thz) frequencies, in *Proceedings of the 10th IFIP International Conference on New Technologies, Mobility and Security (NTMS), Mobility and Wireless Networks Track*, Canary Islands, Spain, Jun 2019
33. C. Chaccour, M. Naderi Soorki, W. Saad, M. Bennis, P. Popovski, Can terahertz provide high-rate reliable low latency communications for wireless VR? arXiv:2005.00536, 2020. [Online]. Available: https://arxiv.org/abs/2005.00536
34. C. Chaccour, M. Naderi Soorki, W. Saad, M. Bennis, P. Popovski, Risk-based optimization of virtual reality over terahertz reconfigurable intelligent surfaces, in *Proceedings of the IEEE International Conference on Communications (ICC), Next-Generation Networking and Internet Symposium*, Dublin, Ireland, Jun 2020
35. M. Chen, W. Saad, C. Yin, Liquid state machine learning for resource and cache management in LTE-U unmanned aerial vehicle (UAV) networks. IEEE Trans. Wireless Commun. **18**(3), 1504–1517 (2019)
36. M. Chen, W. Saad, C. Yin, M. Debbah, Data correlation-aware resource management in wireless virtual reality (VR): an echo state transfer learning approach. IEEE Trans. Commun. **67**(6), 4267–4280 (2019)
37. M. Chen, O. Semiari, W. Saad, X. Lin, C. Yin, Echo-liquid state deep learning for 360 content transmission and caching in wireless VR networks with cellular-connected UAVs. IEEE Trans. Commun. **67**(9), 6386–6400 (2019)
38. M. Chen, W. Saad, C. Yin, Virtual reality over wireless networks: quality-of-service model and learning-based resource management. IEEE Trans. Commun. **66**(11), 5621–5635 (2018)
39. U. Challita, W. Saad, C. Bettstetter, Interference management for cellular-connected UAVs: a deep reinforcement learning approach. IEEE Trans. Wireless Commun. **18**(4), 2125–2140 (2019)
40. M. Chen, W.Saad, S. Cui, H.V. Poor, Convergence time optimization for federated learning over wireless networks, Jan 2020. [Online]. Available: https://arxiv.org/abs/2001.07845
41. A. Ferdowsi, W. Saad, Brainstorming generative adversarial networks (bgans): towards multi-agent generative models with distributed private datasets, Feb 2020. [Online]. Available: https://arxiv.org/abs/2002.00306
42. T. Zeng, O. Semiari, W. Saad, M. Bennis, Joint communication and control for wireless autonomous vehicular platoon systems. IEEE Trans. Commun. **67**(11), 7907–7922 (2019)
43. T. Zeng, O. Semiari, W. Saad, M. Bennis, Joint communication and control system design for connected and autonomous vehicle navigation, in *Proceedings of the International Conference on Communications*, Shanghai, China, May 2019
44. T. Zeng, M. Mozaffari, O. Semiari, W. Saad, M. Bennis, M. Debbah, Wireless communications and control for swarms of cellular-connected UAVs, in *Proceedings of the of the 52nd Asilomar Conference on Signals, Systems and Computers*, Pacific Grove, Nov 2018
45. T. Zeng, O. Semiari, M. Mozaffari, M. Chen, W. Saad, M. Bennis, Federated learning in the sky: joint power allocation and scheduling with UAV swarms, in *Proceedings of the International Conference on Communications*, Dublin, Ireland, Jun 2020
46. B. Cao, L. Zhang, Y. Li, D. Feng, W. Cao, Intelligent offloading in multi-access edge computing: a state-of-the-art review and framework. IEEE Commun. Mag. **57**(3), 56–62 (2019)

Part II
5G New Radio Basics

Chapter 8
A Guide to NG-RAN Architecture

Gino Masini

Acronyms

3GPP	3rd Generation Partnership Project
5GC	5th Generation Core Network
AMF	Access and Mobility Function
BAP	Backhaul Adaptation Protocol
CP	Control Plane
CU	Central unit
DL	Downlink
DRB	Dedicated Radio Bearers
DU	Distributed unit
EPC	Evolved Packet Core
E-UTRA	Evolved Universal Terrestrial Radio Access
E-UTRAN	Enhanced Universal Terrestrial RAN
FFT	Fast Fourier transform
HARQ	Hybrid automatic repeat request
IAB	Integrated Access and Backhaul
iFFT	Inverse FFT
IoT	Internet of Things
LTE	Long-Term Evolution
MAC	Medium access control
MBMS	Multimedia Broadcast Multicast Service
MCE	Multicell/multicast coordination entity
MCG	Master Cell Group

G. Masini (✉)
DU Radio - Systems and Technologies, Ericsson AB, Stockholm, Sweden; 3GPP RAN3 Chairman
e-mail: gino.masini@ericsson.com

X. Lin, N. Lee (eds.), *5G and Beyond*, https://doi.org/10.1007/978-3-030-58197-8_8

MIB	Master Information Block
MIMO	Multiple Input Multiple Output
MME	Mobility management entity
MN	Master Node
MT	Mobile Termination
MTC	Machine-type communications
NB-IoT	Narrowband IoT
NG-RAN	New (G) RAN
NR	New Radio
NRPPa	NR Positioning Protocol A
NSA	Non-standalone
PDCP	Packet Data Convergence Protocol
PDU	Protocol Data Unit
PHY	PHYsical layer
RAN	Radio Access Network
RAT	Radio Access Technology
RF	Radio frequency
RLC	Radio Link Control
RRC	Radio Resource Control
SA	Standalone
SCG	Secondary Cell Group
SDAP	Service Data Adaptation Protocol
SIB	System Information Block
SN	Secondary Node
TP	Transmission point
TR	Technical Report
TRP	Transmission and reception point
TS	Technical Standard
UE	User equipment
UL	Uplink
UMTS	Universal Mobile Telecommunications System
UP	User Plane
UPF	User Plane Function
URLLC	Ultra-Reliable, Low-Latency Communications
V2X	Vehicle-to-everything
WG	Working group

1 Introduction

The standardization process makes it possible to synthesize and specify the best possible technical solution under the condition given by all requirements from the interested parties (especially operators, network infrastructure vendors, and chipset makers). What we call today "NG-RAN architecture" was specified starting from such a mix of very different requirements, while the industry was still in the process of consolidating and managing the success of Long-Term Evolution (LTE). Indeed, NG-RAN architecture, as specified by the 3rd Generation Partnership Program (3GPP), is the result of the following requirement areas:

- Support for enhanced versions of today's mobile broadband services ("more of the same, but better and faster").
- Support for Ultra-Reliable, Low-Latency Communications (URLLC).
- A single architecture to accommodate centralized, distributed, and monolithic deployments – a cornerstone of 5G, supporting the deployment of certain functions in the cloud where it is beneficial.
- In conjunction with the previous point, the possibility to fully separate Control Plane (CP) from User Plane (UP) of a centralized unit, for maximum deployment flexibility – another cornerstone of 5G, given that CP and UP scale differently. This enables the Radio Access Network (RAN) to follow the evolution (and "cloudification") that has happened in the core network in recent years.
- Cooperation and resource sharing with existing Enhanced Universal Terrestrial RAN (E-UTRAN) (LTE) networks, to continue leveraging the operators' installed base – a novel kind of requirement for a new RAT.
- Flexibility to accommodate different migration strategies and "paths" from 4G to 5G, given the uncertain way in which the mobile telecommunications market is evolving.
- Time pressure to release "world class specifications" while accelerating the release process as much as possible; this effectively resulted in three release "drops" (often referred to as "early drop," "regular drop," and "late drop") for the same 3GPP Release 15 – a new way of working in telecommunications standards.

All these factors were already present when 3GPP RAN WG3 ("RAN3") started studying New Radio (NR) architecture and NG-RAN in Rel-14. The resulting Technical Report [1], which only hinted at the massive normative work lying ahead, already contained at least three very distinctive characteristics of NG-RAN: the possibility to operate in both standalone (SA) and non-standalone (NSA) mode, a multitude of architecture options to support various combinations of Dual Connectivity with LTE and different core network types, and the possibility to split the 5G base station into a central unit and one or more distributed units.

This level of flexibility, of course, could have resulted in a considerable cost in terms of complexity. But thanks to the hard work and ingenuity of all the people in 3GPP (especially in RAN3, in charge of RAN architecture, interfaces, and protocols), the gigantic effort to standardize and maintain the NG-RAN architecture has been successful so far.

2 NG-RAN Logical Architecture and Building Blocks

NG-RAN is the "new generation" RAN for 5G, providing both NR and E-UTRA ("LTE") radio access. The building block for the NG-RAN logical architecture is the *NG-RAN node*; the NG-RAN node is a *logical node*.

In general, a *logical node* can be characterized as follows:

- It is defined as a collection of *logical functions*, described in a "Stage 2" type of Technical Standard (TS) – for NG-RAN, this type of description can be found in [2–4].
- It terminates a set of *logical interfaces* toward other logical nodes.
- The logical interfaces it terminates directly depend and are defined based on the set of logical functions of the node. In other words, if the logical functions were to change, so would the logical interfaces.
- The logical interfaces are defined starting from their physical layer, transport requirements, and protocols for both Control Plane (CP) and User Plane (UP).
- Different implementations are possible, completely independently from the logical architecture. This means that the logical architecture does not mandate any specific deployment (e.g., co-located, virtualized, centralized, etc.), but rather it accommodates many such deployment types in a transparent manner. For example, two logical nodes might be implemented as a single physical entity (a single "box") without the need to change anything in the standardized logical architecture: the only tangible effect would be that the logical interface between them would "disappear" inside the physical "box" and would not be visible in the implementation.

With the above definition in mind, an NG-RAN node can be either a gNB (which could be envisaged as a 5G base station) providing NR access or an ng-eNB (which could be envisaged as an enhanced 4G base station, hence the prefix "ng-" to distinguish it), providing E-UTRA access. NG-RAN nodes are connected to the 5G Core Network (5GC) with the NG interface and to one another with the Xn interface.

NG and Xn interfaces are similar to their E-UTRAN counterparts S1 and X2 and are fully described in [5–16].

3 Deployment Flexibility and Architecture Options

The ancient Roman god *Janus Bifrons* was the god of transitions and duality; Janus was always looking at the same time at both past and future, war and peace, beginning and end, so he was always represented with two faces. NR was designed with a similar duality in mind: able to cooperate with existing LTE networks but also capable of operating as a new Radio Access Technology (RAT) with a new core network and flexible enough to support different migration paths from one deployment to the other according to the operator's wishes. The NG-RAN architecture makes this possible. To enable this flexibility while maintaining a single, unified logical architecture, it became necessary to define two different modes of operation.

In *non-standalone (NSA) operation*, gNBs and ng-eNBs tightly interoperate with one another and are connected to the same core network, providing *Dual Connectivity* (DC) toward the same terminal. The core network could be either the EPC (the existing LTE core network), in which case we can refer to it as "NSA within 4G RAN," or the 5GC, in which case we can refer to it as "NSA within NG-RAN." The aim of NSA operation is to provide a higher bit rate to the terminal, thanks to Dual Connectivity (and the concurrent roles of a gNB and an ng-eNB).

In *standalone (SA) operation*, the gNB connects to the 5GC. In this case, both the RAN and the Core Network are "native" 5G. Ideally, this could be considered the point of arrival of an operator's migration path from LTE to NR. In this mode of operation, Dual Connectivity (if present) is provided entirely by NR (referred to as NR-NR DC).

Looking at Figs. 8.1 and 8.2 gives us an idea of the "*Janus*-like" characteristics of our architecture.

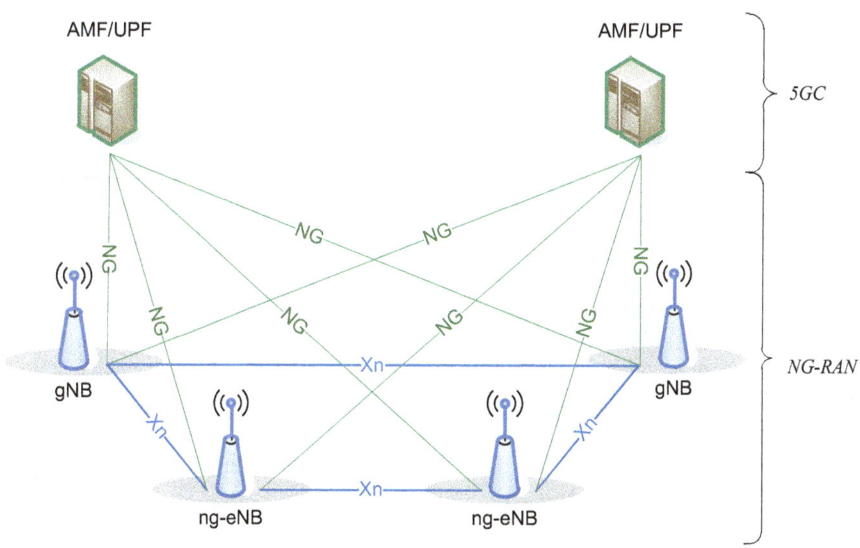

Fig. 8.1 The NG-RAN logical architecture [3]

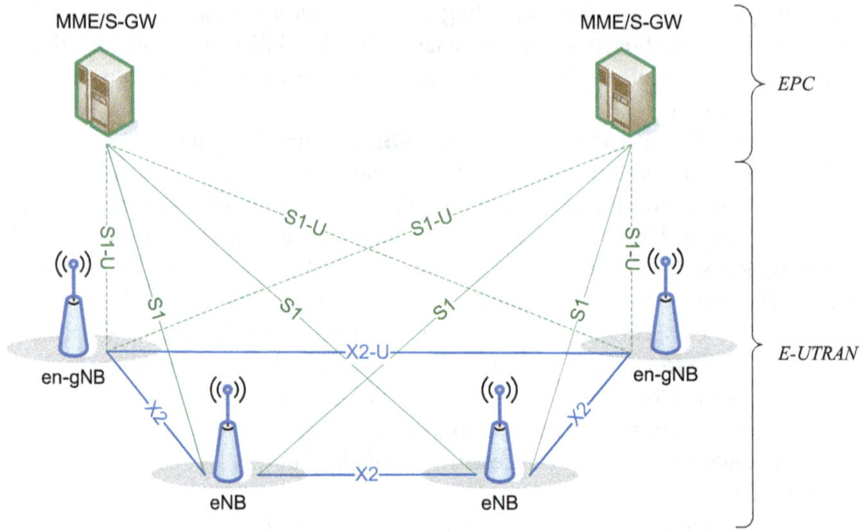

Fig. 8.2 Overall architecture for EN-DC [4]

3.1 A Brief Note on Dual Connectivity (DC)

The concept of *Dual Connectivity* in NG-RAN is derived from the similar functionality specified for LTE in Rel-13: a terminal (UE) capable of multiple reception and transmission may be configured to use resources provided by two different nodes. One node takes the role of Master Node (MN) and the other of Secondary Node (SN). MN and SN are connected via a network interface, and at least the MN is connected to the core network.

From a Control Plane perspective, the UE in DC has a single state in the Radio Resource Control (RRC) protocol, based on the MN RRC and on a single control plane connection toward the core network. Each RAN node has its own RRC entity (E-UTRA RRC if it is an eNB and NR RRC if it is a gNB) which can generate RRC packets (RRC PDUs) to be sent to the UE.

RRC PDUs generated by the SN can be transported via the MN to the UE. The MN always sends the initial SN RRC configuration to the UE, but subsequent reconfigurations may be transported via MN or SN. When transporting RRC PDUs from the SN, the MN does not modify the UE configuration provided by the SN.

Split Dedicated Radio Bearers (split DRBs) are supported for all DC options, allowing duplication of RRC PDUs generated by the MN via the direct path and via the SN.

From a User Plane perspective, three bearer types exist: Master Cell Group (MCG) bearer, Secondary Cell Group (SCG) bearer, and split bearer.

By combining the different alternatives for gNB or ng-eNB as MN and SN with the different core network types (EPC or 5GC), a set of *architecture options* can be

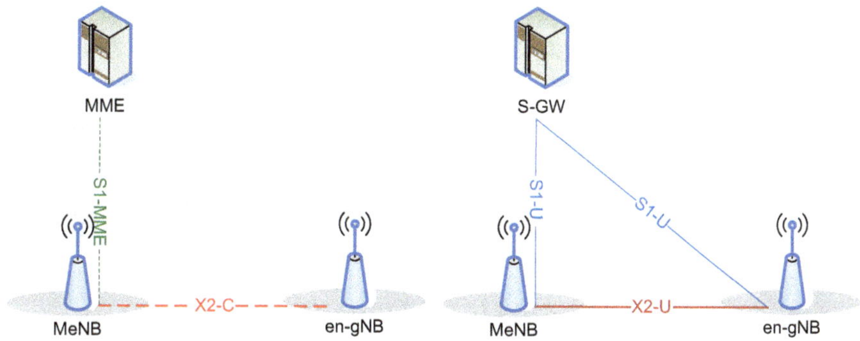

Fig. 8.3 EN-DC logical architecture, showing CP (left) and UP (right) [4]

defined. The various architecture options were defined in the early NR study phase and were first described in [1]. We will briefly describe the most important ones.[1]

3.2 *Option 3 (EN-DC)*

The first architecture option to be introduced in Rel-15 is Option 3 (also referred to as E-UTRAN-NR Dual Connectivity, or EN-DC). EN-DC is the enabler for 5G into an already deployed LTE infrastructure: it leverages Dual Connectivity to enhance the data rate toward the UE for mobile broadband services, using the 5G base station as a capacity booster.

In EN-DC, a UE is connected to one eNB that acts as an MN (MeNB) and one gNB (en-gNB) that acts as an SN. All logical interfaces used are the 4G versions, appropriately enhanced in Rel-15 for this functionality. The eNB is connected to the EPC via the S1 interface and to the en-gNB via the X2 interface. The en-gNB might also be connected to the EPC via the S1-U interface and other en-gNBs via the X2-U interface (Fig. 8.3).

Only a subset of 5G radio functionality is needed for this use: the RRC connection from the RAN to the UE is "owned" by 4G (the eNB). The en-gNB is required to terminate X2-C, X2-U, and S1-U interfaces, which makes it a special kind of gNB, hence, the prefix "en-" to distinguish it.

[1]The numbering of NG-RAN architecture options dates back to the 5G study phase. Nowadays, such numbers are for reference only and are often used interchangeably with the acronyms shown in parentheses in the following sections.

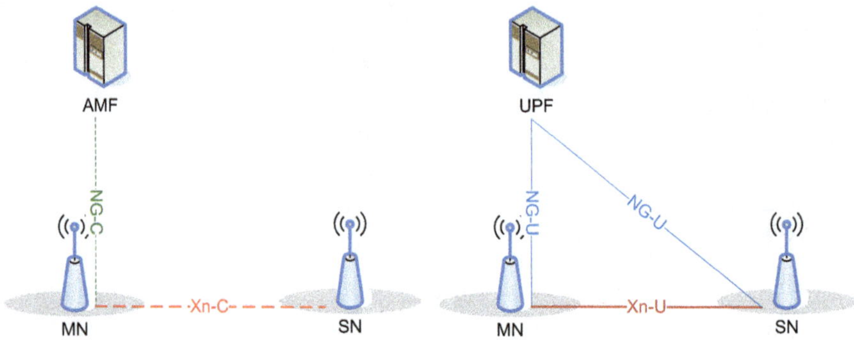

Fig. 8.4 Logical architecture showing CP (left) and UP (right) for NE-DC, NGEN-DC, and NR-NR DC [4]

3.3 *Option 4 (NE-DC)*

Option 4, also referred to as NR-E-UTRAN Dual Connectivity (NE-DC), is part of the "late drop" for Rel-15. With respect to EN-DC, all the 5G logical nodes are swapped with their 4G counterparts, and vice versa: NE-DC requires the 5G Core Network, to which a gNB connects, acting as an MN; a 4G logical node (ng-eNB) acts as SN. The logical interfaces used are the 5G versions, so the en-gNB is required to be upgraded in its functionality, among other things, to terminate Xn-C, Xn-U, and NG-U toward other network nodes (Fig. 8.4).

NE-DC uses 4G as capacity booster in Dual Connectivity while leveraging 5G as coverage anchor.

Option 4 is an addition to Option 2, using Dual Connectivity with an NR anchor. It is primarily relevant when serving mobile broadband traffic via 5GC. The driver for Option 4 is to maximize throughput when the amount of NR spectrum is limited (e.g., if NR is deployed on 700 MHz, 3.5 GHz, and millimeter bands, but the UE is out of coverage of the two higher bands).

This option will typically require new software support in eNB, gNB, and UE, including the necessary interoperability testing. And since it "anchors" the deployment to the presence of the ng-eNBs, its use would require continued investments in such a technology for a significant amount of time.

Focusing on Option 2, on the other hand, will focus the investments on the long-term target architecture (the "end station" of the "journey" to 5G) [17].

3.4 *Option 7 (NGEN-DC)*

NG-RAN Option 7 (NGEN-DC) is the remaining architecture option which leverages multi-RAT Dual Connectivity; it is also introduced as part of the Rel-15 late drop. In this case, the MN role is taken by the eNB, and the SN role is taken by the gNB, similar to Option 3, but the core network required for this option is the 5GC. The set of interfaces which are used are NG and Xn.

Option 7 builds on Option 5 (see Sect. 3.6 below) and cannot exist without it. If Option 5 were to be used, it is very likely that Option 7 would also be supported in areas with NR. It is to be noticed that the driver for Option 7 is the same as for Option 3, namely, to use Dual Connectivity to aggregate NR and LTE bands to enhance capacity, but in this case for a UE connected via eLTE to 5GC [17].

3.5 *Option 2*

The architecture option where both the RAN and the Core Network are "native" 5G is called Option 2; this architecture option identifies the standalone (SA) mode of operation. When Dual Connectivity is provided in this architecture option, both the MN and the SN are gNBs, and it is called NR-NR DC.

3.6 *Option 5*

In Option 5, the upgraded eNB (ng-eNB) is connected to the 5GC. This architecture option is related to Option 7: it really consists of Option 7 without the Dual Connectivity part.

The main driver for deploying Option 5 is to allow devices that move outside the area covered by Option 2 to remain connected to 5GC, which would also increase the 5GC coverage. Option 5 could indeed be used to increase wide-area 5GC coverage, but this would require new UEs, new RAN functionality, and additional system testing (with major impact on both the UE and the network) [17].

3.7 *Migration from 4G to 5G*

From the start of the NR study phase in 3GPP, different business models from a very diverse set of operators resulted in the specification of a very diverse set of architecture options for NG-RAN. Moreover, also the migration from an "all-LTE" to an "all-NR" mobile network can be performed in many ways, at least in theory, depending on the business priorities, strategic choices, and market situation for each operator. In reality, the set of options that an operator will have to "pragmatically" choose from will be much more limited.

Any potential exercise on prioritizing the various architecture options in 3GPP early in the study process proved particularly challenging. Nonetheless, it was

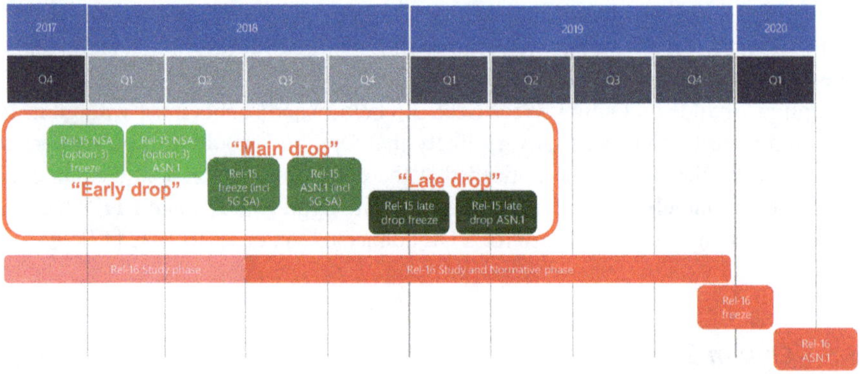

Fig. 8.5 NR specification timeline in 3GPP RAN, highlighting the Rel-15 "early," "main," and "late" drops [18]

possible to group NR architecture options and features in three sets, called "drops," which would be part of the 3GPP Release 15 timeline:

- *Rel-15 Early Drop* – this was the first set of features to be introduced in time; it consists of Option 3 (EN-DC), and it was designed to be the first "stepping stone" into the 5G era. This addressed the most urgent deployment needs for mobile broadband services, leveraging DC.
- *Rel-15 Main Drop* – introduced immediately after the "early drop," the main drop includes NR standalone functionality with the 5G core (Option 2).
- *Rel-15 Late Drop* – this drop includes the remaining DC options (Options 4 and 7 and NR-NR DC), to complete the full set of NR architecture options. This drop contains all remaining options potentially useful for migration from 4G to 5G, so it aims to provide some additional "glue" to bind together some residual scenarios.

But the *specification* timeline for the NR architecture options only shows that NSA was deemed more urgent than SA, which in turn was deemed more urgent than Options 4 and 7. There is no indication of how these options may accommodate *an operator's growth strategy* in time (Fig. 8.5).

In general, an operator's migration path toward 5G depends on many factors, including [19]:

- The operator's business strategy, including the decision on when to deploy the 5GC (thereby introducing the distinctive 5G features, e.g., network slicing)
- Availability of new frequencies for NR in the area considered
- The density of the existing network
- The estimated growth of end-user traffic and demand for services
- The availability of 5G-ready terminals with the appropriate feature set and supporting the right frequency bands
- Other factors

In this ideal journey, we can take as "departure" an LTE RAN connected to the EPC and as "end station" the NR SA deployment (Opt. 2). At least in theory, several "itineraries" are possible for this journey; however, not all of them will be justified by real-life requirements [17]. This journey should be as direct and as profitable as possible for an operator.

EN-DC is likely to be the first stop in this journey. If NR is deployed on higher frequencies (e.g., above 6 GHz), NR coverage is going to be much smaller than on LTE; hence, at least initially, it will be natural to use LTE for coverage and NR for higher capacity in busier areas.

From this first step, it will be possible to introduce the 5GC. As the NR coverage grows and densifies, it will eventually take over as coverage anchor, and an operator's deployment will morph into Option 2. Then the migration toward Option 2 (SA operation), the final stop of the journey, will be complete.

4 Splitting the RAN Node: From the Atom to the *Matryoshka*

Since the early phases of the NR study in 3GPP, it was understood that in order to support the whole set of diverse 5G use cases (including mobile broadband, IoT, URLLC, and ultra-high throughput, among others), the 5G network needed to include the capability to place selected logical functions closer to the network edge or to separate Control Plane and User Plane handling, when needed. Such capability would have additional benefits, including [1, 20]:

- More flexibility in hardware implementation, allowing better scalability and cost-effectiveness
- Better coordination of features and load management and better performance optimization
- Enabling of virtualized deployments and software-defined networking
- Better adaptation to different user density and load demand in a given geographical area
- Better adaptation to variable transport network performance

It was also understood that the NR design should support the possibility to distribute RAN functions between a "centralized" and a "distributed" unit, depending on the above. New dimensions of flexibility in the architecture were then required for 5G.

LTE had been designed with the eNB as the basic building block, and in the standard 3GPP E-UTRAN architecture, the eNB is "atomic" (from the Greek ἄτομος, indivisible). This resulted in a very compact RAN architecture, requiring fewer interactions between logical nodes to be specified with respect to Universal Mobile Telecommunications System (UMTS).

For 5G, it was now required to "split the atom" in a standardized fashion and possibly in more than one way – a challenge within a challenge. Additionally, it was required that "monolithic" and "split" gNBs would coexist and interoperate in the same network and that the rest of the network would ignore whether (or how) a certain network node was split. A gNB, in other words, should always “appear” and “behave” to the Core Network and to all other gNBs in the same way, regardless of whether it was “monolithic” or not. The split architecture, if any, would remain nested inside the gNB, yet it would be fully exposed and specified in the standard to allow full interoperability of all its parts. This was soon recognized as a distinctive feature of the 5G architecture.

The *Matryoshka* (set of traditional Russian wooden dolls, nested one inside the other) might be a good analogy for this concept: the outer *Matryoshka* will always appear the same regardless of what it contains, but it can be opened in order to look at the inner *Matryoshkas*, if present.

4.1 CU-DU Split

Description

The NG-RAN specification supports splitting the gNB into a gNB-CU (central unit) and a gNB-DU (distributed unit). The gNB-CU and the gNB-DU are both logical nodes, connected by the F1 interface (which has both Control and User Plane functions, defined in [16, 21–25]). Figure 8.6 shows the NG-RAN architecture with the CU-DU split visible for the gNB on the right. The gNB-CU terminates the NG and Xn interfaces toward the rest of the network, so the other gNBs and the 5GC see the gNB-CU and the gNB-DUs connected to it as a gNB. Hence, according to our analogy, only the outer *Matryoshka* is visible to the rest of the network regardless of what is inside.

The gNB-CU hosts the RRC, Service Data Adaptation Protocol (SDAP[2]), and Packet Data Convergence Protocol (PDCP) (“high layers”) and may connect to one or more gNB-DUs. The gNB-DU hosts the RLC, MAC, and PHY layers (“low layers”) and may support one or more cells; one cell, however, is supported by only one gNB-DU. A gNB-DU connects to only one gNB-CU.

This functional split between gNB-CU and gNB-DU (“centralized” PDCP and RRC; “decentralized” RLC, MAC, and PHY) closely resembles the protocol stack for Dual Connectivity: for DC, the Master Node and the Secondary Node are split in the same way as the gNB-CU and the gNB-DU.

Among the benefits for this functional distribution between gNB-CU and gNB-DU, identified early in the study phase, are that it will allow centralization of traffic aggregation from NR and E-UTRA transmission points and that it can facilitate traffic load management between such points. Considering the possible migration strategies from LTE to NR, this was considered beneficial. Furthermore, this option “opened up” further possibilities to separate the PDCP for the CP and UP stacks into

[2]SDAP is not present in an en-gNB.

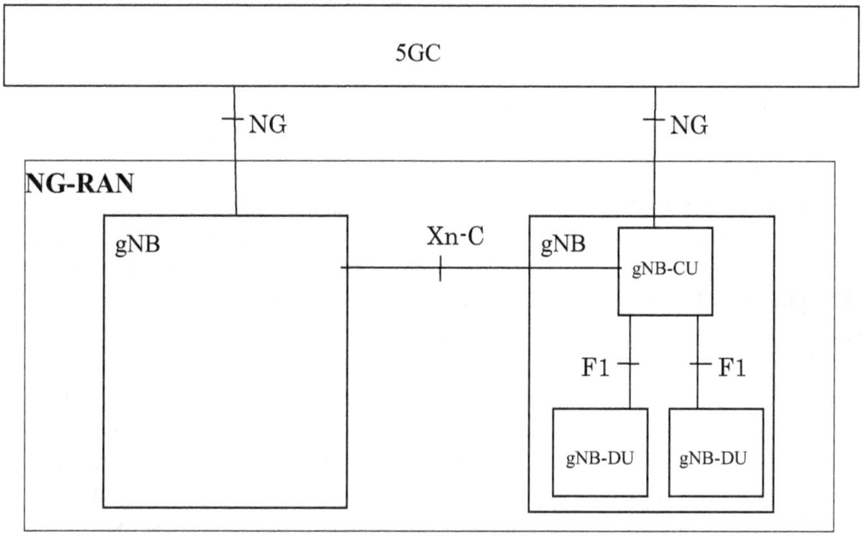

Fig. 8.6 Overall NG-RAN architecture showing the CU-DU split [2]

different central entities, allowing a separate UP while having a centralized RRC. This would bring better scalability according to User Plane traffic load [1]. This possibility eventually became reality, and the CP-UP split (see Sect. 4.2) was also specified in the standard.

The gNB-CU "owns" and manages the UE context (the set of information associated with a UE served by the gNB) and requests the gNB-DU to allocate or modify the required radio resources for the UE. The gNB-DU can accept or reject the request based on admission control criteria (e.g., in case the requested radio resources are not available), and it can also request to modify the resources for an existing UE context. The gNB-DU is responsible for scheduling and broadcasting of system information, and it encodes the NR MIB (Master Information Block) and SIB1 (System Information Block 1); the gNB-CU encodes the other System Information messages.

Consistently with this functional distribution between gNB-CU and gNB-DU, the functions defined for the Control Plane part of the F1 interface include the following [21]:

- *Interface management* – setting up and resetting the F1 interface, updating gNB-CU and gNB-DU configurations, and error indication
- *System information management* – transferring the appropriate information to support broadcasting of SIBs and MIB
- *UE context management* – establishment, modification, and release of the UE context
- *RRC message transfer* – transferring RRC messages from gNB-CU to gNB-DU and vice versa

Impacts of the High Layer Split on Other RAN Functions

The fact that in 5G it is possible to split the gNB requires to further analyze some RAN features if they are to be adopted from 4G. Features such as positioning, multicast/broadcast (called MBMS (Multimedia Broadcast Multicast Service) for E-UTRAN), and others were defined for E-UTRAN where the eNB is "monolithic"; their specification in the presence of a split gNB implies some additional considerations. Either they are adopted from E-UTRAN (with the necessary modifications for NR) assuming a "monolithic" gNB only, or they are also specified for the CU-DU split case. In the first case, inter-vendor interoperability for these features for split gNBs is not guaranteed. In the second case, these features may need to be "broken down" and to some extent redesigned into smaller parts to be assigned to the gNB-CU or to the gNB-DU, and the necessary support needs to be added to the F1 interface. While requiring considerable additional effort, the second case may enable potential enhancements and optimizations to these features due to the split architecture, for example, because of similar functionality already present in the gNB-CU or in the DU. As always, the standardization process is a balance between interoperability, implementation flexibility, and specification effort. When considering the adoption of 4G RAN functionality into a possible 5G equivalent, the balance between reusability and optimization is another critical trade-off in standardization. The result will depend on the different business strategies of operators and equipment vendors.

With that in mind, we will briefly mention possible impacts of the gNB CU-DU split on two popular RAN features: positioning and multicast/broadcast.

Impacts on 5G Positioning Architecture

5G positioning is going to be discussed later in more detail. In general, 5G positioning architecture is derived from the E-UTRAN positioning architecture, with the appropriate modifications due to the different logical nodes involved in the core network.

Figure 8.7 shows the Rel-15 positioning architecture for NG-RAN. Of interest to us at this stage is only the fact that also for positioning, the building block in NG-RAN is the NG-RAN node. Among other things, we can see that the gNB exchanges the necessary positioning information with the core network (this information is conveyed by the NRPPa (NR Positioning Protocol A) protocol [26], transported over the CP of the NG interface [8]). Furthermore, transmission points (TPs) such as remote radio heads, described in this context for positioning purposes, may be part of the ng-eNB. Notice that in Rel-15 the split gNB is not considered for positioning: hence, how to support this feature in the presence of a gNB-CU and a gNB-DU is left to implementation. In Rel-16, however, the gNB-CU and the gNB-DU become part of the standardized positioning architecture, and their set of logical functions include the necessary positioning functionality.

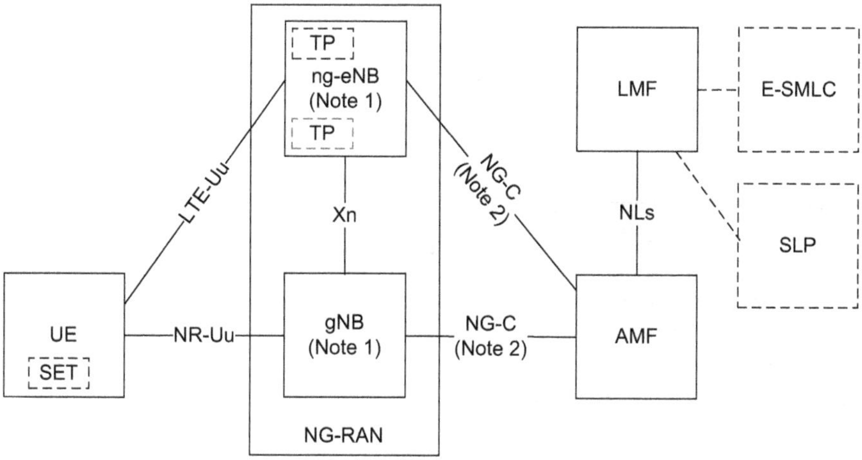

Fig. 8.7 Rel-15 positioning architecture in NG-RAN [27]

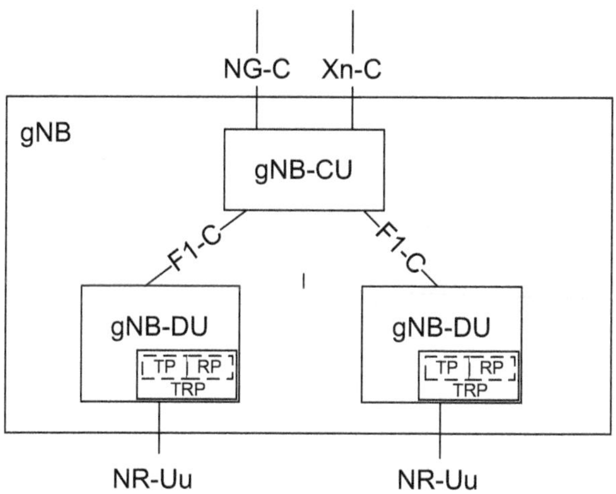

Fig. 8.8 Rel-16 positioning architecture in NG-RAN, showing a split gNB [28]

A possible way to include the gNB-CU and the gNB-DU in the Rel-16 NG-RAN positioning architecture, endorsed by RAN3 at the time of writing, is shown in Fig. 8.8. Let us focus on the gNB: the gNB-CU terminates the NRPPa protocol, and the transmission and reception points (TRPs), if present, are part of the gNB-DU. This is a logical consequence of the fact that the gNB-CU and the gNB-DU already host the necessary functionality (information exchange with the core network and RF functionality, respectively).

The above is only an example of how the gNB split may impact 5G functionality when imported from 4G architecture. Other impacts of CU-DU split on positioning architecture can be expected, for example, on the placement of positioning measurement functionality: if such functionality is to reside in the gNB-DU, the appropriate support may need to be added to the F1 interface.

Impacts on Potential 5G Multicast/Broadcast Functionality

Another popular feature with impact on RAN architecture is cellular multicast/broadcast, called MBMS in E-UTRAN. A full description of MBMS is out of our scope, but for our purposes, it is useful to consider how the corresponding architecture aspects, once adopted in 5G, would be impacted by the CU-DU split.

MBMS architecture in E-UTRAN calls for a logical node, the multicell/multicast coordination entity (MCE), between the eNB and the mobility management entity (MME). Among other things, the MCE is responsible for admission control, for the allocation of radio resources used by an eNB for MBMS transmissions, and for suspending and resuming MBMS sessions [29].

Let us concentrate on the control plane (the left part of Fig. 8.9). As with positioning functionality, if we substitute all E-UTRAN nodes with their 5G counterparts (e.g., the eNB with the gNB, the MME with the Access and Mobility Function (AMF)) and create 5G equivalents of the MCE and of the M2 and M3 interfaces, in principle, the same MBMS architecture could be adopted in 5G, at least for a monolithic gNB. Then, in order to standardize a potential "5G-MBMS" also for a split gNB, the F1 interface would then need to be extended to support the appropriate information exchange.

But if we consider the split gNB from the beginning, we can see that at least some MCE functionality is similar to what can be found in the gNB-DU (e.g., radio resource allocation), while some other is similar to what can be found in the gNB-

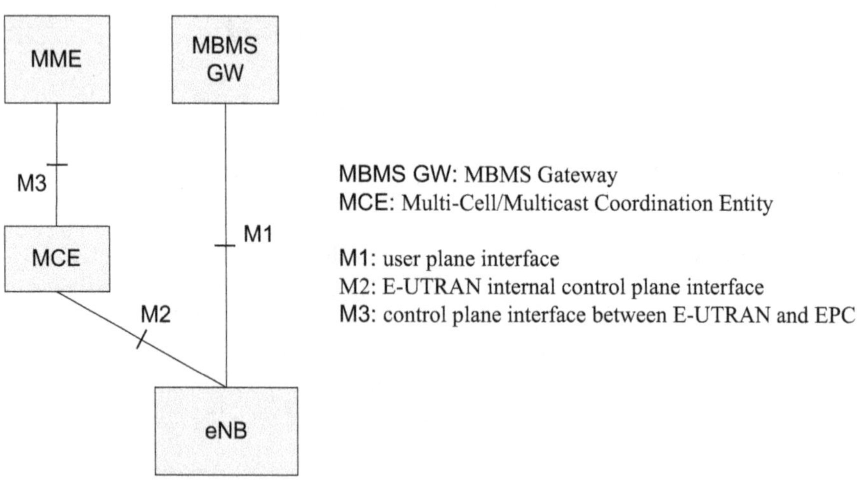

Fig. 8.9 MBMS logical architecture for E-UTRAN [29]

CU (e.g., the exchange of information with the core network). So, standardizing a "5G-MBMS" with the CU-DU split in mind from the beginning might lead to a different architecture altogether, possibly more optimized and more integrated with the gNB-CU and the gNB-DU; even the role and definition of the MCE might need to change. On the other hand, such a strategy might make it difficult to reuse parts of existing 4G implementations for the equivalent 5G feature.

Influences on LTE: eNB Architecture Evolution

The work on NR architecture had shown the benefits of splitting the "atomic" RAN node, at least for the gNB; the influence of such work inspired some companies to propose exploring a similar possibility for 4G. A study was then performed to analyze a possible split for the eNB using a similar architecture as for the gNB high layer split (RRC and PDCP in the eNB-CU; RLC, MAC, and PHY in the eNB-DU; a new interface between them called W1, at first named V1).

One of several challenges of trying to split the eNB is as mentioned, that LTE with all its functionality had been designed from the start with the "atomic" eNB in mind. For this reason, the eNB functionality was never conceived to be split between a gNB-CU and a gNB-DU. By the time of Rel-15, such functionality had grown to include Narrowband IoT (NB-IoT), machine-type communications (MTC), MBMS, vehicle-to-everything (V2X), and others, which had no NR counterpart and therefore was not supported over F1; for this reason, there was no corresponding F1 functionality for a potential "backporting." Furthermore, unlike its NR counterpart, LTE RRC had been optimized for the "atomic" eNB and did not seem well suited for this a posteriori splitting exercise.

The study concluded nonetheless that such a high layer split for the eNB was feasible to be specified for E-UTRAN and/or NG-RAN [30]. Such specification can be expected to be part of 3GPP Rel-16; it is probably too early to tell whether such a high layer split will be feasible in real networks, considering also that LTE is a rather mature technology.

Previous Studies on Low Layer Split for the gNB

Several alternatives for splitting the gNB between its "higher" and "lower" layers were considered during the NR study phase. In fact, the study started from a very general description, listing eight possible options (Fig. 8.10) and even considering additional variants[3]:

Option 1 – RRC in the central unit; PDCP, RLC, MAC, PHY, and RF in the distributed unit

[3]The numbering of gNB split options has no relationship to the numbering of NG-RAN architecture options.

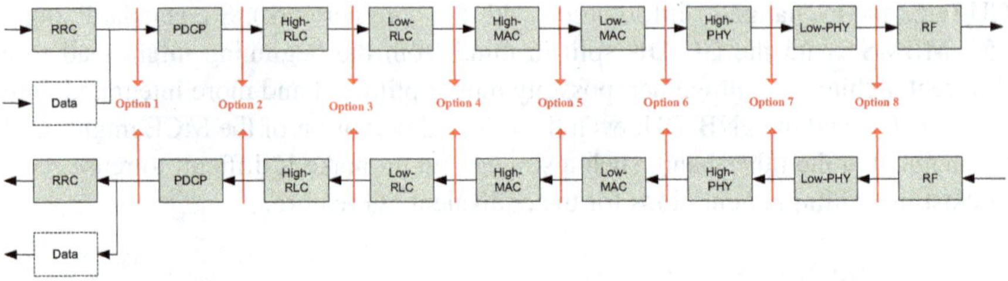

Fig. 8.10 Possible options for the function split between a central and a distributed unit [1]. Option 2 is the one that was standardized

Option 2 (selected for standardization) – RRC and PDCP in the central unit; RLC, MAC, PHY, and RF in the distributed unit
Option 3 (intra-RLC split) – low RLC (partial function of RLC), MAC, PHY, and RF in the distributed unit; PDCP and high RLF (the remaining function of RLC) in the central unit
Option 4 (RLC-MAC split) – PDCP and Radio Link Control in the central unit; MAC, PHY, and RF in the distributed unit
Option 5 (intra-MAC split) – upper layers in the central unit; RF, PHY, and some parts of MAC (e.g., HARQ) in the distributed unit
Option 6 (MAC-PHY split) – upper layers in the central unit; PHY and RF in the distributed unit
Option 7 (intra-PHY split) – upper layers in the central unit; part of PHY and RF in the distributed unit
Option 8 (PHY-RF split) – upper layers in the central unit; RF in the distributed unit

In general, having multiple options in the standard hinders interoperability and requires additional effort at all levels to resolve possible configuration mismatch problems. Therefore, especially for the general architecture, it is critical that a single option is selected for standardization. Option 2 was eventually selected for normative work and standardized, resulting in the current architecture described in Sect. 4.1.1.

Some companies in the operator community, however, felt that some "low layer split" options could have benefited from further study, to better assess their relative technical benefits and feasibility. Their ambition level aimed at potentially specifying an "alternative" split. According to our previous discussion on logical nodes, a central and distributed unit specified according to such an alternative split would be different from the gNB-CU and gNB-DU in "Option 2," and the interface between them would not be the same as F1. In other words, this would result in a *Matryoshka* with inner dolls incompatible with the other set.

Such a study was performed in parallel to 5G specification work, and it aimed to further analyze and refine Options 6 and 7. In order to do this, one of the inner *Matryoshkas* would have to be further opened to look at its components (Fig. 8.11):

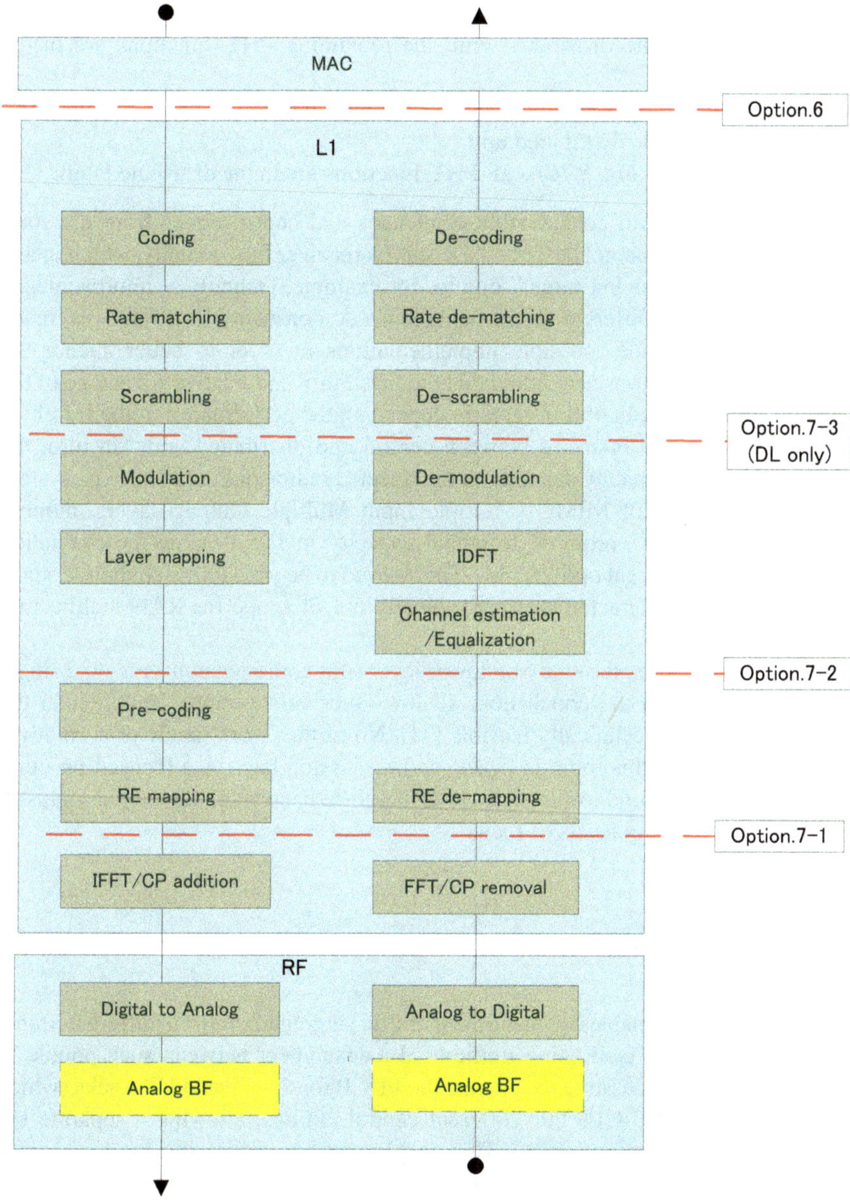

Fig. 8.11 Possible lower layer split options studied by 3GPP RAN3 for DL (left) and UL (right) [31]

Option 7-1 – FFT and Cyclic Prefix removal/addition are in the distributed unit; the remaining PHY functions are in the central unit.

Option 7-2 – FFT, Cyclic Prefix removal/addition, and resource mapping/de-mapping are in the distributed unit; the remaining PHY functions are in the central unit.
Option 7-3 (DL only) – Only the encoder is in the central unit; the remaining PHY functions are in the distributed unit.
Option 6 (same as in Fig. 8.10) – all PHY functions are in the distributed unit.

The study met with considerable challenges and controversies from the start. 3GPP specifications do not describe base station receiver functionality, which in any case is always a "moving target" due to, for example, technology improvements, new features, and different implementations. A considerable effort was made nonetheless to consider possible implementations in order to better anchor the technical analysis. It was also found that the feasibility and relative advantage of the various lower layer split options greatly depend on the performance of the fronthaul used to transport the interface between central and distributed units. In turn, the required fronthaul capacity depends on the selected radio configuration (e.g., system bandwidth, number of MIMO (Multiple Input Multiple Output) layers, number of antenna ports); the required fronthaul capacity in fact became the evaluation criterion for the different options [31]. This proved to be yet another challenge, since transport network characteristics are typically out of scope for RAN architecture specifications.

Given the situation, the study on lower layer split could not converge on a single option: according to the conclusion, all low layer split options identified in the study were deemed technically feasible [31]. No further work (study or normative) could be agreed on this topic in 3GPP, and discussion has since focused on other things. The work around lower layer splits has since been taken up in other industry alliances, with varying levels of focus.

4.2 CP-UP Split

The CU-DU split enables to centralize traffic aggregation from different transmission points, thus facilitating traffic load management between such points. In addition, it makes it possible to separate the PDCP for the CP and UP stacks (which is hosted in the gNB-CU) into different central entities, allowing a separate UP while having a centralized RRC. This enables further optimizing the location of the different RAN functions for better scalability, and it is yet another example of how NG-RAN architecture can support the "cloudification" of the appropriate RAN functions.

This results in the splitting of the gNB-CU into its CP and UP parts, resulting in the gNB-CU-CP and in the gNB-CU-UP, connected by the E1 logical interface. We have thus split also the inner *Matryoshka*.

The gNB-CU-CP hosts RRC and the CP part of the PDCP protocol; it also terminates the CP part of F1 (F1-C) toward the gNB-DU, as well as E1 toward the gNB-CU-UP. The gNB-CU-UP hosts SDAP[4] and the UP part of PDCP; it also terminates the UP part of F1 (F1-U) toward the gNB-DU, as well as E1 toward the gNB-CU-CP.

With this further split, a gNB may consist of one gNB-CU-CP, one or more gNB-CU-UPs, and one or more gNB-DUs. A gNB-CU-UP is connected to only one gNB-CU-CP, but implementations allowing a gNB-CU-UP to connect to multiple gNB-CU-CPs (for redundancy purposes, for example) are not precluded. One gNB-DU can connect to multiple gNB-CU-UPs under the control of the same gNB-CU-CP, and one gNB-CU-UP can connect to multiple gNB-DUs under the control of the same gNB-CU-CP [2] (Fig. 8.12).

The gNB-CU-CP requests the gNB-CU-UP to set up, modify, and release the bearer context; the gNB-CU-UP can accept or reject the setup or modification based on admission control criteria (e.g., in case the requested resources are not available). The gNB-CU-UP can also request a bearer context modification to the gNB-CU-CP, which can accept or reject the request.

Among other things, the gNB-CU-CP is responsible for bearer mapping configuration and security configuration and provides them to the gNB-CU-UP; the gNB-CU-CP is also responsible for notifying the gNB-CU-UP of suspension and resumption of bearer contexts.

The gNB-CU-UP is responsible for notifying the gNB-CU-CP of user inactivity, so that gNB-CU-CP may take further action, and for reporting data volume to the gNB-CU-CP [2].

E1 only has a CP part; it is defined in [32–35]. Its main functions are:

- *Interface management* – setting up and resetting the F1 interface, updating gNB-CU-CP and gNB-CU-UP configurations, and error indication
- *Bearer context management* – setup, modification, and release of the bearer context

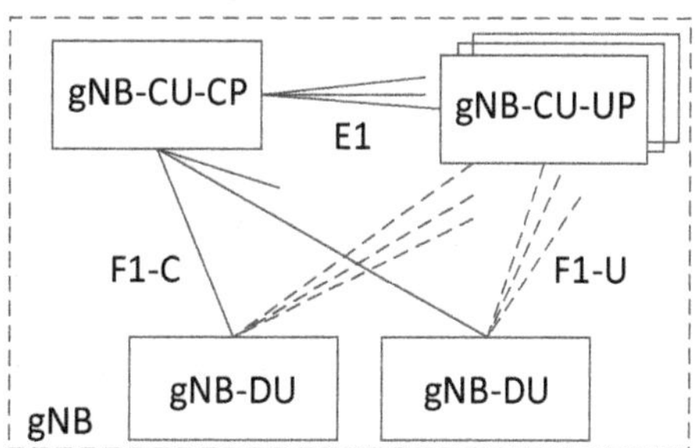

Fig. 8.12 Overall architecture for separation of gNB-CU-CP and gNB-CU-UP [2]

[4]SDAP is not present in an en-gNB.

5 The Unified User Plane

Network interfaces can be typically defined in terms of Control Plane and User Plane. Control Plane carries signaling traffic, and User Plane carries the network user traffic. Apart from few exceptions (like the E1 interface, which only has CP functions), most RAN interfaces include both CP and UP functions.

In Sect. 3, we have seen that DC is one of the key functions provided by NR through NG-RAN. Considering the CU-DU split and the various architecture options, there can be up to three interfaces involved in DC: X2, Xn, and F1 (with both CP and UP). The Xn interface functions, for example, include the following [11]:

Control Plane

- *Interface management and error handling* (including Xn setup, configuration update, error indication, and interface reset and removal)
- *Mobility management* (handover preparation and cancellation, UE context retrieval, paging, data forwarding control)
- *Dual Connectivity* (addition, modification, and release of resources between MN and SN)
- *Energy saving* (indication of cell activation/deactivation)
- *Data volume reporting* (reporting data usage in the secondary RAT to the core network)

User Plane

- *Data transfer* (transferring user data traffic between NG-RAN nodes for mobility or DC)
- *Flow control* (providing feedback information associated with the data flow)
- *Assistance information* (providing information, e.g., related to radio conditions from an NG-RAN node receiving UP data)
- *Fast retransmission* (coordination between two nodes involved in DC in case of outage in one of the nodes, enabling the node in good radio conditions to handle the data previously forwarded to the node in outage)

A similar list can be compiled for F1 [21] (see Sect. 4.1.1) and for X2 [36]. For CP functions, we can observe some similarities between the X2 and Xn interfaces (at least in general terms), while the F1 interface functions significantly differ from the other two. UP functions, however, are remarkably similar for all three interfaces considered: the *flow control function*, in fact, is common to all three in NG-RAN.

This was observed early in the normative phase for NG-RAN, when specifying the details for EN-DC. Although initial discussions on flow control had assumed separate procedures and specifications for X2, Xn, and F1, it was eventually agreed to have a single, unified UP with common flow control for all three interfaces, specified in a single document [16]. If needed, though, specific UP enhancements can be introduced for F1 without breaking this principle. This has become another key feature of the NG-RAN architecture, and it has the following benefits:

- When DC is active between two NG-RAN nodes, it is possible to terminate the UP directly in the end points, regardless of the architecture of the NG-RAN nodes involved. For example, if DC is active between the two NG-RAN nodes of Fig. 8.6, it is possible to terminate the UP in one of the gNB-DUs on the right and in the gNB on the left, bypassing the gNB-CU on the right.[5] This minimizes UP latency.
- A node terminating UP is unaware of whether the remote termination of the UP is in a gNB-CU or a gNB-DU.
- Having the common flow control functionality described in the same specification ensures that the common "language" for the three interfaces is maintained and greatly improves specification maintainability.
- All logical nodes in NG-RAN (e.g., gNBs, eNBs, gNB-CUs, gNB-DUs, gNB-CU-UPs) "speak" exactly the same "language," as far as UP is concerned. This makes it possible for an equipment vendor to reuse the same implementation for the three different interfaces.

6 Building on NG-RAN Architecture: IAB

Integrated Access and Backhauling (IAB) enables NR to act as both access and transport network, increasing the flexibility and versatility of NR for the operator. The main driver for IAB technology is to allow the NR to "backhaul itself" wirelessly, thus providing an alternative to fiber transport. IAB will be discussed in more detail later, but in the scope of this chapter, we can see how the architecture chosen for IAB is anchored in the CU-DU split architecture.

Initial studies for IAB were performed in parallel with the NR specification work, and different architecture options for the IAB node were considered. One such group of architectures ("Architecture Group 1" [37]) had in common the fact that it leverages the CU-DU split architecture. The architecture eventually selected for the normative phase ("Architecture 1a" [37]), expected to be part of Rel-16, is part of this group [37]. The IAB node includes a DU, whose F1 interface is backhauled toward the gNB-CU of a IAB-donor gNB via the NR air interface (Uu). The Uu interface itself is terminated by the Mobile Termination (MT) part of the IAB node (Fig. 8.13).

According to the study conclusion, the architectures of "Architecture Group 1" have advantages in most key performance indicators considered in the study, and the adopted architecture makes it feasible to support the following [37]:

- A physically fixed IAB node
- Both in-band and out-of-band scenarios in both paired and unpaired spectrums
- NR backhauling of NR access traffic

[5]This also applies to X2, so it is valid across all architecture options.

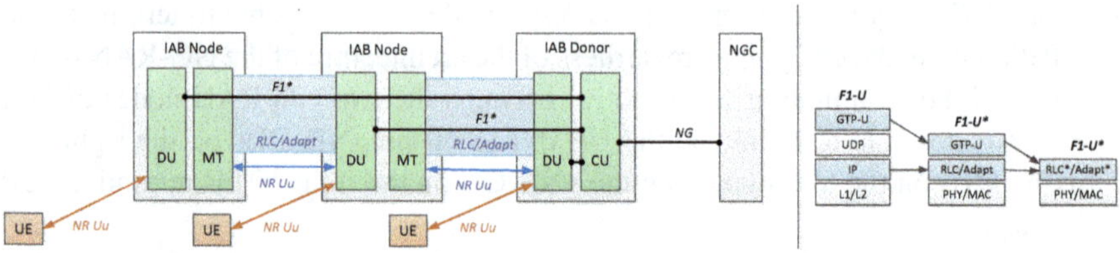

Fig. 8.13 Reference diagram with two IAB nodes and an IAB-donor gNB [37]. (It is worth noting that during the course of the work item leading to the specification of IAB in Rel-16, it was decided to define an IAB-specific protocol (BAP) to enable multi-hop forwarding, according to the description given in [38])

- Both SA and NSA modes for the UE and for the IAB node
- Multi-hop backhauling
- Topology adaptation
- Network synchronization of IAB nodes

The above is yet another example of how the CU-DU split architecture introduced for NG-RAN will be pervasive also for the more future-oriented features of 5G networks.

7 Conclusions

If we had to choose a single slogan to describe and promote the NG-RAN architecture, "future-proof thanks to flexibility" might be a very fitting one.

NG-RAN architecture is all about flexibility. First, thanks to its ambivalence (NSA and SA, split and monolithic, LTE and NR, legacy and future – a modern-day *Janus Bifrons*), it tightly interoperates with existing LTE networks while also existing as a standalone radio access network. Then, since it supports several architecture options, it is designed to adequately address the different deployment scenarios from any operator. Furthermore, additional flexibility has been built into the NG-RAN architecture, thanks to the CU-DU split and the CP-UP split, which have turned the once "atomic" 4G RAN node into a *Matryoshka* of sorts for 5G. The split gNB architecture is so pervasive that it is also influencing how current and future 5G features are being conceived, and it is even being backported to 4G architecture. All such upcoming enhancements are also addressing new requirements beyond mobile broadband (e.g., automated driving, industrial automation, E-health services, and more).

The standardized NG-RAN interfaces and protocols are designed to facilitate the flexible evolution of 4G toward 5G according to the operator's business strategy and to help the update of the 5G core network, thanks to which the distinctive 5G services can be offered to end customers.

Acknowledgments NG-RAN architecture is the result of countless hours of meetings, discussions, and heated but always civilized debates in the 3GPP RAN3 working group. Serving as Chairman of such a talented, dedicated, and passionate group of technologists is an incredible challenge, but it gives one the rare privilege of learning from some of the best in the industry. I must thank the whole group for this privilege.

I am indebted to many wonderful colleagues, especially Martin Israelsson (whom I consider a role model) and all the specialists in the Ericsson RAN3 team. I have received precious feedback and guidance on this material from many colleagues, especially Filip Barac, Joakim Bergström, Angelo Centonza, Elena Myhre, and Per Willars; Yngve Selén also took the time to proofread the draft. My heartfelt thanks go to all of them.

Standards work, including the heavy traveling, takes a massive toll on our families; my wife Daniela and our children Luigi, Lucia, and Giulia have patiently endured this for years. I am deeply grateful to them because nothing would be possible without their love and their support.

References

1. 3GPP TR 38.801: Study on new radio access technology: Radio access architecture and interfaces, v. 14.0.0
2. 3GPP TS 38.401: NG-RAN; Architecture description, v. 15.6.0
3. 3GPP TS 38.300: NR; Overall description; Stage-2, v. 15.7.0
4. 3GPP TS 37.340: NR; Multi-connectivity; Overall description; Stage-2, v. 15.7.0
5. 3GPP TS 38.410: NG-RAN; NG general aspects and principles, v. 15.2.0
6. 3GPP TS 38.411: NG-RAN; NG layer 1, v. 15.0.0
7. 3GPP TS 38.412: NG-RAN; NG signalling transport, v. 15.3.0
8. 3GPP TS 38.413: NG-RAN; NG Application Protocol (NGAP), v. 15.5.0
9. 3GPP TS 38.414: NG-RAN; NG data transport, v. 15.2.0
10. 3GPP TS 38.415: NG-RAN; PDU session user plane protocol, v. 15.2.0
11. 3GPP TS 38.420: NG-RAN; Xn general aspects and principles, v. 15.2.0
12. 3GPP TS 38.421: NG-RAN; Xn layer 1, v. 15.1.0
13. 3GPP TS 38.422: NG-RAN; Xn signalling transport, v. 15.3.0
14. 3GPP TS 38.423: NG-RAN; Xn Application Protocol (XnAP), v. 15.5.0
15. 3GPP TS 38.424: NG-RAN; Xn data transport, v. 15.2.0
16. 3GPP TS 38.425: NG-RAN; NR user plane protocol, v. 15.6.0
17. T. Cagenius, A. Ryde, J. Vikberg, P. Willars, Simplifying the 5G Ecosystem by Reducing Architecture Options, in *Ericsson Technology Review*, 30 Nov 2018. Available: https://www.ericsson.com/en/ericsson-technology-review/archive/2018/simplifying-the-5g-ecosystem-by-reducing-architecture-options
18. B. Bertenyi, Overview of RAN aspects, RWS-180005, *3GPP Workshop on IMT-2020 Submission*, 24–25 Oct 2018, Brussels, Belgium. Available: https://www.3gpp.org/ftp/workshop/2018-10-24_25_WS_on_3GPP_subm_tw_IMT2020/Docs/RWS-180005.zip
19. G. Masini, NR Architecture, RWS-180009, *3GPP Workshop on IMT-2020 Submission*, 24–25 Oct 2018, Brussels, Belgium. Available: https://www.3gpp.org/ftp/workshop/2018-10-24_25_WS_on_3GPP_subm_tw_IMT2020/Docs/RWS-180009.zip
20. B. Bertenyi, R. Burbidge, G. Masini, S. Sirotkin, Y. Gao, NG radio access network (NG-RAN). Journal of ICT Standardization **6**(1–2), 59–76 (2018)
21. 3GPP TS 38.470: NG-RAN; F1 general aspects and principles, v. 15.6.0
22. 3GPP TS 38.471: NG-RAN; F1 layer 1, v. 15.0.0
23. 3GPP TS 38.472: NG-RAN; F1 signalling transport, v. 15.5.0
24. 3GPP TS 38.473: NG-RAN; F1 Application Protocol, v. 15.7.0
25. 3GPP TS 38.474: NG-RAN; F1 data transport, v. 15.3.0
26. 3GPP TS 38.455: NG-RAN; NR Positioning Protocol A (NRPPa), v. 15.2.1

27. 3GPP TS 38.305: NG Radio Access Network (NG-RAN); Stage 2 functional specification of User Equipment (UE) positioning in NG-RAN, v.15.4.0
28. Ericsson, Transmission Measurement Function in NG-RAN, R3-196508, *3GPP TSG-RAN WG3 #106*, Reno, NV, USA, 18–22 Nov 2019. Available: https://www.3gpp.org/ftp/tsg_ran/WG3_Iu/TSGR3_106/Docs/R3-196508.zip
29. 3GPP TS 36.300: Evolved Universal Terrestrial Radio Access (E-UTRA) and Evolved Universal Terrestrial Radio Access Network (E-UTRAN); Overall description; Stage 2, v. 15.7.0
30. 3GPP TR 37.876: Study on eNB(s) Architecture Evolution for E-UTRAN and NG-RAN, v. 15.0.0
31. 3GPP TR 38.816: Study on Central Unit (CU) – Distributed Unit (DU) lower layer split for NR, v. 15.0.0
32. 3GPP TS 38.460: NG-RAN; E1 general aspects and principles, v. 15.4.0
33. 3GPP TS 38.461: NG-RAN; E1 layer 1, v. 15.1.0
34. 3GPP TS 38.462: NG-RAN E1 signalling transport, v. 15.5.0
35. 3GPP TS 38.463: NG-RAN; E1 Application Protocol (E1AP), v. 15.5.0
36. 3GPP TS 36.420: Evolved universal terrestrial radio access network (E-UTRAN); X2 general aspects and principles, v. 15.1.0
37. 3GPP TR 38.874: NR; Study on integrated access and backhaul, v. 16.0.0
38. O. Teyeb, A. Muhammad, G. Mildh, E. Dahlman, F. Barac, B. Makki, Integrated Access Backhauled Networks, in *IEEE Vehicular Technology Conference – Fall 2019*, 22–25 Sept 2019, Honolulu. Available: https://arxiv.org/ftp/arxiv/papers/1906/1906.09298.pdf

Chapter 9
NR Physical Layer Overview

Daniel Chen Larsson

The physical layer of NR is designed to support a large set of use cases from day one and operate tightly with Long Term Evolution (LTE). In addition, NR is designed so that new use case and design can be added on in later releases. It is further designed to support performances in terms of throughputs, latency, energy efficiency, deployment flexibility and different spectrums. The design of NR will support the general technology development during the 2020 decade and 2030 decade, wherein connectivity is a key enabler for all sorts of applications and uses.

NR shares some basics with LTE, in that they are both based on OFDM in downlink (DL) and NR supports both transform precoded Orthogonal Frequency-Division Multiplexing (OFDM) (same as LTE) and OFDM in uplink (UL). NR, however, is designed to support a larger spectrum range and wider carriers. NR supports spectrum ranges from around 500 MHz to up to 52.6 GHz in its first release, and work is being done to expand this beyond 52.6 GHz. A key aspect in this design is the supported subcarrier spacings, wherein NR supports a large set of numerologies in its first release which are 15, 30, 60, 120 and 240 kHz.

NR is designed so that it can tightly coexist with LTE in different levels, where we have the high-layer protocol aspects defining the dual connectivity support. Another aspect is that NR is designed so that an NR carrier can operate in the same spectrum as LTE and even overlap with an LTE carrier.

Many key features from LTE are further supported in NR, such as carrier aggregations. The carrier aggregations are supported from the first release of NR.

When it comes to data rate and particularly latency, NR has taken a large step compared to LTE. The design is done on the physical layer to allow quicker decoding and transmissions of data and control channels. This is, for example, related to placements of reference signals within the data and control channel transmissions, which significantly reduces the minimum latency on LTE Rel-8

D. C. Larsson (✉)
Ericsson AB, Lund, Sweden
e-mail: daniel.chen.larsson@ericsson.com

X. Lin, N. Lee (eds.), *5G and Beyond*, https://doi.org/10.1007/978-3-030-58197-8_9

physical layer, e.g. being 4 ms to something much smaller in value, e.g. in some cases reaching below 1 ms latency. The channel coding schemes have further gotten a slight update where the polar code is used for control signalling and low-density parity-check code (LDPC) is used for data channels.

Energy efficiency has been considered not only on the UE side but also on the network side. On the network side, this enables the base station to go into micro sleep because the base station does not always have to transmit channels and signals. This has a further performance benefit that it creates less interference among base stations and hence the system performance increases. For example, a key factor in NR is the non-existence of general reference symbols, e.g. Cell-specific Reference Signals (CRS) being transmitted, and a large periodicity of the synchronization signals.

Another key aspect built in from the start in NR is the support for beams and the associated active antennas. The beam management aspect has introduced a complete new set of features of both the physical layer and the control plane to be able to handle multiple beams and move the UE between different beams. The MIMO support is further significantly enhanced taking a step from the later LTE releases wherein support for active antennas has been introduced.

Within this chapter, we outline the physical layer highlighting aspects around waveforms and numerologies, bandwidth parts, downlink and uplink control information, downlink and uplink data channels, NR-LTE interworking on the physical layer, power control and UE capabilities.

1 Waveform and Basic Structure of NR

NR has OFDM as the waveform in DL, and for UL, NR supports both OFDM and transform precoded OFDM. The support for transform precoded OFDM is limited in UL to a single layer, while the OFDM-based waveform supports both one and multiple layers. In addition to OFDM in UL, the underlying reason to support transform precoded OFDM is to be able to achieve better coverage in UL, in which scenarios peak to average power ratio (PAPR) becomes influential on the coverage.

As described in the introduction, there is the concept of the general frequency ranges, i.e. frequency range 1 and frequency range 2, in which frequency range 1 covers the spectrum up to 6 GHz. Frequency range 2 covers the spectrum from 24.25 GHz up to 52.6 GHz. Obviously, both frequency division duplexing (FDD) and time division duplexing (TDD) spectrums are supported. The NR is even more generically defined than LTE in terms of FDD and TDD spectrums on the physical layer. There is work ongoing in 3rd Generation Partnership Project (3GPP) to expand the supported spectrum beyond 52.6 GHz, which is targeted for future releases of NR.

There are multiple sets of numerologies that can be used, which are 15, 30, 60, 120 and 240 kHz. Further, NR supports normal cyclic prefix for both waveforms

Table 9.1 Subcarrier spacing together with cyclic prefix and its associated frequency range

Subcarrier spacing	Cyclic prefix	Frequency range
15 kHz	Normal	FR1
30 kHz	Normal	FR1
60 kHz	Extended	FR1
60 kHz	Normal	FR2
120 kHz	Normal	FR2
240 kHz	Normal	FR2

Table 9.2 Supported subcarrier spacings and cyclic prefix for data and SS/PBCH block

Subcarrier spacing	Cyclic prefix	Supports
15 kHz	Normal	Data, SS/PBCH
30 kHz	Normal	Data, SS/PBCH
60 kHz	Extended	Data
60 kHz	Normal	Data
120 kHz	Normal	Data, SS/PBCH
240 kHz	Normal	SS/PBCH

and extended cyclic prefix. Normal cyclic prefix is the main cyclic prefix to be used in practice, and extended cyclic prefix is introduced as an additional option for supporting Industrial IoT application in specific use cases. Industrial IoT application can, of course, be deployed using normal cyclic prefix. Typically, the subcarrier spacing and the cyclic prefix are chosen to be applicable for the given spectrums and deployment scenarios. Since NR supports a large set of subcarrier spacings, it is sufficient in all scenarios currently to deploy normal cyclic prefix and set the subcarrier spacing accordingly. The applicability of different numerologies and cyclic prefix differs between the different frequency ranges as highlighted in Table 9.1.

For a given frequency band regardless of whether it is in frequency range 1 or frequency range 2, the supported subcarrier spacings are given in [1].

There are some limitations in terms of subcarrier spacings. For initial access, there are some limitations in terms of subcarrier spacing and cyclic prefix supported. In addition, not all the subcarrier spacings are supported for data transmissions. The supported subcarrier spacings together with their cyclic prefixes and applicability are shown in Table 9.2. More details are described in Chap. 11 on cell search and random access.

The downlink and uplink transmissions are organized into frames. The frame duration is 10 ms. Each such frame consists of ten subframes, wherein each subframe is 1 ms. Each frame is in addition divided into two equally sized half frames, i.e. five subframes each. The first five subframes in the first half frame and the last five subframes in the last half frame are given in Fig. 9.1. The half-frame concept is used during the initial access procedure. There is one set of frames for downlink and one set for uplink on a carrier.

Each frame is also divided into a number of slots. The number of slots depends on the subcarrier spacing. A slot has a certain number of OFDM symbols, where

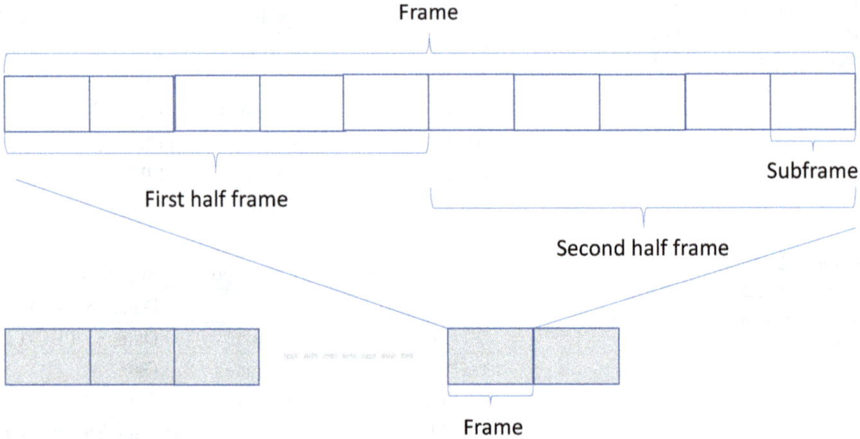

Fig. 9.1 Frame and half frames

Table 9.3 Number of slots per frame and slots per subframe for normal cyclic prefix

Subcarrier spacing	Number of slots per frame	Number of slots per subframe
15 kHz	10	1
30 kHz	20	2
60 kHz	40	4
120 kHz	80	8
240 kHz	160	16

the amount is different between normal and extended cyclic prefixes. For normal cyclic prefix, a slot has 14 OFDM symbols, while for extended cyclic prefix, a slot has 12. The difference is to make the alignment work between subframe lengths, slots lengths and across different numerologies. On the physical layer, the slots are typically used as timing reference together with the frame when applicable. The number of slots per frame and subframe for normal cyclic prefix is given in Table 9.3. It is notable that the subframe definition in NR differs from LTE, wherein the subframe definition from a physical layer perspective in LTE is more similar to the slot definition in NR. The slot lengths are further illustrated in Fig. 9.2, wherein each box is a slot.

The basic transmission unit is a resource element. A resource element is defined by an antenna port, a single subcarrier (and its associated subcarrier spacing) and one OFDM symbol. It is, hence, uniquely defined in the frequency and time domain based on its antenna port and subcarrier spacing.

A resource block is built up by 12 consecutive subcarriers in frequency domain and hence does not have any time domain component as, for example, a resource element does. The relationship between resource block, slot and resource element is illustrated in Fig. 9.3.

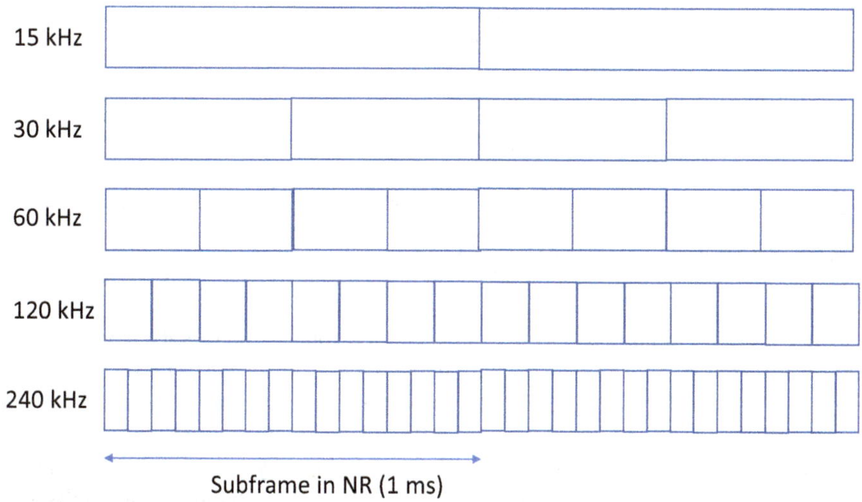

Fig. 9.2 Slot lengths and subframe lengths in NR for different subcarrier spacing

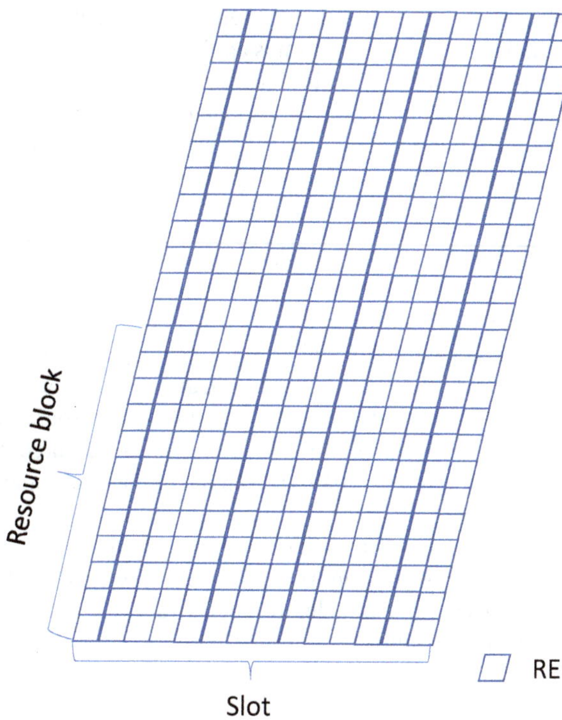

Fig. 9.3 Resource block, slot and resource element

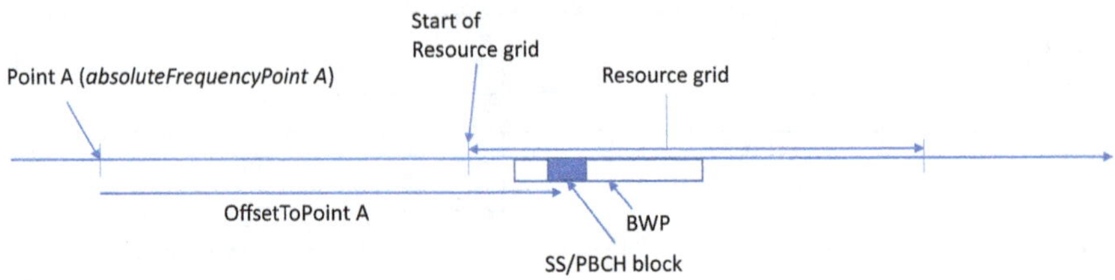

Fig. 9.4 Resource grid and relation to point A and carriers

Each carrier and numerology are placed on a resource grid of subcarriers. There is one resource grid for downlink and one resource grid for UL. The resource grid is defined to cater for the possibility to support multiple numerologies and different bandwidths between different UEs or at different stages at the access to the network by the same UE. The resource grid is defined to be very large so that it is large enough to be future compatible in terms of providing large enough bandwidths and placements. The location of the grid is signalled in relation to point A. There are two ways by which the location of point A can be derived. One is that the UE is aware of the absolute frequency of point A that is signalled as absolute radio-frequency channel number (ARFCN). The mechanism is applicable for the primary cell on the downlink wherein the offset of point A is signalled. The offset is defined in relation to the lowest subcarrier of the lowest resource block that overlaps with the SS/PBCH block used by the UE for initial access. The information is provided in 15 kHz scale for frequency range 1 and 60 kHz for frequency range 2 (Fig. 9.4).

There are three different types of resource blocks.

- Common resource blocks can be translated into an absolute frequency due to their connection to point A. The common resource blocks are numbered starting from 0 to upwards. Common resource block 0 is coinciding with point A.
- Physical resource blocks are defined within a bandwidth part. The bandwidth parts are further described in Sect. 2. The physical resource blocks are numbered from 0 to the size of the bandwidth part. The physical resource block can be converted to a common resource block by the physical resource block number plus the start of bandwidth part (BWP) (given in common resource blocks) for a given subcarrier spacing.
- Virtual resource blocks are also defined within a bandwidth part. The virtual resource blocks are numbered from 0 to the size of the bandwidth part. Virtual resource blocks are used, for example, to support various randomizing functions of the resource element mapping.

As stated above, there is one frame structure in downlink direction and one in uplink direction, and NR supports both FDD and TDD spectrums. A given spectrum is either FDD or TDD spectrum which is known from the band definitions in [1]. For FDD spectrum, whether both DL and UL directions can be utilized simultaneously or not depends on if the UE supports the transmission and reception at the same time. If not, the transmission direction needs to be controlled. For TDD spectrum, the transmission direction obviously needs to be controlled. For FDD spectrum, the DL and UL frames are not on the same carrier frequency and hence not overlapping in frequency, while for TDD spectrum, the DL and UL frames are on the same frequency and are hence overlapping. There are a set of different mechanisms to control it. This is controlled by assigning the slots or OFDM symbols in both the DL and UL frames to be one type of 'downlink', 'uplink' or 'flexible' symbol/slot. Their definitions are as follows:

- An OFDM symbol/slot that is defined as 'downlink' can contain downlink transmissions.
- An OFDM symbol/slot that is defined as 'uplink' can contain uplink transmissions.
- An OFDM symbol/slot that is defined as 'flexible' can contain either uplink or downlink transmissions, depending on which frame the OFDM symbol is associated with. OFDM symbols are assigned as 'flexible' in both the downlink and uplink frames at the same time occasion. This can be set as fast as scheduling allows it.

It is then given that a DL frame can only contain DL transmissions and a UL frame can only contain UL transmissions.

The assignment procedure for a given type of OFDM symbols/slots is as follows. For initial access purpose, the UE needs to assume that all resources are 'flexible'. When receiving the system information, it is possible for the gNB to include information about a fixed pattern on a cell level for the OFDM symbols being either 'downlink', 'uplink', 'flexible' or none. None is signalled by not being assigned to any direction. It is further possible to signal a UE specific modification of this signalling after the cell access is completed. The configurability with the cell level signalling is large as it is generically defined for both frequency range 1 and frequency range 2. For example, for frequency range 1, it is, of course, possible to configure it so that an NR system can operate adjacent or co-channel with an LTE TDD system. The NR system has either 15 or 30 kHz subcarrier spacing. If no cell level signalling is provided by the gNB to the UE, the UE assumes that all OFDM symbols are flexible.

It is further possible through a downlink control information (DCI) message to indicate a specific resource direction on a more short-term basis of the flexible resources. This is referred to as slot format indicator (SFI).

Depending on how faraway a UE is from the base station's antenna, the propagation time to and from it will be different. In an OFDM-based system, all the UE transmissions need to reach the base station roughly at the same time for the base station to be able to effectively demodulate them. In order to achieve this, the UEs that are further away from the base station need to start to transmit earlier than the UEs closer to the base station so that the corresponding transmissions reach the base station at the same time. The mechanism to control this is referred to as timing advance. Adjustments of the timing advance are named timing advance commands. Adjustments are mostly based on the fact that the UE is moving in relation to the base station. The timing advance commands are sent on medium access control (MAC) level to the UE. The UE uses this to adjust when it transmits its uplink transmissions in relation to when it receives the DL transmissions. Except for the initial timing advance value, the timing advance command is accumulated, i.e. it sets relative adjustment to the current utilized value. The initial timing advanced command is an absolute timing advance. The gNB can provide this based on measuring on the arrival timing of the physical random access channel (PRACH). The PRACH is the only channel/signal the UE transmits without applying a timing advance.

2 Bandwidth Part (BWP)

In NR, the concept of carrier and bandwidth part exists. A carrier is defined by a carrier bandwidth for a subcarrier spacing and a starting location for the subcarrier spacing. The carrier bandwidths are used to define aspects such as Radio Frequency (RF) filters and set requirements on the UE in terms of out-of-band emissions and so forth. The network signals to the UE a set of carrier bandwidths for each applicable numerology it is operating. The bandwidth part concept is introduced for multiple purposes where some of them are to be able to introduce power saving capability in case of very large bandwidths of carriers and handle different bandwidth capabilities between different UEs and not common spectrum sizes. Further on, we will describe in detail about the use case, but initially we describe the functionality of bandwidth part. The bandwidth part defines the current allowed transmission/reception bandwidth and not the carriers' bandwidth.

The UE will operate with what is referred to as an initial bandwidth part when it accesses the system. The initial bandwidth part is the same bandwidth and positioned as the CORESET #0 for DL. For UL, the initial bandwidth part is provided to the UE in terms of system information. The initial bandwidth part is typically used for the purposes of basic functionality and is common for all UEs in the system such as receiving system information, paging and the random-access procedure. The UE, in addition to the initial bandwidth part, can be configured with up to four UE specific bandwidth parts (in pairs for DL and UL) wherein only a single one can be active at any given time. If the UE is reading system information for the cell, any configured bandwidth parts need to encapsulate with the initial bandwidth part. Hence, any UE specific bandwidth part is large in bandwidth than initial bandwidth part. If the UE has an active bandwidth part that is different than

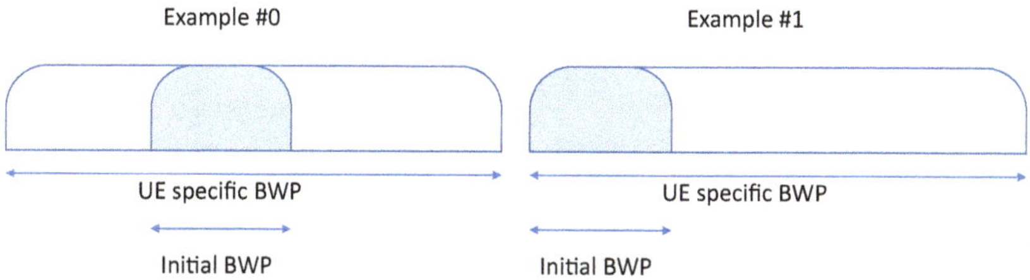

Fig. 9.5 Initial bandwidth part and UE specific bandwidth part relation

the initial bandwidth part, the UE does not switch bandwidth part when receiving information related to the initial bandwidth part. Instead, this is done by an offset in the resource grid and adjusting the resource allocation so that it fits within the initial bandwidth part (Fig. 9.5).

The configuration of the bandwidth parts is performed over radio resource control (RRC). The procedure to activate a bandwidth part can either be done by RRC, timer or DCI message. For the RRC-based approach, an Radio Resource Control RRC message is sent to the UE indicating which bandwidth part is the active bandwidth part. It is further also possible to configure a default bandwidth part (among the several bandwidth parts). The default bandwidth part configuration is used in connection with the timer so that the UE switches to the default bandwidth part if it does not transmit or receive any information on the current active bandwidth until the timer runs out. The last approach is that the UE can switch the active bandwidth part by a field in the DCI message wherein the field indicates one out of maximumly four bandwidth part pairs the UE should switch to (assuming both UL and DL bandwidth parts). The DCI message is being sent by the base station.

To illustrate the above functionality, we here take an example. The example is chosen for the case when the system is operating a larger bandwidth, and that bandwidth is only in use when the UE either needs to receive or transmit large amount of data; otherwise, the UE can use a smaller bandwidth. The approach is that the UE operates on small bandwidth like the initial bandwidth part or slightly large bandwidth until a large amount of data is received in the base station. At this point, the base station will start to schedule the UE. The base station indicates to the UE via a DCI message or RRC to switch to a bandwidth part of a significantly larger bandwidth. The base station continues to schedule the UE on that large bandwidth until it has emptied its data buffer to the UE; after that, the base station either indicates to the UE to switch bandwidth part or the timer in the UE which makes it switch back to a smaller bandwidth part. This example is illustrated in Fig. 9.6.

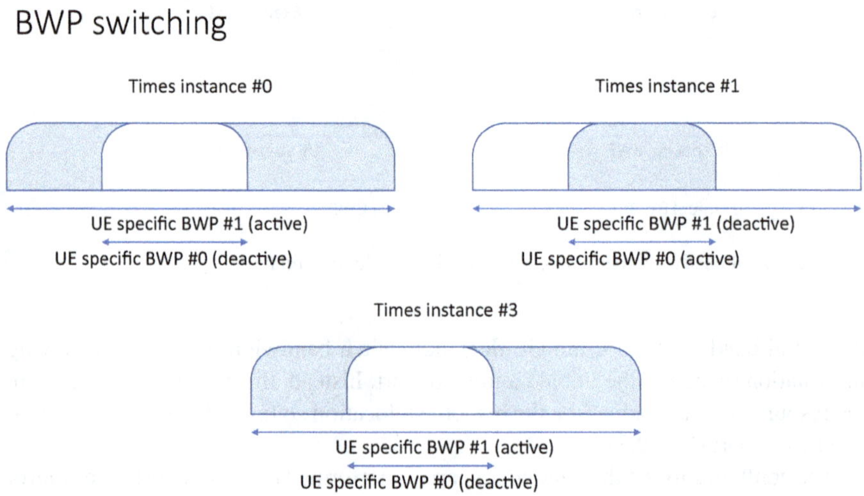

Fig. 9.6 Bandwidth part switching from a bandwidth part with a small bandwidth to a bandwidth part with a large bandwidth and back again to the bandwidth part with the smaller bandwidth

3 Downlink Control Information

Control information is provided on the physical layer, MAC and RRC level. On RRC, it is provided by the system information and the dedicated UE configurations in terms of RRC signalling. On MAC level, it is provided in terms of MAC control elements (CE). On the physical layer, the downlink control information is provided by the physical downlink control channel (PDCCH). A PDCCH message is the encoded version of downlink control information (DCI) message. The structure and content of the DCI messages are described further below.

The UE will attempt to decode a number of PDCCHs in a given number of search spaces per slot in active BWP. The term search space comes from that the UE does not know prior to decode a PDCCH message if a PDCCH message is sent by the gNB or not in the given search location in the search space. Hence, a search space is a space wherein the UE searches for PDCCH message. Each attempt to decode a PDCCH message is referred to as a blind decode. Each search space is further associated with a Control Resource Set (CORESET). The CORESET provides bandwidth in terms of number of RBs, number of consecutive OFDM symbols and a set of properties on how the CORESET is generated. A search space is always linked to a CORESET. The additional properties are quasi-co-location properties, reference signalling scrambling properties, control channel element (CCE) to resource element group (REG) mapping and transmission configuration indicator (TCI) present in the DCI. A CORESET can be one, two or three OFDM symbols long. The bandwidth of a CORESET is possible to allocate in the lengths of six physical resource blocks (PRBs) and can cover up to the full bandwidth of the BWP.

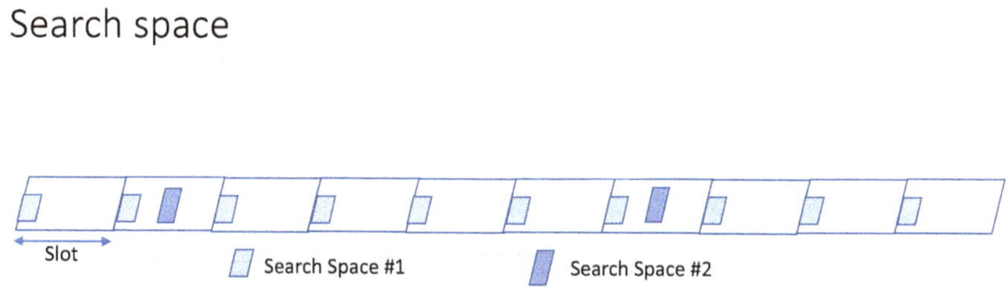

Fig. 9.7 Search spacing monitoring periodicity and monitoring within a slot

The specific search space properties on top of the CORESET are the periodicity of the search space in slots, the applicable OFDM symbol(s) to monitor the search space within a slot, a duration in the number of slots where the search space exists (if signalled), number of PDCCH candidates, search space type and the applicable DCI formats to search for. The periodicity of a search space that is signalled per range of slot can be from every slot to up to every 256 slots. With the signalling of the location of the search space within a slot, it is possible to configure multiple occasions for a single search space. A UE can maximally support three CORESETs and ten search spaces (Fig. 9.7).

A PDCCH message is the encoded message of a DCI. There are several different DCI formats. More details are provided further down with the DCI formats. It can be noticed that some of the DCI formats can have the same size. A PDCCH message is mapped out to one or several CCEs. The UE will look blindly for one or more PDCCH candidates in a search space set that covers one or several aggregations levels of CCEs. A unique candidate corresponds to a given aggregation level and a certain PDCCH size. In case there are several DCI formats of the same size which are available at the same aggregation level, the same blind decoded candidate is shared.

The CCEs are continuous in logical domain, but on the mapping towards physical resources, the CCEs do not need to be mapped continuously in time or frequency domain. The supported aggregation levels, i.e. number of CCEs for the UE to attempt blind decoding, are one, two, four, eight and sixteen CCEs. An exemplified search space with five different aggregation levels, i.e. number of CCEs, is shown in Fig. 9.8.

A CCE is mapped to resource elements in the following manner. A CCE consists of six REGs. Each REG is located in one resource block during one OFDM symbol. REGs are mapped out in a time manner first starting from the first OFDM symbol and lowest resource block in the control resource set. The CCE to REG mapping can either be interleaved or non-interleaved. The mapping is done in a REG bundle. A REG bundle consists of a few consecutive REGs. For non-interleaved mapping, the REG bundle size is six. The CCE is mapped out directly in increasing numbering order to REG bundles. For interleaved mapping, the REG bundle size can be either two or six, depending on configuration. The CCE to REG bundles are interleaved when mapped out. In Fig. 9.9, the mapping of REGs to two REG bundles is

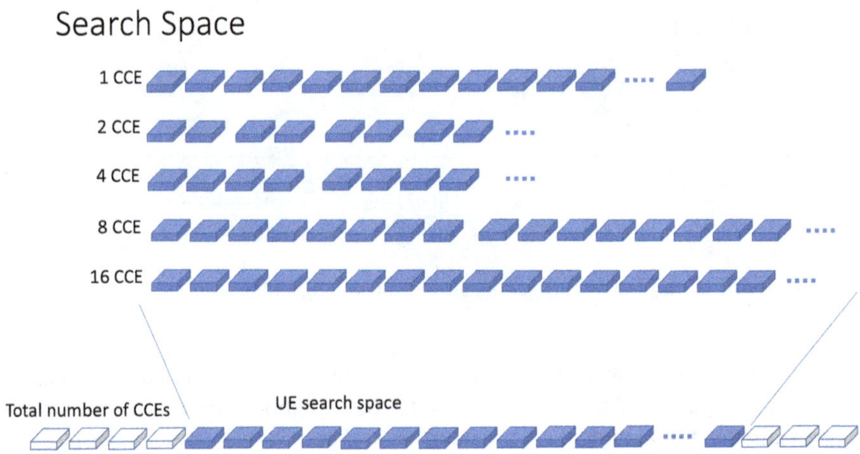

Fig. 9.8 Search space and CCE aggregation levels

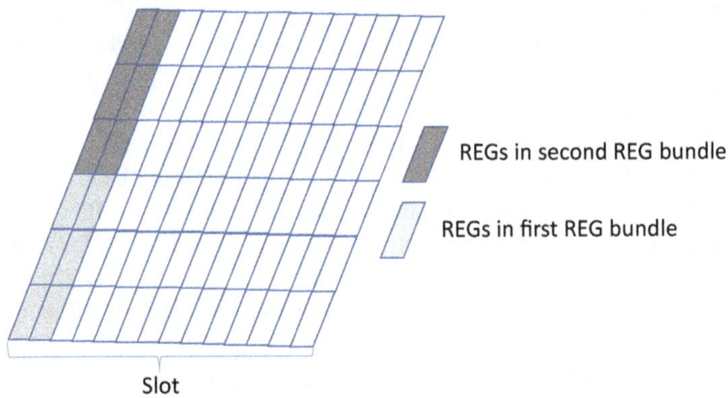

Fig. 9.9 REGs mapped to two REG bundles for non-interleaved mapping and six REGs per REG bundle

illustrated for the case of six REGs per REG bundle and non-interleaved mapping. Further in Fig. 9.10, the mapping of REGs to N REGs is illustrated for the case of six REGs per REG bundle and interleaved mapping.

Reference signals are mapped out to every OFDM symbol and to every fourth resource element. The mapping of the data and reference symbols of PDCCH are shown in Fig. 9.11 for one REG.

There are two types of search spaces: common search spaces and UE-specific search spaces. The common search spaces are intended to be used for control information that is intended for multiple UEs or all UEs operating in the cell, e.g. scheduling messages for system information, paging, random access, etc. In addition, the common search space has the possibility to include scheduling message to individual UEs. A UE-specific search space is used to send scheduling messages intended for one specific UE. Since multiple UEs are using the carrier spectrum, it

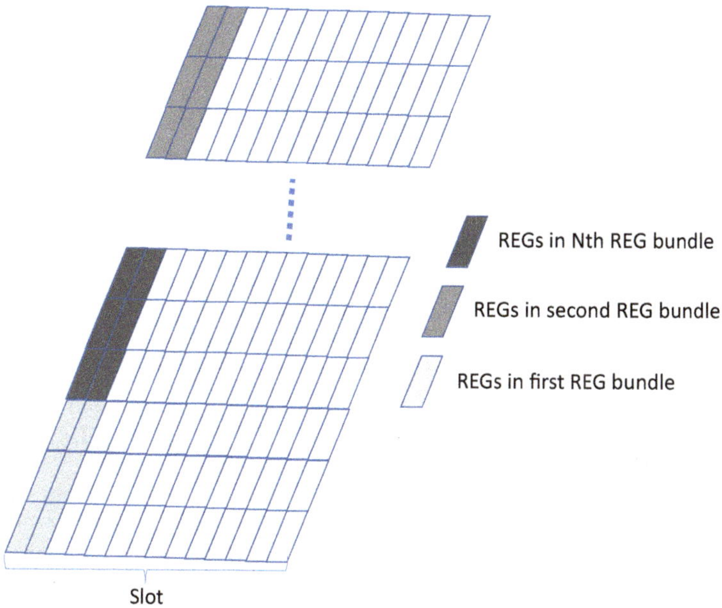

Fig. 9.10 REGs mapped to N REG bundles for interleaved mapping and six REGs per REG bundle

Fig. 9.11 Data and RS mapping for one REG for PDCCH

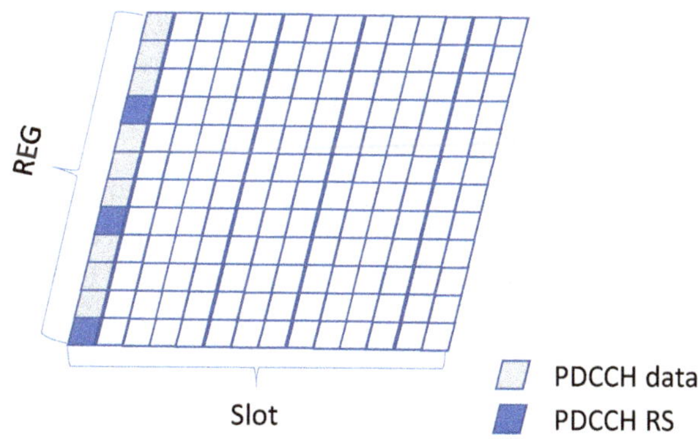

is very likely that the UE-specific search spaces of different UEs will overlap with each other. The search spaces are defined so that the CCEs they cover can change over time, which ensures that multiple UEs can coexist with overlapping search spaces and not always block each other. It could also ensure that different search spaces to the same UE do not block each other (Fig. 9.12).

The number of candidates per aggregation level is either configurable or fixed by the specification. For the common search space for system information block 1 (SIB1) reception, the CCE aggregation levels and number of candidates are fixed in the specification as given in Table 9.4.

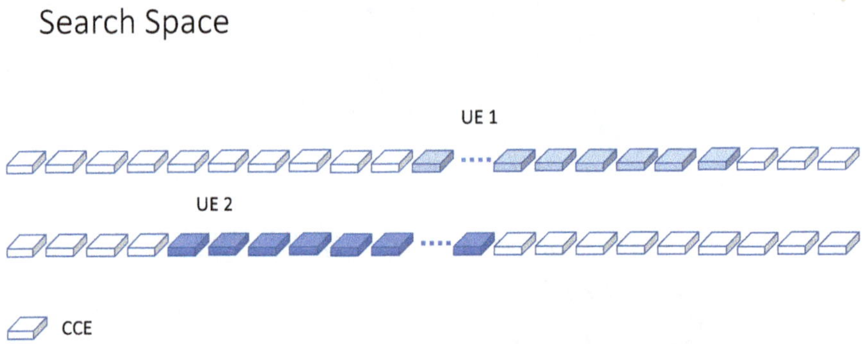

Fig. 9.12 Different UE search spaces not overlapping in CCE domain

Table 9.4 Number of candidates per CCE aggregation level for the common search space for SIB1 reception

CCE aggregation level	Number of candidates
4	4
8	2
16	1

For other search spaces, the number of candidates and the corresponding aggregation levels are configurable. In the standalone version of NR, there is at least one common search space that is referred to as the Type0-PDCCH CCS set, wherein the UE is receiving scheduling messages scrambled with SI-RNTI, P-RNTI, RA-RNTI, TC-RNTI and C-RNTI. It is possible to configure unique common search spaces for specific purposes. These purposes and the corresponding common search space names are as follows:

- If Type0A-PDCCH CSS set is configured, it is used for reception of other system information, i.e. all system information except SIB1 and master information block (MIB). If it is not configured, Type0-PDCCH CCS set is used for this purpose.
- If Type1-PDCCH CSS set is configured, it is used for reception of PDCCH in which cyclic redundancy check (CRC) is scrambled with RA-RNTI and TC-RNTI. If it is not configured, Type0-PDCCH CCS set is used for this purpose.
- If Type2-PDCCH CSS set is configured, it is used for reception of PDCCH in which CRC is scrambled with P-RNTI. If it is not configured, Type0-PDCCH CCS set is used for this purpose.
- If Type3-PDCCH CSS set is configured, it is used for reception of PDCCH in which CRC is scrambled with INT-RNTI, SFI-RNTI, TPC-PUSCH-RNTI and TPC-SRS-RNTI. In addition, it supports unicast scheduling messages on the primary cell. If it is not configured, Type0-PDCCH CCS set is used for this purpose.

The UE can be configured with one or more UE-specific search spaces. The total number of PDCCH candidates the UE can handle depends on the UE processing limitation and is defined per cell. There are two sets of limitations, and both are

Table 9.5 Maximum number of PDCCH candidates per slot for a given subcarrier spacing

Subcarrier spacing	Maximum number of PDCCH candidates
15 kHz	44
30 kHz	36
60 kHz	22
120 kHz	20

Table 9.6 Maximum number of non-overlapping CCEs per slot for a given subcarrier spacing

Subcarrier spacing	Maximum number of non-overlapping CCEs
15 kHz	56
30 kHz	56
60 kHz	48
120 kHz	32

subcarrier spacing dependent. One is the number of unique PDCCH candidates per slot and per serving cell, as in Table 9.5. Another one is the maximum number of non-overlapping CCEs per slot, as in Table 9.6. The reason for the two limitations is as said due to the processing limitations in the UE. In practice, the first limit means the number of PDCCH candidates that can be decoded, and the second one is further related to channel estimation limitations in the UE. The second one is possible by configuration from the network to partly overcome with overlaying different search spaces on top of each other in time and frequency domain.

In a given slot, if the number of PDCCH candidates or the maximum number of non-overlapping CCEs is more than what can be supported by the UE, the UE will prioritize the search spaces to perform blind decodes within. In order, the UE will prioritize first the common search space sets and after that assign blind decodes to the UE-specific search space sets with an increasing search space ID number starting from the lowest assigned number.

The search spaces can be located on the same carrier as the physical downlink shared channel (PDSCH) when it is scheduled, or they can be also located on another carrier. If they are located on another carrier, it is referred to as cross-carrier scheduling. The carrier that is supposed to be scheduled is indicated by a bit field in the DCI. In case of cross-carrier scheduling, to maintain the scheduling flexibility, it is possible to configure additional search spaces on the carrier which provides them, i.e. one can view it as the search spaces have moved from one carrier to another one but still scheduling the same carrier. In Fig. 9.13, an example of two DL cells utilizing cross-carrier scheduling is illustrated. In the figure, the cell #1 is cross-carrier scheduling cell #2. Cell #1 is hence the scheduling cell, while cell #2 is the scheduled cell. In case of the scheduling for cell #1 purely, cell #1 is both the scheduled and scheduling cell.

Each PDCCH messages CRC is scrambled by a Radio Network Temporary Identity (RNTI). The RNTI acts as an identifier for which type of PDCCH message it is. There are RNTIs that are unique to a single UE, and there are RNTIs that are common for all UEs in the cell. The different RNTIs and their purposes are given in Table 9.7.

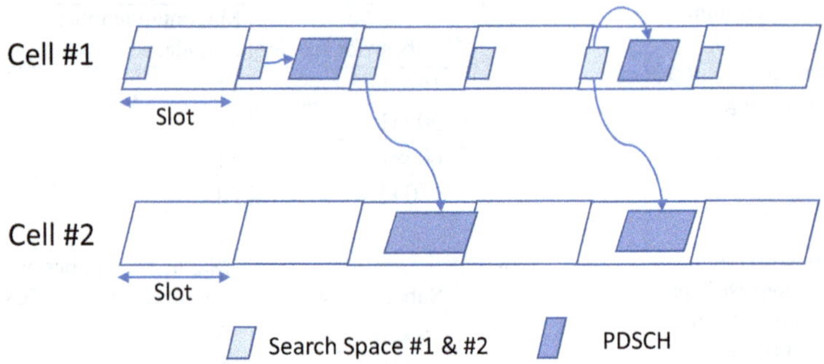

Fig. 9.13 Cross-carrier scheduling for two DL cells

Table 9.7 RNTI and usage

RNTI	Usage
P-RNTI	Paging and system information change notification
SI-RNTI	Broadcast of system information, either being SIB1 or other system information
RA-RNTI	Random access response
Temporary C-RNTI	Contention resolution (when no valid C-RNTI is available)
Temporary C-RNTI	Msg3 transmission
C-RNTI, MCS-C-RNTI	Dynamically scheduled unicast transmission, triggering of PDCCH ordered random access
CS-RNTI	Configured scheduled unicast transmission (activation, reactivation and retransmission)
CS-RNTI	Configured scheduled unicast transmission (deactivation)
TPC-PUCCH-RNTI	PUCCH power control
TPC-PUSCH-RNTI	PUSCH power control
TPC-SRS-RNTI	SRS trigger and power control
INT-RNTI	Indication pre-emption in DL
SFI-RNTI	Slot format indication on the given cell
SP-CSI-RNTI	Activation of semi-persistent CSI reporting on PUSCH

Each RNTI is further associated with one or several DCI formats. The DCI formats provide the control information, which can consist of a scheduling or other types of messages. The DCI formats and their associated RNTIs and usage are shown in Table 9.8.

To exemplify the content of a DCI message, we use here DCI format 0_0 if the CRC is scrambled with C-RNTI that schedules unicast PUSCH and DCI format 1_0 if the CRC is scrambled with C-RNTI that schedules a unicast PDSCH.

DCI Format 0_0

- Identifier for DCI formats – 1 bit
  - Identifies whether it is DCI format 0_0 or DCI format 1_0

Table 9.8 DCI formats, RNTIs and usage

DCI formats	RNTIs	Usage
DCI format 0_0	C-RNTI, CS-RNTI, MCS-C-RNTI and TC-RNTI	Basic format for scheduling of PUSCH in one cell
DCI format 0_1	C-RNTI, CS-RNTI, SP-CSI-RNTI and MCS-C-RNTI	Scheduling of PUSCH in one cell
DCI format 1_0	C-RNTI, CS-RNTI, MCS-C-RNTI, SI-RNTI, P-RNTI-RA-RNTI and TC-RNTI	Basic format for scheduling of PDSCH in one cell
DCI format 1_1	C-RNTI, CS-RNTI, and MCS-C-RNTI	Scheduling of PDSCH in one cell
DCI format 2_0	SFI-RNTI	Notifying a group of UEs of the slot format
DCI format 2_1	INT-RNTI	Notifying a group of UEs of the PRB(s) and OFDM symbol(s) where UE may assume no transmission is intended for the UE
DCI format 2_2	TPC-PUSCH-RNTI and TPC-PUCCH-RNTI	TPC commands for PUCCH and PUSCH for a group of UEs or a single UE
DCI format 2_3	TPC-SRS-RNTI	TPC commands for SRS transmissions for a group of UEs or a single UE

- Frequency domain resource allocation field
  - The frequency domain resource for the PUSCH that is being scheduled
- Time domain resource assignment field
  - The slot the PUSCH is scheduled in and the specific OFDM symbols within that slot. Further defines the mapping type of the PUSCH
- Frequency hopping flag
  - Indicates whether frequency hopping is to be applied or not
- Modulation and coding scheme
  - Modulation order and the MCS
- New data indicator
  - Indicates if the UE should retransmit the transport block associated with the indicated HARQ process or generate a new transport block
- HARQ process number
  - The HARQ process number of the transmitted transport block

- Redundancy version
  - The redundancy version to use when transmitting the transport block
- TPC common for scheduled PUSCH
  - Power control command for PUSCH
- Padding bits, if required
  - Padding bits are added to either align or misalign DCI formats. Alignment could be so that DCI formats 0_0 and 1_0 have the same size. Misalignment can be added so that the DCI format 0_0 of a certain aggregation level becomes unique.
- UL/SUL indicator (if configured)
  - Indicates if the PUSCH is transmitted on the UL carrier or the SUL carrier.

DCI Format 1_0

- Identifier for DCI formats – 1 bit
  - Identifies whether it is DCI format 0_0 or DCI format 1_0
- Frequency domain resource allocation field
  - The frequency domain resource for the PDSCH that is being scheduled
- Time domain resource assignment
  - The slot the PDSCH is scheduled in and the specific OFDM symbols within that slot. Further defines the mapping type of the PDSCH
- VRB-to-PRB mapping
  - Iindicates whether VRB-to-PRB mapping is non-interleaved or interleaved
- Modulation and coding scheme
  - Modulation order and the MCS
- New data indicator
  - Indicates if the UE should retransmit the transport block associated with the indicated HARQ process or generate a new transport block
- Redundancy version
  - The redundancy version to use when transmitting the transport block
- HARQ process number
  - The HARQ process number of the received transport block
- Downlink assignment index
  - Indicates the number of scheduled PDSCH within this HARQ-ACK reporting window

- TPC common for scheduled PUCCH
  - Power control command for PUCCH
- PUCCH resource indicator
  - Gives a point to a PUCCH resource; see Sect. 4 for more details
- PDSCH-to-HARQ_feedback timing indicator
  - Indicates in which slot the PUCCH should be transmitted. Counted from the slot the PDSCH was received.

There are some specific control messages sent on PDCCH that are worth highlighting in this section. For example, 'dynamic' OFDM symbols/slots that are discussed in Sect. 1 are assigned to be either DL or UL based on the PDCCH scheduling content; for example, if the PDCCH schedules a PDSCH, the OFDM symbols are in DL direction, and if the PDCCH schedules a PUSCH, then the OFDM symbols are in UL direction. This further applies to other channel and signals such as SRS or PUCCH, wherein the direction would be UL only.

In addition to this, there is the possibility to use the SFI, which uses a DCI format 2_0 to signal a certain pattern of DL and UL transmission in the coming future slots.

4 Uplink Control Information

The uplink control information on the physical layer consists of Hybrid ARQ-acknowledgement (HARQ-ACK), channel state information (CSI) and scheduling request (SR). This section will describe how the HARQ-ACK and SR are generated and in which forms they are transmitted from the UE. The different forms of how CSI is transmitted from the UE to the network will be further described.

The uplink control information or UCI in common can generally be transmitted on either PUSCH or the physical uplink control channel (PUCCH). All three forms of UCI can be transmitted on PUCCH, while on PUSCH, only CSI and HARQ-ACK can be transmitted. The reason for not transmitting SR on PUSCH is that data transmitted on PUSCH can contain a buffer status report (BSR). This replaces the need for transmitting a scheduling request on PUSCH.

The HARQ-ACK reporting is done on a single UL carrier within a PUCCH group. This single UL carrier is either the primary SpCell or one of the cells that has a PUSCH scheduled that is overlapping according to a set of rules with the HARQ-ACK reporting time. This is illustrated in Fig. 9.14, which shows a PUCCH group of three UL cells. The SpCell is cell #0. It is assumed that the PUCCH is transmitted towards the end of the slot. Hence, if there is a PUCCH transmission and there is no PUSCH overlapping with it, the PUCCH transmission will be transmitted. If there is PUSCH transmission overlapping in time with the PUCCH transmission, the PUCCH will not be transmitted. Instead, the content of the PUCCH will be multiplexed into the PUSCH transmission. If there are multiple

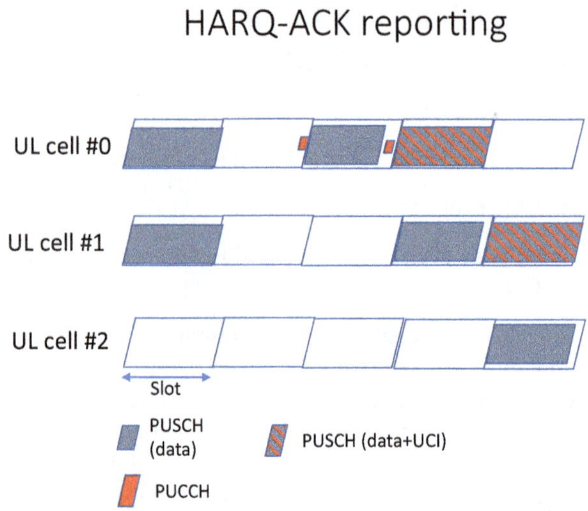

Fig. 9.14 UCI on PUSCH and PUCCH

PUSCH transmissions overlapping with the same PUCCH transmission. The UCI will be multiplexed into one of the PUSCHs. If the UCI is purely HARQ-ACK information or periodic CSI reporting, the UCI will be multiplexed into the PUSCH on the cell with the lowest cell ID.

The UE generates one or two HARQ-ACK bits per transport block it receives from PDSCH. When reporting the HARQ-ACK bits, the UE can report for multiple scheduled transport blocks at once. This is due to the reason that either multiple carriers are used and the UE is scheduled on them simultaneously or the UE is scheduled several PDSCH transport blocks one after each other on the same carrier. The latter is typical for TDD spectrum. Within the DCI message that schedules the PDSCH, there are multiple fields related to the reporting of the HARQ-ACK bits. One of these fields indicates the number of slots into the future wherein the HARQ-ACK bits should be sent. This field is measured from the slot in which the PDSCH is received within. The field is named 'PDSCH-to-HARQ_feedback timing indicator'. All the HARQ-ACK bits that are pointed out to be reported in the same slot are all jointly reported together because only a single HARQ-ACK report per slot is supported. An example of this is illustrated in Fig. 9.15, wherein there are three cells utilizing the TDD spectrum. In total, there are six PDSCHs scheduled, and the UE will send a PUCCH with HARQ-ACK information related to all the six PDSCHs.

Below we further outline the different PUCCH formats that can be used for transmission on PUCCH followed by how to derive the number of HARQ-ACK bits that are transmitted by the UE. After that, the UCI on PUSCH follows.

On PUCCH, there are in total five different PUCCH formats defined.

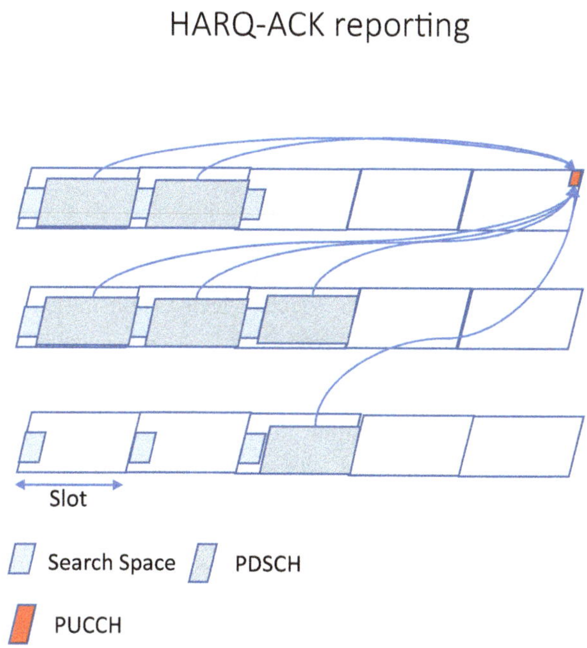

Fig. 9.15 HARQ-ACK reporting time based on PDCCH messages

PUCCH format 0 supports a payload size of 1 or 2 bits and covers a length of either one or two OFDM symbols. PUCCH format covers one resource block in frequency domain. If the length of the PUCCH format is two OFDM symbols, it is possible to support frequency hopping. If frequency hopping is used, the two OFDM symbols are not transmitted on the same frequency. The design of PUCCH format 0 is based on sequence selection. The sequence can be cyclically shifted in order to support randomization between cells and within cells. The cyclic shift and the sequence for the PUCCH transmission are selected based on at least the HARQ-ACK payload, the cell ID, slot number and OFDM symbol numbers. Different UEs are separated by PUCCH format being allocated to different frequency locations or time. Alternatively, if they are overlapping in time and frequency, they are then separated by different cyclic shifts and sequences.

PUCCH format 1 supports a payload of 1 or 2 bits and covers a length of 4–14 OFDM symbols. PUCCH format covers one resource block in frequency. PUCCH format 1 is based on a similar structure as PUCCH format 0, wherein it is based on sequence selection and cyclic shift. It is similar to PUCCH format 0 to support the frequency hopping. Frequency hopping is supported. Different UEs are separated by PUCCH formats being allocated to different frequency locations or time. Alternatively, if they are overlapping in time and frequency, they are then separated by different cyclic shifts and sequences.

PUCCH format 2 supports a payload size of above 2 bits and covers a length of either one or two OFDM symbols. PUCCH format covers one or more resource blocks in frequency domain. It is based on a structure wherein encoded UCI and demodulation reference signals (DMRS) are assigned to different subcarriers. Different UEs are supported by assigning them to different resources, i.e. non-overlapping within the same cell.

PUCCH format 3 supports a payload size of above 2 bits and covers a length of 4–14 OFDM symbols. PUCCH format covers one or more resource blocks in frequency domain. It is based on a structure wherein encoded UCI and DMRS are assigned to OFDM symbols. The encoded UCI is transmitted with transform precoded OFDM, and the DMRS are generated in a similar manner as DMRS for transform precoded OFDM. There are two OFDM symbols used for DMRS, and the remaining are for encoded UCI. The DMRSs are placed roughly evenly throughout the transmission length (exact position depends on the length of the PUCCH transmission). Frequency hopping is supported. Different UEs are supported by assigning them to different resources, i.e. non-overlapping within the same cell.

PUCCH format 4 supports a payload size of above 2 bits and covers a length of 4–14 OFDM symbols. PUCCH format covers one or more resource blocks in frequency domain. PUCCH format 4 is very similar to PUCCH format 3 with the difference that an orthogonal cover code (OCC) is added to allow multiplexing of up to four UEs.

The specific PUCCH format to be assigned for a transmission differs from the type of transmitted UCI and some other factors. To start with, we describe how the PUCCH format is determined for an HARQ-ACK feedback, and after that, the pure SR and CSI transmission follows.

For HARQ-ACK feedback, we start with the situation when the UE has been configured with dedicated configuration by the network, and after these, other cases are described.

The UE can be configured with up to four PUCCH sets. A PUCCH set is used for a given range of HARQ-ACK payload bits. To each PUCCH resource set, a number of PUCCH resources up to eight are associated, except the first PUCCH resource set that can have a maximum of 16. Each PUCCH resource consists of a PUCCH format, and it is associated with configurations, e.g. frequency and OFDM symbol placement and frequency hopping or not. The UE picks which PUCCH set to use based on the number of HARQ-ACK bits it is supposed to transmit. The PUCCH resource within the PUCCH set is picked directly based on the indicated resource by the DCI field ‘PUCCH resource indicator’ (in the DL DCI). Alternatively, for the case that the first PUCCH resource set is used and if the number of PUCCH resources is more than eight, a combination of the indicated DCI field ‘PUCCH resource indicator’ and factors such as the CCE number of the last received DL DCI that is associated with HARQ reporting. For both cases, the DCI message to use is the last received DCI message for the UE associated with that HARQ-ACK report to determine the ‘PUCCH resource indicator’. In Fig. 9.16, the PUCCH resource set selection is illustrated where each PUCCH resource set contains configurations for eight PUCCH resource indicators (PRI). The boundary between the PUCCH

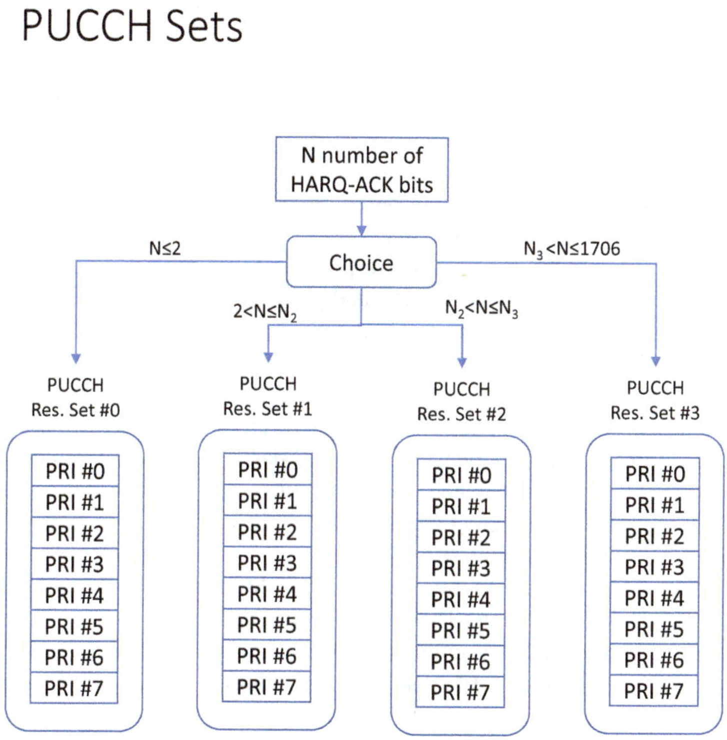

Fig. 9.16 PUCCH resource sets

resource sets, i.e. N_2 and N_3, is configurable by the network. If, for example, there are only two PUCCH resource sets, then N_2 corresponds to 1706 which is the maximum supported UCI payload size.

Before RRC has configured specific PUCCH resources, there is a significant limitation in what the UE can report in number of HARQ-ACK bits which is limited to 2. The UE can further only use PUCCH format 0 or 1 for the HARQ-ACK reporting. The UE uses these PUCCH formats in a preconfigured fashion using a default operation, wherein RRC signalling defines which PUCCH format to use and the length and the starting position of it in time. The DCI format further then defines the exact position of it in frequency domain based on the field 'PUCCH resource indicator' and the starting CCE of the last received associated PDCCH message.

One ACK/NACK bit is generated per transport block on PDSCH per scheduling occasion. A scheduling occasion is one PDCCH message. For up to four layers, one transport block is supported, and for above four layers, two transport blocks are supported. There is also a possibility based on a specific UE capability to report more HARQ-ACK bits per transport block if the UE is configured to report HARQ-ACK feedback per code block based group.

The HARQ-ACK bits are reported using one of two codebooks. The codebooks are a mechanism to generate the actual number of HARQ-ACK bits the UE sends to the gNB.

The two available codebooks are dynamic and semi-static HARQ-ACK codebooks.

The dynamic HARQ-ACK codebook is the simplest form where its number of reported HARQ-ACK bits is directly based on how many scheduled PDSCHs the gNB has sent to the UE. In a design like such, one must tackle the problem that the UE may miss some PDCCH and its corresponding PDSCH so there may then be a different understanding between the UE and gNB on the number of PDSCHs that have been scheduled and its corresponding number of HARQ-ACK bits. This is not possible to deduce by the gNB from the sent PUCCH format, and hence, the error will propagate above the physical layer. To handle this issue, two bit fields have been introduced in the DCI message, which are referred to as downlink assignment index (DAI) jointly. Specifically, there are two separate fields being named counter DAI and total DAI. This becomes specifically applicable when carrier aggregation is used, i.e. the UE is aggregating multiple DL carriers, which is the technique very much used to increase downlink throughput.

The counter DAI represents the number of PDSCH that is being scheduled at the same scheduling occasion. The same scheduling occasion spans in multiple carriers. The counter DAI is accumulated counter of the total number of PDSCH being scheduled and each DCI having a unique number (leaving out limitation in number of bits in the DCI message). With this, the UE can detect whether it has missed a specific PDCCH message during a specific scheduling occasion, if it is not the last one. The other field is the total DAI, which represents the total number of scheduled PDSCHs up to that specific scheduling occasion. This will handle the case of all PDCCH missed at a certain scheduling occasion if it is not the last one. As there is a limitation on the number of bits the DAI field has, they wrap around at specific lengths. The maximum value for each field is 4. It is, however, envisioned that it is unlikely to have so many missed PDCCH one after the other for this to be a problem.

If the UE has found that it has missed a specific PDCCH message, the corresponding HARQ-ACK bit to that PDSCH is set to NACK. The remaining HARQ-ACK bits that are then associated with a PDSCH that has not been missed are set to either ACK or NACK based on whether the associated transport block has been successfully decoded or not.

The number of bits derived out of the codebook is used to determine the used PUCCH set as previously described. In Fig. 9.17, the counter and total DAI are shown, C meaning counter DAI and T meaning total DAI. Here, the number of scheduled PDSCH takes into account a modulo operation and hence the value would not exceed four.

The semi-static codebook uses fixed codebook as the name implies. It does so by providing HARQ-ACK bits for all possible scheduling occasions across all carriers to the indicated HARQ-ACK reporting time. All possible scheduling occasions here are derived out from all the PDCCH that can schedule a specific slot for

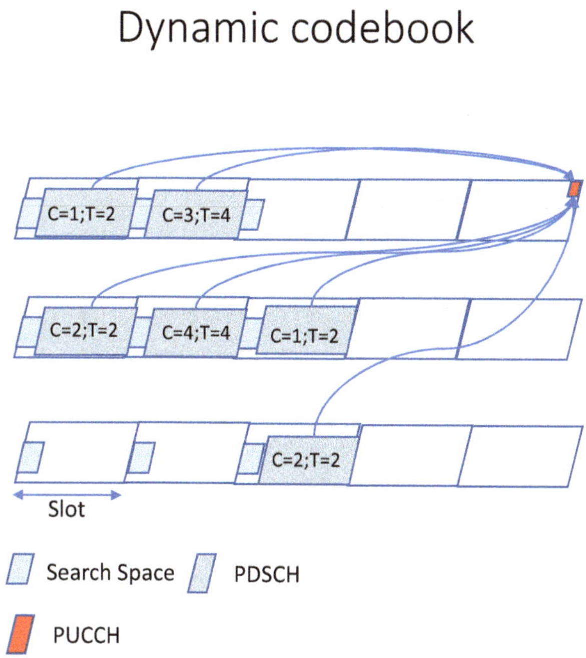

Fig. 9.17 Dynamic HARQ-ACK codebook

HARQ-ACK reporting. The UE will generate the HARQ codebook for all these possible scheduling occasions and set all bits to NACK. For the occasion wherein a transport block has been successfully decoded that is associated with a specific bit in the HARQ-ACK codebook, it will be set to ACK. This is illustrated in Fig. 9.18, wherein there are nine different PDCCH scheduling occasions for the single HARQ-ACK reporting occasion. Therefore, nine bits are generated for the HARQ-ACK codebook assuming that the maximum number of schedulable layers does not exceed four.

The advantage of the dynamic codebook is that it provides a smaller HARQ-ACK codebook because the number of bits it generates only represents the number of scheduled occasions. This can be observed from the example in Figs. 9.17 and 9.18, wherein the dynamic HARQ-ACK codebook generates 6 bits while the semi-static codebook generates nine bits. It hence does not require as good UL coverage as the semi-static codebook. On the other hand, the semi-static codebook is very robust in what is being reported and is therefore error-free in terms of the missed PDCCHs.

Described shortly above was the possibility to report multiple HARQ-ACK information bits per transport block. This is referred to as reporting HARQ-ACK for a code block group and their multiple code block groups typically within a transport block. The basic principle is that the UE is configured with the number of code block groups to report. The UE will based on that report HARQ-ACK information for that number of given code block groups. Each transport block is constructed by

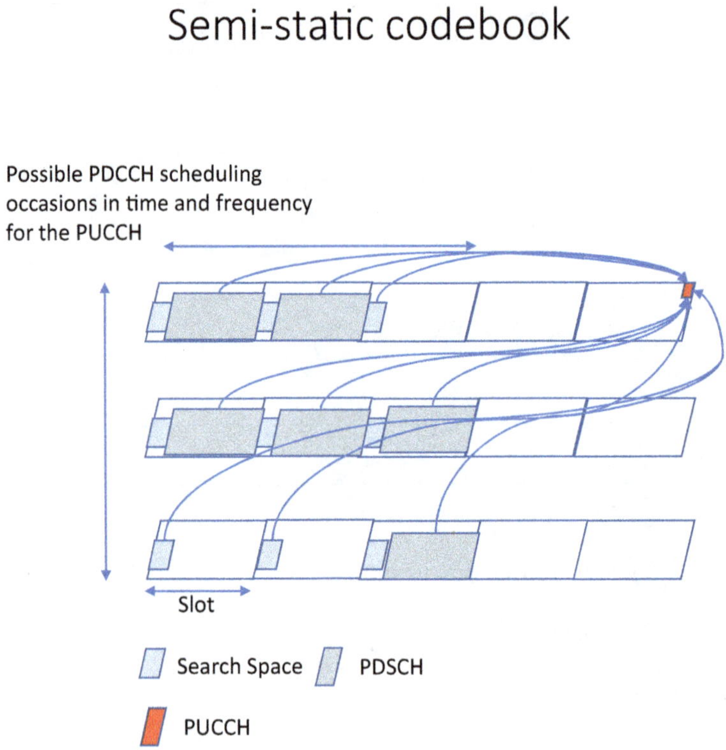

Fig. 9.18 Semi-static HARQ-ACK codebook

a number of code blocks. Each code block group consists of a number of groups of code blocks for which the HARQ-ACK is bundled. Hence, if one of the code blocks in a specific code block group is not successfully decoded, then that specific code block group HARQ-ACK bit is set to NACK, and if all of the code blocks within that group are successfully decoded, then that specific HARQ-ACK bit is set to ACK. In case of retransmission of the transport block, it is possible for the gNB to just retransmit specific code block groups. This avoids retransmitting the whole transport block. This will generate significantly more HARQ-ACK information bits to transmit from the UE and is by that not a generally applicable feature, but if the UE is operating in a scenario wherein it is very likely that parts of its transport blocks are consistently unsuccessfully decoded, the feature is suitable to use.

In addition to transmitting UCI on PUCCH, UCI can also be transmitted on PUSCH. The UCI on PUSCH is HARQ-ACK or CSI. To support the CSI transmission in an effective manner, the CSI is split up in two parts: CSI part 1 and CSI part 2. The CSI part 1 is typically smaller in size, and CSI part 2 is larger and typically contains more frequency selective information.

The UCI on PUSCH can be transmitted in two different ways. The first one is when it is multiplexed with UL-SCH, i.e. data. The second one is UCI only on

PUSCH. The first case would occur, for example, if HARQ-ACK feedback is to be transmitted at the same time as PUSCH with UL-SCH granted or a grant indicates that PUSCH should include an aperiodic CSI report together with UL-SCH. The other alternative happens when semi-persistent CSI report is to be transmitted on PUSCH or a grant indicates that a PUSCH should be transmitted with only aperiodic CSI report(s).

In case UCI is transmitted on PUSCH together with UL-SCH, the amount of coded UCI bits is controlled with a set of offsets from the code rate of the data. There is one or several offsets per UCI type. The reason for the offsets is that typically there is a need for higher coding protection for the UCI as it does not have any retransmissions, which is valid for normal mobile broadband traffic. The offsets can either be configured to a fixed value or it is possible for the gNB to configure another mode wherein the gNB can select one out of four possible offsets per UCI type in the PDCCH message granting the PUSCH. The latter one targets for the scenario wherein one has data types of different error protection needs, i.e. not constantly the same. It is typically same for mobile broadband case, so the feature is to rather target for a scenario with mixture of mobile broadband traffic and, for example, Industrial IoT application (which may need higher protection). At the same time as the data error protection is adjusted, the UCI error protection could also be adjusted because the example CSI feedback may not be equally critical to be correct as the data (so it may not need to follow data).

Independent of the above procedure, the order of mapping of UCI on PUSCH is the same. The HARQ-ACK is mapped starting from the OFDM symbol right after the first DMRS OFDM symbols. If the amount of HARQ-ACK information bits is 2 or less, the coded HARQ-ACK bits are puncturing the information that is prior mapped to the location wherein the HARQ-ACK bits are placed. If, however, the HARQ-ACK information bits are more than 2, it is considered together with mapping out all other UCI and UL-SCH, in a non-puncturing manner, referred to as rate-matching in 3GPP. The reason for the split between the HARQ-ACK bit mapping is the 2 information bits of HARQ-ACK corresponding to either one or two PDCCH which could have been missed by the UE. Puncturing the encoded HARQ-ACK bits into PUSCH provides less errors in the decoding of the other UCI and data on PUSCH compared to the wrong understanding of the number of HARQ-ACK bits between the gNB and the UE. To further assist to avoid the wrong understanding between the UE and the gNB on the number of HARQ-ACK bits, the DCI format 0_1 contains a UL DAI. The UL DAI for fixed codebook is 1 bit and that for dynamic codebook is 2 bits. The DAI is used to indicate either that PDSCHs have been scheduled (for the semi-static HARQ-ACK codebook) or the total number of PDSCHs scheduled (for the dynamic HARQ-ACK codebook). However, DCI format 0_0 does not contain UL DAI, and hence, the above aspect with puncturing provides error protection.

The CSI part 1 is mapped to the first OFDM symbol in the PUSCH and onwards, excluding the OFDM symbols containing the DMRS. The CSI part 2 is mapped starting in the empty resource elements in the last OFDM symbol of the CSI part 1. Finally, after that, the remaining resource elements contain encoded UL-SCH. The

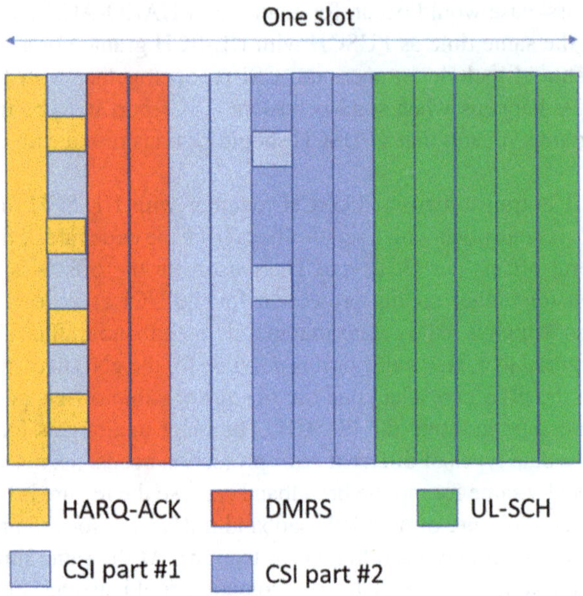

Fig. 9.19 UCI mapping on PUSCH with UL-SCH

above condition is valid with the exception that the CSI part 1, part 2 and UL-SCH are not mapped to resource element that will contain encoded HARQ-ACK bits if HARQ-ACK with more than 2 information bits is to be multiplexed on PUSCH (Fig. 9.19).

In case there is no UL-SCH but only UCI on PUSCH, the last UCI type given above fills out the rest of the PUSCH, so there is no remaining space available on it, which is in the order of HARQ-ACK, CSI part 1 and CSI part 2. Note that UCI on PUSCH without UL-SCH can only occur together with both CSI part 1 and CSI part 2 or only with CSI part 1 (Fig. 9.20).

An SR is a transmission indicating that the UE has new data or new data of higher priority to be transmitted to the gNB. The SR consists of either an indication of a single bit (certain state of the bit indicates positive scheduling request) or the transmission of a specific format on PUCCH on a specific resource at a specific time. The possible SR occasions are configured by the gNB with a range which can go down to every second OFDM symbol or up to the order of hundreds of milliseconds. It is possible to configure multiple SR processes, in which each has its own reporting time. Each SR report is connected to one or several buffers in the UE. Each buffer has data of different priorities. The possible reporting periodicity that is selected is based on how urgent is the need for the gNB to know that the UE has data or new data in its buffer, which is then connected to what data are in the buffers and how time critical they are.

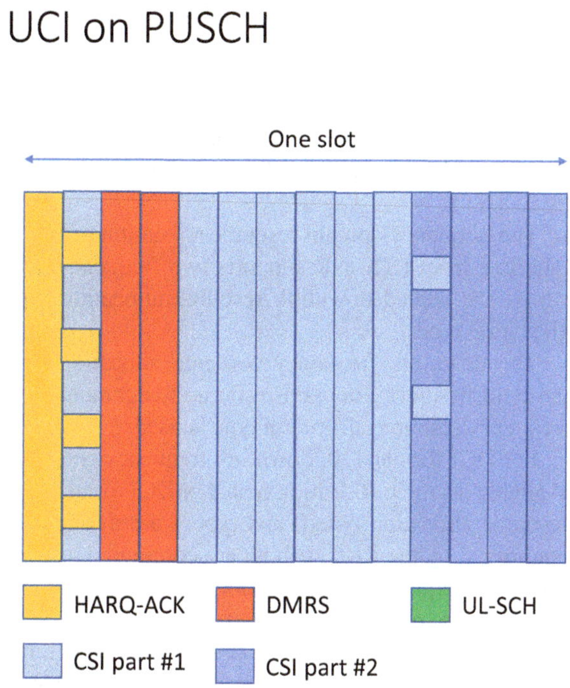

Fig. 9.20 UCI mapping on PUSCH without UL-SCH

5 Data Channels

The data channels for NR are Physical Downlink Shared Channel (PDSCH) in DL and the Physical Uplink Shared Channel (PUSCH) in UL. The PDSCH supports up to 256QAM in modulation and up to eight MIMO layers, while the PUSCH supports up to 256QAM modulation and up to four MIMO layers. The PUSCH supports both OFDM-based waveform and a transform precoded OFDM. The latter is similar in waveform structure as in LTE in UL. For the case of an OFDM-based waveform, the mapping to physical resources is very similar between PDSCH and PUSCH.

5.1 *Physical Downlink Shared Channel (PDSCH)*

From a PDSCH scheduling perspective, two scheduling principles are supported. The first is that the UE receives a DL assignment in the form of PDCCH which is based on a certain DCI format. The DL assignment assigns resource in either the same slot as the PDCCH is received or in the slot following that slot. The other mechanism is that the UE is set up with a semi-static scheduling. The UE would in such a case be assigned in a certain PDSCH resource with a certain periodicity, for example, 10 or 20 ms. This would be used if the UE is connected to a service that generates packets to it with a fixed periodicity, and by doing this, it can reduce the load on PDCCH. The most typical manner to assign the UE with resources is to dynamically send the UE a PDCCH message and further apply the assignment in the same slot where the PDCCH is received. The UE could also receive multiple PDCCH contents in the same slot for different RNTIs. In addition, if it supports certain capabilities, it could also do so for the UE specific data, e.g. PDSCH assigned by C-RNTI.

The frequency domain resource allocation schemes for PDSCH and PUSCH are inherited from LTE. NR supports two frequency allocation schemes in downlink, which are named downlink resource allocation type 0 and downlink frequency allocation type 1.

The downlink frequency resource allocation is given in the DCI format by the field frequency domain resource assignment, independent of which downlink frequency resource allocation type is used.

For DCI format 1_0, downlink frequency resource allocation type 1 is the only available resource allocation type. For DCI format 1_1, both the downlink frequency resource allocation type 0 and type 1 are available, where either one or both are configured by the RRC. If both are configured, the one used is selected by a single bit in the frequency resource allocation field.

Downlink resource allocation type 0 is available for scheduling by a PDCCH with DCI format 1_1. The resource allocation scheme consists of a bit map in which each bit is set representing a group of allocated resource blocks. The group of resource blocks is referred to as resource block groups (RBGs). Downlink resource allocation type 0 is illustrated in Fig. 9.21.

The number of resource blocks per RBG is dependent on the active bandwidth size, wherein the number of resource blocks in an RBG increases with the increasing of the size of the bandwidth part. There are two different possible configurations of the number of resource blocks per RBG given in the standard, which are simply referred to as configuration 1 and configuration 2. The possible RBG sizes for these

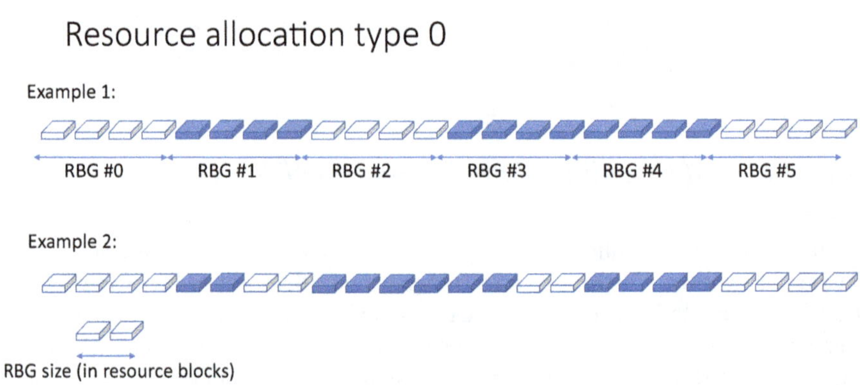

Fig. 9.21 Downlink resource allocation type 0

Table 9.9 RBG configuration for the bandwidth part size

Bandwidth part size	Configuration 1	Configuration 2
1–36	2	4
37–72	4	8
73–144	8	16
145–275	16	16

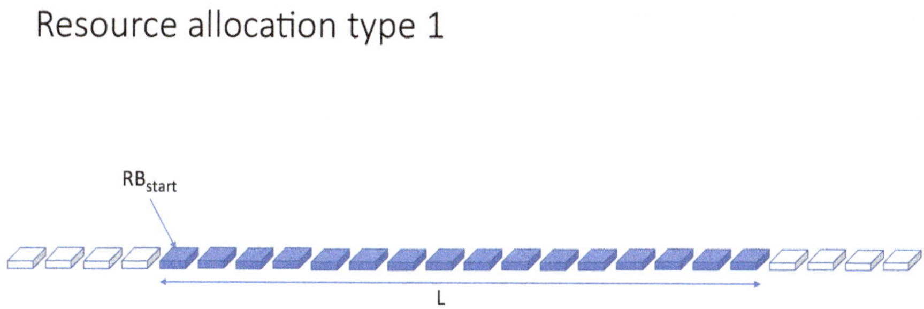

Fig. 9.22 Downlink resource allocation type 1

configurations are given in Table 9.9. The configuration for which RBG size to use is given by RRC signalling and is configured per UE in a dedicated manner.

Downlink resource allocation type 1 is the only available frequency resource allocation scheme from a PDCCH with DCI format 1_0. It is further available if the UE has been configured with it for DCI format 1_1. The basic purpose of the frequency resource allocation scheme is to provide a compact frequency resource allocation. It is constructed so that it provides a starting resource block with the length of the allocated resource blocks within the scheduled bandwidth part. The length in number of resource block is signalled with the granularity of a resource block except in the case of a DCI format 1_0 in a UE-specific search space set, which has its size derived from the common search space set. For that specific case, more specifically, if there is an active bandwidth part that is not the initial bandwidth part, the starting resource block location and the length in number of resource blocks are given by an integer, which is the ratio of the size in resource block between the active bandwidth part and the initial bandwidth part. Hence, by using such mechanism, it will provide a continuous number of resource blocks for the frequency allocation (Fig. 9.22).

The time domain resource allocation is selected by the field 'Time domain resource assignment' in DCI format 1_0 or DCI format 1_1. That field may not exist if only a single time domain resource allocation scheme is configured by RRC for DCI format 1_1. The time domain resource allocation assignment field points into a row of a table. The table will give the PDSCH mapping type, which slot the PDSCH is allocated in, the starting OFDM symbol index in the slot the PDSCH is allocated and the length of the OFDM symbols of the PDSCH.

The PDSCH mapping type corresponds to the indexes of OFDM symbols which contain DMRS. There are two types defined, i.e. type A and type B. The starting

OFDM symbol of the DMRS for mapping type A is the OFDM symbol of the first DMRS in case of type A mapping. The slot for which the PDSCH is allocated to is referred to as K_0. The starting OFDM symbol for the allocation is referred to as S and the length is referred to as L.

The table of the time domain resource allocation assignment could be either given by the specification or configured through broadcast signalling or dedicated RRC signalling depending on the applicable RNTI that the PDCCH CRC is scrambled with. This is further outlined in Table 9.10.

For SI-RNTI and specifically SIB1, which the UE receives after PBCH, there is no signalling possibility of either a broadcasted or dedicated resource allocation table. Hence, there are three default tables specified, which are named Default A, Default B and Default C. Which of the tables the UE should use depends on the SS/PBCH block and CORESET multiplexing pattern. The SS/PBCH block and CORESET multiplexing pattern is given by a frequency resource allocation field for CORESET #0 in MIB. There are three SS/PBCH block and CORESET multiplexing patterns 1, 2 and 3. The applicable ones to choose from are subcarrier spacing dependent, specifically the subcarrier spacing of the SS/PBCH block and the CORESET #0. For a given subcarrier spacing combination, there are at most two different options to pick from, in which for some subcarrier spacing combination there is only one single option available. The most common SS/PBCH block and CORESET multiplexing pattern is pattern 1.

Within SIB1, it is possible to configure a broadcasted time domain resource allocation table for PDSCH that would be applicable for all the RNTIs, which is given in Table 9.10. If such table is not broadcasted, one of the default tables A, B and C would be used, depending on the SS/PBCH block and CORESET multiplexing pattern.

Further, when a UE is in connected state, it is possible to configure the UE with a dedicated time domain resource allocation table. That table would be applicable for the unicast messages that are addressed with the C-RNTI, MCS-C-RNTI or CS-RNTI.

The PDSCH mapping types described previously give the locations of the DMRS within the PDSCH allocations. The different schemes are defined as type A and type B. Type A has the first DMRS allocated a bit into the PDSCH allocation to give the possibility to derive channel estimation in a more precise manner if the channel is varying during the PDSCH reception for the UE. On the other hand, the decoding of the PDSCH needs to wait until the first DMRS is received. To allow faster processing, PDSCH mapping type B has the first DMRS allocated in the first OFDM symbol of the PDSCH. In addition, there is the possibility to allocate more than one symbol for DMRS to allow higher protection against channel variations in time. Both mapping type A and mapping type B are illustrated in Figs. 9.23 and 9.24, respectively.

The mapping from virtual resource block to physical resource block can either be interleaved or non-interleaved. For the non-interleaved mapping, the order of the PRBs are fixed, but depending on the scheduling search space, the starting PRB may be adjusted. If the scheduling search space is a common search space associated with

Table 9.10 Applicable PDSCH time domain resource allocation

RNTI	PDCCH search space	SS/PBCH block and CORESET multiplexing pattern and PDSCH time domain resource allocation	*If no dedicated and broadcasted configurable table provided*
SI-RNTI	Type0 common	1 default A for normal CP	–
		2 default B	–
		3 default C	–
SI-RNTI	Type0A common	1 default A	No
		2 default B	No
		3 default C	No
		1, 2, 3	Yes, both broadcasted table
RA-RNTI, TC-RNTI	Type1 common	1, 2, 3 default A	No
		1, 2, 3	Yes, both broadcasted table
P-RNTI	Type2 common	1 default A	No
		2 default B	No
		3 default C	No
		1, 2, 3	Yes, broadcasted table
C-RNTI, MCS-C-RNTI, CS-RNTI	Any common search space associated with CORESET 0	1, 2, 3 default A	No
		1, 2, 3	Yes, both broadcasted and dedicated table
C-RNTI, MCS-C-RNTI, CS-RNTI	Any common search space not associated with CORESET 0 UE-specific search space	1,2,3 default A	No
		1, 2, 3	Yes, both broadcasted and dedicated table

Fig. 9.23 Mapping type A of PDSCH

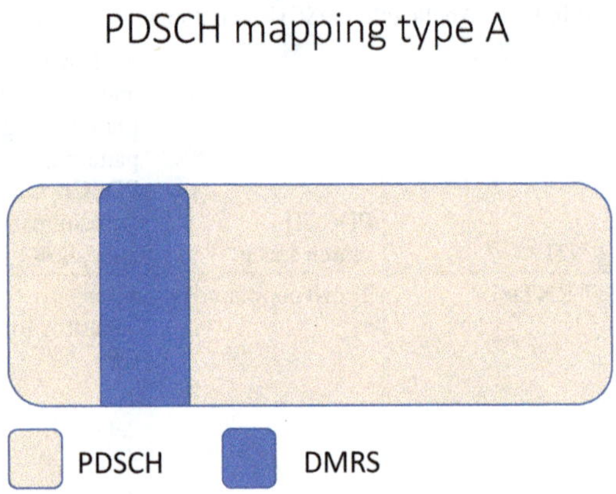

Fig. 9.24 Mapping type B of PDSCH

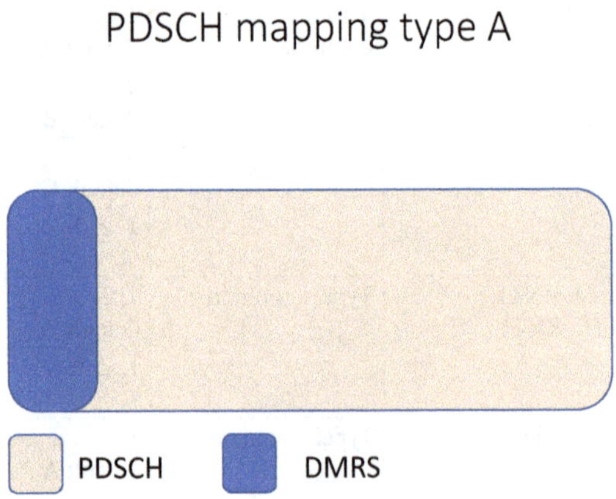

CORESET #0, the allocation starts from the start of initial bandwidth part. If the scheduling is from a UE-specific search space, the allocation starts from the start of the scheduled bandwidth part. This is to align the allocation within the correct bandwidth part.

The modulation and code rate are selected by the DL assignment by a field in the corresponding DCI that indicates a given modulation and code rate. The field is 5 bits long and an entry points into a table. The table used is configured to the UE. There are three modulation and coding scheme (MCS) tables supported. Two of them are targeting for mobile broadband use cases. The difference is that one supports a maximum modulation order of 64QAM and the other supports a maximum modulation order of 256QAM. The reason for having two tables is that certain spectrum may not be suitable for 256QAM and deployments. Together with supporting a higher modulation order, it would also yield to higher UE complexity. The other supported modulation orders are QPSK and 16QAM. The third MCS table is designed for ultra-low-latency high reliability, i.e. IIoT (Industrial Internet of Things) deployment wherein there is a need for a very low code rate. Note that not all IIoT deployment may need this stringent low code rate, but some may need it.

Based on the indicated code rate, the frequency and time resource allocation, the selected modulation order and the number of scheduled layers all together, a transport block size can be calculated.

The minimum processing times are defined as the time between the end of the PDSCH and the start of the HARQ-ACK transmissions. At a high level, there are two categories of UE processing times defined as cap #1 and cap #2. The faster one targeting PDSCH and the corresponding HARQ-ACK in the same slot is referred to as processing time cap #2. The processing capability #1 is the basic processing capability of an NR UE. Typically, the processing time requirement is chosen so that it allows the HARQ-ACK to be reported in the next or second next slot after a PDSCH was received. For both categories described here, we assume no timing advance, and hence, if there is timing advance, it is added on top of the processing time of the UE. There is one key aspect that can decide the processing time, which is the used subcarrier spacing. There are some other factors that could also control the processing time such as the selected mapping type of PDSCH. In most cases, if subcarrier spacing becomes larger, the processing time becomes shorter. This certain functionality scaled with the subcarrier spacing would then allow for quicker processing times. It is, however, important to note that not all components are scaled with the processing time, so this is not a linear dependency.

5.2 *Physical Uplink Shared Channel (PUSCH)*

From a PUSCH scheduling perspective, three scheduling principles are supported. The first is that the UE receives an uplink grant in the form of PDCCH based on a certain DCI format, which grants resource in a given slot as indicated by the UL grant. The other two mechanisms are referred to as configured UL grant type 1 and type 2 and are both based on the same principle. It is in principle an adaption of semi-persistent scheduling for UL (very similar to DL and LTE). For a configured UL grant type 1, everything is configured by RRC. Further, configured UL grant type 1 is activated/deactivated by RRC. For configured UL grant type 2, most aspects are configured by RRC, but the activation/deactivation is done by a PDCCH message. The configured UL grant type 2 provides greater and faster flexibility for the scheduling to activate and deactivate than the configured UL grant type 1. Note as well that UL SPS scheme in LTE is very similar to UL grant type 2 in this respect. The configured UL grant is used similar to the DL for periodic service usage which could also be used for the UE in order to be able to quickly send data when it has new data in its buffer. And by doing that, it avoids the need for the UE to send a scheduling request and then be granted by the base station before sending the data. It is worth noticing that this type of operation would only work for a limited amount of data as the reliable link adaption for configured grant needs a rather limited amount of grant bits to transmit and the overhead of such a scheme will be large if there are hundreds of UE operating under a certain base station and each UE would have a periodic configured grant resource. Hence, the main scheduling mechanism is similar to DL dynamically granting resource based on PDCCH, together with the configured UL grant which provides a good complement to this.

On PUSCH, both OFDM and transform precoded OFDM are supported. The OFDM flavour is very similar to the PDSCH version with some differences, for example, the total amount of supported layers that is limited to four in UL compared to eight in DL. For a single layer, both transform precoded OFDM and OFDM are supported. The UE is configured to operate one of the waveforms and can also dynamically change between the two based on the granted resource, which allows the operation of both one layer with transform precoded OFDM and more than one layer with OFDM. The merit with transform precoded OFDM is the lower PAPR than OFDM, and hence, it will have a larger coverage than an OFDM-based waveform.

Similar to PDSCH, both resource allocation type 0 and resource allocation type 1 are supported for PUSCH. However, there is a very limited set of requirements being specified for resource allocation type 0 due to the impact of PAPR of that scheme on the UL transmission.

The uplink frequency resource allocation is given in the DCI format by the field frequency domain resource assignment, independent of which uplink frequency resource allocation type is used.

For DCI format 0_0, uplink frequency resource allocation type 1 is the only available resource allocation type. For DCI format 0_1, both the uplink frequency resource allocation type 0 and type 1 are available, where only one or both could be configured, which is decided by dedicated RRC signalling. If both are configured, the one used is selected by a single bit in the frequency resource allocation scheme. As the functionality of the resource allocation types of PUSCH and PDSCH is similar, more details on the resource allocation type 0 and 1 are available in the PDSCH section.

The time domain resource allocation is selected by the field 'Time domain resource assignment' in DCI format 1_0 or DCI format 1_1, which may not exist if only a single time domain resource allocation scheme is configured by RRC for DCI format 1_1. The time domain resource allocation assignment field points into a row in a table. The table will give the PUSCH mapping type, which slot the PUSCH is allocated in, the starting OFDM symbol in the slot the PUSCH is allocated and the length of OFDM symbols of the PUSCH.

The PUSCH mapping type corresponds to the OFDM symbols which contain DMRS. There are two types defined, i.e. type A and type B. The starting OFDM symbol of the DMRS for mapping type A is the OFDM symbol of the first DMRS in case of type A mapping. The slot for which the PUSCH is allocated to is referred to as K_2. The starting OFDM symbol for the allocation is referred to as S and the length is referred to as L. The PUSCH is always allocated in contiguous OFDM symbols.

The table that is used for time domain resource allocation assignment could be either given by the specification or configured through broadcast signalling or dedicated RRC signalling depending on the applicable RNTI that the PDCCH

Table 9.11 Applicable PUSCH time domain resource allocation

RNTI	*If no dedicated and broadcasted configurable table provided*
PUSCH for RAR	Yes, both broadcasted and dedicated table
C-RNTI, MCS-C-RNTI, CS-RNTI	Yes, both broadcasted and dedicated table

CRC is scrambled with. This is further outlined in Table 9.11. The difference from downlink is that there is only one single default table for normal cyclic prefix.

Further, when a UE is in connected state, it is possible to configure the UE with a dedicated time domain resource allocation table. That table would be applicable for the unicast messages that are addressed with the C-RNTI, MCS-C-RNTI or CS-RNTI.

How the PUSCH mapping types are defined is very similar to the case on PDSCH, and hence, we refer to the mapping type A and mapping type B PDSCH for more details.

For PUSCH, frequency hopping is supported wherein the allocated frequency resource hops either in the middle of an allocation or between allocations in time which achieves a better diversity. The hopping is done with one out of four frequency offsets.

The modulation and code rate are selected by the UL grant by a field in the corresponding DCI that indicates a given modulation and code rate. The field is 5 bits long and an entry points into a table. The table used is configured to the UE. If the OFDM-based waveform is used, then the same tables as for downlink are available to configure the UE with. For more details, see the PDSCH section. If the UE is configured with transform precoded OFDM, there are two MCS tables to configure the UE with. One is targeting for a mobile broadband scenario, and the other one is targeting for IIoT scenarios. Both tables support the modulations orders Pi/2-BPSK, QPSK, 16QAM and 64QAM.

Based on the indicated code rate, the frequency and time resource allocation, the selected modulation order and the number of scheduled layers all together, a transport block size is calculated. This is in a similar manner as for PDSCH.

The minimum processing times are defined as the time between the end of the PDCCH giving the UL grant and the start of the PUSCH. At a high level, two categories of UE processing times are defined. This is also similar as for PDSCH to HARQ. Processing capability #1 for PUSCH is the basic UE capability and supports PUSCH being granted in the following slot after the PDCCH, while UE capability #2 supports this in the same slot. For both of the capabilities described here, we assume no timing advance, and hence, if there is timing advance, it is added on top of the processing time of the UE. There is one key aspect that can decide the processing time, which is the used subcarrier spacing. There are some other factors that could also control the processing time such as the selected mapping type of PUSCH. This certain functionality scaled with the subcarrier spacing would then allow for quicker processing times. It is, however, important to note that not all components are scaled with the processing time, so this is not a linear dependency.

6 Power Control

The UL power control in NR is very similar in many respects to LTE but with some differences. At a high level, the power control includes an open loop component and a closed loop component based on TPC commands. There are up to two power control loops running per carrier. One power control loop is for PUSCH and SRS. If there is no PUSCH on the carrier, then it is solely used for SRS. In addition, if there is PUCCH configured for the carrier, there is a power control loop for PUCCH, which is separated from the PUSCH/SRS power control loop.

The open loop component in the power control loop is either the SS/PBCH block which is used to obtain MIB for accessing the cell or a configured reference signal, which can be either another SS/PBCH block or a specific CSI-RS. The aspect of having different reference for open loop component differs from LTE.

For the closed loop power control, the accumulated component is different based on the associated SRI, i.e. the associated SRS resource for the PUSCH, which is an additional aspect designed to handle UL and DL beam management.

In addition to the above, power headroom reports (PHR) are defined for each specific power control loop and can be reported.

Another key aspect with NR is the operation of dual connectivity together with LTE, which is included from the start and is a key feature in the market. To support this, there are many features introduced which have multiple options based on the UE support on how to operate when connected to both NR and LTE.

The UE is configured with a total maximum output power for both NR and LTE. This is applicable when NR is within FR1 frequency range if the total output power of both NR and LTE is configured to be less or equal to the total UE power. The output power is statically split between the NR and LTE connection. This is the simplest aspect to implement and ensure connectivity on both NR and LTE. If the coverage becomes limiting, the NR connection is deconfigured by the network and the UE is configured with full output power on the LTE connection only. If the UE is more advanced, the UE can handle the case when the maximum output power of both NR and LTE connection goes beyond the maximum output power of the UE. If the scheduling does not collide in time, each power control is operated independently, and if they collide, the NR power is adjusted down (due to that the NR connection is on the secondary cell group (SCG) and assumed to be of lower priority). Instead of using this option, other less favourable options are introduced wherein a scheduling restriction in time is introduced and wherein the scheduler on the NR side is not allowed to schedule the UE in certain time instance on the NR side as this is solely dedicated to LTE.

In addition to all the above, the UE can still be configured with a total maximum output power for all operations within FR1 which leaves the remaining output power for FR2.

7 NR-LTE Interworking

A key aspect of NR is the interworking with LTE. This can be done in different flavours. One is that the UE is connected to both LTE and NR at the same time. This is described more in other parts of this book and is referred to as EN-DC (E-UTRAN-NR Dual Connectivity) or as non-standalone operation of NR. On the physical layer, the large impact of this is on the UE power control, as the power needs to be shared between NR and LTE in UL. There are significant impacts on higher layers due to this.

An important feature in NR is the future compatibility aspects that were described in the introduction to this chapter. Some of the features that are put in place to support future compatibility can also be used from day one. Specifically, the features can be used to enable an NR UE to handle the case where an NR carrier is overlapping with an LTE carrier, LTE-M carrier and/or NB-IoT carrier (Fig. 9.25).

To be able to handle migration of NR UE over several years, it is important to allow operation of an NR carrier overlapping with an LTE carrier. This is needed to handle a mixture of different types of UE in the network wherein some only support LTE and some support both LTE and NR. If the operators switch off the LTE carrier and directly migrate the network to NR, the UEs that only support LTE may get less quality in the network (in terms of bit rate or coverage). The NR UEs are informed to take away some resource elements that are used for LTE CRS and some general configured blocks of resource elements (which may be used for PDCCH, PSS/SSS, and PBCH on LTE). This is done by configuring a puncturing pattern on which the UE assumed for LTE channels where the PDSCH is not mapped onto. Such a pattern is a given CRS configuration. This configuration may be covering the full bandwidth of the currently used bandwidth part or only parts of it. NR is rather flexibly designed and hence the PDCCH on the NR side can be configured so that it does not overlap with any CRS on an overlaying LTE carrier. An example of such

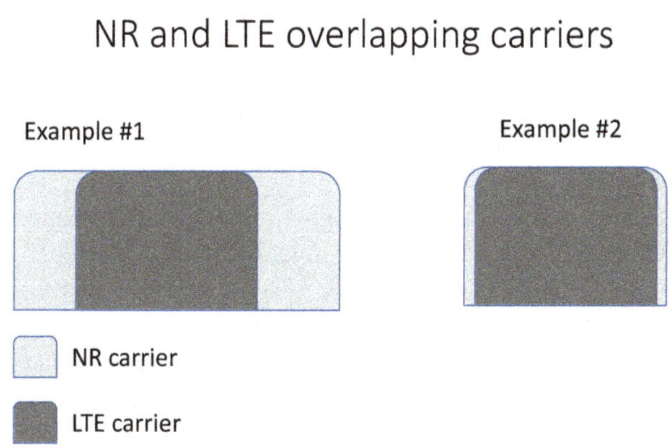

Fig. 9.25 Overlapping NR and LTE carriers in examples #1 and #2

Fig. 9.26 CRS rate matching of a PDSCH

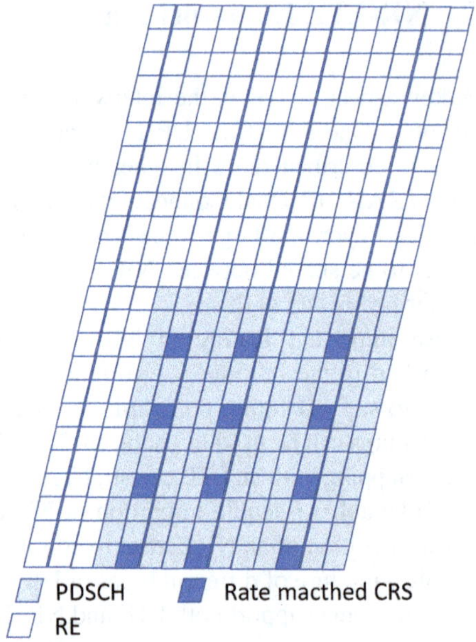

a pattern is shown in Fig. 9.26 wherein the PDSCH is allocated to a certain number of OFDM symbols and one resource block. Note that the CRS are rate matched and not transmitted from an NR perspective. Since the PDSCH is rate matched, the CRS pattern is not illustrated outside the PDSCH allocation as it is not applicable there.

Further, as mentioned, a generic rate matching pattern is also defined. This is defined so that the UE rate matches PDSCH around frequency domain on PRB level and in time domain on an OFDM symbol level. The pattern can either be configured semistatically by RRC and is then always available with a given periodicity in slots. Alternatively, a given pattern can be indicated to be used by the bit field in the scheduling DCI and hence for that specific PDSCH, a given pattern is used. An example of such a pattern is illustrated in Fig. 9.27. The figure illustrates a few different rate-matching examples wherein a specific RB at a specific OFDM symbol is rate matched. Furthermore, an example is given that an RB is rate matched in all OFDM symbols of a slot and that all RBs in a given OFDM symbol are rate matched.

In addition, it is also possible to configure the UE with a static rate-matching pattern for a CORESET or configure it dynamically by DCI. This is to avoid PDSCH to be mapped onto a PDCCH that is not addressed to the assigned UE.

Additional aspects which have also been introduced to support the overlaying of LTE and NR carriers are specific DMRS patterns on the NR side that can match the location better than the LTE and the possibility to align the subcarrier spacing grid in UL between NR and LTE.

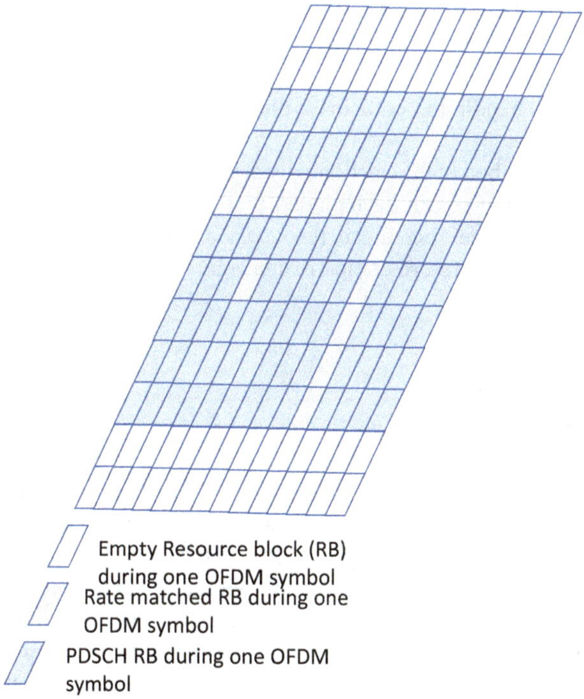

Fig. 9.27 Rate-matching pattern

All these rate matching patterns and some of the configurability options of the slots in PDSCH, PDCCH and so forth can be placed and are designed to allow LTE to be overlaid with NR, from a UE perspective.

For all the above functions, it is assumed that the scheduler operating for NR and the one operating for LTE are coordinated in some manner, so that they do not assign overlapping resources between NR and LTE. The speed at how often they coordinate is an implementation aspect at the network side, so it would not be directly visible to the UEs operated in the network.

In addition, it is also important to allow LTE-M carriers and/or NB-IoT carriers to overlap with an NR carrier. Here, the reason for this is that the corresponding functionality is not there on NR and the devices using these technologies are envisioned to be out in the market for 10+ years or so. Hence, such devices may be in the market as long as NR networks are operated, and a migration may not be considered as an option.

A side topic within this area is the ability to run two UL carriers towards a single DL carrier. One of the UL carriers is referred to as supplementary uplink (SUL). This is envisioned to be used by a UE that does not directly use CA but switches its uplink transmitter chain between these UL carriers. Typically, one of the uplinks would be in a lower LTE band, while the other one is in the same band as the corresponding DL. The UE will then be able to operate with larger uplink coverage. The same approach can be achieved by configuring the UE with DL CA with two DL carriers and a single UL in the lower spectrum and then using the mechanism described above to handle the collision with LTE DL resource, so that they do not overlap.

The features described in this section pave way for future development within NR wherein new technologies can be included that are not yet foreseen, and existing UEs can just avoid receiving those resources assigned to them. By using this mechanism, it provides a future compatible design.

8 UE Capabilities

NR supports a very large set of features that span over a large set of use cases, just like the later releases of LTE. It is necessary for NR to allow certain types of use cases to implement a specific feature set, to allow introduction of features over time, to allow different complex UEs on the market and to provide a filtering of features to the market. To allow this, a flexible structure with a wide possibility of capabilities is introduced. Here we do not describe the details of the UE capability signalling but focus on why there is UE capability signalling. Capability signalling is an indication to the network whether a certain feature is implemented and tested in the UE. If there is no capability signalling, it is assumed that the UE has implemented and tested the features already in the real network deployments.

The first set of features are the ones that do not have any capability signalling associated with them and are mandatory for the UEs to implement. These should be features that are basic and need to be supported by all UEs. It could be, for example, cell search-related features or to receive and transmit basic control channels and data. This only works if both networks and UEs implement the features. If the features are not implemented by all UEs, it would be difficult to use the features in practice as there is no mechanism for the network to know if the UEs support the features or not.

The second set of features are features that are associated with capability signalling and are mandatory for the UE to implement. The background to this is that the feature is seen as key to be implemented by the UEs on a long-term basis but may not be applicable to all bands, initial deployments and so forth; hence, the network needs the possibility to be able to know if the feature in question is implemented or not in the UE. UE features that do not provide an end user merit (e.g. higher throughput or lower latency) but rather a system benefit are suitable to be categorized as part of the first or second set of features. Note that some UE features that provide an end user merit are also suitable for being categorized as part of the first or second set of features.

The third set of features are features that are optional with capability signalling. The set of features typically defined here is not key for the basic support in the UE but could give a clear end user merit. Good examples of these are features that give higher throughput or lower latency, for example, the number of supported carriers in carrier aggregation.

The last set of features are features that are optional without capability signalling. This category is used for features that the network does not need to be aware of if the UE supports them or not. An example from LTE on this is basic LTE broadcast, i.e. eMBMS support, or idle mode-related feature as the network does not need to be aware if UEs support it or not.

With the above framework of capability signalling, it is possible for UEs to be designed for specific use cases. For example, a high-end NR UE may support different feature sets than an NR UE towards an Industrial IoT application. The capability signalling is defined further so that there are band-specific components, TDD/FDD components and FR1 or FR2 components. It is further possible for the UE to signal different sets of capabilities it supports in combination, which allows the network to utilize the UE capabilities in the most efficient way. For example, the UE can indicate that either it supports a large number of carriers but with fewer number of spatial layers or it can indicate the opposite. The network can choose how to configure the UE in its optimized way.

Reference

1. TS 38.101, NR; User Equipment (UE) radio transmission and reception

Chapter 10
Channel Coding in NR

Yufei Blankenship, Dennis Hui, and Mattias Andersson

To achieve the demanding performance targets of 5G NR, new channel coding techniques are introduced for both the data channels and the control channels. Compared to fourth-generation (4G) long-term evolution (LTE), low-density parity-check (LDPC) codes are introduced for the data channels, replacing the turbo codes of LTE. Similarly, polar codes are introduced for the control channels, replacing the tail-biting convolutional codes (TBCC) of LTE.

1 LDPC Coding in NR

1.1 Introduction

Robert Gallager introduced LDPC codes in his doctoral thesis [1] in 1963. Interest was renewed following the success of turbo codes in the early 1990s, and they were studied by several authors [2, 3]. Since then they have been incorporated in several standards such as DVB-S2, IEEE 802.16e, IEEE 802.11n, etc. LDPC codes were also considered during standardization of 4G LTE more than 10 years ago, though turbo codes with new QPP interleavers were ultimately adopted instead [4].

When channel coding techniques were investigated for 5G NR, it was decided that LDPC codes should be selected to replace the turbo code, considering the

Y. Blankenship
Ericsson Inc., Business Area Networks, Schaumburg, IL, USA

D. Hui (✉)
Ericsson Inc., Ericsson Research, Santa Clara, CA, USA
e-mail: dennis.hui@ericsson.com

M. Andersson
Ericsson AB, Ericsson Research, Stockholm, Sweden

X. Lin, N. Lee (eds.), *5G and Beyond*, https://doi.org/10.1007/978-3-030-58197-8_10

stringent performance requirements of 5G. 5G NR targets very high throughputs with peak data rates of 20 Gbps in the downlink and 10 Gbps in the uplink, as well as ultra-reliable low-latency communication with block error rate (BLER) targets down to 1e-5. Compared to the 4G LTE turbo codes, the 5G NR LDPC codes offer the following advantages:

- Increased throughput, both in terms of area efficiency (as measured by Gbps/mm^2) and peak throughput
- Higher degree of parallelization leading to reduced decoding complexity and latency, especially at high code rates
- Improved performance in the error floor region, with no error floors above BLER 1e-5 regardless of code size or code rate

1.2 Coding Chain of NR Data Channel

In NR, LDPC codes are used in the downlink and uplink data channels (i.e., PDSCH and PUSCH). The NR LDPC coding chain includes code block (CB) segmentation, cyclic redundancy check (CRC) attachment, LDPC encoding, rate matching, and systematic-bit-priority channel interleaving; see Fig. 10.1. Specifically, code block segmentation allows very large transport blocks to be split into multiple smaller-sized code blocks which can be efficiently processed by the LDPC encoder/decoder. CRC bits are then attached to each code block for error detection purposes. Combined with the built-in error detection of LDPC codes through the parity-check equations, very low probability of undetected errors can be achieved for each code block. After applying the CRC attachment, LDPC encoding, rate matching, and bit interleaving steps to each code block individually, the coded bits of all code blocks are concatenated into a single bit sequence for transmission.

CRC Attachment

As shown in Fig. 10.1, two levels of CRC attachment are applied to a transport block: first CRC attachment to the entire transport block and then CRC attachment to each of the code blocks individually after code block segmentation.

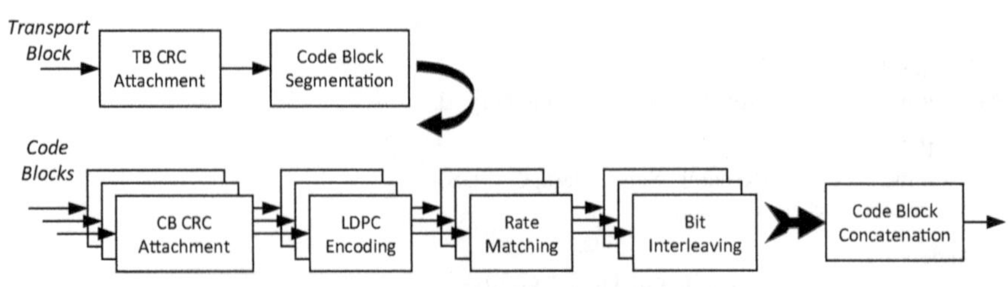

Fig. 10.1 NR LDPC coding chain at transmitter side

At the transport block level, CRC polynomials of two different degrees are used when generating CRC bits for a transport block. To reduce overhead, a CRC polynomial $g_{\mathrm{CRC16}}(D) = [D^{16} + D^{12} + D^{5} + 1]$ of degree 16 is used for transport blocks shorter than or equal to 3824 bits. For larger transport blocks, a degree-24 polynomial $g_{\mathrm{CRC24A}}(D) = [D^{24} + D^{23} + D^{18} + D^{17} + D^{14} + D^{11} + D^{10} + D^{7} + D^{6} + D^{5} + D^{4} + D^{3} + D + 1]$ is used.

In addition, when a transport block is segmented into two or more code blocks, 24 CRC bits generated using the polynomial $g_{\mathrm{CRC24B}}(D) = [D^{24} + D^{23} + D^{6} + D^{5} + D + 1]$ are attached to each CB. The addition of CRC bits to individual CBs allows for detailed hybrid automatic repeat request (HARQ) feedback at the CB level, which can be used to save radio resources via partial retransmission of a transport block. For example, if the HARQ feedback indicates that only a subset of the CBs of a given TB is incorrectly decoded, then the scheduler can request retransmissions of incorrectly decoded CBs only, without retransmitting the correctly received CBs belonging to the same TB.

Code Block Segmentation

Code block segmentation is depicted in Fig. 10.2. Let A (bits) be the transport block size generated by higher layer and B (bits) be the size of the transport block after CRC attachment. If B is smaller than the largest code block size K_{cb} for the given base graph, the transport block is not segmented.

Otherwise, the transport block is segmented into $C = \lceil B/(K_{\mathrm{cb}} - 24) \rceil$ equal-sized code blocks[1]. The 24 in the denominator accounts for the 24 CB CRC bits attached to each CB after segmentation. The size $K_{\mathrm{cb}} = 8448$ (bits) for base graph #1, and

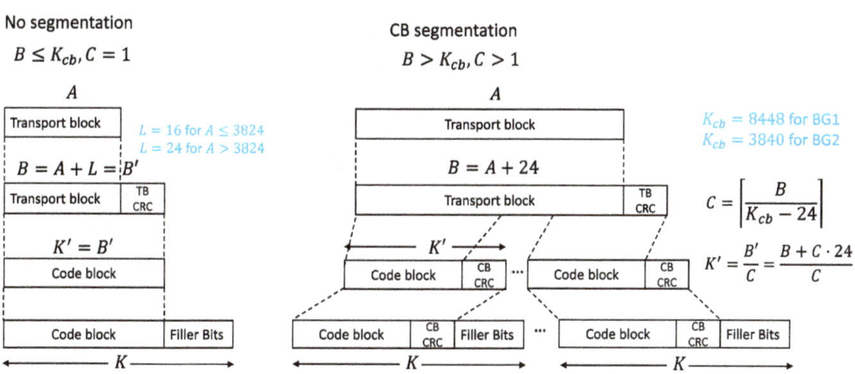

Fig. 10.2 Code block segmentation of NR data channel

[1]The transport block sizes are chosen in such a way that equal-sized code blocks are guaranteed.

$K_{cb} = 3840$ (bits) for base graph #2. The number of information bits in each CB after segmentation and CB CRC attachment is $K' = B/C + 24$.

NR LDPC Structure

NR LDPC codes are quasi-cyclic codes, where the parity-check matrix (PCM) H is defined by a small bipartite base graph represented by a binary matrix together with a set of shift coefficients [5]. To obtain the parity-check matrix, the base graph is expanded by replacing each entry by a $Z \times Z$ matrix. Each entry with value zero in the base graph is replaced by a $Z \times Z$ zero matrix, and each entry with value one is replaced by a shifted $Z \times Z$ identity matrix. The identity matrix is cyclically shifted to the right by a cyclic shift corresponding to the shift coefficient associated with the entry.

The base graph is represented by a binary matrix with M_b rows and N_b columns. The first K_b columns, where $K_b = N_b - M_b$, correspond to information bits and are sometimes denoted as systematic columns. It has been noticed that LDPC code performance can be improved by including a small fraction of punctured variable nodes[2] with high degree. In NR, the $2 \times Z$ systematic bits corresponding to the first two columns in the base graph are always punctured.

After expansion, the parity-check matrix H has size of $M_b \times Z$ rows, $N_b \times Z$ columns. Using this H matrix, the number of information bits to be encoded is $K = K_b \times Z$ (bits). After encoding, the number of coded bits available for transmission is $N = (N_b - 2) \times Z$ (bits), since the first $2 \times Z$ systematic bits are always punctured. Thus, without considering further rate-matching operations, the native code rate based on the H matrix is $R = K/N = K_b/(N_b - 2) = (N_b - M_b)/(N_b - 2)$. The size of the base graph ($M_b \times N_b$) hence determines the native code rate.

The NR LDPC codes use a large number of degree-one, or single parity check, parity bits as can be seen by the identity sub-matrix of the base graph in the right part of Figs. 10.3 and 10.4, similar to the codes proposed in [6]. Puncturing a different number of these parity bits generates code words of different rates. This is especially useful for communication systems employing HARQ, allowing for the use of incremental redundancy instead of Chase combining for retransmissions. Rows and columns corresponding to punctured degree-one parity bits can be removed from the parity-check matrix when decoding, hence making the decoding complexity and latency smaller for higher code rates. This contrasts with the LTE turbo codes which have constant decoding complexity and latency irrespective of the code rate.

[2]Punctured variable nodes are sometimes referred to as state variable nodes.

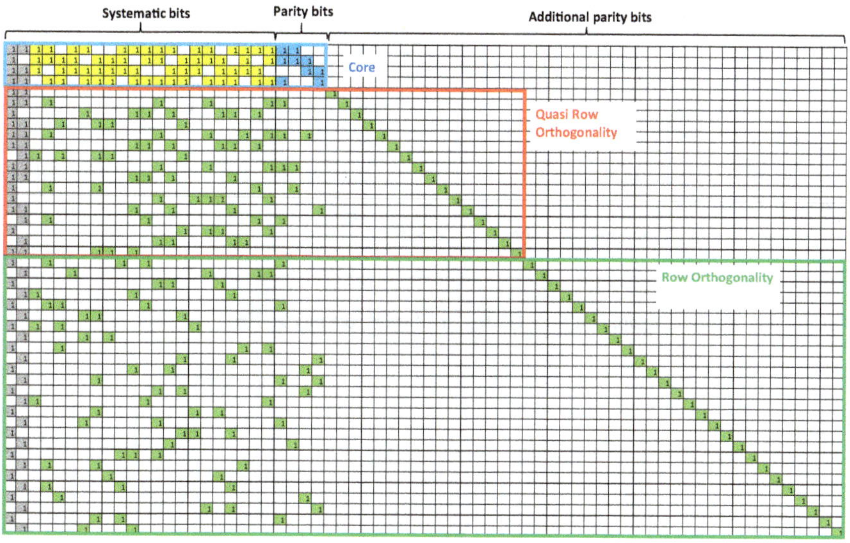

Fig. 10.3 Base graph #1

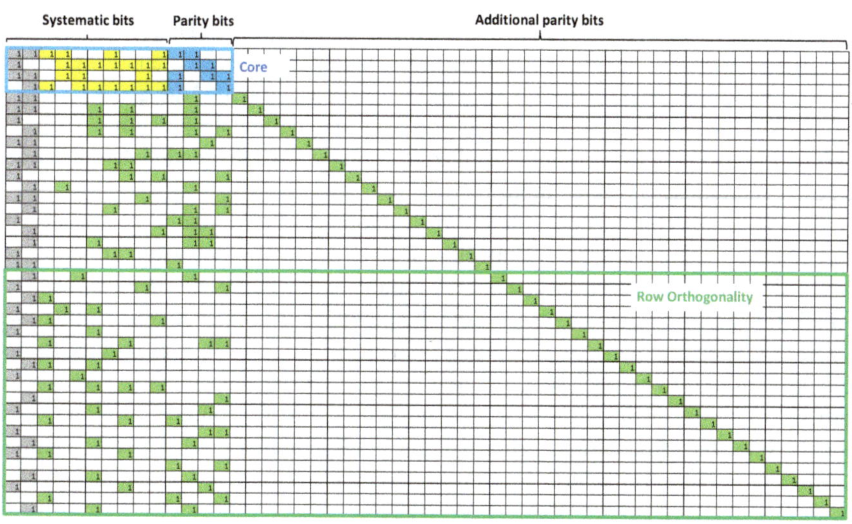

Fig. 10.4 Base graph #2

Two Base Graphs

In order to efficiently support a wide range of use cases with various throughput and reliability requirements, two different base graphs are adopted in NR. This allows NR to efficiently support mobile broadband use cases with large information block

Table 10.1 NR LDPC base graph parameters

Parameter	Base graph #1	Base graph #2
Native code rate of the base graph	1/3	1/5
Base graph size ($M_b \times N_b$)	46 × 68	42 × 52
Number of systematic columns (K_b)	22	10
Maximum information block size K (=$K_b \times Z_{max}$)	8448 bits(=22 × 384)	3840 bits(=10 × 384)
Number of nonzero elements in the base graph	316	197

Table 10.2 Shortening applied for BG2 for small transport block sizes

TB size B (bits) after CRC attachment	K_b	Range of Z
$640 < B$	10	72 <= Z <= 384
$560 < B <= 640$	9	64 <= Z <= 72
$192 < B <= 560$	8	26 <= Z <=72
$B <= 192$	6	7 <= Z <= 32

size and high code rate, as well as ultra-reliable low-latency communications with low code rate and small information block size.

Base graph #1 (BG#1) is designed for larger block lengths and higher code rates, while base graph #2 (BG#2) is designed for smaller block lengths and lower code rates. The design parameters of base graph #1 and base graph #2 are given in Table 10.1. Note that different lifting sizes Z as well as additional operations, such as shortening and puncturing, can be applied to obtain many other code rates and code sizes beyond those shown in Table 10.1.

To improve performance and to allow for more parallelization in the decoder, it can be beneficial to use a larger Z together with a smaller K_b for a given information block size. For this reason, shortening is additionally applied when BG#2 is used for small transport block sizes. This is done by choosing a smaller K_b and setting the value of the remaining $(10 - K_b) \times Z$ systematic bits to zero when encoding. These bits are referred to as filler bits and are discarded before transmission. The whole procedure is equivalent to removing the unused systematic columns from the base graph. Since the effect of using shortening to achieve higher level of parallelization is most prominent for the smallest information block sizes, the shortening scheme is applied to BG#2 but not BG#1 (Table 10.2).

Turning now to the BG#1 and BG#2 matrices illustrated in Figs. 10.3 and 10.4, it is observed that the matrices contain several carefully selected design features. These features allow the base graphs to simultaneously achieve the goals of superior code performance, simple encoder and decoder architecture, low decoding complexity, and low decoding latency.

Each base graph contains a very small core matrix of dense connection (i.e., dense ‘1’s), together with extension parts of sparse connection (i.e., sparse ‘1’s). We refer to the sub-matrix containing the top-left four rows and $(K_b + 4)$ columns in BG#1 and BG#2 as the core matrix. The column weight of the parity bits that are not in the core matrix is 1, and the corresponding bottom-right sub-matrix is the identity matrix. This structure implies that the parity-check matrix of the LDPC code obtained by puncturing some of the additional parity bits is given by the sub-matrix that does not include the corresponding punctured parity bits and the parity-check equations that they are involved in. Hence the corresponding decoding operations are no longer necessary. The core matrix of BG#1 is the 4 row-by-26 column sub-matrix at the top-left corner, i.e., $M_{\mathrm{b},0}^{(1)} = 4$, $N_{\mathrm{b},0}^{(1)} = 26$. Hence the highest code rate for which BG#1 can be used without extension[3] is given by:

$$R_{1,\mathrm{core}} = \frac{\left(N_{\mathrm{b},0}^{(1)} - M_{\mathrm{b},0}^{(1)}\right)}{\left(N_{\mathrm{b},0}^{(1)} - 2\right)} = \frac{22}{24} = 0.92.$$

Similarly, the core matrix of BG#2 is the 4 row-by-14 column sub-matrix at the top-left corner, i.e., $M_{\mathrm{b},0}^{(2)} = 4$, $N_{\mathrm{b},0}^{(2)} = 14$. Thus, the highest code rate that BG#2 can be used without extension is given by:

$$R_{2,\mathrm{core}} = \frac{\left(N_{\mathrm{b},0}^{(2)} - M_{\mathrm{b},0}^{(2)}\right)}{\left(N_{\mathrm{b},0}^{(2)} - 2\right)} = \frac{10}{12} = 0.83.$$

Larger sub-matrices are obtained by successively extending the core matrix toward the bottom-right corner, thus achieving lower code rates. For BG#1, if d_{b} additional number of rows and columns are used, the sub-matrix is composed of the top $\left(M_{\mathrm{b},0}^{(1)} + d_{\mathrm{b}}\right)$ rows and the left $\left(N_{\mathrm{b},0}^{(1)} + d_{\mathrm{b}}\right)$ columns of the BG#1 base graph, resulting in the code rate $\frac{\left(N_{\mathrm{b},0}^{(1)} - M_{\mathrm{b},0}^{(1)}\right)}{\left(N_{\mathrm{b},0}^{(1)} + d_{\mathrm{b}} - 2\right)}$. A similar procedure can be applied to BG#2. Example sub-matrix dimension and code rates are shown in Tables 10.3 and 10.4 for BG#1 and BG#2, respectively, assuming no additional shortening for BG#2. This demonstrates that when higher code rates are used, a smaller sub-matrix is used for encoding and decoding, leading to reduced complexity for encoding and decoding.

The code rates illustrated in Tables 10.3 and 10.4 are achieved by using different sub-matrices of the base graph. In these examples, the number of coded bits is an integer multiple of Z bits, for example, $\left(N_{\mathrm{b},0}^{(1)} + d_{\mathrm{b}} - 2\right) \times Z$ bits for BG#1. To provide full flexibility for NR, circular buffer-based rate-matching algorithm is applied to the output of LDPC decoder. The rate matcher can provide an arbitrary number of bits for transmission, corresponding to the finest granularity of code rate.

[3]During the code search process, BG#1 and BG#2 were designed for maximum code rates of 8/9 and 2/3, respectively. The sub-matrices corresponding to these code rates are referred to as the kernel of the matrices and have size 5×27 and 7×17 for the two base graphs. Additionally, code rates up to 0.95 can be achieved by puncturing parity bits in the core matrix.

Table 10.3 Example code rates provided by the sub-matrix of BG1 base graph, which is composed of the top $\left(M_{\mathrm{b},0}^{(1)} + d_{\mathrm{b}}\right)$ rows and the left $\left(N_{\mathrm{b},0}^{(1)} + d_{\mathrm{b}}\right)$ columns

d_{b} additional number of rows, columns	Top $\left(M_{\mathrm{b},0}^{(1)} + d_{\mathrm{b}}\right)$ rows	Left $\left(N_{\mathrm{b},0}^{(1)} + d_{\mathrm{b}}\right)$ columns	Code rate of the (sub-)matrix
1	5	27	22/25
9	13	35	2/3
20	24	46	1/2
42	46	68	1/3

Table 10.4 Example code rates provided by the sub-matrix of BG2 base graph, which is composed of the top $\left(M_{\mathrm{b},0}^{(2)} + d_{\mathrm{b}}\right)$ rows and the left $\left(N_{\mathrm{b},0}^{(2)} + d_{\mathrm{b}}\right)$ columns

d_{b} additional number of rows, columns	Top $\left(M_{\mathrm{b},0}^{(2)} + d_{\mathrm{b}}\right)$ rows	Left $\left(N_{\mathrm{b},0}^{(2)} + d_{\mathrm{b}}\right)$ columns	Code rate of the (sub-)matrix
3	7	17	2/3
8	12	22	1/2
18	22	32	1/3
28	32	42	1/4
38	42	52	1/5

Base graph #1, shown in Fig. 10.3, is optimized for larger information block sizes and higher code rates. It is designed for a maximum code rate of 8/9 and may be used for code rates up to $R = 0.95$. As discussed, the systematic bits corresponding to the first two columns (shown in gray) are never transmitted. It is observed that for the top four rows, the 4-by-4 parity block (shown in blue) has a dual-diagonal structure for easy encoding as well as good performance. For parity columns beyond the dual-diagonal structure, the simple diagonal structure (i.e., the green diagonal extending to bottom-right corner) allows for even simpler encoding. For the rows below the core (i.e., row 5 to row 46), the design strives to achieve low connection density (i.e., low density of '1's) and row orthogonality for efficient decoding without compromising the superior BLER performance. Lower connection density implies fewer decoding operations. Row orthogonality allows more parallel decoder processing, hence facilitating higher throughput LDPC decoder implementation. For base graph #1, row 5 to row 20 (shown in red box) have the property of quasi-row orthogonality, i.e., when excluding the first two columns, row i and row $i + 1$ of the base graph are orthogonal, $i = 5, \ldots, 20$. Row 21 to row 46 have the property of row orthogonality, i.e., row i and row $i + 1$ of the base graph are orthogonal (including the first two columns), $i = 21, \ldots 45$.

Base graph #2, shown in Fig. 10.4, is optimized for smaller information block sizes and lower code rates than base graph #1. It is designed with a smaller number of systematic columns than base graph #1, which gives a smaller information block size for the same Z. On the other hand, base graph #2 can reach code rate 1/5 without repeating coded bits. This is significantly lower than base graph #1 and the LTE turbo codes, which have design code rates of 1/3. To reach code rates below the design code rate, repetition is used, resulting in worse performance compared to using a code with lower design code rate. Base graph #2, and its additional coding gain at low code rates, is suitable for use cases that require very high reliability. Similar to base graph #1, the base graph #2 design carefully incorporates several features to achieve easy encoding, high-throughput decoding, as well as optimized code performance. For easy encoding, the parity portion contains the 4-by-4 dual-diagonal structure (shown in blue) as well as the simple diagonal structure beyond (i.e., the green diagonal extending to bottom-right corner). To facilitate high-throughput decoding, for the rows below the core (i.e., row 5 to row 42), row orthogonality is applied as much as possible without degrading code performance. In particular, row 21 to row 42 exhibit row orthogonality, i.e., row i and row $i + 1$ of the base graph are orthogonal, $i = 21, , \ldots 41$.

There is a need to support a large number of different information block sizes for a cellular system to support the wide range of use cases. As discussed, the base graphs are designed to encode information blocks with size $K_b \times Z$ (bits), where K_b is the number of systematic columns and Z is the lifting size. To support arbitrary, smaller, information block sizes than $K_b \times Z$, shortening is applied, where the last information bits are set to zero and not transmitted. In principle, shortening can be used to support any information block size smaller than $K_b \times Z$, but excessive shortening leads to performance degradation. Therefore, 51 different lifting sizes Z are used to define 51 different parity-check matrices for each base graph, with each Z supporting a different information block size. Shortening is still used to support intermediate information block sizes. In total, there are 102 parity-check matrices defined for the NR data channels. For comparison, we note that IEEE 802.11n only specifies 12 PCMs with 4 different code rates and 3 different information block sizes. The lifting sizes are of the form $Z = a \times 2^j$ to facilitate hardware implementation of arbitrary circular shifts of the $Z \times Z$ identity matrix. The possible Z values are listed in Table 10.5, ranging from 2 to 384.

The supported information block size range and code rate range by base graph #1 and #2 overlap, as can be seen in Table 10.1. In the overlapping region, one base graph, from the two candidate base graphs, needs to be selected for actual transmission of a transport block. The selection criteria are largely based on the error-correcting performance of the candidate LDPC codes, including the effect of

Table 10.5 Parameters for lifting sizes of the LDPC codes

		a							
Z		2	3	5	7	9	11	13	15
j	0	2	3	5	7	9	11	13	15
	1	4	6	10	14	18	22	26	30
	2	8	12	20	28	36	44	52	60
	3	16	24	40	56	72	88	104	120
	4	32	48	80	112	144	176	208	240
	5	64	96	160	224	288	352		
	6	128	192	320					
	7	256	384						

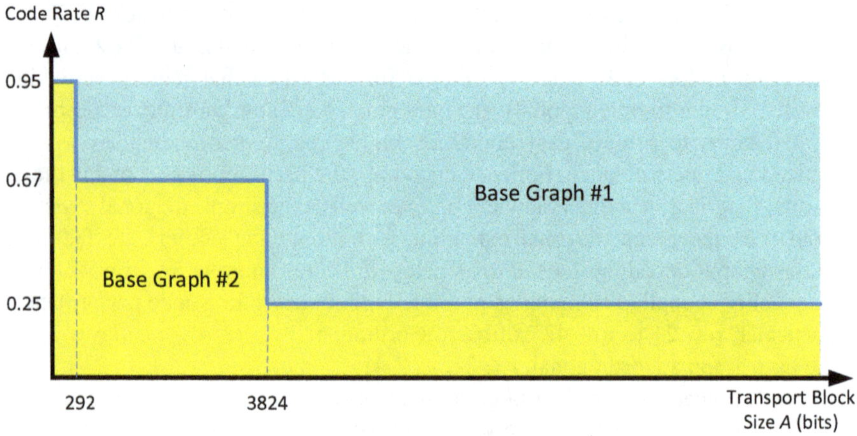

Fig. 10.5 Base graph selection according to transport block size A and code rate

puncturing, repetition, code block segmentation, etc. Overall, the range of transport block sizes A and code rates R covered by each of the two base graphs is shown in Fig. 10.5.

In general, base graph #1 is used for higher code rates and base graph #2 for lower code rates. The information block size is also taken into account when determining the switching point, due to performance differences between the two base graphs for different information block sizes. Base graph #2 is used for all code rates for $A \leq 292$ bits. For $292 < A \leq 3824$, BG#1 is used for code rates above 2/3, with BG#2 used for code rates lower than 2/3. For $A > 3824$ bits, the maximum information block size of $K = 3840$ for BG#2 is reached, when taking into account the 16 TB CRC bits. Hence BG#1 is used for all combinations of $A > 3824$ and $R > 1/4$. BG#2 is used for all cases with $R \leq 1/4$, due to the additional coding gain available from the design rate of 1/5 for BG#2. BG#1 has a design rate of 1/3, but the switching point is at code rate 1/4 due to considerations from code block segmentation. Base graph #1 supports a larger maximum information block size of $K = 8448$, so code block segmentation results in fewer code blocks for a given transport block size, when BG#1 is used. This offsets the additional coding gain from BG#2 for $1/4 < R < 1/3$. That is, for a given transport block size, BG#1 with repetition to reach down to $R = 1/4$ gives better TB-level BLER performance than BG#2, where TB-level decoding success requires all of its code blocks to be successful.

The rate-matching simulation studies show that BG#1 performance is good for 256 QAM up to code rate 0.9375 and good for QPSK up to code rate 0.9565. Hence it was decided that UE can skip decoding with either BG#1 or BG#2 when the effective code rate is greater than 0.95, where the effective code rate refers to the code rate obtained when only actually transmitted coded bits are counted.

Table 10.6 Starting point in the circular buffer for the four RV indices

RV	Starting coded bit index for BG#1	Starting coded bit index for BG#2
0	0	0
1	$17 \times Z$	$13 \times Z$
2	$33 \times Z$	$25 \times Z$
3	$56 \times Z$	$43 \times Z$

1.3 Rate Matching for LDPC Codes

To be able to provide an arbitrary number of coded bits (hence arbitrary code rate) as needed, the circular buffer-based rate-matching algorithm is defined. Coded bits at the output of LDPC encoder are entered into the circular buffer and read out consecutively starting at a predefined point on the circular buffer. The predefined point is according to the redundancy version (RV) index, where the RV takes value of 0–3. In Table 10.6, the predefined starting point for each of the RV indices is shown for both base graphs. Note that the first $2 \times Z$ punctured systematic bits are never entered into the circular buffer; hence the first $2 \times Z$ systematic bits are never transmitted regardless of the code rate.

The full circular buffer size is $66 \times Z$ for BG#1, and $50 \times Z$ for BG#2. RV0, RV1, and RV2 are evenly distributed on the circular buffer (i.e., 0/4, 1/4, 2/4 of the full circular buffer size). The exception is RV3, where RV3 is shifted closer to the starting point of the circular buffer so that a transmission using RV3 is self-decodable even for high code rate since it allows quick wrap around to pick up systematic bits. Thus, together with RV0, two RV indices support self-decodability at high code rate.

The default redundancy version (RV) sequence for HARQ (re-)transmission is {RV0, RV2, RV3, RV1}, which has been shown by simulation to provide the best performance across all block sizes and all code rates. This is illustrated in Fig. 10.6.

1.4 Bit-Level Channel Interleaver for LDPC Codes

The systematic-bit-priority mapping (HSPA-like) was used in the bit-level channel interleaver. The rectangular interleaver improves performance by making systematic bits more reliable than parity bits for the initial transmission of the code blocks. Specifically, the interleaver is realized via a table with the number of rows equal to the modulation order, with the number of columns then determined by the number of coded bits.

The coded bits at the output of the LDPC rate matcher are written into the table row-by-row starting from the top-left corner and read out from the table column-by-column starting from the top-left corner. When the bits are read out and used

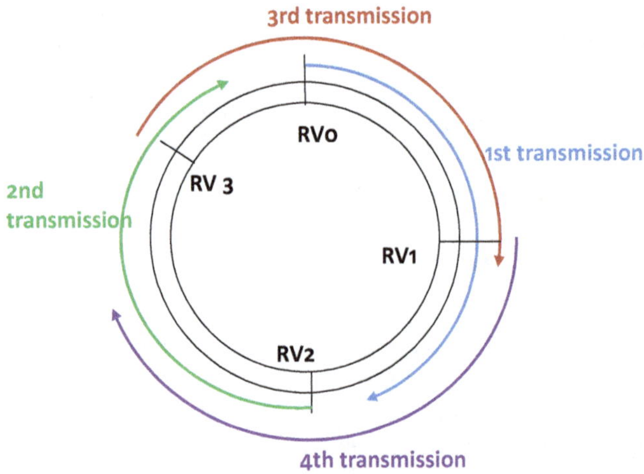

Fig. 10.6 Extraction of coded bits from the circular buffer for transmission assuming the default RV sequence of {RV0, RV2, RV3, RV1}

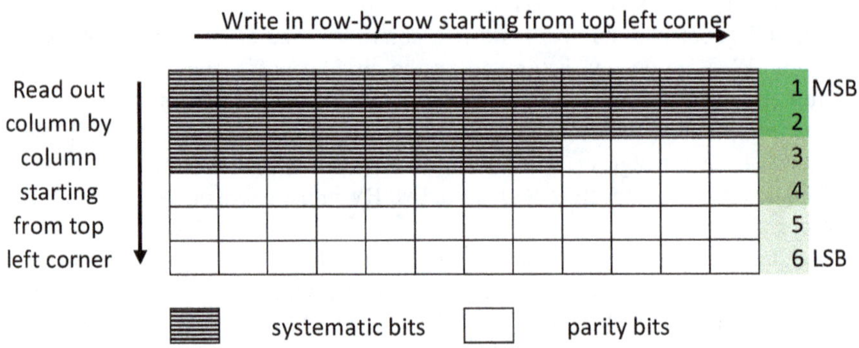

Fig. 10.7 Illustration of bit-level channel interleaver for LDPC codes assuming modulation of 64-QAM

to generate modulation symbols, the bits at the top row are mapped to the most significant bit (MSB) of the modulation symbols, providing them with the highest bit reliability. Similarly, the bits at the bottom row are mapped to the least significant bit (LSB) of the modulation symbol, providing them with the lowest bit reliability. This effect exists for modulation order higher than QPSK. This tends to improve the LDPC decoding performance since the systematic bits are located toward the MSB when the redundancy version is 0, which the initial transmission of a transport block uses. This is illustrated in Fig. 10.7 assuming 64-QAM.

1.5 Performance of NR LDPC Codes

The performance of NR LDPC codes over an AWGN channel has been evaluated using a normalized min-sum decoder, layered scheduling, and a maximum of 20 decoder iterations.

Figure 10.8 shows the SNR required to achieve certain BLER targets as a function of information block size K for code rate 1/2 and QPSK modulation. The results show that NR LDPC codes provide consistently good performance over the full range of block sizes. According to the base graph selection rules, BG#2 is used for $K \leq 3840$ (including CRC bits), and BG# 1 is used for $K > 3840$ (bits). This accounts for the small jump in performance at $K = 3840$.

In Fig. 10.9, the 5G NR LDPC codes are compared with 4G LTE turbo codes for $K = 6144$ (bits), the largest information block size defined for the turbo codes. BG# 2 with two code blocks is used for rate 1/5, and BG#1 is used for the other code rates. The two code families show similar performance, except at high code rates where the LTE turbo codes have a tendency of an error floor. This can be seen in Fig. 10.9 where the LTE turbo code is 0.06 dB worse than the NR LDPC code at BLER = 1e-4. This error floor becomes higher at higher code rates; see [7] for a thorough evaluation of the LTE turbo codes for a large range of code rates and information block sizes.

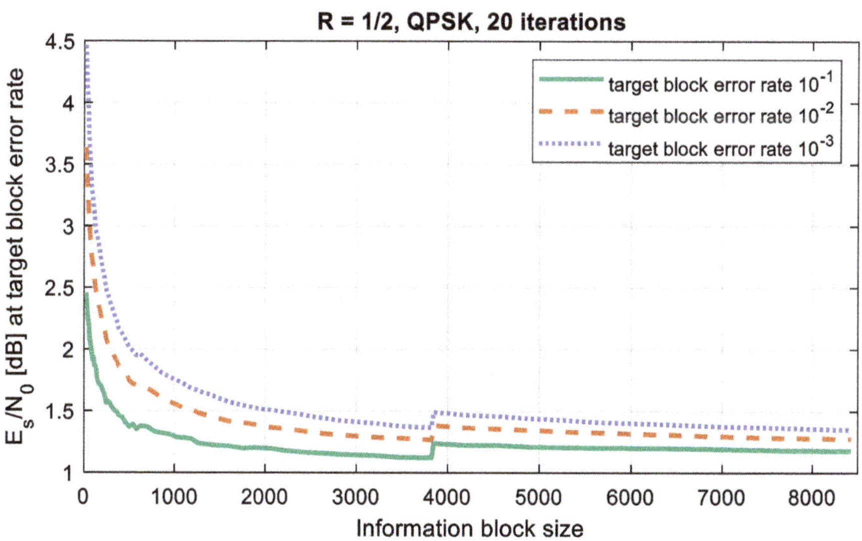

Fig. 10.8 Performance of NR LDPC codes at code rate 1/2 for QPSK modulation

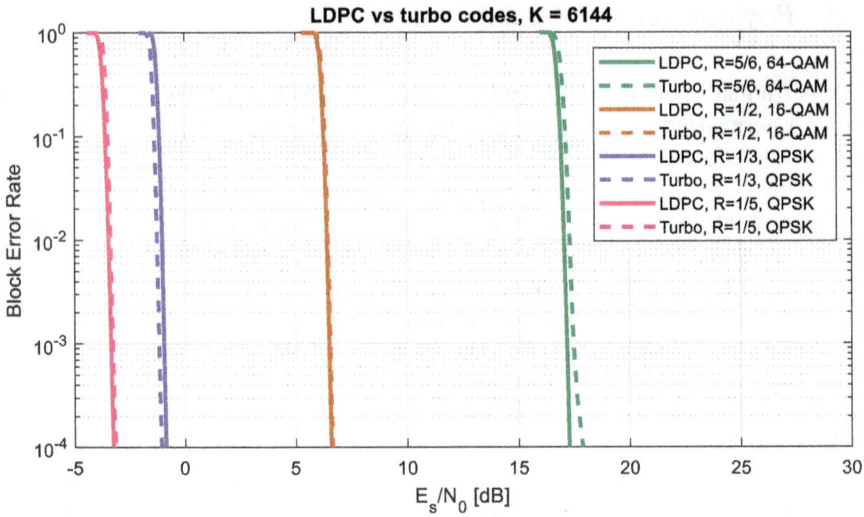

Fig. 10.9 Performance comparison between NR LDPC codes (solid) and LTE turbo codes (dashed)

2 Polar Coding in NR

2.1 Introduction

Polar codes were first introduced by Arıkan [8] who proved constructively that they can achieve the symmetric (Shannon) capacity of binary-input discrete memoryless channel using a low-complexity decoder, namely, a successive cancellation (SC) decoder. However, in practice, polar codes of finite size often exhibit a poor minimum distance property when used alone. As a result, despite being capacity-achieving, they do not perform well at high SNR regime even with exhaustive-search ML decoding [9]. Competitive performance is achieved only in concatenation with an outer code, such as a CRC code (or any parity-check code in general), together with the use of a successive cancellation list (SCL) decoder.

In NR, polar codes are used to protect the downlink control information (DCI), the uplink control information (UCI), as well as the system information in the physical broadcast channel (PBCH). DCI is transmitted over PDCCH, while UCI may be transmitted over PUCCH or PUSCH. As illustrated in the following sections, the NR polar code is significantly more complex than the tail-biting convolutional code (TBCC) used in 4G LTE control channels. The main advantage of polar codes over the TBCC, as well as the turbo codes used in 4G LTE and the NR LDPC codes, is that polar codes with SCL and CRC outer code typically yield better performance at moderate payload sizes (in the order of $K = 250$ bits or less). While not appropriate for data channels, small to moderate payload sizes are typically sufficient for transmitting control information and system information.

Polarization Theory

The core idea of polar coding is based on the transformation of a pair of identical binary-input channels into two distinct binary-input channels of different qualities. One of them is better, while the other is worse, than the original binary-input channel. To understand this transformation, consider the use of a polar code of length $N = 2$ to encode a pair of binary inputs, X_1 and X_2, into two coded bits, Z_1 and Z_2, that are sent over two identical channels with binary input to yield a pair of channel output, Y_1 and Y_2, as shown in Fig. 10.10.

The two inputs, X_1 and X_2, can be viewed as being transmitted individually over two distinct channels when a successive cancellation (SC) decoder is used. On the one hand, since the input X_1 is decoded before the input X_2, the input X_2 is unknown and acts as additional "noise" when the input X_1 is decoded, as depicted in the lower-left figure in Fig. 10.10. Because of this additional "noise," the input X_1 is effectively sent over a channel that has a worse reliability than the original binary-input channel. On the other hand, when decoding the input X_2 with a SC decoder, the input X_1 is already known (and presumably correctly decoded), and its impact can thus be nullified. As a result, the input X_2 can be viewed as being sent over a diversity channel with two outputs, Y_1 and Y_2, as illustrated in the lower-right figure in Fig. 10.10. The diversity channel clearly has a better reliability than the original binary-input channel with a single output, Y_2. Hence, when a SC decoder is used, a polar code of length two essentially "polarizes" two identical binary-input channels into a pair of distinct binary-input channels, commonly referred to as "bit-channels," with two different reliabilities.

Now consider the use of a polar code of length $N = 4$ over four identical binary-input channels, as depicted in the left figure in Fig. 10.11. Using the transformation illustrated above for polar code of length $N = 2$, one can view a polar code of length $N = 4$ as two independent polar codes of length $N = 2$ applied over two kinds of binary-input channel with different reliabilities, as shown in Fig. 10.11. The first polar code of length $N = 2$ has two inputs, X_1 and X_3, and is applied over a worse pair of identical binary-input channels ("channel –"). The second polar code has

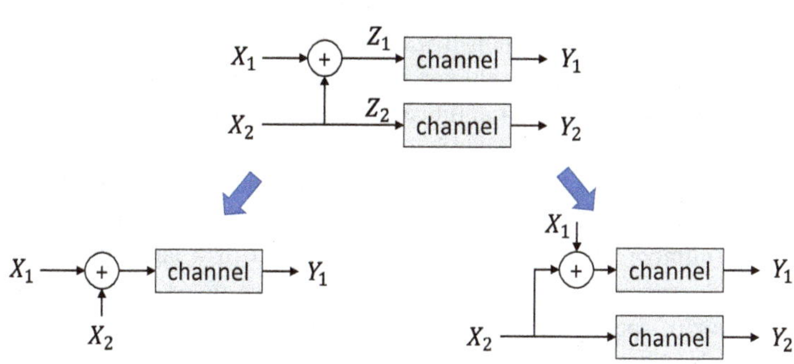

Fig. 10.10 Transformation of two identical channels into a better channel and a worse channel

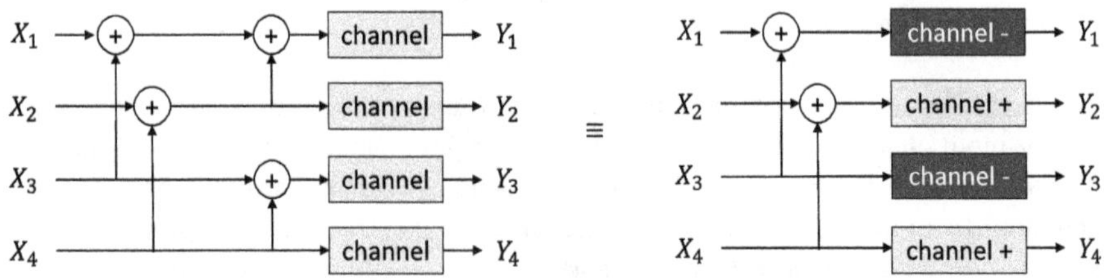

Fig. 10.11 Use of polar code of length four as two uses of polar code of length two, each applied over a different channel

Fig. 10.12 Four bit-channels of different reliabilities transformed from four identical binary-input channels via a length-4 polar code

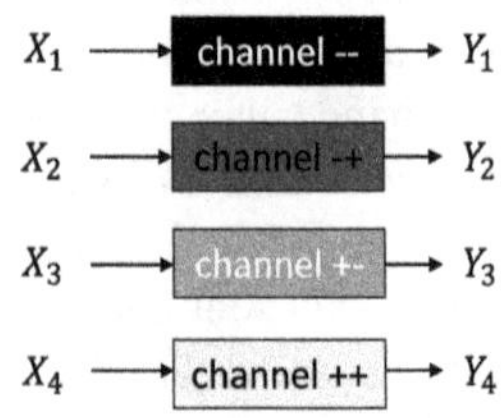

two inputs, X_2 and X_4, and is applied over a better pair of identical binary-input channel ("channel +"). Applying the transformation again on each of these polar codes of length $N = 2$, one transforms the four original binary-input channels into four distinct bit-channels with four different reliabilities, as depicted in Fig. 10.12.

Such a pairwise polarizing operation can be repeated on a set of $N = 2^n$ independent uses of a binary-input channel for any given integer n to obtain a set of 2^n bit-channels of varying reliabilities. When n increases, some of these bit-channels become nearly perfect (i.e., error-free) while the others become nearly useless (i.e., totally noisy). Arıkan [8] showed that the fraction of nearly perfect bit-channels is exactly equal to the capacity of the original binary-input channel as n approaches infinity. The key idea of polar coding is to transmit data on the nearly perfect channels while fixing the input to the useless channels to fixed or frozen values (e.g., 0) that are known to the receiver. Frozen bits and non-frozen (or information) bits are commonly used to refer to the input bits to the nearly useless and the nearly perfect bit-channels, respectively. Information are only carried on the non-frozen bits. The set of non-frozen bit locations is often referred to as the *information set* and has a direct impact on the performance of a polar code. The number of data bits K to be communicated over the N independent binary-input channels determines the size of the information set.

Since a more reliable bit-channel is clearly more desirable than a less reliable bit-channel in terms of carrying information, a smaller information set (i.e., that for a smaller number of data bits) is always a subset of a larger information set (i.e., that for a larger number of data bits) for the same polar code. Consequently, an efficient way of specifying the information sets for different sizes is a sequence of indices to the bit-channels, termed an *information sequence*, which ranks the bit-channels in an ascending order of reliabilities. The order with which the bit-channels should be used to carry data can be derived from such an information sequence. For instance, the bit-channels corresponding to the last K indices in the information sequence should be used for carrying a given set of K data bits and so forth. Note that puncturing (or shortening) of coded bits can cause some bit-channels to become highly unreliable (i.e., incapable of information-carrying) and thus alter the order with which the bit-channels should be used (as described in more details later). Hence, the puncturing (or shortening) patterns and the information sequence should be designed jointly for polar codes.

2.2 *Coding for Downlink Control Information*

The core components in coding of DCI in NR are the cyclic-redundancy-check (CRC) encoder, the CRC interleaver, the polar encoding kernel, and the rate matcher, as illustrated in Fig. 10.13. All except the CRC interleaver are also included in the coding chain for uplink control information (UCI) in NR, as described in the next section.

Compared to the polar coding chain for UCI, the polar coding chain for DCI does not have a bit-level channel interleaver. This is due to the existence of interleaver for REG bundles before assigning the DCI coded symbols to the time-frequency resources. Furthermore, since the polar code is not required to support input size larger than 164 bits for DCI, segmentation procedure is not necessary for DCI coding chain.

CRC Encoding for DCI

CRC encoding is an important part of wireless communications. Traditionally, the CRC bits generated by a CRC encoder are used by the receiver to perform error detection in order to maintain a false-detection or false-alarm rate (FAR) below a certain target $P_{\text{FAR}} \approx 2^{-n_{\text{FAR}}}$. Operating with low FAR allows the network to selectively communicate different messages to a large group of different users through blind decoding over a common downlink control channel by scrambling the CRC bits differently. With polar coding, however, the CRC bits are also used for error correction to eliminate decoding paths in the list that fail CRC at the end of SCL decoding. With a target list size $L = 2^{n_L}$ in SCL decoding, the CRC length (i.e., the number of CRC bits) is given by $n_{\text{CRC}} = n_{\text{FAR}} + n_L$. For downlink control channel as well as uplink control channel, the target SCL list size is implicitly assumed to be $L = 8$ (i.e., $n_L = 3$) in the specifications of 5G NR [10].

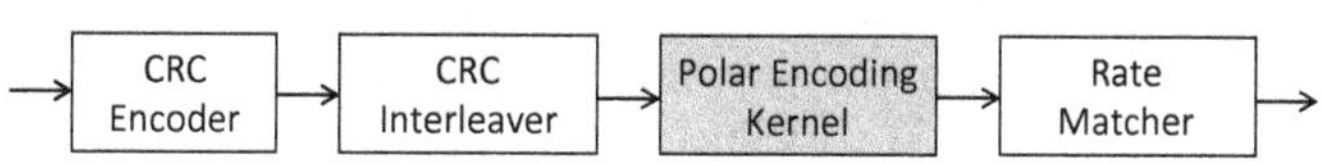

Fig. 10.13 NR coding chain for DCI

Since the number of blind decoding attempts of DCI in NR has been increased compared to LTE, the CRC for DCI has been lengthened to $n_{\mathrm{CRC}} = 24$ bits in NR. Accounting for the additional CRC length of 3 (i.e., $n_L = 3$) required for list decoding, one obtains an improved error detection capability of 2^{-21} (i.e., $n_{\mathrm{FAR}} = 21$) for DCI in NR, in comparison to 2^{-16} of LTE. The CRC polynomial for DCI is $g_{\mathrm{CRC24C}}(D) = [D^{24} + D^{23} + D^{21} + D^{20} + D^{17} + D^{15} + D^{13} + D^{12} + D^{8} + D^{4} + D^{2} + D + 1]$. Note that, among the 24 CRC bits, up to 7 of them are distributed among the information bits, and the remaining CRC bits are clustered at the end. To additionally provide identification (i.e., RNTI) of the UE being targeted by the DCI, 16 RNTI bits are scrambled onto the last 16 CRC bits, such that only the UE with the target identity can successfully pass the CRC check when performing the blind decoding.

The CRC bits are generated as usual by performing a long division of the data polynomial by a CRC polynomial, which is typically implemented by shift registers. When CRC is to be generated for DCI, the input vector is prepended with n_{CRC} ones, which achieves the effect of initializing the shift register by all ones.

CRC Interleaver

As in LTE, all CRC bits are typically clustered together and placed after the corresponding information block [9]. In NR, however, CRC bits are distributed more evenly among the data (non-frozen) and frozen bits using a CRC interleaver in downlink. The main purpose of this interleaver is to facilitate early error detection and the resulting early termination of the decoding process that reduces the latency and the average energy consumption of a polar decoder without performance degradation.

To enable early error detection during SC/SCL decoding, the value of each CRC bit must be computable using only the values of the data bits that come before the CRC bit. Hence, CRC interleaver must be designed under the constraint that each CRC bit is placed after all the data bits that the CRC bit depends on. CRC checks can then be performed at any of these CRC bits during the decoding process so that decoding can be terminated earlier when all surviving paths in the list fails a CRC check. Alternatively, these distributed CRC bits can be used to improve error correction capability. For example, any of these CRC bits can be used as dynamic frozen bits [11] to trim the list of surviving paths during SCL decoding for an improved error-correcting performance. However, each CRC bit can only be used either for early error detection or for improved error correction, but not for both.

CRC interleaving is the only component in the entire NR polar coding chain that is performed only in downlink but not in uplink. The size of the CRC interleaver in NR supports a maximum DCI payload size of 140 bits. Hence the maximum number of information bits at the input of the polar encoder of DCI is 164 bits, including 140 payload or data bits and 24 CRC bits.

Polar Encoding Kernel

Polar encoding kernel performs the basic polar encoding for a given mother code size, $N = 2^n$, that is a power of two and for an information set chosen as in [8] except without the bit-reversal permutation. More precisely, the output of NR polar encoding kernel is given by

$$\boldsymbol{x} = \boldsymbol{G}^{\otimes n}\boldsymbol{u},$$

where $\boldsymbol{G}$ is the 2-by-2 Arıkan kernel matrix given by

$$\boldsymbol{G} = \begin{bmatrix} 1 & 0 \\ 1 & 1 \end{bmatrix},$$

where $\otimes n$ denotes the n-time Kronecker power and $\boldsymbol{u}$ is the input to polar encoding kernel.

Mother Code Size

In principle, the power index n can simply be chosen such that the mother code size N is just large enough for the desired number of coded bits E (e.g., $n = \lceil \log_2 E \rceil$). In NR, however, a number of additional practical considerations have to be addressed, and the power index n is chosen based not only on E but also on the number of information bits K at the input of polar encoder, where K includes the CRC bits attached to the payload (data) bits, i.e., $K = A + n_{\mathrm{CRC}}$ with A being the payload size. First, when the code rate $R = K/E$ is small, and when E is only slightly higher than a power of two, it is desirable to choose a smaller n (i.e., $n = \lceil \log_2 E \rceil - 1$) and then repeat some of the coded bits, so that a smaller mother code size of 2^n (rather than twice as large, 2^{n+1}) is used for the polar coding kernel. This helps to reduce the latency and complexity of the encoder and decoder. The exact condition under which the smaller n is selected is when $E \leq 9/8 \times 2^{\lceil \log_2 E \rceil - 1}$ and the code rate $R = K/E < 9/16$. Second, at a very low code rate, using a smaller n with repetition of code bits yields similar performance as using a larger n. In NR, a lower bound of $R_{\min} = 1/8$ on code rate is imposed in the determination of the mother code size. Note that $R_{\min}$ is only used in the determination of n but does not limit the choice of code length E. Third, due to hardware limitations, there must be an upper limit on the mother code size. The upper bound on n is $n_{\max} = 9$ (i.e., mother code size $N_{\max} = 512$) for downlink and $n_{\max} = 10$ (i.e., mother code size $N_{\max} = 1024$) for uplink, as a base station can typically afford a higher computational load than a user equipment. Lastly, a lower limit of $n_{\min} = 5$ (i.e., mother code size $N_{\min} = 32$) is also imposed on the mother code size to avoid over-specification with little performance difference.

In summary, the Kronecker power index n is calculated as

$$n = \max\{n_{\min}, \min\{n_1, n_2, n_{\max}\}\}$$

where

$$n_1 = \begin{cases} \lceil \log_2 E \rceil - 1, \text{ if } E \le \left(\frac{9}{8}\right) 2^{\lceil \log_2 E \rceil - 1} \text{and } \frac{K}{E} < 9/16 \\ \lceil \log_2 E \rceil, & \text{otherwise.} \end{cases}$$

$$n_2 = \lceil \log_2 (K/R_{\min}) \rceil,$$

and where $R_{\min} = 1/8$, $n_{\min} = 5$, and $n_{\max} = 9$ for downlink while $n_{\max} = 10$ for uplink.

Information Sequence

A polar code of mother code size N provides a maximum of N bit-channels with different reliabilities. In NR, the relative reliabilities of bit-channels for $N = 1024$ are captured in an information sequence of length $N = 1024$ that lists the indices (from 0 to 1023) of bit-channels in an ascending order of reliabilities. The corresponding information sequence for a shorter length is nested within this information sequence of length $N = 1024$ in the sense that the sequence of length N', where $N' < N$, can be obtained by removing the indices of values higher or equal to N' from the sequence of length N. Thus, the information sequence of any length smaller than $N = 1024$ can be derived from the information sequence of length $N = 1024$ specified in [10]. For example, the information sequence of length $N = 64$ is given by

0,1,2,4,8,16,32,3,5,9,6,17,10,18,12,33,20,34,24,36,7,11,40,19,13,48,14,21,35,26,37,25,22,38,41,28,42,49,44,50, 15,52,23,56,27,39,29,43,30,45,51,46,53,54,57,58,60,31,47,55,59,61,62,63.

Getting rid of the indices larger than or equal to 32 (in color red) yields the information sequence of length $N = 32$ as

$$0, 1, 2, 4, 8, 16, 3, 5, 9, 6, 17, 10, 18, 12, 20, 24, 7, 11, 19, 13, 14,$$
$$21, 26, 25, 22, 28, 15, 23, 27, 29, 30, 31.$$

For DCI, the maximum mother code size is $N = 512$. Its information sequence is derived from the $N = 1024$ sequence used for UCI by removing all indices larger than or equal to 512. Note that, in general, the best information sequence for a smaller mother code size may not be nested within that for a larger mother code size. However, it was found that the nested structure simplifies the specification with little performance impact.

The information sequence of length N is used to derive the locations of frozen and non-frozen bits of the polar encoding core of mother code size N. Specifically, the bit-channels corresponding to the last K indices in the information sequence that do not correspond to any punctured or shortened code-bit indices should be used for carrying a given set of K information bits.

Rate Matcher

The rate matcher is used to match the number of coded bits at the input of rate matcher (i.e., a coded bit sequence of length $N = 2^n$) with the amount of available radio resource. It in effect adjusts the code length from N bits to E bits, the number of bits that can be carried by the available radio resource. Three methods of code length adjustment, namely, puncturing, shortening, and repetition, are used in the rate matcher for NR polar codes. Similar to LDPC codes, puncturing refers to discarding coded bits that are generated based on the (unknown) information bits and do not have known values. Hence the punctured coded bits should be treated as unknown by setting the log-likelihood ration (LLR) to zero at the decoder input. Shortening refers to assigning known values (e.g., 0) to a selected set of non-frozen bit positions of the polar code so that a corresponding set of coded bits also have known values (e.g., 0), which are then discarded before transmission. Hence the shortened (i.e., discarded) coded bits should be treated as known (e.g., LLR $= \infty$) at the decoder input. Repetition refers to repeating a selected set of coded bits to arrive at a longer sequence of coded bits. Puncturing typically yields better performance at low code rates (i.e., lower K/E) while shortening performs better at high code rates (i.e., higher K/E). Repetition is used when the desired code length E is longer than the mother code size $N = 2^n$.

A size-N subblock interleaver and a circular buffer, as depicted in Fig. 10.14, are used to implement all three methods of code length adjustment.

The subblock interleaver is used to arrange the coded bits before placing them into the circular buffer so that they are discarded or admitted in a certain desired order. The interleaver simply takes the 5 most significant bits of the binary representation of the indices of the polar encoded bits and permutes them according to a pre-determined integer sequence of length 32 as illustrated in Fig.

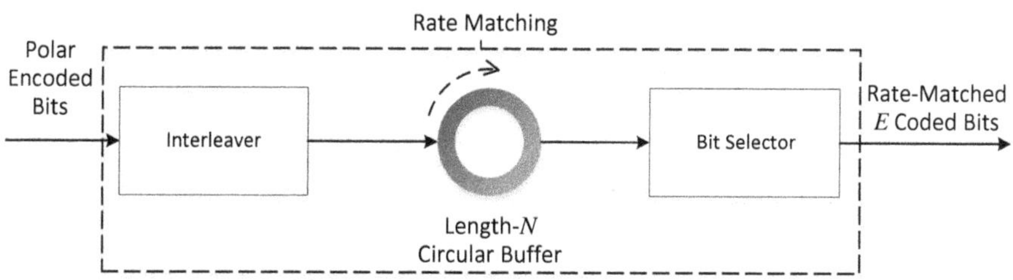

Fig. 10.14 Rate-matching process with interleaver, circular buffer, and bit selector

10.15. Equivalently, such a permutation first divides the N coded bits into 32 subblocks, each of size $N/32$, and then rearranges these subblocks according to the predetermined integer sequence.

The coded bits are selected from the circular buffer according to the relative sizes of E, N, and $A + n_{\mathrm{CRC}}$, as described in Table 10.7 and illustrated in Fig. 10.16, where n_{CRC} is the number of CRC bits and $R_{\mathrm{ps}} = 7/16$ is a threshold for determining whether puncturing or shortening should be used.

As mentioned earlier, discarding coded bits through puncturing or shortening can cause some bit-channels of a polar code to become highly unreliable in carrying information. Hence, depending on the number of discarded coded bits and the method of rate matching, a corresponding set of bit-channel indices needs to be skipped (or so-called pre-frozen) from the information sequence when forming the information set. For example, the indices of those coded bits that are discarded through puncturing or shortening are skipped from the information sequence when determining the information set.

Polar Coding for DCI

For DCI, the number of coded bits is determined by the grid of aggregation levels (AL) for transmitting DCI and can be calculated as follows. A resource element

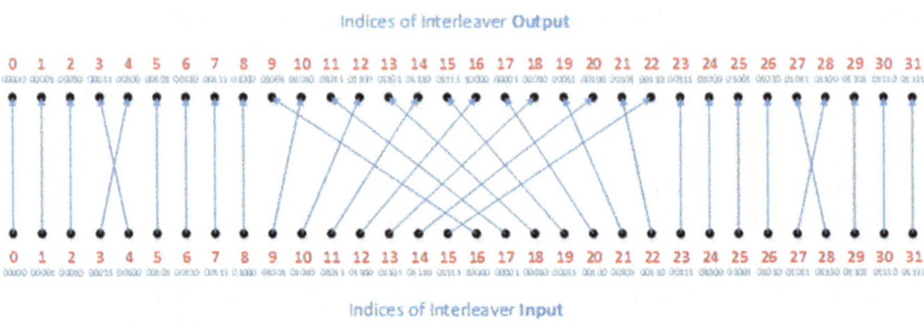

Fig. 10.15 Bipartite graph of rate-matching interleaver

Table 10.7 Bit selection from circular buffer for different rate-matching methods

Method	Condition	Operation
None	$E = N$	Extract all N bits from circular buffer starting from the first position
Puncturing	$E < N$ and $\frac{A+n_{\mathrm{CRC}}}{E} \le R_{\mathrm{ps}}$	Extract E consecutive bits starting from the $(N - E + 1)$-th position of circular buffer
Shortening	$E < N$ and $\frac{A+n_{\mathrm{CRC}}}{E} > R_{\mathrm{ps}}$	Extract E consecutive bits starting from the first position of circular buffer
Repetition	$E > N$	Extract E consecutive bits starting from the first position of circular buffer and wrapping around when reaching the end of circular buffer

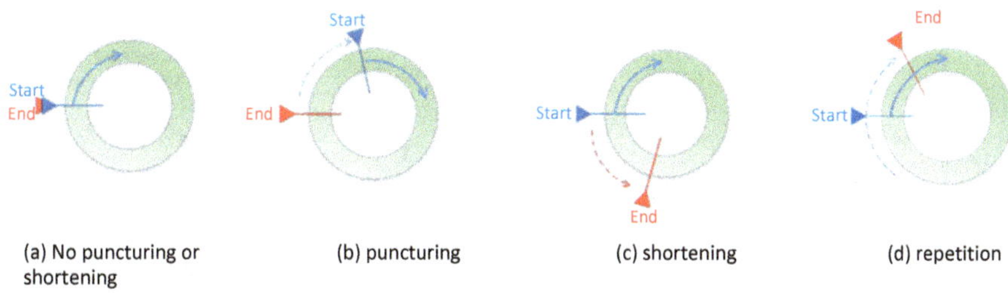

Fig. 10.16 Bit selection from circular buffer

Table 10.8 Number of coded bits and polar code mother size for polar code of DCI

Aggregation level (AL)	Number of coded bits E	Polar code mother code size N	Rate-matching method
1	108	128	Puncturing/shortening
2	216	256	
4	432	512	
8	864	512	Repetition
16	1728	512	

group (REG) consists of 12 resource elements (REs). A quarter of these REs in a REG are occupied by DMRS. Since QPSK modulation is used, the number of coded bits in each REG is thus $(12 \cdot 3/4) \cdot 2 = 18$ bits. For a control channel element (CCE), which is composed of 6 REGs, the number of coded bits is $6 \cdot 18 = 108$ bits. For a DCI candidate of aggregation level AL, a set of AL CCEs are associated with the corresponding time-frequency resources, and the number of coded bits E in the set is $E = 108 \cdot AL$ (bits). Table 10.8 shows the number of coded bits E for the different aggregation levels together with the polar code mother code size N and whether shortening/puncturing or repetition is applied per the rate-matching procedure. Since for DCI transmission $N_{\max} = 2^{n_{\max}} = 512$, repetition is always applied if $E > 512$ is needed.

As mentioned in Sect. 2.2, $R_{\min} = 1/8$ is used in determining the polar code mother code size. For $A_{\min} = 12$ with 24 CRC bits attached, the mother code size corresponding to $R_{\min} = 1/8$ is $N_2 = 2^{n_2} = 512$, where

$$n_2 = \lceil \log_2((12 + 24)/R_{\min}) \rceil = 9.$$

Thus, $R_{\min} = 1/8$ is not limiting in the determination of the DCI mother code size.

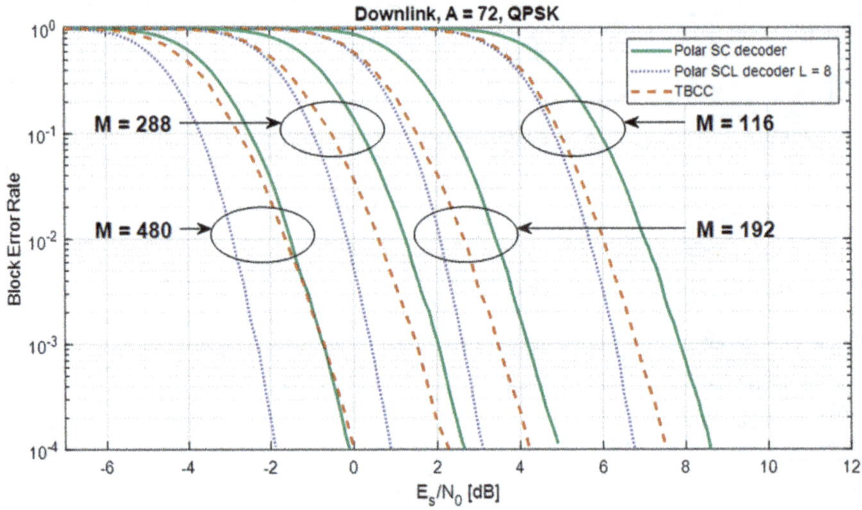

Fig. 10.17 Performance comparison of NR polar code in downlink with 24 CRC bits and LTE TBCC with 21 CRC bits for 72 payload bits

Performance of NR Polar Codes in Downlink

The performance of NR polar code in terms of block error rate (BLER) versus E_s/N_0 is shown in Fig. 10.17 for 72 payload bits in downlink. The performance of LTE TBCC for the same values of A and E is also shown, where the LTE TBCC is used in conjunction with a 21-bit CRC, instead of the 16-bit CRC in LTE DCI, for a fair comparison. NR polar code with SCL decoding outperforms LTE TBCC, especially at low code rates, as shown in the Fig. 10.17.

2.3 *Coding for Uplink Control Information*

When considering the full range of UCI bits supported in Rel-15, four types of channel coding schemes are used, as illustrated in Fig. 10.18. The parameters are:

- A: number of information bits excluding CRC parity bits
- K: number of information bits including CRC parity bits ($K = A + n_{\mathrm{CRC}}$)
- N: number of coded bits at the output of Polar encoding kernel
- E: rate-matching output size

Polar coding is applied when the number of UCI payload bits A is 12 bits or more. Channel coding schemes as defined in LTE are reused when UCI payload bits is 11 bits or fewer. Specifically, when the UCI payload size is between 3 and 11, the LTE Reed Muller code (32, K) is reused. A repetition code and a simplex code are used for payload size of 1 and 2, respectively.

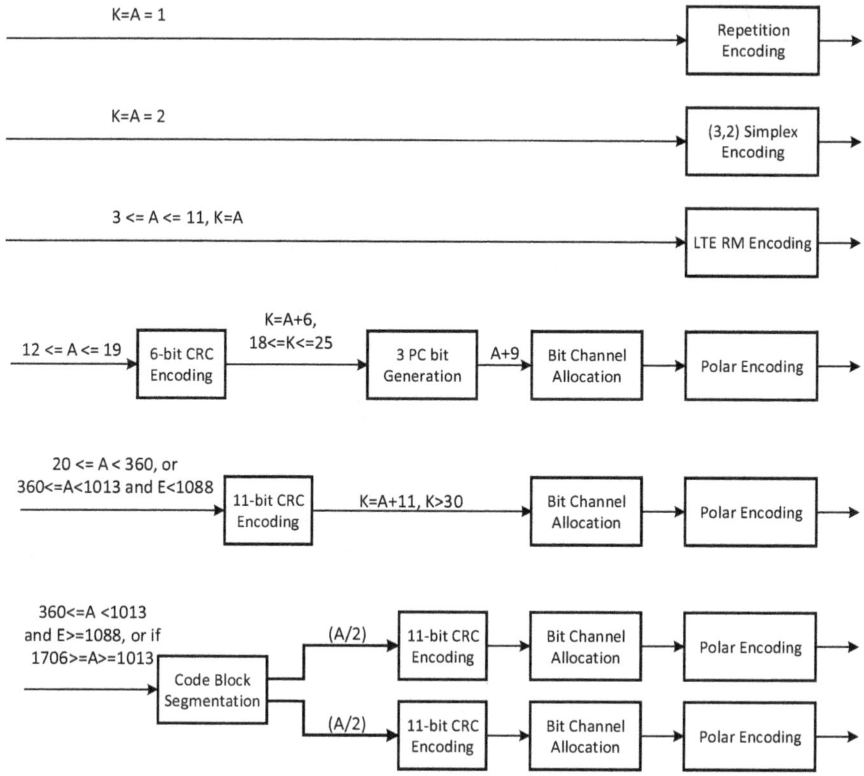

Fig. 10.18 Coding schemes used for UCI

Table 10.9 CRC polynomials for UCI

Range of K	Number of CRC bits	Nominal FAR	CRC Polynomial
$12 \leq A \leq 19$	6	2^{-3}	$g_{CRC}(D) = D^6 + D^5 + 1$
$A \geq 20$	11	2^{-8}	$g_{CRC}(D) = D^{11} + D^{10} + D^9 + D^5 + 1$

CRC Encoding for UCI

The CRC bits are generated as usual by performing a long division of the polynomial associated with the data bits by an CRC polynomial, which is typically implemented by shift registers. The CRC polynomials used for UCI for different payload sizes are shown Table 10.9. The shift registers are initialized to all zeros for UCI encoding. Since the target SCL list size of $L = 8$ (i.e., $n_L = 3$) is implicitly assumed, the nominal FAR is $2^{-n_{FAR}} = 2^{-(n_{CRC}-3)}$. There is no CRC interleaving in uplink. The CRC bits are appended to the data bits, and they are collectively allocated to the bit-channels of the polar code according to the information set derived from the information sequence.

Parity-Check (PC) Bits

For conventional polar codes, frozen bits typically take a constant value (commonly, 0) that is independent of the data and is known to the decoder. However, by making the frozen bit values dependent on the values of the non-frozen (i.e., information) bits, the error-correcting performance of NR polar code in uplink can be improved slightly in most cases. The use of this kind of data-dependent frozen bits is similar to the use of distributed CRC bits as dynamic frozen bits [11] in downlink as described before. Apart from dynamic frozen bits, they are also referred to as parity-check (PC) frozen bits, or simply PC bits, which is the terminology adopted in 5G NR specifications. The addition of these PC bits essentially forms another layer of outer code between the CRC outer code and the polar code.

To facilitate SCL decoding, the value of these PC bits must be readily determinable during successive decoding from previously decoded bits for each hypothesized decoding path in the list. Hence, each PC bit must be placed, according to the successive decoding order, after all the information bits that the PC bit depends on.

The addition of PC bits in principle should increase the minimum distance of the overall code. However, with successive cancellation decoding, it does not provide significant improvement in error-correcting performance.

Only when UCI has size satisfying $12 \le A \le 19$ in 5G NR, $n_{\text{PC}} = 3$ PC bits are added at the input of the polar encoding kernel. No PC bits are added outside this range of A.

Placement of PC Bits

In vast majority cases, the PC bits reside in the least reliable positions among the $(A + n_{\text{CRC}} + n_{\text{PC}})$ most reliable positions at the input of the polar encoder. The only exception occurs when $E - (A + n_{\text{CRC}} - n_L) > 192$ or, equivalently, $E > 195 + A$ (since $n_L = 3$ and $n_{\text{CRC}} = 6$ when the number of PC bits is nonzero). In this case, two of the three PC bits are placed in the least reliable positions among the $(A + n_{\text{CRC}} + n_{\text{PC}})$ most reliable positions, while one PC bit is placed elsewhere as specified later.

Replacing the conventional frozen bits with PC (frozen) bits in the least reliable positions among the aggregate of $(A + n_{\text{CRC}} + n_{\text{PC}})$ ensures no performance degradation in successive cancellation decoders. However, placing a PC bit elsewhere in a higher reliability position pushes another data bit or CRC bit to a lower reliable position and can thus cause performance degradation. In fact, it has been verified that there is a slight performance degradation when $E - A > 195$ precisely due to the fact that one of the three PC bits is not place in the lowest reliable positions among all data-dependent bits at the input of the polar encoder.

Now for the case when $E - A > 195$, two of the three PC bits are placed in the least reliable positions among the $(A + n_{\text{CRC}} + n_{\text{PC}})$ most reliable positions, while the remaining one bit is placed at the most reliable position among those positions with indices whose binary representations have the smallest weight (i.e.,

Table 10.10 Placement of PC Bits

Range of A and E	Number of PC bits	Placement of PC bits
$12 \le A \le 19$ and $E \le 195 + A$	3	The 3 least reliable positions of the $(A + n_{\mathrm{CRC}} + n_{\mathrm{PC}})$ most reliable positions
$12 \le A \le 19$ and $E > 195 + A$	3	The 2 least reliable positions of the $(A + n_{\mathrm{CRC}} + n_{\mathrm{PC}})$ most reliable positions, plus the most reliable position among all positions with indices in the set $I = \{i \in \{0, 1, \cdots, N-1\} : w(i) = w_{\min}(A + n_{\mathrm{CRC}})\}$, where $w(i)$ denotes the weight of the binary representation of index i, and $w_{\min}(m)$ denotes the minimum weight among the binary representations of the m most reliable positions
$A \ge 20$	0	N/A

the number of ones) among the $(A + n_{\mathrm{CRC}})$ most reliable positions at the input of the polar encoder.

The placement of PC bits is summarized in Table 10.10.

Computation of PC Bits

For each PC bit location $k \in I_{\mathrm{PC}}$, where I_{PC} denotes the set of the three PC bit locations if applicable, the value of the PC bit is computed based on the previously decoded information bits, frozen bits, or PC bits as follows:

$$u_k = \sum_{i \in S_k} u_i$$

where

$$S_k = \{i < k : i \notin I_{PC} \text{ and } \mathrm{mod}\,(k - i, 5) = 0\},$$

u_i denotes the value of the bit at position i, and the summation is modulo-2 (i.e., binary addition or XOR). In other words, the PC bit value is simply the modulo-2 cumulative addition of every 5th bit value in front of the PC bit, except for the values of other PC bits. The computation of PC bits can be carried out by a length-5 cyclic shift register.

Polar Coding for UCI

In this subsection, we focus on polar coding only. Without considering the special case of PC bits, the polar coding chain is illustrated in Fig. 10.19.

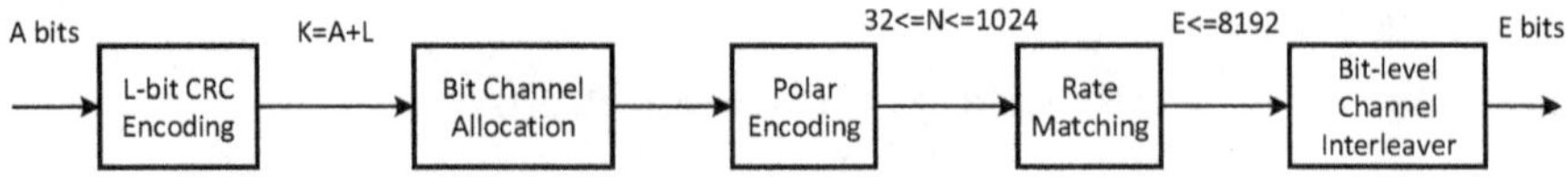

Fig. 10.19 Polar coding for UCI when PC bits are not applied

When the UCI is carried by PUCCH, the maximum UCI payload size supported by polar code is $A_{\max} = 2048 \times \left(\frac{5}{6}\right) \cong 1706$ bits, assuming the highest supported code rate of 5/6. This occurs when code block segmentation is applied, where each of the two segments is polar encoded with code rate 5/6 to $N_{\max} = 1024$ bits.

When UCI is carried by PUSCH, the same set of modulation orders and the same range of modulation symbols apply as UL data. Thus, the number of coded bits E for UCI can be very large, compared to DCI.

For each code block, the maximum number of coded bits after rate matching is $E_{\max} = 8192$ bits. This is due to the need to limit the maximum channel interleaver size, since the complexity of the triangular channel interleaver increases with the number of coded bits E. Considering that two segments are allowed, the total number of coded bits for the entire UCI payload is $2 \times E_{\max} = 2 \times 8192 = 16384$ bits. Since the maximum polar code size is $N_{\max} = 1024$ bits, repetition (part of rate matching) is applied, where the output of polar encoder is repeated eight times to arrive at $E_{\max} = 8192$ bits for each code block.

The supported modulation orders for UCI are the same as uplink data when UCI is carried by PUSCH. When OFDM is used for PUSCH, the modulation order can be QPSK, 16-QAM, or 64-QAM when the maximum modulation order is 64-QAM. The modulation order can be QPSK, 16-QAM, 64-QAM, or 256-QAM when the maximum modulation order is 256-QAM. When transform precoding is used for PUSCH with the maximum modulation order being 64-QAM, pi/2-BPSK is also supported for PUSCH.

Overall, the following applies:

1. When $12 \le A \le 19$ bits, 6-bit CRC is applied. Additionally, three PC bits are applied as PC-frozen bits to assist with polar decoding.
2. When $A \ge 20$ bits, 11-bit CRC is applied without PC bits.
    (a) When $20 \le A < 360$ or $\{306 \le A < 1013$, and $E < 1088\}$, code block segmentation is not applied. Only a single polar coding chain is applied.
    (b) When $\{360 \le A < 1013$ and $E \ge 1088\}$ or $1706 \ge A \ge 1013$, code block segmentation is applied. The UCI payload is subdivided into two segments as evenly as possible. Each segment is encoded separately in parallel according to the coding chain illustrated in Fig. 10.20. At the end of channel interleaving, the two segments are concatenated without interlacing.

For polar coding chain of UCI, the rate-matching algorithm is the same as that of DCI. While no bit-level channel interleaver is applied for polar coding chain of DCI, a triangular channel interleaver is applied to polar coding chain of UCI.

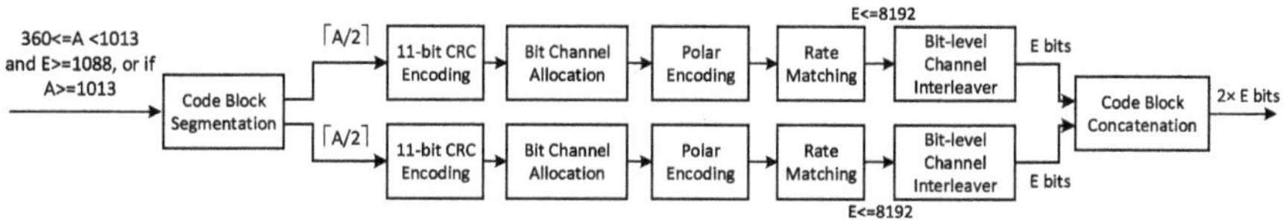

Fig. 10.20 Polar coding for UCI when segmentation is applied

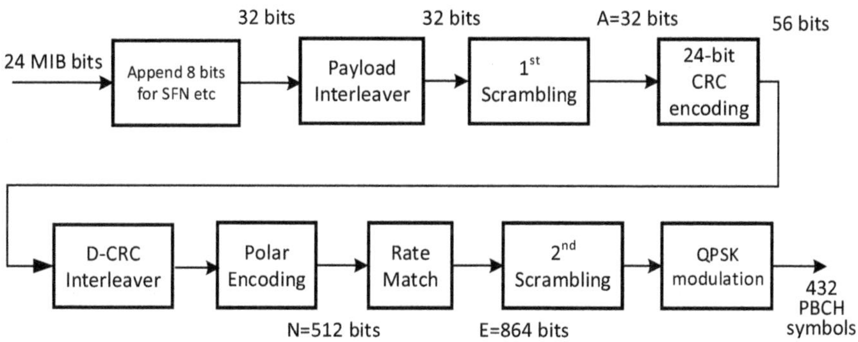

Fig. 10.21 Transmitter procedure for generating the modulation symbols of PBCH

2.4 *Polar Coding for PBCH*

Apart from DCI and UCI, polar code is also used for encoding of physical broadcast channel (PBCH) payload bits. The same polar code construction as that of physical downlink control channel (PDCCH) is used, with the mother code size of $N_{\max} = 512$ (bits). Similarly, the 24-bit CRC code and the associated interleaver designed for PDCCH is also adopted in PBCH. This allows the UE to implement a single polar decoder to cover both PDCCH and PBCH. The only difference between the encoding procedure of PBCH and PDCCH is that an interleaver is additionally applied to the payload of PBCH. The payload interleaver takes advantage of the unequal bit error probability (or unequal bit-channel reliabilities) provided by the polar code and match that with the unequal error protection needs of the payload bits. For instance, the system frame number (SFN) bits can be known under certain conditions; therefore they are mapped to least reliable information bit positions in the polar encoder and treated as frozen bits (value can be 1 or 0 depending on actual SFN bit values) if known a priori at the polar decoder.

The PBCH payload contains the Master Information Block (MIB). The total number of information bits at the input of encoding procedure is 32 bits. With 24-bit CRC attachment, the number of input bits at the polar encoder is 56 bits. The number of coded bits after polar encoding is 864 bits, which are then used to generate 432 modulation symbols with QPSK. The PBCH is broadcast periodically as a component of the SS/PBCH block. The procedure is illustrated in Fig. 10.21.

Since polar code of $N_{\max} = 512$ is used (the same as that of DCI), repetition is applied to generate the 864 output bits. The approximate polar code rate (including CRC bits) is 56/864 $\cong$ 1/15. Such low code rate is necessary for PBCH because MIB is the most fundamental system information that a device needs to obtain to gain initial access to the system.

References

1. R.G. Gallager, "Low density parity-check codes" (MIT Press, Cambridge, MA, 1963)
2. D. MacKay, R. Neal, "Good codes based on very sparse matrices," Cryptography and coding, 5th IMA Conf., C. Boyd, Ed., Lecture Notes in Computer Science, Oct. 1995
3. N. Alon, M. Luby, A linear time erasure-resilient code with nearly optimal recovery. *IEEE Transactions on Information Theory* **42**, 1732–1736 (1996)
4. ETSI TS 136 212 v12.2.0 (2014-10) LTE Release 12. Available: http://www.etsi.org/deliver/etsi_ts/136200_136299/136212/12.02.00_60/ts_136212v120200p.pdf
5. G. Liva, W.E. Ryan, M. Chiani, Quasi-cyclic generalized LDPC codes with low error floors. *IEEE Transactions on Communications* **56**(1), 49–57 (2008)
6. T.Y. Chen, K. Vakilinia, D. Divsalar, R.D. Wesel, Protograph-based raptor-like LDPC codes. *IEEE Transactions on Communications* **63**(5), 1522–1532 (2015)
7. Ericsson, "Performance evaluation of turbo codes and LDPC codes at higher code rates," 3GPP TSG RAN WG1 Meeting #85, doc. no. R1-164359, Nanjing, China, 23rd–27th May 2016. Available: http://www.3gpp.org/ftp/TSG_RAN/WG1_RL1/TSGR1_85/Docs/R1-164359.zip
8. E. Arıkan, Channel polarization: a method for constructing capacity-achieving codes for symmetric binary-input memoryless channels. *IEEE Transactions on Information Theory* **55**, 3051–3073 (2009)
9. I. Tal, A. Vardy, "List decoding of polar codes," in *Proceedings of IEEE Symposium of Information Theory*, pp. 1–5, 2011
10. 3GPP 5G NR Specification, "Multiplexing and channel coding", TS 38.212, v15.0.0, 2018-01-03, Release15. Available: http://www.3gpp.org/ftp//Specs/archive/38_series/38.212/38212-f20.zip
11. P. Trifonov, V. Miloslavskaya, "Polar codes with dynamic frozen symbols and their decoding by directed search," *Proceedings of IEEE Information Theory Workshop*, pp. 1–5, Sept. 2013

Chapter 11
5G NR Cell Search and Random Access

Jingya Li

Before a UE can properly communicate within a network, it must carry out cell search to find, synchronize, and identify a cell. Then, it can acquire basic system information and perform random-access procedure to establish a connection to the cell.

For non-standalone (NSA) NR deployment, a UE uses the LTE network as an anchor for control plane communications; thus, the UE performs the legacy LTE cell search and random-access procedure to initially access the network. On the other hand, for standalone (SA) NR deployment, initial access must be carried out on NR carriers using the newly designed NR signals.

In this chapter, we introduce the details of NR design on cell search, basic system information acquisition, and random access, with a focus on the SA NR deployment.

1 Cell Search

For SA NR deployments, initial cell search is carried out based on SS/PBCH block, a new concept introduced in NR to support beam-sweeping for SS/PBCH block transmission. Beamforming is important for improving the coverage of SS/PBCH block transmission, especially for compensating the high path loss in high carrier frequency bands.

Cell search can be used to enable a UE to access new cells when moving in the network or switch to new beams when performing beam management.

J. Li (✉)
Department of Ericsson Research, Ericsson, Gothenburg, Sweden
e-mail: jingya.li@ericsson.com

X. Lin, N. Lee (eds.), *5G and Beyond*, https://doi.org/10.1007/978-3-030-58197-8_11

1.1 SS/PBCH Block

A SS/PBCH block comprises a pair of synchronization signals (SSs) and physical broadcast channel (PBCH) with associated demodulation reference signal (DMRS). By detecting SSs, a UE can obtain the physical layer cell identity (PCI) and acquire DL timing of the cell in both time and frequency domain. By detecting PBCH DMRS and decoding PBCH, a UE knows the SS/PBCH block index (the beam index if a SS/PBCH block is associated with a beam), derives the system frame timing, and obtains the very basic system information that is needed for acquiring the remaining minimum system information for performing subsequent random-access procedure.

Structure of SS/PBCH Block

Similar to LTE, NR SSs consist of primary synchronization signal (PSS) and secondary synchronization signal (SSS). The PSS, SSS, PBCH, and DMRS for PBCH within a SS/PBCH block have the same cyclic prefix length and the same subcarrier spacing (SCS). The SCS of a SS/PBCH block can be 15 or 30 kHz in FR1 and 120 or 240 kHz in FR2. To limit the UE cell search complexity, the SCS of a SS/PBCH block in most cases is uniquely identified by the operating frequency band [1, 2].

A SS/PBCH block is mapped to 4 consecutive OFDM symbols in the time domain and 240 contiguous subcarriers (20 RBs) in the frequency domain, as illustrated in Fig. 11.1 [3]. The sequence of symbols constituting the PSS is mapped to the middle 127 subcarriers in the first symbol of the SS/PBCH block. The same set of subcarriers in the third symbol of the SS/PBCH block are used for SSS. PBCH and its associated DMRS are mapped to all subcarriers in the second and fourth symbols of the SS/PBCH block, together with 96 subcarriers in the third symbol (4 RBs on each side of SSS) of the SS/PBCH block. The resource elements that are not used for SS, PBCH, and DMRS for PBCH within a SS/PBCH block are transmitted with zero power.

Time Domain Configuration for SS/PBCH Block

To support beamforming and beam-sweeping for SS/PBCH block transmission, in NR, a cell can transmit multiple SS/PBCH blocks in different narrow beams in a time-multiplexed fashion, as illustrated in Fig. 11.2. The transmission of these SS/PBCH blocks is confined to a half-frame time interval (5 ms). Note that it is also possible to configure a cell to transmit multiple SS/PBCH blocks in a single wide-beam with multiple repetitions. The design of beamforming parameters for each of the SS/PBCH blocks within a half-frame is up to network implementation.

The same SCS (i.e., 15 or 30 kHz in FR1 and 120 kHz or 240 kHz in FR2) is applied for all SS/PBCH block transmissions from a cell. A larger SCS enables a

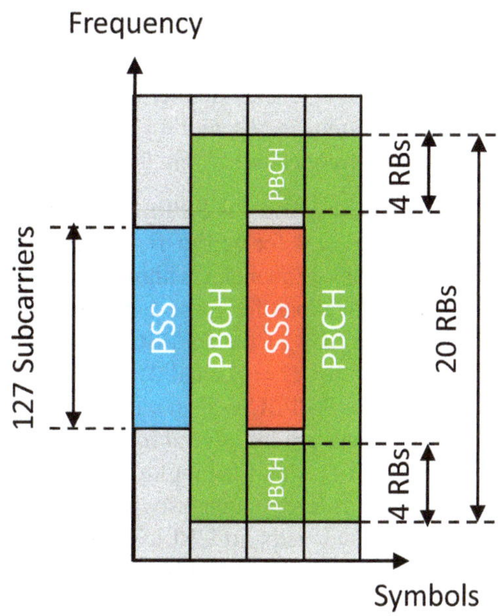

Fig. 11.1 NR SS/PBCH block structure

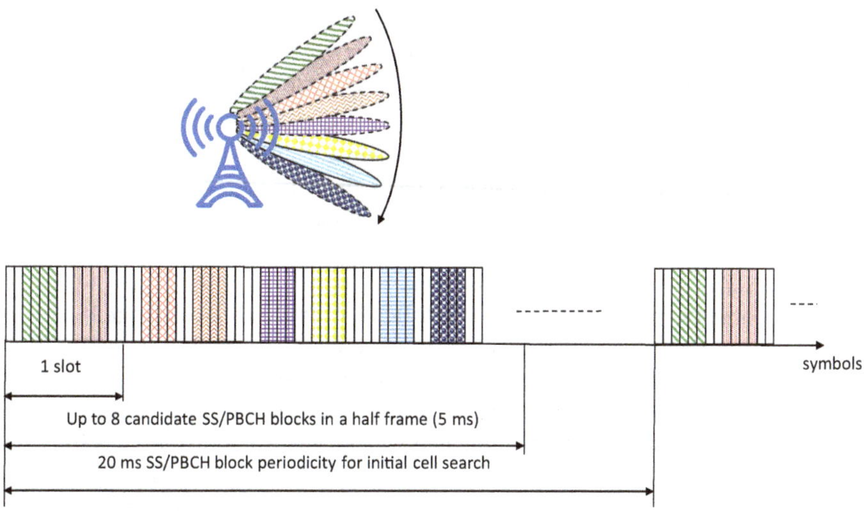

Fig. 11.2 An example of beam-sweeping for SS/PBCH block transmission with SCS of 15 kHz

faster beam-sweeping for SS/PBCH block transmissions, which is useful for cell search in higher carrier frequencies, since more narrow beams are required in a beam-sweep in order to provide good coverage of SS/PBCH block transmissions in all directions. Therefore, in contrast to LTE, where the time locations for SS and PBCH transmissions are always fixed, NR supports different patterns of possible SS/PBCH block time locations for different carrier frequencies.

The maximum number of SS/PBCH blocks within a half frame, denoted by L, depends on the frequency band, and it is defined as follows [3]:

- For carrier frequencies smaller than or equal to 3 GHz, $L = 4$.
- For carrier frequencies within FR1 larger than 3 GHz, $L = 8$.
- For carrier frequencies within FR2, $L = 64$.

Note: An exception for SS/PBCH block transmission with SCS of 30 kHz in TDD, where $L = 4$ for carrier frequencies smaller than or equal to 2.4 GHz and $L = 8$ for carrier frequencies within FR1 larger than 2.4 GHz.

The time locations for these L candidate SS/PBCH blocks within a half-frame depend on the SCS of the SS/PBCH blocks. As an example, Fig. 11.2 shows the time locations for $L = 8$ candidate SS/PBCH blocks within a half frame, when the SCS of the SS/PBCH blocks is set to 15 kHz. The gaps between consecutive SS/PBCH blocks are introduced to support transmission of DL control signaling in the beginning of a slot, time multiplexing the transmissions of SS/PBCH block and PDSCH with different numerologies within a slot, and short UL transmissions in the end of a slot to support URLLC.

The L candidate SS/PBCH blocks within a half-frame are indexed in an ascending order in time from 0 to L-1. For $L = 4$ or 8, a candidate SS/PBCH block index per half-frame is implicitly indicated by the DMRS sequence index associated with the PBCH; for $L = 64$, the 3 least significant bits (LSB) of a candidate SS/PBCH block index is implicitly indicated by the PBCH DMRS sequence index, while the 3 most significant bits (MSB) of the candidate SS/PBCH block index is obtained from the PBCH payload bits generated at physical layer.

The combination of the SCS and the time location pattern for candidate SS/PBCH blocks within a half-frame is uniquely defined for a given operating frequency band [4]. By successfully detecting PBCH, a UE knows the SS/PBCH block index, from which the UE can further derive the relative time location of the detected SS/PBCH block within a half frame.

A cell does not necessarily transmit SS/PBCH blocks in all L candidate locations in a half frame, and the resource of the un-used candidate positions can be used for the transmission of data or control signaling instead. It is up to network implementation to decide which candidate time locations to select for SS/PBCH block transmission within a half-frame and which beam to use for each SS/PBCH block transmission. For a UE carrying out initial cell search, it assumes SS/PBCH transmission at all L candidate locations. Once the UE successfully detects a SS/PBCH block, the time locations of the actual transmitted SS/PBCH block in a half-frame can be derived from the higher-layer parameter, *ssb-PositionsInBurst*, in *SIB1*.

Similar to the transmissions of SS and PBCH in LTE, the SS/PBCH blocks within a half-frame are broadcasted periodically from each cell. The periodicity of the half frames with SS/PBCH blocks is referred to as *SS/PBCH block periodicity*. For initial cell search, a UE assumes a SS/PBCH block periodicity of 20 ms, which is four times longer than the periodicity of LTE SS transmission. This follows the NR ultra-lean design principle to minimize always-on transmissions, thereby reducing network energy consumption. The actual SS/PBCH block periodicity for a serving cell is indicated by SIB1, and the periodicity can be configured to be 5 ms, 10 ms, 20 ms, 40 ms, 80 ms, or 160 ms. A shorter SS/PBCH block periodicity allows for a faster cell search, while a longer periodicity than 20 ms can further improve the network energy efficiency, and it can be configured for UEs connected to a secondary cell.

Frequency Domain Configuration for SS/PBCH Block

In LTE, the granularity of the *channel raster* is fixed as 100 kHz for all frequency bands, which means that the LTE carrier center frequency must be an integer multiple of 100 kHz [5]. The LTE SSs are always placed in the middle of the carrier bandwidth. When performing initial cell search in an LTE operating band, a UE must search for SSs at a set of candidate carrier frequencies on the channel raster within the operating band. Once the SSs are found, the position of the carrier center frequency is also known.

In NR, the granularity of the channel raster is 100 kHz for most of the operating bands in FR1 and 60 or 120 kHz for all defined operating bands in FR2 [4]. Due to very large channel bandwidth for operation bands in FR2, if reusing LTE principle and placing SS/PBCH block at the center of the carrier, there can be significant large number of candidate frequency locations on NR channel raster that a UE has to search, which increases the UE complexity. In addition, as discussed above, the default SS/PBCH block periodicity for initial cell search is four times longer than that for LTE SS transmission; this implies that a UE must stay for a longer time to perform SS/PBCH block detection at each candidate frequency location.

To limit UE complexity for initial cell search and to compensate the sparser SS/PBCH block periodicity, NR defines a sparser raster, referred to as *synchronization raster*, to indicate the candidate frequency positions of SS/PBCH block transmission [4]. The granularity of the synchronization raster is defined as 1.2 MHz for frequencies below 3GHz, 1.44 MHz for frequencies from 3GHz to 24.25 GHz, and 17.28 GHz for frequencies from 24.25 GHz to 100 GHz.

In NR, each candidate frequency from 0 to 100 GHz on the synchronization raster is designated by a global synchronization channel number (GSCN). For each NR operating band, the range of GSCN (a set of candidate frequencies) is defined based on the granularity of the synchronization raster. There is a one-to-one mapping between a GSCN and the corresponding frequency position of the SS/PBCH block, that is, the frequency designated by a GSCN corresponds to the lowest subcarrier (subcarrier 0) of the 11th RB (RB 10) of the corresponding SS/PBCH block, i.e., the center of SS/PBCH block bandwidth, as shown in Fig. 11.3. When carrying out initial cell search in an NR operating band, a UE only needs to search for SS/PBCH blocks at frequency locations on the synchronization raster that are within the defined range of GSCN for this operating band.

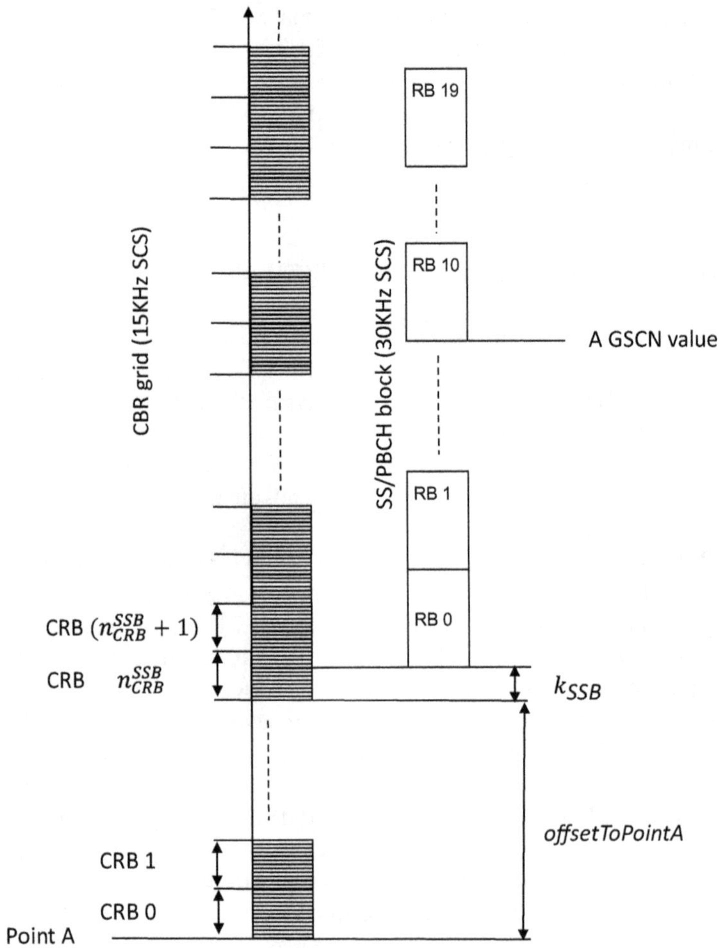

Fig. 11.3 Frequency location of SS/PBCH block

The exact frequency location of SS/PBCH block expressed in terms of the carrier common resource block (CRB) grid is configured by two higher-layer parameters, *offsetToPointA* and *ssb-SubcarrierOffset.*

The parameter, *offsetToPointA*, defines the frequency offset (denoted by $n_{\mathrm{CRB}}^{\mathrm{SSB}}$ in Fig. 11.3) between Point A and the lowest subcarrier of the lowest CRB, which overlaps with the SS/PBCH block used by the UE for initial cell search [6]. Here, Point A is defined as the center of the lowest subcarrier of the lowest CRB in the system bandwidth. The frequency offset value, $n_{\mathrm{CRB}}^{\mathrm{SSB}}$, is expressed in units of RBs assuming 15 kHz SCS for FR1 and 60 kHz SCS for FR2.

Relative to the common resource block grid, the entire SS/PBCH block can have a subcarrier-level offset (denoted by k_{SSB} in Fig. 11.3). The value of the subcarrier offset is provided by the system information carried on PBCH [3, 7], and the details will be discussed in section "Information Carried on PBCH".

Details of PSS, SSS, and PBCH Design

In this section, we discuss the details of the PSS, SSS, and PBCH design.

PSS and SSS

In NR, PSS and SSS together can be used to indicate in total 1008 different PCIs, which is twice as much as compared to LTE. More PCIs are needed in NR in order to support more dense network deployment scenarios. A PCI can be derived by $N_{\mathrm{ID}}^{\mathrm{cell}} = 3N_{\mathrm{ID}}^{(1)} + N_{\mathrm{ID}}^{(2)}$, where $N_{\mathrm{ID}}^{(1)} \in \{0, 1, \ldots, 335\}$ and $N_{\mathrm{ID}}^{(2)} \in \{0, 1, 2\}$.

Like in LTE, the PSS is the first signal searched by a UE when conducting initial cell search. In LTE, PSS is generated as a Zadoff-Chu (ZC) sequence with a length of 63. In NR, due to the possible low-cost terminals in higher carrier frequencies and the absence of tracking from the frequent-static reference signals (CRS in LTE), there could be larger initial frequency errors between the gNB and UEs as compared to LTE. Therefore, the PSS in NR is generated by a 127-length BPSK modulated M-sequence in frequency domain, in order to fix the time/frequency offset ambiguity problem of traditional ZC sequence-based PSS in LTE and to have a unified design of NR PSS sequence for different numerologies [6]. Three cyclic shift values, $m = 43N_{\mathrm{ID}}^{(2)}$, were selected according to the PCI for generating three different PSS sequences, with each sequence defined as

$$d_{\mathrm{PSS}}(n) = 1 - 2x\left((n + m) \bmod 127\right), 0 \leq n \leq 127,$$

where

$$x\left(i + 7\right) = \left(x\left(i + 4\right) + x(i)\right) \bmod 2,$$

and

$$[x(6)\ \ x(5)\ \ x(4)\ \ x(3)\ \ x(2)\ \ x(1)\ \ x(0)] = \left[1\ 1\ 1\ 0\ 1\ 1\ 0\right].$$

A UE performs matched filtering to find PSS. By detecting PSS, a UE acquires initial DL OFDM symbol boundary synchronization to the cell and achieves a coarse frequency synchronization by evaluating several hypotheses of the frequency error. It also obtains the timing of SSS and PBCH based on the fixed structure of SS/PBCH block as shown in Fig. 11.1.

The UE can then continue to detect SSS in the frequency domain. In NR, SSS is generated by using a BPSK modulated Gold sequence with a length of 127 [6].

The cyclic shift values of the two m-sequences, m_0 and m_1, used for deriving a Gold sequence for SSS are jointly determined by the PCI as $m_0 = 15\left\lfloor \frac{N_{\mathrm{ID}}^{(1)}}{112} \right\rfloor + 5N_{\mathrm{ID}}^{(2)}$ and $m_1 = N_{\mathrm{ID}}^{(1)} \bmod 112$. The 127-length Gold sequence for SSS is defined by

$$d_{\mathrm{SSS}}(n) = S_0(n)S_1(n), 0 \le n \le 127,$$

where

$$S_j(n) = 1 - 2x_j\left((n + m_i) \bmod 127\right), j \in \{0, 1\}$$

and

$$x_0\left(i + 7\right) = \left(x_0\left(i + 4\right) + x_0(i)\right) \bmod 2$$

$$x_1\left(i + 7\right) = \left(x_1\left(i + 1\right) + x_1(i)\right) \bmod 2,$$

and the initial value of both two m-sequences is

$$\left[x_j(6)\; x_j(5)\; x_j(4)\; x_j(3)\; x_j(2)\; x_j(1)\; x_j(0)\right] = [0\,0\,0\,0\,0\,0\,1]\,.$$

By detecting SSS, a UE acquires the PCI, and it can also refine the residual frequency error after PSS detection based on the phase rotation between the detected PSS and SSS symbols.

PBCH with Associated DMRS

Following the NR lean design principle, the always-on CRS, which can be used in LTE for PBCH demodulation, does not exist in NR any longer. Therefore, in NR, a dedicated DMRS is introduced for PBCH. Similar to the DMRS for other physical channels, a pseudo-random sequence is used as a basis for generating DMRS for PBCH. Unique for PBCH DMRS, the initialization for the sequence depends on the PCI and its associated SS/PBCH block index.

The sequence of the DMRS for PBCH is mapped to each PBCH symbol within a SS/PBCH block, i.e., the second, third, and fourth OFDM symbols shown in Fig. 11.1. In the frequency domain, the PBCH DMRS sequence is evenly mapped to the subcarriers with a density of 3 REs per RB, which gives a good trade-off between DMRS overhead and channel estimation accuracy. The subcarriers used for PBCH DMRS transmission are aligned for the PBCH symbols within a SS/PBCH block. The aligned mapping pattern makes it easy for a UE to use the DMRS for frequency error refinement after initial synchronization from SS.

Similar to LTE CRS, a frequency shift is defined for determining the exact set of subcarriers to use for mapping PBCH DMRS within a SS/PBCH block. The frequency shifts are PCI dependent. The PCI in a network can be planned such that different frequency shifts are used for neighboring cells. This can be beneficial to improve the SIR of the PBCH DMRS if neighboring cells are configured with the same time-frequency resources for SS/PBCH block transmission and power boosting is configured for PBCH DMRS.

In NR, the PBCH content consists of 32 bits, with 8 bits generated at physical layer and 24 bits generated at higher layer. The details of the information carried on PBCH will be explained in the next section. The NR PBCH physical processing procedure is shown in Fig. 11.4 [6, 8]. Firstly, the 32 information bits are scrambled based on the PCI and the second and third LSBs of the system frame number (SFN) on which the SS/PBCH block is transmitted. Secondly, the scrambled 32-bit sequence is used to generate 24 CRC parity bits, and the CRC attachment is performed, resulting in a 56-bit sequence. Thirdly, the 56 bits are Polar encoded and rate matched, resulting in a block of 864 coded bits. Then, a second scrambling is applied on these 864 coded bits, based on the PCI and the 3 LSBs of the SS/PBCH block index. Finally, QPSK modulation is performed, resulting in a block of 432 modulation symbols. These symbols are mapped to the time-frequency resources for PBCH in the associated SS/PBCH block.

When detecting a SS/PBCH block, a UE assumes that the same antenna port is used for transmitting PSS, SSS, PBCH, and DMRS for PBCH, and it also assumes

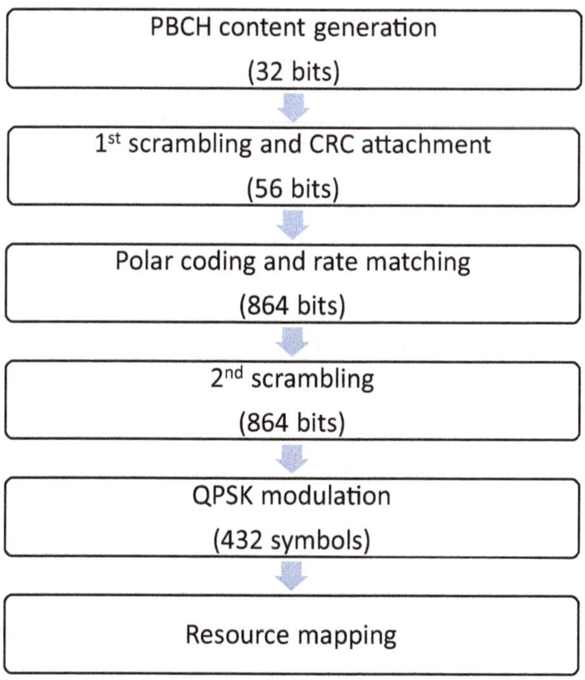

Fig. 11.4 PBCH physical layer processing procedure

that the SSS, PBCH, and DMRS for PBCH have the same energy per resource element (EPRE). The compact SS/PBCH block structure and the design of mapping PBCH close to SSS in time make it possible for a UE to jointly use the SSS and the DMRS for PBCH to improve the channel estimation accuracy for PBCH demodulation [9].

After successfully decoding of PBCH, a UE acquires the SSB index, from which it can derive the slot timing by utilizing the pre-defined the SS/PBCH block time location pattern on the operating band. In addition, the UE can acquire the half-frame number from the PBCH content, and in combination with the SSB index, the UE can also obtain the frame timing.

1.2 Basic System Information Acquisition

Basic system information is acquired by decoding the information contained in PBCH and the remaining minimum system information, referred to as SIB 1, carried on a normal PDSCH.

Information Carried on PBCH

The information carried on PBCH are summarized in Table 11.1. As discussed above, in NR, PBCH payload includes two parts, 24 information bits from the higher layer and 8 bits generated at the physical layer.

Table 11.1 Information carried on PBCH

Information carried on PBCH		Number of bits
BCCH-BCH message (1 bit)	CHOICE	1
MIB (23 bits)	systemFrameNumber	6
	subCarrierSpacingCommon	1
	ssb-SubcarrierOffset	4
	dmrs-TypeA-Position	1
	pdcch-ConfigSIB1	8
	cellBarred	1
	intraFreqReselection	1
	spare	1
L1 generated (8 bits)	3 MSB of SS/PBCH Block Index for FR2 1 MSB of ssb-SubcarrierOffset for FR1	3
	Half-frame indication	1
	4 LSB of systemFrameNumber	4

The 24-bit higher-layer information include 23-bit master information block (MIB) and 1-bit CHOICE parameter. The CHOICE parameter indicates the broadcast control channel (BCCH) message type, i.e., MIB or messageClassExtension. One broadcast channel (BCH) transport block, containing CHOICE and MIB, is transmitted every 80 ms, which implies that MIB on PBCH is the same within an 80 ms time interval.

The 23-bit MIB carries the following basic system information [7]:

- The 6-bit *systemFrameNumber* field indicates the 6 MSB of the 10-bit SFN. The 4 LSBs of the SFN are carried on the physical layer-generated PBCH payload.
- The 1-bit *subCarrierSpacingCommon* field indicates the SCS for PDSCH carrying SIB1, message 2, and message 4 for initial access, paging, and other broadcast SIB-messages. The SCS can be either 15 or 30 kHz for FR1 and 60 or 120 kHz for FR2.
- The 4-*bit ssb-SubcarrierOffset* field indicates the frequency offset in the unit of subcarriers, k_{SSB}, between the lowest subcarrier of SSB and the lowest subcarrier of the lowest CBR overlaps with the SS/PBCH block, as explained in section "Frequency Domain Configuration for SS/PBCH Block" and illustrated in Fig. 11.3. For FR1, $k_{\text{SSB}} \in \{0, 1, 2, \ldots, 23\}$ and 5 bits are needed to indicate the subcarrier offset value. Thus, an additional MSB bit is carried out on the physical layer-generated PBCH payload. For FR2, $k_{\text{SSB}} \in \{0, 1, 2, \ldots, 11\}$, the 4-*bit ssb-SubcarrierOffset* field is enough to indicate the offset. This field may indicate that this cell does not provide SIB1 by setting $k_{\text{SSB}} > 23$ for FR1 or if $k_{\text{SSB}} > 11$ for FR2.
- The 1-*bit dmrs-TypeA-Position* defines the time domain position of the first DMRS for PDSCH and PUSCH.
- The 8-*bit pdcch-ConfigSIB1* determines the *ControlResourceSet* (CORSET), a common search space (CCS), and necessary PDCCH parameters that are needed for acquiring SIB1 (see section "SIB 1").
- The 1-*bit cellBarred* field indicates whether UEs are allowed to access the cell. This barring indication can be configured to prevent UEs from accessing a NR cell in an NSA deployment where the initial cell search should be carried out on the LTE carrier.
- The 1-*bit intraFreqReselection* field indicates whether UEs can access other intra-frequency cells.

The eight information bits generated at the physical layer can be different for different SS/PBCH blocks, and they include:

- 3 bits for indicating the 3 MSB of the associated SSB index for operating band in FR2. For FR1, the SSB index is indicated by the PBCH DMRS sequence index; thus, one of these 3 bits is used for indicating the subcarrier offset together with the 4-bit *ssb-SubcarrierOffset* field in MIB, while the rest two bits are reserved.
- 1 bit for indicating the half radio frame index, i.e., the first or second half of a radio frame.
- 4 bits for indicating the 4 LSB of the SFN.

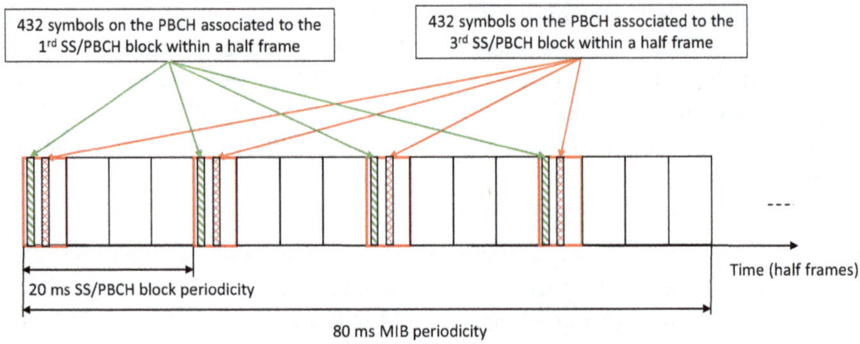

Fig. 11.5 An example of time domain resource mapping for PBCH

Considering the example of an SS/PBCH block time location pattern shown in Fig. 11.2, the Fig. 11.5 illustrates the time domain resource mapping for the PBCH transmissions that are associated with the first and the third SS/PBCH blocks within a half frame.

SIB 1

SIB1 carries the remaining minimum system information that is needed for a UE to be able to perform subsequent random-access procedure. *SIB1* is carried out on normal PDSCH, which is scheduled by a physical downlink control channel (PDCCH) carrying a downlink control information (DCI) format 1_0 with CRC scrambled by a SI-RNTI. The set of PDCCH candidates for a UE to monitor for acquiring *SIB1* is defined as *Type0-PDCCH common search space (CSS).* Upon detecting a SS/PBCH block, if the PBCH content indicates that *SIB1* is present, a UE can further obtain from *MIB* the configuration of a *CORESET* (referred to as *CORESET 0*) and monitoring occasions for *Type0-PDCCH CSS.*

NR supports three different SS/PBCK block and *COREST 0* multiplexing patterns, as illustrated in Fig. 11.6. In pattern 1, SS/PBCH block and *CORESET 0* are time domain multiplexed, which is needed when the system bandwidth is limited. In multiplexing patterns 2 and 3, SS/PBCH block and *CORESET 0* are frequency domain multiplexed, which enables fast beam-sweeping for initial cell search by transmitting SS/PBCH block and *SIB1* in the same beam at the same time. Patterns 2 and 3 require higher system bandwidth, and they are only supported for FR2.

In NR, for each given combination of SS/PBCH block SCS and *SIB1* SCS, a table is pre-defined for its *CORESET 0* configuration, with each row of the table indicating a set of different configuration parameters, including the multiplexing pattern, the size of *CORESET 0* (the number of consecutive RBs in frequency and

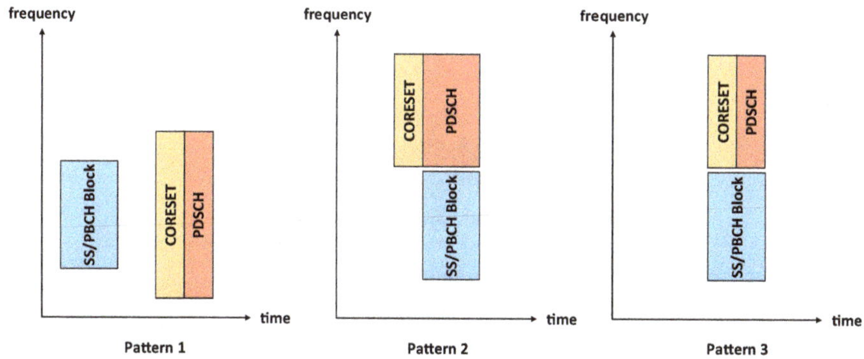

Fig. 11.6 NR SS/PBCH block and CORESET 0 multiplexing patterns

the number of consecutive OFDM symbols in time), as well as the frequency offset of the lowest RB index of *CORESET 0* with respect to the lowest RB index of the SS/PBCH block. The row index to use for a UE to acquire the *CORESET 0* configuration is indicated by the four MSB of the *pdcch-ConfigSIB1* field in *MIB*.

The PDCCH monitoring occasions for *Type0-PDCCH CSS* are configured by the four LSB of the *pdcch-ConfigSIB1* field in *MIB*. For multiplexing pattern 1, the PDCCH monitoring window for *Type0-PDCCH CSS* is two consecutive slots with *SIB1* SCS, and up to two monitoring occasions can be configured for each slot. For multiplexing patterns 2 and 3, the duration of the monitoring window is one slot. The monitoring occasion associated with a certain SS/PBCH block index occurs either earlier than the SS/PBCH block in the same slot or one slot before (pattern 2), or it is the same as the starting symbol of the SS/PBCH block (pattern 3).

After detecting a PDCCH scheduling physical downlink shared channel (PDSCH) carrying SIB1, a UE can further decode SIB1 and obtain the remaining minimum system information, e.g., actually transmitted SS/PBCH blocks, initial UL bandwidth part (BWP) configuration, random access channel (RACH) configuration, etc., which are necessary for the UE to perform random-access procedure to establish connection to the cell.

2 Random Access

After initial cell search and system information acquisition, a UE can perform random access to access the cell. NR supports two types of random-access procedure: *Type-1 random-access procedure* introduced in NR Rel-15 and similar to the one adopted in LTE, and *Type-2 random-access procedure* introduced in NR Rel-16 [10]. This section focus on the NR design of type-1 random-access procedure, whose design concepts are reused in a large extent for type-2 random access.

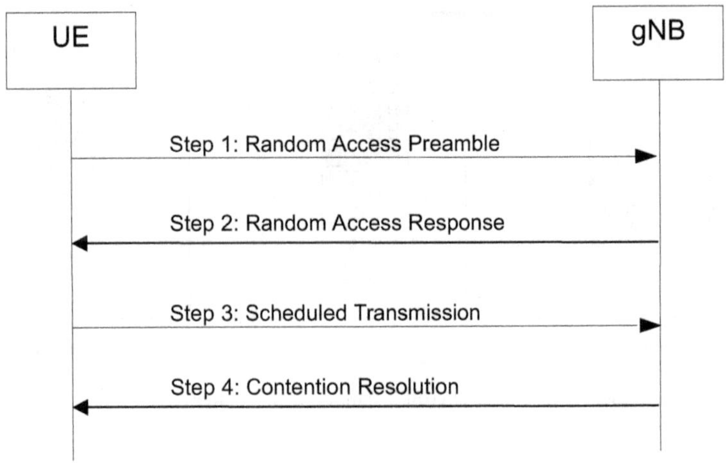

Fig. 11.7 Type-1 contention-based random-access procedure

Figure 11.7 illustrates the type-1 contention-based random-access procedure, which consists of four steps. In the first step, a UE initiates the random-access procedure by transmitting in UL a random-access preamble (Msg1) in a physical random-access channel (PRACH). After detecting the Msg1, in the second step, the gNB will respond by transmitting in DL a random-access response (RAR) with a PDSCH (Msg2). In the third step, after successfully decoding Msg2, the UE continues the procedure by transmitting in UL a PUSCH (Msg3) for terminal identification and RRC connection establishment request. In the last step of the procedure, the gNB transmits in DL a PDSCH (Msg4) for contention resolution.

In step 1, there is a possibility that multiple UEs select the same preamble and transmit the preamble on the same PRACH resource. This preamble collision is referred to as contention, and one of the main purposes of applying step 3 and step 4 is to resolve such potential contention. Therefore, the four-step random-access procedure described above is called type-1 contention-based random-access procedure. For a UE in the connected mode, it is possible to configure it with a dedicated preamble and PRACH resource for its Msg1 transmission. In this case, only the first two steps shown in Fig. 11.7 are needed, and this procedure is called type-1 contention-free random access.

In NR Rel-16, a type-2 random access procedure was introduced. The type-2 random access consists of only two steps. In step 1, a UE transmits in UL a MsgA, which is a combination of Msg 1 (preamble on PRACH) and Msg 3 (payload on PUSCH). In the second step, the gNB responses by transmitting a MsgB for contention resolution. The reduced number of handshaking between a UE and a gNB can provide benefits of reduced access latency and control signaling overhead. This makes type-2 random access procedure fit for NR unlicensed operations, where the UE or the gNB are typically required to perform listen-before-talk (LBT) to determine whether it can transmit or not.

Besides initial access, random-access procedure can also be used for other cases, like UL synchronization re-establishment, scheduling request in case no PUCCH resources are available, radio link failure recovery, beam failure recovery, RRC connection resume when transitioning from RRC inactive state to RRC connected state, and request of additional SI. The latter three use cases are NR specific.

2.1 Random-Access Preamble

PRACH is used to transmit a random-access preamble from a UE to indicate to the gNB a random-access attempt and to assist the network to adjust the UL timing of the UE.

Preamble Sequence Design

Like in LTE, Zadoff-Chu sequences are used for generating NR random-access preambles. To support the wide range of deployments for which NR is designed, NR supports two random-access preamble sequence lengths with different format configurations, as examples shown in Fig. 11.8.

For the long sequence of length 839, four preamble formats are supported. These formats are designed for large cell deployment scenarios and can only be used in FR1. Preamble formats 0, 1, and 2 have a subcarrier spacing of 1.25 kHz, while a SCS of 5 kHz is used for preamble format 3.

For the short sequence of length 139, nine different preamble formats are introduced in NR Rel-15, mainly targeting the small/normal cell and indoor deployment

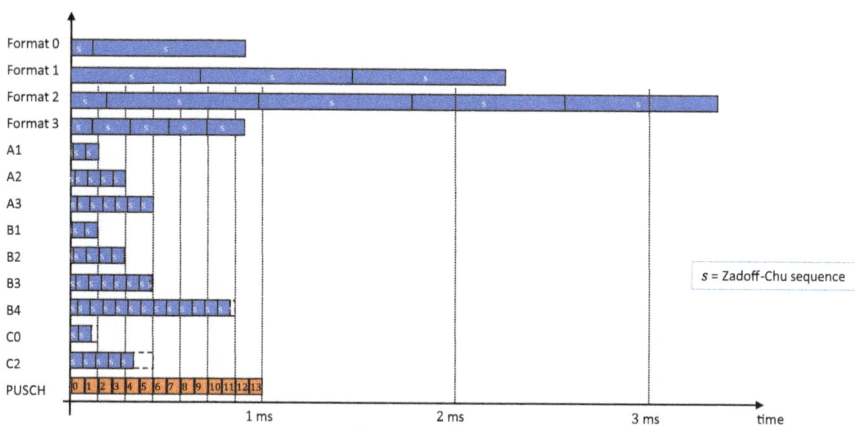

Fig. 11.8 Illustration of NR random-access preamble formats

scenarios. The short preamble formats can be used in both FR1 with SCS of 15 or 30 kHz and FR2 with SCS of 60 or 120 kHz. The examples shown in Fig. 11.8 assume SCS of 15 kHz for both the short preamble formats and PUSCH. The SCS of the short sequence preamble is signaled by the higher-layer parameter *msg1-SubcarrierSpacing* in *RACH-ConfigCommon* in *SIB1*.

The basic design principle for PRACH preamble is that the last part of each preamble OFDM symbol acts as a cyclic prefix (CP) for the next OFDM symbol. In contrast to LTE, for the design of the short preamble formats, the length of a preamble OFDM symbol equals the length of data OFDM symbols. This new design allows the gNB receiver to use the same fast Fourier transform (FFT) for data and random-access preamble detection. In addition, due to the composition of multiple shorter OFDM symbols per PRACH preamble, depending on the algorithms used for PRACH detection, the new short preamble formats can be more robust against time-varying channels and frequency errors.

PRACH Configuration

In NR, a *PRACH slot* is defined as one slot with the SCS of the random-access preamble. The time and frequency resource on which a random-access preamble is transmitted is defined as a *PRACH occasion*. There can be multiple PRACH occasions configured within a PRACH slot, and multiple frequency-multiplexed PRACH occasions configured in one-time instance.

In this section, we introduce the details of the NR design on PRACH configuration for type-1 random-access procedure, including the resource mapping of PRACH occasions in the time and frequency domain, as well as the association between SS/PBCH blocks and random-access preamble transmissions.

Time Domain PRACH Configuration

Three different random-access configuration tables are defined for the cases of FR1 with paired spectrum (i.e., FDD operation), FR1 with unpaired spectrum (i.e., TDD operation), and FR2 with unpaired spectrum, respectively [6]. There is no configuration table defined for FR 2 with paired spectrum, since no NR spectrum has been identified for this case yet.

The configuration of candidate time domain PRACH occasions and the preamble format are jointly indicated by a higher-layer parameter *prach-ConfigurationIndex* in *SIB1*, which indicates a row in the associated random-access configuration table. A subset of rows of the random-access configuration table specified in NR for FR1 unpaired spectrum is copied in Table 11.2.

For long preamble sequence, the time domain PRACH occasions are determined by the following parameters in the associated random-access configuration table:

Table 11.2 Random-access configurations for FR1 and unpaired spectrum

PRACH Configuration index	Preamble format	$n_{SFN} \bmod x = y$		Subframe number	Starting symbol	Number of PRACH slots within a subframe	$N_t^{RA,slot}$, number of time domain PRACH occasions within a PRACH slot	N_{dur}^{RA}, PRACH duration
		x	y					
0	0	16	1	1	0	–	–	0
1	0	16	1	4	0	–	–	0
2	0	16	1	7	0	–	–	0
68	A1	8	1	9	0	2	6	2
69	A1	4	1	9	0	1	6	2
70	A1	2	1	9	0	1	6	2
71	A1	2	1	4,9	7	1	3	2

- The PRACH configuration periodicity, indicated by the values in the unit of system frame in the column "*x*". The PRACH configuration periodicity can be configured to be 10, 20, 40, 80, or 160 ms.
- The system frame within each PRACH configuration period, on which the PRACH occasions are configured. This information is indicated by the values in the unit of system frame in the column "*y*". For instance, if y is set to 0, then it means that PRACH occasions are only configured in the first system frame of each PRACH configuration period.
- The subframes within a system frame that are configured with PRACH occasion, are indicated by the values in the unit of subframe in the column "subframe number." For FR2, instead of subframes, the slots that are configured with PRACH occasions are indicated by the values in the column "slot number," and the unit is slot with SCS of 60 kHz.

For short preamble sequences, the following additional parameters are configured:

- Number of PRACH slots within a subframe (for FR1) or a 60 kHz slot (for FR2). This configuration is only valid when the preamble SCS is configured to be 30 kHz for FR1 or 120 kHz for FR2. If the value in the corresponding column is 2, it indicates that both PRACH slots are configured with PRACH occasions; otherwise, only the second PRACH slot is configured with PRACH occasions.
- Number of time domain PRACH occasions within a PRACH slot. If configured, these PRACH occasions are consecutive in time.
- Starting symbol of the first PRACH occasion within a PRACH slot, indicated by the values in the column "starting symbol" in Table 11.2. For TDD systems, NR supports PRACH occasion configurations with later starting symbol than symbol 0 in a PRACH slot. These configurations can be used to create the guard period (GP) between DL and UL so that the preambles transmitted on the first PRACH occasion within a PRACH slot will not be interfered by the DL transmissions from other BSs. These configurations can also be used to support the transmission of DL control signaling in the beginning of a PRACH slot.
- The duration of one PRACH occasion in the unit of OFDM symbols of preamble SCS.

For FDD, all time domain PRACH occasions defined in the PRACH configuration table are valid. In the case of TDD, the validity of these time domain PRACH occasions is determined by the cell-specific TDD pattern and the actually transmitted SS/PBCH blocks. More specifically, a PRACH occasion defined in the PRACH configuration table is valid if it is within the UL part of the cell-specific TDD pattern or when it is within the flexible part of the TDD pattern and it does not precede an SS/PBCH block in the RACH slot. For short format preambles, an additional condition is applied, that is, the PRACH occasion should be at least two symbols after the DL part of the TDD pattern and at least two symbols after the last symbol of an actually transmitted SS/PBCH block [3].

Frequency Domain PRACH Configuration

In the frequency domain, multiple PRACH occasions can be frequency-multiplexed in one-time instance. This can be used to create more PRACH resources in frequency so that random-access attempts from more UEs can be supported within a time domain PRACH occasion.

The number of PRACH occasions that are frequency-multiplexed in one time domain PRACH occasion can be 1, 2, 4, or 8, and the value is indicated by the parameter *msg1-FDM* in *SIB1*. The frequency-multiplexed PRACH occasions are consecutive in frequency, and they are configured within the initial active UL BWP for initial access. The start position in frequency is indicated by the parameter *msg1-FrequencyStart* in *SIB1*, which provides a frequency offset of the lowest PRACH occasion in frequency with respective to the lowest RB of the configured initial UL BWP.

Association Between SS/PBCH Block and Preamble Transmission

In NR, each SS/PBCH block index is associated with a set of valid PRACH occasions and a set of preambles. If different SS/PBCH blocks are transmitted in different DL beams, then, by detecting a random-access preamble in one PRACH occasion transmitted from a UE, a gNB can acquire the information about the DL beam selected by the UE after cell search. Therefore, the gNB can use this beam for subsequent DL transmissions to the UE. In addition, if there is a beam correspondence between the DL beam used to transmit a SS/PBCH block and the UL beam used to receive the associated random-access preambles, then, as shown in Fig. 11.9, by associating all frequency-multiplexed PRACH occasions at one-time instance with one SS/PBCH block, fast analog beam-sweeping can be applied at the gNB for preamble reception in UL.

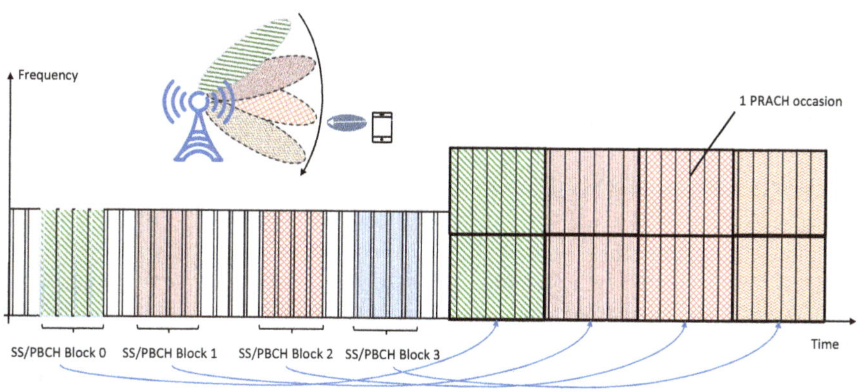

Fig. 11.9 An example of association between SS/PBCH blocks and PRACH occasions

NR supports one-to-one, one-to-many, and many-to-one association between SS/PBCH blocks and valid PRACH occasions. For each PRACH occasion, there are 64 preambles defined. The total number of preambles per PRACH occasion used for contention-free and contention-based random access is provided by *totalNumberOfRA-Preambles* in *SIB1*. The number of SS/PBCH blocks associated with one PRACH occasion and the contention-based preambles assigned per SS/PBCH block per valid PRACH occasion is jointly indicated by the parameter *ssb-perRACH-OccasionAndCB-PreamblesPerSSB* in *SIB1* [3].

The SS/PBCH block indexes are mapped to preambles in valid PRACH occasions in the following order:

- First, in increasing order of preamble indexes within a single PRACH occasion
- Second, in increasing order of frequency resource indexes for frequency-multiplexed PRACH occasions
- Third, in increasing order of time resource indexes for time-multiplexed PRACH occasions within a PRACH slot
- Fourth, in increasing order of indexes for PRACH slots

For initial access to a cell that is configured with a supplementary UL carrier, an reference signals received power (RSRP) threshold, *rsrp-ThresholdSSB-SUL*, is provided in *SIB1* for the selection between the non-supplementary UL carrier and the supplementary UL carrier [11]. If the RSRP of the DL pathloss reference is less than *rsrp-ThresholdSSB-SUL*, a UE will select the supplementary UL carrier for performing all uplink transmissions of the random-access procedure in order to improve the coverage in UL; otherwise, the non-supplementary UL carrier will be selected.

2.2 Random-Access Response

According to the type-1 random access procedure shown in Fig. 11.7, upon reception of a random-access preamble in a PRACH occasion, a gNB will transmit in DL a RAR using a normal PDSCH (Msg2). As discussed above, by detecting a random-access preamble from a UE, a gNB can acquire the information about the DL beam selected by the UE, thereby using the same beam for Msg2 transmission to this UE.

The PDSCH carrying Msg2 has the same SCS as for PDSCH carrying *SIB1*, and it is scheduled by a PDCCH carrying a DCI format 1_0 with CRC scrambled by a RA-RNTI. The RA-RNTI is associated with the time-frequency PRACH occasion in which the preamble is transmitted. If multiple random-access preambles from multiple UEs are detected on one time-frequency PRACH occasion, then the Msg2 carried on a PDSCH will consist of multiple RARs, with each RAR corresponding to an individual detected preamble [11]. Multiplexing of multiple RARs into one Msg2 reduces the overhead for both scheduling and transmitting random-access responses.

The content of the RAR for a detected random-access preamble includes:

- Random-access preamble ID (RAPID), which indicates the received preamble index
- Timing advance command, which indicates the index value to be used for adjusting the timing for the subsequent UL transmissions
- UL grant, also referred to as RAR UL grant, which is used for scheduling PUSCH carrying Msg3
- Temporary C-RNTI (TC-RNTI), which indicates the temporary ID that a UE can use for Msg3 transmission

A UE derives its corresponding RA-RNTI based on the time-frequency PRACH occasion selected for its preamble transmission. After transmitting a random-access preamble, the UE will start monitoring the PDCCH candidates in the *Type1-PDCCH CSS* set for a DCI format 1_0 with CRC scrambled by its corresponding RA-RNTI within a time window. The time window starts from the first symbol of the earliest CORESEST for Type*1-PDCCH CSS* set. The length of the time window is provided by the parameter *ra-ResponseWindow* in *SIB1*, which can be configured to be {1, 2, 4, 8, 10, 20, 40, 80} slots in Msg2 SCS, with a maximum length of 10 ms. The short RAR time windows have been introduced in NR to support low latency use cases that require fast connection establishment.

If a UE detects a PDCCH with its corresponding RA-RNTI within the RAR window and if it successfully decodes the associated PDSCH carrying Msg2, then the UE will check whether the RAPID contained in a RAR in the received Msg2 matches with the transmitted preamble index. If the UE identifies the RAPID, it will consider this RAR reception successful. Otherwise, the UE will consider this RAR reception not successful, and it can start a new random-access attempt according to the information signaled by the network.

For each new random-access attempt, if the selected PRACH occasion is associated with the same SS/PRACH block (the same beam) as for the last random-access procedure, the UE will transmit a preamble with an increased transmit power according to the configured power ramping step size for PRACH, as long as the increased transmit power does not exceed the UE configured maximum output power [1, 2]. If a different SS/PBCH block is selected for the new random-access attempt, then the same transmit power as for the last random-access preamble transmission is used for this new attempt. Multiple random-access attempts can be performed until the UE successfully received a RAR from the gNB or until it reaches the maximum allowed number of preamble transmissions provided by the parameter *preambleTransMax* in *SIB1*.

If multiple UEs transmit preambles in the same PRACH occasion, they will monitor the PDCCH candidates with the same RA-RNTI and receive the same Msg2 from the gNB. If different preamble indexes are selected by different UEs, then no preamble collision occurs, and each UE can identify the corresponding RAR based on the RAPID field of the multiple RARs contained in Msg2. However, there is a possibility that multiple UEs select the same preamble and transmit the preamble on the same PRACH resource. Step 3 and step 4 of the random-access procedure are used to resolve such potential contention.

2.3 Scheduled Msg3 Transmission

For type-1 contention-free random access, after the successful reception of the corresponding RAR, a UE will adjust its uplink transmission timing according to the timing advance command received from the RAR, and then, it will perform initial Msg3 transmission based on the UL grant in the RAR.

Msg3 is carried out on a normal PUSCH. The SCS of PUSCH carrying Msg3 is configured by *subcarrierSpacing* in *SIB1*. For FR1, the SCS can be either 15 or 30 kHz; for FR2, the SCS can be either 60 or 120 kHz. The waveform to use for Msg3 transmission, i.e., OFDM or DFT-precoded OFDM, is provided by the parameter *msg3-transformPrecoder* in *SIB1*. To improve the coverage of Msg3 transmission, a gNB can schedule a Msg3 PUSCH retransmission using a DCI format 0_0 with CRC scrambled by the TC-RNTI provided in the corresponding RAR.

The content of the Msg3 in a contention-free random access differs for different triggering events. For initial access, a UE includes the *rrcSetupRequest* message in Msg3, which consists of the following information [7]:

- *establishmentCause*, which indicates the establishment cause that triggers the access request, e.g., emergency call, mission critical services, multimedia priority services, etc. This information can be used by the network to prioritize the access requests from the high-priority UEs in high network load situation.
- *ue-Identity*, which provides an initial UE identity to facilitate contention resolution in step 4.

For the cases where a UE is already assigned a unique UE identity (C-RNTI) by the network, e.g., when the random-access procedure was initialed by a PDCCH order for UL synchronization re-establishment, the UE will include the assigned C-RNTI in its MAC CE in its Msg3 transmission.

If multiple UEs select the same preamble and transmit the preamble on the same PRACH resource, these UEs will read the same RAR content. Thus, they will transmit their individual Msg3 using the same time and frequency resources based on the UL grant received from the RAR. However, only one of these UEs can connect to the network during this random-access procedure, and this is enabled by the last step of the procedure, which is called contention resolution.

2.4 Contention Resolution

In response to a Msg3 reception from a UE that has not been assigned a C-RNTI, a gNB transmits a PDSCH (Msg4) carrying a UE contention resolution identity. The PDSCH carrying Msg4 has the same SCS as for PDSCH carrying SIB1, and it is scheduled by a PDCCH carrying a DCI format 1_0 with CRC scrambled by a TC-RNTI associated with the received Msg3.

After transmitting a Msg3 PUSCH scrambled with TC-RNTI, a UE starts monitoring the PDCCH candidates in the *Type1-PDCCH CSS* set for a DCI format 1_0 with CRC scrambled by its corresponding TC-RNTI within a time window.

If the UE detects a PDCCH with the TC-RNTI within the time window, and if the UE successfully decodes the associated PDSCH carrying Msg4, the UE will check the UE contention resolution identity contained in the MAC CE in the received Msg4 [11]. If the UE contention resolution identity matches with the *ue-Identity* transmitted in Msg3, the UE will consider this random-access procedure successful, and it will then send an ACK in a PUCCH to the gNB and set the C-RNTI to the value of the TC-RNTI. Otherwise, the UE considers this random-access procedure not successful, and it will start a new random-access attempt.

For the cases where the detected Msg3 is from a UE that has a C-RNTI, a gNB will transmit a PDCCH carrying a DCI format 1_0 with CRC scrambled by the C-RNTI. Within the contention resolution time window, upon a reception of a PDCCH transmission that is addressed to the C-RNTI, the UE considers the random-access procedure successfully completed. Otherwise, the UE considers this random-access procedure not successful, and it will start a new random-access attempt.

References

1. 3GPP TS 38.101-1, "NR; User Equipment (UE) radio transmission and reception Part I: Range 1 Standalone", v16.1.0, Sept. 2019
2. 3GPP TS 38.101-2, "NR; User Equipment (UE) radio transmission and reception Part II: Range 2 Standalone", v16.1.0, Sept. 2019
3. 3GPP TS 38.213, "NR; Physical layer procedures for control," V15.7.0, Sept. 2019
4. 3GPP TS 38.104, "NR; Base Station (BS) radio transmission and reception", v16.1.0, Sept. 2019
5. 3GPP TS 36.104, "Evolved Universal Terrestrial Radio Access (E-UTRA); Base Station (BS) radio transmission and reception," V15.7.0, Sept. 2019
6. 3GPP TS 38.211, "NR; Physical channels and modulation," V15.7.0, Sept. 2019
7. 3GPP TS 38.331, "NR; Radio Resource Control (RRC); Protocol specification," V15.7.0, Sept. 2019
8. 3GPP TS 38.212, "NR; Multiplexing and channel coding," V15.7.0, Sept. 2019
9. Z. Lin, J. Li, Y. Zheng, etc., "SS/PBCH Block Design in 5G New Radio (NR)," *IEEE Globecom Workshops (GC Wkshps)*, Abu Dhabi, 2018, pp. 1–6
10. 3GPP TS 38.300, "NR; NR and NG-RAN overall description; Stage 2" V15.7.0, Sept. 2019
11. 3GPP TS 38.321, "NR; Medium Access Control (MAC) protocol specification," V15.7.0, Sept. 2019

Chapter 12
A Primer on Bandwidth Parts in 5G New Radio

Xingqin Lin, Dongsheng Yu, and Henning Wiemann

1 Introduction

The 3rd Generation Partnership Project (3GPP) has developed a new radio-access technology known as New Radio (NR) in its Release 15 and continues to evolve NR to further improve performance and address new use cases in the fifth generation (5G) era [1]. Compared to the previous generations of radio-access technologies, NR introduces many new features to support a wide range of services, devices, and deployments. In this chapter, we focus on one of the basic NR features – bandwidth part (BWP).

To develop a preliminary understanding of BWP, we first review the hierarchy of spectrum management in NR, which is illustrated in Fig. 12.1. At a high level, NR defines frequency ranges (FRs). In 3GPP Release 15 there are two FRs defined: the first is FR1 ranging from 410 to 7125 MHz [2] and the second is FR2 ranging from 24.25 to 52.6 GHz [3]. 3GPP further defines operating bands in each FR. An operating band is a frequency band associated with a certain set of radio frequency (RF) requirements. Bandwidths of different operating bands can vary from several MHz to a few GHz. Different operators may have different amounts of spectrum within an operating band. To accommodate diverse spectrum scenarios while limiting implementation complexity, NR supports a range of channel bandwidths

X. Lin (✉)
Ericsson, Santa Clara, CA, USA
e-mail: xingqin.lin@ericsson.com

D. Yu
Ericsson, Ottawa, Canada
e-mail: dongsheng.yu@ericsson.com

H. Wiemann
Ericsson, Aachen, Germany
e-mail: henning.wiemann@ericsson.com

X. Lin, N. Lee (eds.), *5G and Beyond*, https://doi.org/10.1007/978-3-030-58197-8_12

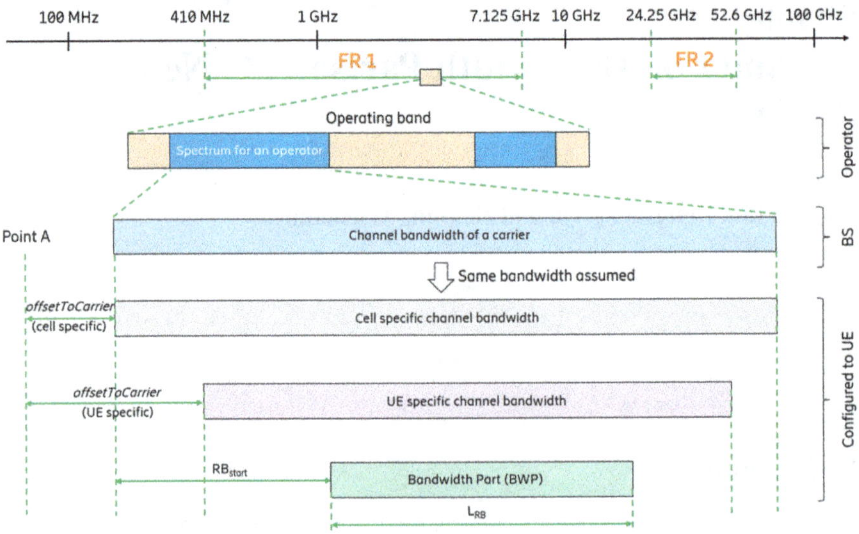

Fig. 12.1 An illustration of the 5G NR spectrum management and configuration

from 5 to 400 MHz, where a channel bandwidth refers to the bandwidth of an NR carrier. The number of resource blocks (RBs) that may be configured in a channel bandwidth, known as transmission bandwidth configuration, shall meet the specified minimum guard band requirements [2, 3]. Base station (BS) and user equipment (UE) can support different channel bandwidths. Like in Long-Term Evolution (LTE), a UE camps on and connects to a cell. The UE is made aware of the channel bandwidth of the cell. In addition to the cell bandwidth, the network informs the UE about the position and width of a BWP. Loosely speaking, a BWP is hence a set of contiguous RBs configured inside a channel bandwidth. The width of a BWP may be smaller than or equal to the cell bandwidth.

One motivation of introducing BWP in NR is to support UE bandwidth adaptation to help reduce device power consumption [4]. The main idea is that a UE may use a wide bandwidth when a large amount of data is scheduled, while being active on a narrow bandwidth for the remaining time. Another motivation is to support devices of different bandwidth capabilities by configuring the devices with different BWPs. A BS may support a very wide channel bandwidth which may not be supported by some UEs. BWP provides a mechanism to flexibly assign radio resources such that the signals for a UE are confined in a portion of BS channel bandwidth that the UE can support.

BWP, as a basic concept in NR, spans across different 3GPP specifications. Understanding how BWP operates is vital to developing a good knowledge of NR. A high-level introduction to BWP can be found in [4]. The white paper [5] provides further introduction to BWP concepts with a focus on UE power consumption. In contrast, the objective of this chapter is to delve into the detailed NR technical specifications to provide a complete overview of BWP design, while keeping the overall contents at a level accessible to an audience working in the wireless communications and networking communities. Besides, we take a network-centric approach and provide insights into NR deployments using BWPs.

The remainder of this chapter is organized as follows. In Sect. 2, we introduce the basic concepts of BWPs. Then we describe how a network may configure BWPs in Sect. 3 and BWP switching mechanisms in Sect. 4. We discuss UE capabilities of supporting BWPs in Sect. 5. Several use cases of BWPs for NR deployments are described in Sect. 6, followed by our concluding remarks in Sect. 7.

2 Basic Concepts of Bandwidth Parts

2.1 Fundamentals of Bandwidth Parts

NR defines scalable orthogonal frequency division multiplexing (OFDM) numerologies using subcarrier spacing (SCS) of $2^{\mu} \cdot 15$ kHz ($\mu = 0, 1, \ldots, 4$) [6]. An RB consists of 12 consecutive subcarriers in the frequency domain. NR uses "Point A" as a common reference point for RB grids. "Point A" is illustrated in Fig. 12.1.

As illustrated in Fig. 12.1, a BWP starts at a certain common RB and consists of a set of contiguous RBs with a given numerology (SCS and cyclic prefix) on a given carrier. For each serving cell of a UE, the network configures at least one downlink (DL) BWP (i.e., the initial DL BWP). The network may configure the UE with up to four DL BWPs, but only one DL BWP can be active at a given time. If the serving cell is configured with an uplink (UL), the network configures at least one UL BWP. Similar to the DL, the network may configure the UE with up to four UL BWPs, but only one UL BWP can be active at a given time. NR also supports a so-called supplementary UL (SUL), on which UL BWP(s) can be similarly configured as on a normal UL.

For paired spectrum, i.e., frequency division duplex (FDD), DL BWPs and UL BWPs are configured separately. For unpaired spectrum, i.e., time division duplex (TDD), a DL BWP is linked to a UL BWP when the indices of the two BWPs are the same. In this case, the paired DL BWP and UL BWP must share the same center frequency, but they can have different bandwidths.

In general, a UE only receives physical downlink shared channel (PDSCH), physical downlink control channel (PDCCH), or channel state information reference signal (CSI-RS) inside an active DL BWP. But the UE may need to perform radio resource management (RRM) measurements outside the active DL BWP via measurement gaps. Similarly, the UE only transmits physical uplink shared channel (PUSCH) or physical uplink control channel (PUCCH) inside an active UL BWP, and for an active serving cell, the UE does not transmit sounding reference signal (SRS) outside an active UL BWP.

2.2 Bandwidth Part Types

Activating an inactive BWP and deactivating an active BWP are called BWP switching to enforce that it is not possible to deactivate all BWPs or to activate more than one. For paired spectrum, DL BWPs and UL BWPs can be switched separately. For unpaired spectrum, the paired DL BWP and UL BWP are switched together. The detailed BWP switching mechanisms are described in Sect. 4. In this subsection, we describe the types of BWPs that may be active at a given time.

Initial DL/UL BWP The initial DL and UL BWPs are used at least for initial access before radio resource control (RRC) connection is established. An initial BWP has index zero and is referred to as BWP #0. During the initial access, the UE performs cell search based on synchronization signal block (SSB) composed of primary synchronization signal (PSS), secondary synchronization signal (SSS), and physical broadcast channel (PBCH). To access the system, the UE needs to further read system information block 1 (SIB1) which carries important information including the initial DL/UL BWP configuration. The SIB1 is transmitted on the PDSCH, which is scheduled by downlink control information (DCI) on the PDCCH using the control resource set with index zero (CORESET #0) [7, 8].

Before the UE reads the SIB1, the UE's initial DL BWP has the same frequency range and numerology as those of CORESET #0. After reading the SIB1, the UE follows the initial DL/UL BWP configuration in the SIB1 and uses them to carry out random-access procedure to request the setup of RRC connection. The network should configure the frequency domain location and bandwidth of the initial DL BWP in the SIB1 so that the initial DL BWP contains the entire CORESET #0 in the frequency domain.

First Active DL/UL BWP The first active DL and UL BWPs may be configured for a Special Cell (SpCell) or a secondary cell (SCell). In a master cell group (MCG), the SpCell refers to the primary cell (PCell) in which the UE performs the connection (re-)establishment procedure. In a secondary cell group (SCG), the SpCell refers to the primary SCG cell (PSCell) in which the UE performs random access for RRC (re-)configuration. An SCell provides additional radio resources on top of an SpCell in a cell group. The first active DL and UL BWPs are the active DL and UL BWPs upon RRC (re-)configuration for an SpCell or activation of an SCell.

Default BWP For a serving cell, the network may configure the UE with a BWP inactivity timer. The expiration of this timer may, for example, indicate that the UE has no scheduled transmission and reception for a while on the currently active BWP. Thus, the UE can switch its active BWP to a default BWP to save power. The default DL BWP can be configured. If not configured, the UE uses the initial DL BWP as the default DL BWP. For unpaired spectrum, when the UE switches its

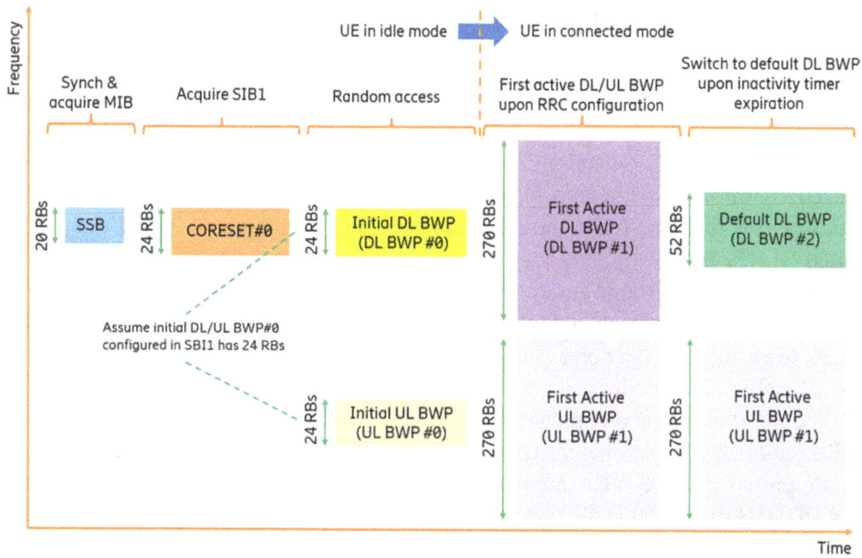

Fig. 12.2 An illustration of UE BWP adaptation from idle mode to connected mode

active DL BWP to the default DL BWP, the active UL BWP is switched accordingly since the BWP switching for TDD is common for both DL and UL.

Figure 12.2 provides an illustration of the aforementioned BWP types from a UE processing perspective. The UE first performs downlink synchronization and acquires PBCH based on 20-RB SSB. Assuming the CORESET #0 configured in the MIB has 24 RBs, the UE may assume that the initial DL BWP is 24 RBs wide and proceeds to acquire SIB1, which in this example also configures 24 RBs for both initial DL and UL BWPs. The UE then performs random-access procedure with the small initial DL and UL BWPs. After the random access, the UE reports that it is capable of supporting multiple BWPs. With dedicated RRC signaling, the network configures the UE with large DL/UL BWP #1 (270 RBs), small DL/UL BWP #2 (52 RBs), and BWP inactivity timer. The network sets the large DL/UL BWP #1 as the first active DL/UL BWP and the small DL BWP #2 as the default DL BWP. Upon RRC configuration, the first active DL and UL BWPs (i.e., DL/UL BWP #1) become activated and are used for scheduling a large amount of data. After that, the UE does not have traffic demand and has no scheduled transmission. As a result, the BWP inactivity timer expires, upon which the UE switches its active DL BWP to the default DL BWP (i.e., DL BWP #2). Note that the active UL BWP does not need to switch to UL BWP #2 because Fig. 12.2 illustrates an FDD system in which DL and UL BWPs are switched separately.

3 Bandwidth Part Configurations

3.1 Configuration of a Bandwidth Part with a Nonzero Index

In this subsection, we discuss how to configure a BWP with a nonzero index. A DL/UL BWP with a nonzero index is a non-initial DL/UL BWP (recall that the index zero is reserved for initial DL/UL BWP) and is configured in addition to the initial DL/UL BWP.

The DL/UL BWP configurations are divided into common and dedicated parameters. The BWP-common parameters are cell specific, implying that the network needs to ensure that the corresponding parameters are appropriately aligned across the UEs. The BWP-dedicated parameters are UE specific.

The BWP-common parameters for a DL BWP with a nonzero index include basic cell-specific BWP parameters (frequency domain location, bandwidth, SCS, and cyclic prefix of this BWP) and additional cell-specific parameters for the PDCCH and PDSCH of this DL BWP. The BWP-dedicated parameters for a DL BWP with a nonzero index include UE-specific parameters for the PDCCH, PDSCH, semi-persistent scheduling, and radio link monitoring configurations of this DL BWP. The BWP-common parameters for a UL BWP with a nonzero index include basic BWP parameters and cell-specific parameters for the random access, PUCCH, and PUSCH of this UL BWP. The BWP-dedicated parameters for a UL BWP with a nonzero index include UE-specific parameters for the PUCCH, PUSCH, SRS, configured grant, and beam failure recovery configurations of this UL BWP.

3.2 Configuration of a Bandwidth Part with Index Zero

There are two options for configuring a BWP with index zero (i.e., the initial BWP):

- Option 1: Configure the BWP #0 with cell-specific parameters only.
- Option 2: Configure the BWP #0 with both cell-specific and UE-specific parameters.

The DL/UL BWP #0 configured by Option 1 does not have the dedicated parameters and thus has limited functionality. In this case, the DL/UL BWP #0 mainly plays a temporary role and is used by the UE, for example, during the initial-access procedure. To set up a fully operational connection, the network should also configure the UE with an additional full-featured DL/UL BWP equipped with both cell-specific and UE-specific parameters.

The DL/UL BWP #0 configured by Option 2 is a full-featured BWP equipped with both cell-specific and UE-specific parameters. The UE may obtain the cell-specific and UE-specific parameters via different signaling messages. For example, during the initial access, the UE can obtain the cell-specific parameters of the

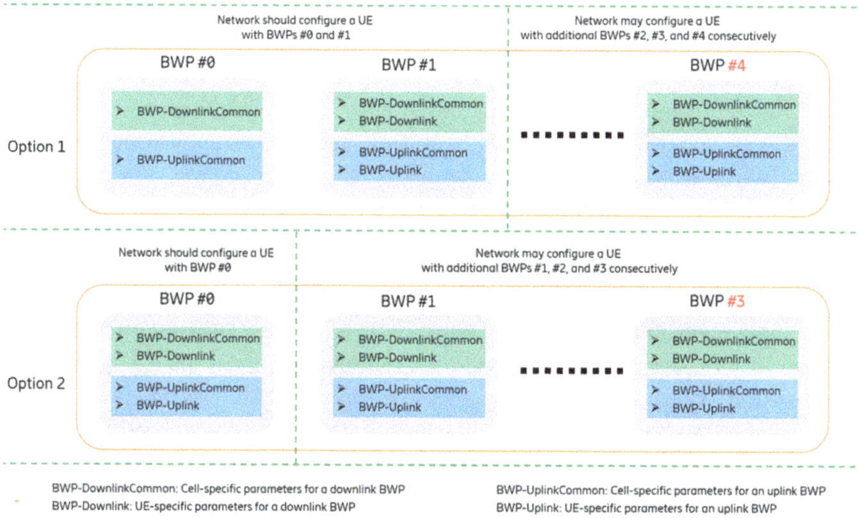

Fig. 12.3 An illustration of the 5G NR bandwidth part configuration options

DL/UL BWP #0 by reading the SIB1. The UE can further obtain the UE-specific parameters upon RRC configuration after the initial access. Option 2 is appealing in the deployments where multiple DL/UL BWPs are not needed. In this case, the network can set up a fully operational connection with a UE by only configuring DL/UL BWP #0 using Option 2.

NR supports configurations of up to four "RRC-configured" DL/UL BWPs. The DL/UL BWP #0 configured by Option 1 only has cell-specific parameters and is not counted as an "RRC-configured" BWP. Therefore, additional four DL/UL BWPs #1, #2, #3, and #4 may be consecutively configured. The DL/UL BWP #0 configured by Option 2 has both cell-specific and UE-specific parameters and thus is counted as an "RRC-configured" BWP. Therefore, additional three DL/UL BWPs #1, #2, and #3 may be consecutively configured. Figure 12.3 provides an illustration of the BWP configuration options in NR.

4 Bandwidth Part Switch

4.1 RRC Reconfiguration-Based Bandwidth Part Switch

When more than one UE-specific DL/UL BWP is configured to the UE on a serving cell, the first active DL/UL BWP, if configured, indicates the DL/UL BWP to be activated upon RRC (re-)configuration for an SpCell and upon activation of an SCell. If the first active BWP is not configured, there is no BWP switch upon RRC (re-)configuration. The first active DL/UL BWP is always configured upon SCell addition, upon PCell change in MCG, and PSCell addition or change in SCG.

For BWP configuration Option 1, switch from the initial DL/UL BWP to another DL/UL BWP requires RRC reconfiguration since only DCI format 1_0/0_0 can be used with initial DL/UL BWP without dedicated configuration which does not support DCI-based BWP switch.

For RRC-based BWP switch, there is a delay of receiving (for DL active BWP switch) or transmitting (for UL active BWP switch) on the new BWP on the serving cell after the UE receives RRC reconfiguration involving active BWP switch or parameter change of its active BWP. The delay requirement for RRC-based BWP switch, within which UE shall complete the switch of active DL and/or UL BWP, is the sum of processing delay for RRC procedure and the delay for UE to perform BWP switch. The processing delay requirements for RRC procedure are in the range of 5–80 ms and differ among connection control procedures [9]. The delay requirement for UE to perform RRC-based BWP switch is 6 ms [11].

4.2 DCI-Based Bandwidth Part Switch

With initial DL/UL BWP and one or more additional DL/UL BWPs being configured to a UE, the network can schedule the UE to switch the active DL/UL BWP from one configured BWP to another using BWP indicator in DCI format 1_1/0_1. The possibility of DCI-based BWP switch involving BWP #0 is dependent on BWP configuration option, as described in Table 12.1. DCI format 1_1 and DCI format 0_1 are non-fallback DCI formats for downlink assignment and uplink grant, respectively [10]. They support the full set of NR features, and their fields are largely configurable. On the other hand, fallback DCI formats 1_0 and 0_0, used respectively for downlink assignment and uplink grant, do not contain BWP indicator field and thus do not support DCI-based BWP switch.

BWP field in DCI format 1_1/0_1 has a bitwidth of 0–2. The exact value is determined by the number of RRC configured DL/UL BWPs, excluding the initial DL/UL BWP. Table 12.1 provides the interpretation of BWP indicator field for DCI-based BWP switch.

There is a transmission/reception delay between network and UE associated with DCI-based BWP switch. UE shall complete the switch of active DL and/or UL BWP within the required BWP switch delay. BWP switch delay requirements are listed in Table 12.2 for both DCI- and timer-based BWP switch [11]. The switch delay denoted by $T_{BWPswitchDelay}$ for DCI-based BWP switch is defined as the slot offset between the DL slot in which the UE received switch request and the first slot in which the UE shall be able to receive PDSCH (for DL active BWP switch) or transmit PUSCH (for UL active BWP switch) on the new BWP. There are two levels of BWP switch delay requirement, type 1 and type 2, as given in Table 12.2. The UE indicates in its UE capabilities which of the two types it supports (see section 5).

Table 12.1 Interpretation of BWP indicator field in DCI format 0_1/1_1 for DCI-based BWP switch

# of RRC configured BWPs, excluding the initial BWP	Bitwidth of BWP indicator field	BWP indicator in DCI 0_1/1_1	BWP Id	Comments	
0	0	Absent	0	For BWP configuration option 1, DCI 0_1/1_1 is not applicable for operation on BWP#0	For BWP configuration option 2, this is single BWP operation on BWP #0
1	1	0, 1	0, 1	Switch among BWP 0, 1	For BWP configuration option 1, once switched to BWP#0, the active BWP can no longer be switched to another BWP with index other than zero.
2	2	00, 01, 10	0, 1, 2	Switch among BWP 0, 1, 2	
3	2	00, 01, 10, 11	0, 1, 2, 3	Switch among BWP 0, 1, 2, 3	
4	2	00, 01, 10, 11	1, 2, 3, 4	Switch among BWP 1, 2, 3, 4	

Table 12.2 DCI- and timer-based BWP switch delay requirements

SCS (kHz)	NR slot length (ms)	BWP switch delay requirement $T_{BWPswitchDelay}$ (slots)	
		Type 1	Type 2
15	1	1	3
30	0.5	2	5
60	0.25	3	9
120	0.125	6	18

The UE is not required to transmit UL signals or receive DL signals during the time duration $T_{BWPswitchDelay}$ on the serving cell where DCI-based BWP switch occurs.

Note that the BWP switch delay is dependent on SCS. If the BWP switch happens between BWPs of different SCS values, the switch delay requirement is determined by the smaller SCS.

4.3 Timer-Based Bandwidth Part Switch

The network may configure a UE with a BWP inactivity timer and a default DL BWP on a serving cell. The default DL BWP is one of the DL BWPs configured to the UE and becomes the active DL BWP upon expiry of the inactivity timer. If no default DL BWP is configured, the default DL BWP is the initial DL BWP. As mentioned in Sect. 2, for unpaired spectrum (TDD), a DL BWP and a UL BWP with the same indices are linked and switched together. Thus, a DL BWP is effectively a DL/UL BWP pair in this case.

The granularity of the timer is 1 ms (i.e., one subframe) for FR1 and 0.5 ms for FR2. When the timer is running, the UE decrements the timer at the end of each subframe for FR1 or at the end of each half-subframe for FR2. The values for the BWP inactivity timer have the range of 2–2560 ms. The maximum value for the BWP inactivity timer matches the maximum value of discontinuous reception (DRX) inactivity timer, which allows for a configuration that prevents the timer from expiring while the DRX inactivity timer is running.

A UE starts the BWP inactivity timer of a serving cell, if configured, when it activates a DL BWP other than the default DL BWP. A UE restarts the BWP inactivity timer of the serving cell when it decodes a DCI with downlink assignment for the active DL BWP in paired spectrum, or when it decodes a DCI with downlink assignment or uplink grant for its active DL/UL BWP pair in unpaired spectrum. A UE shall also start/restart the BWP inactivity timer when a PDCCH for DCI-based BWP switch is received. BWP inactivity timer can only be started or restarted when there is no ongoing random-access procedure associated with the serving cell.

For timer-based BWP switch, the BWP switch transition time duration is from the subframe/half-subframe for FR1/FR2 immediately after a BWP inactivity timer expires until the beginning of a slot where the UE can receive or transmit. The UE is not required to receive or transmit on the serving cell during the transition. Timer-based BWP switch shares the same BWP switch delay requirements as DCI-based BWP switch, as shown in Table 12.2.

Switching between configured BWPs may also happen when random-access procedure is initiated on a serving cell. UL BWP is switched to the initial UL BWP if the physical random access channel (PRACH) occasions are not configured for the active UL BWP of the serving cell. If the serving cell is SpCell, the active DL BWP needs to be switched to the one with the same BWP index as the active UL BWP.

5 UE Capabilities of Bandwidth Part Support

UEs typically support only a subset of the specified radio access features due to implementation constraints and test limitation. The UE sends its capability parameters to the network, and the network shall configure and schedule the UE accordingly. In this section, we describe the BWP-related UE capabilities and the corresponding parameters [12, 13].

As mentioned in Sect. 2, a UE only receives PDCCH and PDSCH in an active DL BWP and transmits PUCCH and PUSCH in an active UL BWP per serving cell. It is mandatory for a UE to support the basic BWP operation of one RRC configured DL BWP and one RRC configured UL BWP.

For the initial access, a UE needs to perform cell search and downlink synchronization by detecting SSB and acquire SIB1 by decoding DCI transmitted in CORESET #0. All UEs support configuration of a BWP if the BWP contains the SSB and CORESET #0 in frequency domain. Only if the UE indicates that it supports ‘BWP operation without bandwidth restrictions’, the network may configure a BWP that does not comprise the SSB and CORESET #0.

Supporting bandwidth adaptation with more than one DL/UL RRC configured BWP and switching among BWPs are optional. UE may support bandwidth adaptation with up to two or four RRC configured DL and/or UL BWPs with the same numerology per serving cell or with up to four RRC configured DL and/or UL BWPs with different numerologies per serving cell.

RRC-based BWP switch is a default function supported by all UEs. DCI- and timer-based BWP switches, which enable efficient bandwidth adaptation, are applicable to the UE supporting more than one RRC configured BWP. The UE can also report which of the two switch delay requirements listed in Table 12.2 it supports.

BWP is introduced to NR as one option to support flexible bandwidth operation by decoupling the channel bandwidth of a carrier from the UE channel bandwidth. From the physical layer design perspective, the bandwidth of a BWP spans from 1 RB to 275 RBs, although BWP sizes smaller than the resource block group (RBG) size or the precoding resource block group (PRG) size are not supported in Release 15 [14].

6 Use Cases of Bandwidth Parts

6.1 Flexible Bandwidth Support

Basic bandwidth flexibility has been introduced since LTE by supporting multiple carrier bandwidths and enabling carrier aggregation and is adopted in NR. A normal LTE device is required to transmit and receive on the full carrier bandwidth of the frequency band supported by the device. LTE machine type communication (LTE-M) and Narrowband Internet of Things (NB-IoT) have been developed to relax this constraint. There has been growing demand for higher bandwidth flexibility in NR due to several reasons:

- NR should support network operation in a much wider range of spectrum with wider carrier bandwidth than LTE.
- NR should support a wide range of services and applications. They may have different requirements on throughput, latency, and reliability.
- UE devices of different bandwidth capabilities should be supported in the same NR network.

Besides carrier aggregation, the configuration of BWPs is one of the main building blocks to meet the new requirements of bandwidth flexibility in NR, though the current BWP scheme is not as flexible as LTE-M/NB-IoT when it comes to narrow bandwidth and low UE complexity. With DL/UL BWPs, along with DL/UL UE-specific channel bandwidth, configured to a UE (see Fig. 12.1), the reception and transmission bandwidths for the UE are decoupled from each other and decoupled from carrier bandwidth.

6.2 UE Power Saving

Power efficiency is an important design consideration for UE. Several basic UE power-saving schemes from LTE are adopted in NR [15], including wake-up-sleep management for adaptation to traffic load in time and fast activation/deactivation of SCell for adaptation to traffic load in frequency. Wake-up-sleep management such as connected mode DRX (cDRX) is beneficial for UE handling bursty data traffic by switching between network access mode and power efficient mode. Fast activation/deactivation of SCell helps UE to achieve power saving by adjusting bandwidth processing requirements at the granularity of component carrier level.

BWP-based bandwidth adaptation is introduced in NR to improve UE power efficiency by finer-granularity adaptation to traffic variation in frequency dimension [5]. Bandwidth adaptation is typically achieved by configuring the UE with multiple BWPs and dynamically switching the UE's active BWP among the configured BWPs. For maximizing UE power-saving gain, BWP-based bandwidth adaptation is usually applied in conjunction with cDRX and/or fast activation/deactivation of SCell.

6.3 Fast Change of UE Configuration

From a UE's perspective, all physical channels and most physical signals are configured per BWP by the network. Switching among multiple BWPs for the UE usually occurs with UE configuration change as well. Each BWP has specific physical characteristics including frequency location, bandwidth, SCS, and cyclic prefix. UE configuration needs to, at least, convey the physical characteristics of the associated BWP. The network can also configure BWPs to a UE with the same (or similar) physical characteristics but with different UE configurations. For example, two BWPs with the same physical characteristics (e.g., same bandwidth, position, SCS) may be configured to a UE with different uplink waveforms: One BWP is configured with cyclic-prefix OFDM (CP-OFDM) waveform, and the other BWP is configured with discrete Fourier transform spread OFDM (DFT-s-OFDM) waveform.

By applying DCI-based BWP switching among such BWPs, the network may "reconfigure" the UE within 1–3 ms (cf. Table 12.2), which is faster, by at least one order of magnitude, than the legacy RRC reconfiguration procedure. DCI-based BWP switch for fast change of UE configuration is a complementary approach limited by a maximum of four RRC configured DL/UL BWPs in Release 15.

Other use cases and application scenarios can be derived based on BWP concept in NR. Network may provide services with different levels of quality of service (QoS) to the same UE or different UEs. BWPs configured with different configurations can be applied to accommodate different service requirements.

7 Conclusions

BWP is a basic concept in 5G NR. This chapter provides an overview of the essentials of BWP in the NR technical specifications, including the fundamental BWP concepts, BWP configuration methods, BWP switch mechanisms, and UE capabilities in terms of BWP support. As highlighted in this chapter, BWP may have the potential of enabling more flexible bandwidth support, reducing UE power consumption, and achieving fast change of UE configuration, among others. As 5G rollout is happening, it will be interesting to see how BWP will be used in the real networks.

References

1. X. Lin et al., 5G new radio: Unveiling the essentials of the next generation wireless access technology. IEEE Communications Standards Magazine **3**(3), 30–37 (2019)
2. 3GPP TS 38.101-1, NR; User Equipment (UE) radio transmission and reception; Part 1: Range 1 Standalone
3. 3GPP TS 38.101-2, NR; User Equipment (UE) radio transmission and reception; Part 2: Range 2 Standalone
4. E. Dahlman, S. Parkvall, J. Scold, *5G NR: The Next Generation Wireless Access Technology* (Academic Press, Cambridge, MA, 2018)
5. MediaTek, Bandwidth part adaptation; 5G NR user experience & power consumption enhancements, *White paper*, 2018. Available at https://d86o2zu8ugzlg.cloudfront.net/mediatek-craft/documents/Bandwidth-Part-Adaptation-White-Paper-PDFBPAWPA4.pdf
6. 3GPP TS 38.211, NR; Physical channels and modulation
7. K. Takeda, H. Xu, T. Kim, K. Schober, X. Lin, Understanding the heart of the 5G air interface: An overview of physical downlink control channel for 5G New Radio, in IEEE Communications Standards Magazine, **4**(3), 22–29 (2020)
8. 3GPP TS 38.213, NR; Physical layer procedures for control
9. 3GPP TS 38.331, NR; Radio Resource Control (RRC) protocol specification
10. 3GPP TS 38.212, NR; Multiplexing and channel coding
11. 3GPP TS 38.133, NR; Requirements for support of radio resource management
12. 3GPP TS 38.306, NR; User Equipment (UE) radio access capabilities
13. 3GPP TR 38.822, NR; User Equipment (UE) feature list
14. 3GPP TSG RAN WG1, R2-1912026, Reply LS on supported BW for initial BWP, 3GPP TSG RAN WG2#107bis, Oct 2019
15. 3GPP TS 38.300, NR; NR and NG-RAN overall description; Stage 2

Part III
5G New Radio Evolution

Chapter 13
Support of Ultra-reliable and Low-Latency Communications (URLLC) in NR

Sigen Ye

1 Introduction

The design of NR access technology is targeted to support a broad range of use cases with various levels of requirements in terms of data rate, latency, and reliability. Ultra-reliable and low-latency communication (URLLC) is one of the most important features for NR, which can be used to support a large set of applications with stringent end-to-end latency and/or reliability requirements. Significant amount of effort has been spent in different organizations in order to identify the potential URLLC use cases for NR and define the corresponding performance requirements.

To support the stringent end-to-end performance requirements of URLLC applications, special consideration is required in every aspect of the network design, from system architecture all the way down to physical layer design. This includes but is not limited to the Quality-of-Service (QoS) differentiation at the core network which is passed down to the radio access network (RAN), mobile edge computing that reduces the network delay, always-on Protocol Data Unit (PDU) session, a new inactive state in Radio Resource Control (RRC) protocol, packet duplication to improve the reliability, and various techniques to improve latency and reliability in physical layer.

In this chapter, the focus is on the physical layer design that enables the low latency and high reliability over the air interface. The fundamental NR design in Rel-15 already considered URLLC requirements, which led to a flexible framework that includes a set of tools that can be used to support URLLC. This includes the fundamentals such as the flexible numerology and frame structure, flexible control and data channel structure, and much more. Further enhancements were introduced in Rel-16, which provides more optimization for URLLC on different aspects. The

S. Ye (✉)
Apple Inc., Whitehouse Station, NJ, USA

X. Lin, N. Lee (eds.), *5G and Beyond*, https://doi.org/10.1007/978-3-030-58197-8_13

overall design also considers the mixed traffic scenario, where either a particular UE or the network may carry both URLLC and non-URLLC traffic, in which case mechanisms are available to prioritize the URLLC transmission in order to guarantee the performance.

In Sect. 2, we provide an overview of potential URLLC use cases for 5G, together with the performance requirements for some of the use cases. The physical layer design considerations for URLLC in Release 15, the first release of NR, are discussed in Sect. 3. Further URLLC enhancements in Release 16 are summarized in Sect. 4. It is concluded in Sect. 5 with the outlook of future URLLC work in the 3rd Generation Partnership Project (3GPP).

2 Use Cases and Requirements

The wireless industry, together with the industry verticals, has spent significant amount of effort in identifying use cases and requirements for 5G. URLLC has been one area that has attracted a lot of interest.

A few organizations outside 3GPP have established a wide set of use cases and requirements with the intention to influence the related standards bodies including 3GPP so that the identified use cases can be taken into account in standardization. Among some of the early efforts, NGMN (Next Generation Mobile Networks) Alliance included extreme real-time communications and ultra-reliable communications among other 5G use cases in their 5G white paper issued in 2015 [1]. The 5G Infrastructure Public Private Partnership (5G PPP), a joint initiative between the European Commission and European Information and communications technology industry (ICT manufacturers, telecommunications operators, service providers, small and medium-sized enterprises, and research institutions), issued a series of white papers [2–6] on 5G use cases in different sectors since 2015.

More recently, NGMN Alliance issued another white paper on URLLC use cases and requirements specifically [7]. Among all the verticals, Industrial IoT (Internet of Things), or industrial automation, and transport industry have been very active in recent years, with the involvement from the important stakeholders from the vertical domains. For example, 5G Alliance for Connected Industries and Automation (5G-ACIA) is developing requirements for Industrial IoT [8], with the vision of realizing the so-called Industry 4.0 on 5G. 5G Automotive Association (5GAA), on the other hand, focuses on the transport industry [9] and aims to develop end-to-end solutions for future mobility and transportation services.

From 3GPP perspective, the 5G/NR Rel-15 work started with identifying use cases and specifying the corresponding requirements in SA1 [10, 11], taking into account the input from various industry groups and the verticals. Later on, the use cases and requirements for V2X applications were developed and documented in [12], and those for cyber-physical control applications in vertical domains were developed and documented in [13].

Some of the typical examples of vertical domains and applications for URLLC are summarized in Table 13.1.

These applications typically have stringent latency and/or reliability requirements and may additionally have other requirements such as the data rate or packet size, the number of served UEs, clock synchronization, and security. Positioning accuracy is also important for some applications that require the location of the UEs. The performance requirements for some example typical use cases, specified in [12, 13], are provided in Tables 13.2, 13.3, and 13.4 for industrial automation, energy industry, and transport industry. In the remainder of this chapter, the physical layer design discussion mainly focuses on the latency and reliability aspects.

3 URLLC Support in NR Rel-15

Different from LTE, URLLC support in NR is not an add-on feature on top of an existing system design. It has been taken into account in the very first release of NR, Rel-15, and it is an integral part of basic NR design. This allowed a much better and cleaner system design because there was no legacy system that needed to be retained and accommodated, and the design could be done from scratch. With the target to accommodate a wide range of applications including URLLC with various performance requirements, NR Rel-15 provides a lot of flexibility and configurability that allows the service provider to choose based on the targeted services to provide.

The stringent requirements of URLLC use cases on latency and reliability create significant challenges for wireless cellular networks. The radio frequency (RF) condition of a UE fluctuates significantly and is typically not completely predictable due to interference variation, channel fading in wireless channels, and UE movements. These inherent characteristics of the wireless environment make it really challenging to guarantee very high reliability, especially when it comes together with the low-latency requirement. In some sense, latency and reliability are two conflicting requirements as improving one metric would typically negatively impact the other metric. In this section, we will discuss the enabling technologies in NR PHY and MAC design that allows the support of low latency and high reliability.

3.1 Support of Low Latency

Some URLLC use cases have very stringent latency requirements, on the order of a few milliseconds (or even less than 1 millisecond in some extreme cases), for end-to-end latency as shown in Sect. 2. Taking into account the delay caused by the other components of the end-to-end network, the latency budget for the air interface transmission can be very limited. The design target in NR Rel-15 URLLC was 1 ms for the air interface delay. As a reference, the typical delay that can be achieved

Table 13.1 Examples of vertical domains and applications for URLLC

Vertical domains	Applications	Brief description
Industrial/ factory automation	Motion control	Control of moving and/or rotating parts of machines in a well-defined manner, for example, in printing machines, machine tools, or packaging machines
	Mobile robots	A programmable machine that can follow the instruction to move within the industrial environment and complete a large variety of tasks such as transferring goods
	Process automation	Automation for (reactive) flows, e.g., refineries and water distribution networks
	Remote control	Remote operation of a UE, either by a human or a computer
Energy	Electric power distribution (smart grid)	Operations that improve efficiency, controllability, and predictability of the power network, e.g., power reliability and quality, grid resiliency, power usage optimization, and safety and security
	Central power generation	The centralized conversion of chemical energy and other forms of energy into electrical energy, including the planning and installation of respective equipment and plants as well as the operation, monitoring, and maintenance of these plants
Transport industry	Remote driving	Support of a remote driver or application to operate a remote vehicle for those passengers who cannot drive themselves or a remote vehicle located in dangerous environments
	Intelligent transport systems	Automation solutions for the infrastructure supporting street-based traffic. It addresses the connection of the road-side infrastructure (e.g., road-side units) with other infrastructure (e.g., a traffic guidance system)
	Advanced driving	Semi-automated or fully automated driving that requires vehicles and/or road-side units in proximity to exchange real-time information for safe driving, collision avoidance, and improved traffic efficiency
Entertainment	AR/VR	Augmented reality (AR) is an interactive experience where the objects in a real world are enhanced by computer-generated perceptual information. Virtual reality (VR) is a simulated experience that can be similar or completely different from the real world. AR and VR are widely used in video gaming in entertainment
	Online gaming	A video game that is either partially or primarily played through the Internet, which typically requires low latency for satisfactory user experience
Healthcare	Remote medical care	Automated monitoring of patients' health data and triggering of the critical real-time alarm in emergency scenarios
	Remote surgery	Remote surgery in mobile scenarios, disaster situations, or remote areas that requires precise control and feedback mechanisms with high requirements on both latency and reliability
Other	Remote computing	Remote computing (or cloud computing) refers to the on-demand availability of computer system resources such as data storage and computing power that are available over the Internet

Table 13.2 Example performance requirements for URLLC use cases in industrial and factory automation

	Characteristic parameter			Influence quantity							
Use case	Communication service availability: target value in %	Communication service reliability: mean time between failures	End-to-end latency: maximum	Message size [byte]	Transfer interval: lower bound	Target transfer interval	Transfer interval: upper bound	Survival time	UE speed	# of UEs	Service area (note)
Motion control #1	99,999–99,99,999	~10 years	< Transfer interval value	50	500 μs–500 ns		500 μs + 500 ns	500 μs	≤72 km/h	≤20	50 m × 10 m × 10 m
Motion control #2	99,9999–99,999,999	~10 years	< Transfer interval value	40	1 ms–500 ns		1 ms + 500 ns	1 ms	≤72 km/h	≤50	50 m × 10 m × 10 m
Motion control #3	99,9999–99,999,999	~10 years	< Transfer interval value	20	2 ms–500 ns		2 ms + 500 ns	2 ms	≤72 km/h	≤100	50 m × 10 m × 10 m
Mobile robots #1	>99,9999	~10 years	< Target transfer interval value	40–250	− < 25% of target transfer interval value	1–50 ms	+ < 25% of target transfer interval value	Target transfer interval value	≤50 km/h	≤100	≤1 km^2
Mobile robots #2	>99,9999	~1 year	< Target transfer interval value	15 k–250 k	− < 25% of target transfer interval value	10–100 ms	+ < 25% of target transfer interval value	Target transfer interval value	≤50 km/h	≤100	≤1 km^2
Process automa-tion #1	99,9999–99,999,999	≥1 year	< Target transfer interval value	20	−5% of target transfer interval value	≥10 ms	+5% of target transfer interval value	0	Typically stationary	Typically 10–20	Typically ≤100 m × 100 m × 50 m

Table 13.3 Example performance requirements for URLLC use cases in energy industry

Example use case	Characteristic parameter				Influence quantity				
	Communication service availability: target value in %	Communication service reliability: mean time between failures	End-to-end latency: maximum	Service bitrate: user experienced data rate	Message size [byte]	Transfer interval: target value	UE speed	# of UEs	Service area
Electric power distribution #1	99,999		~50 ms		~100	~50 ms		≤100, 000	Several km^2 up to 100,000 km^2
Electric power distribution #2	99,9999	–	<5 ms	1 kbit/s (steady state) 1, 5 Mbit/s (fault case)	<1500	<60 s (steady state) ≥1 ms (fault case)	Stationary	20	30 km × 20 km
Central power generation	99,9,999,999 (packet error rate < 10^{-9})	~10 years	16 ms				Stationary		Several km^2

Table 13.4 Example performance requirements for URLLC use cases in transport industry

Example use case	Payload (bytes)	Tx rate (message/sec)	Max end-to-end latency (ms)	Reliability (%)	Data rate (Mbps)	Min required communication range (meters)
Cooperative collision avoidance between UEs	2000	100	10	99.99	10	
Emergency trajectory alignment between UEs supporting V2X application	2000		3	99.999	30	500
Cooperative lane change between UEs	12,000		10	99.99		
Remote driving			5	99.999	UL: 25 DL: 1	

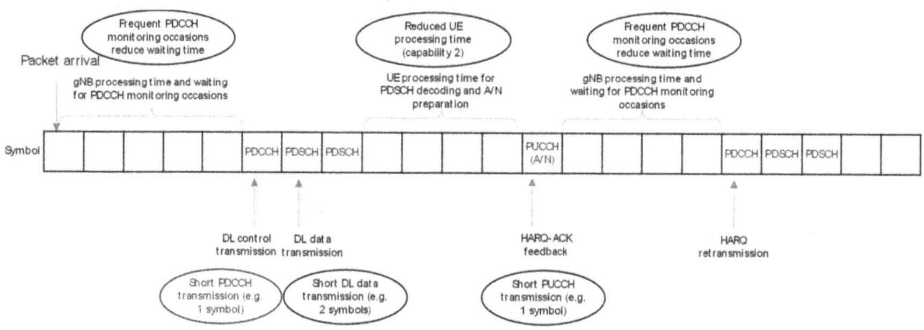

Fig. 13.1 Latency components of a DL transmission and the mechanisms to reduce the latency

in LTE is on the order of 4 ms without hybrid automatic repeat request (HARQ) retransmissions (before short Transmission Time Interval (TTI) and short processing time were introduced). To achieve such low-latency target in NR over the air, each delay component in the data delivery procedure needs to be optimized.

As a high-level overview, Fig. 13.1 illustrates the steps of a downlink data transmission, the delay components, and the corresponding mechanisms to reduce each of the delay components. Figures 13.2 and 13.3 provide illustrations for a grant-based and grant-free (also called configured grant in NR) uplink data transmission, respectively.

Here are some key features in NR Rel-15 design that enables low latency:

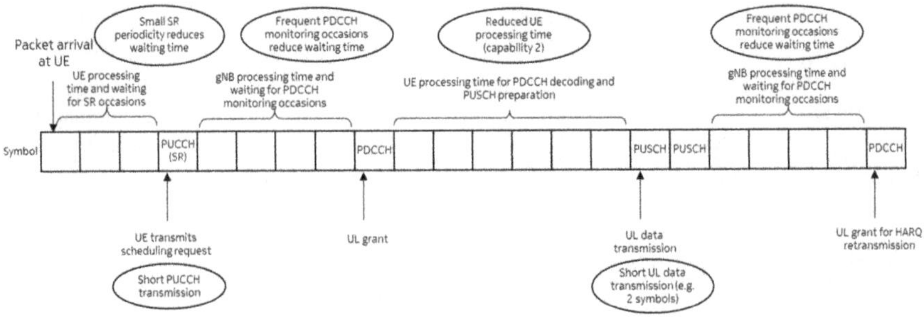

Fig. 13.2 Latency components of a grant-based UL transmission and the mechanisms to reduce the latency

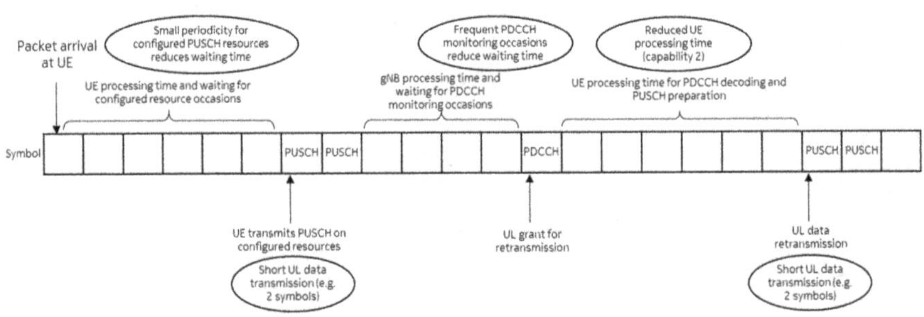

Fig. 13.3 Latency components of a grant-free UL transmission and the mechanisms to reduce the latency

- Scalable numerology, with shorter symbol duration for higher subcarrier spacing
- Short transmission duration for DL and UL control channels, as short as one symbol
- Short transmission duration for DL and UL data channels, as short as one or two symbols
- Multiple DL control monitoring occasions within a slot, which reduces the waiting time to schedule a DL or UL data transmission
- Grant-free UL data transmissions, which allow the UE to transmit data in configured resources without the need to send scheduling request and wait for UL grant
- Flexible Time Division Duplex (TDD) frame structure, which allows the gNB to flexibly change the DL and UL direction to accommodate the traffic need
- Optimized (i.e., significantly reduced) UE processing time for both DL and UL

Note that gNB processing time is also expected to be greatly reduced compared to LTE in order to support URLLC, even though it is not specified in standards.

Scalable Numerology Scalable numerology, one of the basic features for NR, provides substantial benefit for URLLC. With higher subcarrier spacing, the symbol duration, which is the basic time unit for all the transmissions, becomes shorter. With shorter symbol duration, a lot of delay components also scale down correspondingly.

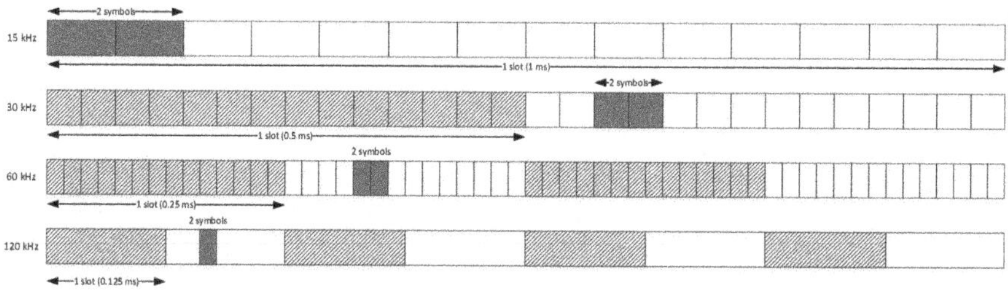

Fig. 13.4 Illustration of slot structure and two-symbol duration for different SCSs

Short/Flexible Transmission Duration for Control and Data Channel The actual transmission time of different channels is a significant component of the overall latency. In order to reduce the transmission time, NR supports the short transmission duration for all channels, including DL and UL control and data channels. This is achieved on one hand by using larger subcarrier spacing when possible and on the other hand by using small number of symbols for transmission.

At the same time, flexible transmission duration is supported which allows the gNB to adapt to different payload sizes and different RF conditions. For example, long transmission duration can be used for data channels for larger payload size. For UL, as the UE reaches the transmit power limit, longer duration can be used to achieve the intended coverage and/or reliability.

Figure 13.4 shows the slot structure for different SCSs and the corresponding two-symbol transmission duration. With all the flexibility, the gNB can choose the minimum transmission duration possible to accommodate the URLLC data packet. This provides significant latency reduction compared to slot-based scheduling with relatively small subcarrier space (SCS).

Frequent PDCCH Monitoring Occasions Physical Downlink Control Channel (PDCCH) carries the scheduling information for DL and UL data transmissions. How often a UE monitors the PDCCH determines the waiting time needed to schedule a UE. Traditionally, a UE does not monitor PDCCH too frequently to reduce power consumption, under the condition that latency is not as critical as in URLLC. To achieve the latency requirements of URLLC, the UE may be required to monitor PDCCH more often (e.g., on the sub-slot level). NR supports the PDCCH monitoring as frequent as every symbol to minimize the waiting time. Practically speaking, it should be avoided to set the monitoring occasions to be unnecessarily frequent because it is very power consuming for the UE.

Frequent Transmission Occasions for Scheduling Request (SR) For UL, in case grant-based transmission is used (with the procedure illustration in Fig. 13.2), the transmission occasions for SR determine how long a UE needs to wait before having the chance to transmit SR, which triggers the gNB to issue a UL grant. NR supports SR periodicity as low as 2 symbols.

Flexible Scheduling Timing Between PDCCH and PDSCH/PUSCH NR allows Physical Downlink Shared Channel (PDSCH) and Physical Uplink Shared Channel (PUSCH) to be scheduled with flexible scheduling delay in terms of both slot offset and the actual time domain resource within the slot. For example, the time domain resource indication is so flexible, and it can start at any symbol in a slot. This allows the gNB to schedule a packet as soon as the resource becomes available, which is especially important for TDD systems where DL and UL are time-division multiplexed.

Reduced UE (and gNB) Processing Time The processing time for different channels itself is one key component of latency, as shown in Figs. 13.1, 13.2, and 13.3. It is even more critical when hybrid automatic repeat request (HARQ) retransmission is considered. HARQ is very important for achieving high reliability while still providing good spectral efficiency. Without any HARQ retransmissions, the link adaptation has to be extremely conservative to ensure very high reliability level such as 10^{-5} or 10^{-6} with one-shot transmission. With HARQ retransmission(s), relatively higher modulation and coding scheme (MCS) can be scheduled for the first HARQ transmission(s), and more conservative MCS can be used when it is getting closer to the latency bound, which improves the overall spectral efficiency significantly. However, to allow HARQ retransmission(s) within the tight latency budget, short HARQ round-trip time (RTT) is required, which puts very stringent requirements on both UE and gNB processing time.

To allow fast processing, especially efficient pipelining processing, at the UE and gNB, many aspects had been considered, with a few highlights as follows:

- *Channel coding*: Low density parity check (LDPC) code, which was adopted in NR for PDSCH and PUSCH, allows efficient parallelization of decoding process to reduce the latency.
- *Front-loaded demodulation reference signal (DMRS)*: DMRS is used for channel estimation, which needs to be done before the demodulation of data symbols. Therefore, the location of the DMRS symbols is important for decoding latency. Unlike in LTE where DMRS symbols are always spread out in time, NR also supports front-loaded DMRS, meaning that DMRS is transmitted at the beginning of the transmission. This is illustrated in Fig. 13.5, where DMRS is carried in the first symbol of a 7-symbol data transmission. Front-loaded DMRS allows the channel estimation to be performed right after receiving the first symbol(s), and the data decoding can start immediately after channel estimation, without waiting for the later symbols in the transmission.
- *Data mapping*: In NR, when the data symbols are mapped to the resource elements, it follows frequency-first mapping, without any time domain interleaving. This is also illustrated in Fig. 13.5. With this kind of data mapping, the UE does not need to wait until receiving all the data symbols for de-interleaving. It allows symbol-by-symbol processing; that is, each data symbol can be processed once it is received. This is very critical for reducing the decoding time.

All these design considerations allow NR to significantly reduce the processing time at both the UE and gNB. From specification point of view, only UE processing

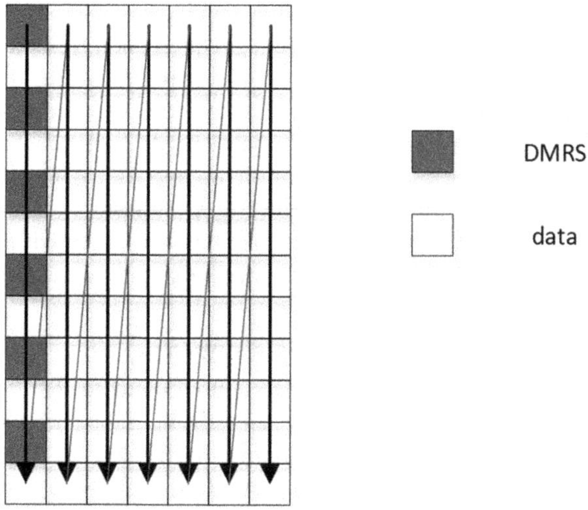

Fig. 13.5 Example of front-loaded DMRS and data mapping

Table 13.5 UE PDSCH and PUSCH processing time in NR

	PDSCH processing time capability 1		PDSCH processing time capability 2		PUSCH preparation time capability 1		PUSCH preparation time capability 2	
SCS (kHz)	Symbol	μs	Symbol	μs	Symbol	μs	Symbol	μs
15	8	571	3	214	10	713	5	357
30	10	357	4.5	161	12	428	5.5	196
60	17	303	9	161	23	410	11	196

time is defined, while gNB processing time is left to implementation. Table 13.5 provides a summary of UE processing time for PDSCH and PUSCH in NR. There are two capabilities defined, with capability 2 targeting for low-latency applications and capability 1 targeting for non-latency critical applications. Here, PDSCH processing time is defined as the minimum time needed from the end of the PDSCH to the beginning of HARQ-ACK transmission, which consists of PDCCH decoding time, PDSCH decoding time, and preparation time for HARQ-ACK transmission. PUSCH preparation time is defined as the minimum time needed from the end of PDCCH carrying UL grant to the beginning of PUSCH transmission, which consists of PDCCH decoding time and preparation time for PUSCH transmission. As can be seen, even the baseline capability 1 already provides significantly smaller processing time compared to LTE without short TTI (~3 ms). Capability 2 requires more optimization at the UE implementation to further reduce the processing time compared to capability 1.

With the defined capability 2, the RTT can be less than 1 ms for 15 kHz SCS and less than 0.5 ms for 30 kHz SCS, assuming the gNB processing time is roughly the same as UE processing time. This is significantly shorter than the 8 ms RTT in LTE without short TTI.

Grant-Free UL Transmissions With the regular grant-based UL transmission, the UE needs to transmit a scheduling request upon UL data arrival and wait for the gNB to issue a UL grant for data transmission. This extra step of handshaking between UE and gNB creates additional delay. Even with the greatly reduced timeline in NR, this extra step can still take a non-negligible portion of the very tight latency budget (e.g., 1 ms or less) and may push the overall latency beyond the budget. Grant-free (also called configured grant in NR) UL transmission allows the UE to skip this step and transmit directly on the configured resource without waiting for the dynamic UL grant. This becomes necessary for some use cases in order to satisfy the latency requirement.

Flexible Frame Structure for TDD Low latency of URLLC is especially challenging for TDD. In TDD, DL and UL need to share the channel in time domain, which naturally incurs more delay in the scheduling and transmission. If a semi-static TDD pattern is configured, the DL and UL directions of a slot or symbol cannot be changed, which prevents the gNB, for example, from using the DL resources for UL transmissions even if there is no (or less important) DL traffic at the time. With flexible frame structure, also called dynamic TDD, the gNB can dynamically determine the DL or UL direction of a slot or a symbol based on the traffic to be transmitted. This is achieved by using slot format indicator (SFI) carried in DCI format 2_0, which can be configured to a UE to monitor and provides the slot format (in terms of which symbols are DL or UL) for one or more upcoming slots.

3.2 *Support of High Reliability*

Traditionally, HARQ has been a very effective and also an efficient way in wireless systems to achieve low block error rate (BLER), or high reliability, after multiple HARQ transmissions. For URLLC applications, when the high reliability and the low-latency requirements need to be satisfied at the same time, HARQ may not always be so useful for high reliability because there may be very limited opportunities for HARQ retransmissions within the latency budget. In extreme cases where no HARQ retransmission can fit in the latency budget, a packet needs to be transmitted with a single-shot transmission with very high reliability requirement (e.g., 10^{-5} or even lower). Therefore, each channel in the system should be designed to be able to achieve high reliability target in order to accommodate these use cases.

In NR, there are some basic features that are useful for achieving high reliability, such as:

- *Channel coding design*: In the process of the channel coding design, efficient HARQ support has been one consideration, which is mainly to improve the efficiency. At the same time, low error floor was another consideration factor, which allows very low BLER to be achievable.

- *Time diversity*: This can be achieved by HARQ retransmissions if the latency budget allows, but for URLLC, the time diversity gain is typically very limited because the HARQ transmissions cannot span over a long duration.
- *Frequency diversity*: This can be achieved by distributed frequency resource allocation (for data and control channel in case of CP-OFDM) or frequency hopping in case of contiguous frequency resource allocation, e.g., for PUSCH and Physical Uplink Control Channel (PUCCH).
- *Spatial diversity*: This can be achieved by spatial diversity transmission schemes (e.g., precoder cycling) in Rel-15.
- *High aggregation level for PDCCH*: For PDCCH, the maximum aggregation level of 16 is supported, compared to the maximum aggregation level of 8 in LTE. Higher aggregation level reduces the effective code rate, which allows higher reliability.
- *Slot-based repetitions*: Repetition is a common approach to improve the coverage and reliability. Slot-based repetition (transmission in several consecutive slots) is supported for PDSCH, PUSCH, and PUCCH. From reliability perspective, slot-based repetition for data channels is especially useful when the latency budget does not allow the transmitter to wait for the feedback to initiate HARQ retransmission. This is sometimes called blind repetition.

In addition, some enhancements have been introduced particularly to support the high reliability requirements from URLLC.

- *Channel State Information (CSI) reporting enhancements*: Traditionally for non-URLLC traffic, the BLER target for CSI reporting is 10%. For URLLC, the initial BLER target is likely to be significantly lower than 10%. If the UE only reports CSI with 10% BLER target, it is quite difficult for the gNB to accurately determine the corresponding MCS for a lower BLER target. Therefore, a new BLER target of 10^{-5} for CSI reporting has been introduced and can be configured for a CSI report. Given that lower BLER target naturally means lower spectral efficiency, a new channel quality indicator (CQI) table with lower spectral efficiency entries has been defined accordingly.
- *MCS table enhancements*: Tied together with the CSI reporting enhancements to support lower BLER, a new MCS table with lower spectral efficiency entries was introduced. The lowest spectral efficiency supported becomes 0.0586 bps/Hz instead of 0.2344 bps/Hz in regular MCS tables.
- *PDCP packet duplication*: Reliability can be additionally improved by enabling higher layer packet duplication. In NR, this is done by Packet Data Convergence Protocol (PDCP) packet duplication, which enables a packet to get transmitted with two independent radio paths (e.g., in two different carriers) over the air interface.

3.3 DL Pre-emption

The features we have discussed so far consider the performance from a single UE perspective, without taking into account how the traffic from other UEs may affect the latency and reliability performance. However, in a real network, it is typically expected that URLLC traffic coexists with non-URLLC traffic due to, for example, UE traffic characteristics, limited spectrum resources, or more efficient resource utilization. When different types of traffic coexist in the network, it needs to be considered how non-URLLC traffic would affect the performance of URLLC traffic and what would be the way to minimize the impact of non-URLLC traffic in order to guarantee the URLLC performance while not sacrificing the spectral efficiency unnecessarily.

To avoid URLLC performance being impacted by non-URLLC traffic, one implementation approach at the gNB is to partition the available resources into two dedicated portions, one for URLLC and one for non-URLLC. The drawback, however, is that the URLLC and non-URLLC traffic cannot dynamically share the resources, and the trunking efficiency is lost. Moreover, given the high performance requirements of URLLC, the resources allocated to URLLC traffic need to be budgeted for the worst case. This means that a significant part of the resources may remain unused through the time and the resource efficiency can be very low. Therefore, it would be desirable to have a mechanism that allows the dynamic resource sharing between URLLC and non-URLLC traffic while still guaranteeing the URLLC performance. For DL, this is achieved by DL pre-emption in Rel-15, while for UL, this is achieved by UL cancellation, which is introduced in Rel-16 and will be discussed in further detail later.

In case resources are dynamically shared among different traffic types, when URLLC traffic comes, there may not be sufficient resources available immediately or soon enough to serve it because the resources have already been allocated to other traffic. DL pre-emption means that the gNB uses the resources that had previously been allocated to other traffic to schedule the more urgent traffic. This is illustrated in Fig. 13.6, where PDSCH #2, a URLLC packet that comes later, is scheduled on some of the resources that were allocated to PDSCH #1.

The scheduling decision can be done by the gNB itself without notifying the UE. However, if the UE being pre-empted is not aware of the situation, it would still assume the received signals on the pre-empted resources are for itself and use them for decoding its PDSCH. This results in unsuccessful decoding and corrupted soft buffer, which may or may not be recovered even with HARQ retransmission. Therefore, in NR Rel-15, the DL pre-emption indication (called interrupted transmission indication in NR specifications) was introduced to notify the UEs about the time-frequency region where the DL transmission was pre-empted so that the pre-empted UE knows the signals received on these resources are not valid. If the pre-empted PDSCH has large transport block size, using code block group-based HARQ retransmission is especially useful for improving the efficiency because it can enable the retransmission of only the code block groups

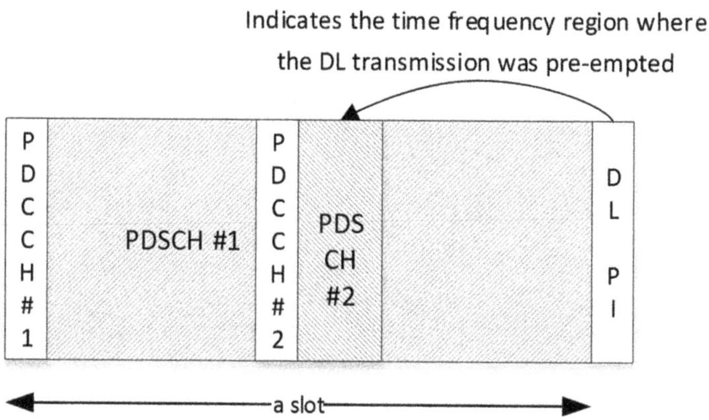

Fig. 13.6 DL pre-emption and pre-emption indication

in error (not the entire transport block) based on HARQ-ACK feedback. The pre-emption indication is carried by a group common PDCCH via DCI format 2_1 in the common search space, and it can be configured such that one message can reach a group of UEs.

One design issue was how often the UE should monitor for the pre-emption indication. The ideal scenario is that the pre-emption indication is sent before or at the time the pre-emption occurs, e.g., sent together with the PDCCH for the pre-empting PDSCH. This allows the pre-empted UE to receive the information earlier so that it can be taken into account in the decoding process earlier. However, the pre-emption typically happens when the PDCCH monitoring occasions are configured on sub-slot level for PDSCH URLLC, because otherwise both URLLC and non-URLLC traffic can be considered in scheduling at the beginning of a slot and no pre-emption would be necessary at all. This means that the pre-empted UE would need to monitor the pre-emption indication on sub-slot level also if we want to deliver it to the UE before or at the time the pre-emption occurs. This frequent PDCCH monitoring is very power consuming for a UE and is highly undesirable especially for a non-URLLC UE. So in NR Rel-15, the pre-emption indication is transmitted in the next monitoring occasion after the pre-emption occurs, e.g., at the start of the next slot. This typically overlaps with the monitoring occasions for the UE's own traffic and does not increase the monitoring frequency of the UE.

Each DL pre-emption indication consists of a 14-bit bitmap. Each bit corresponds to one partition of a reference time-frequency region and indicates whether the corresponding partition is pre-empted or not. The reference region in time covers the symbols from the previous monitoring occasion to the current one, and the reference region in frequency covers the entire DL bandwidth part. Two patterns are specified to define how the reference time-frequency region is divided into 14 partitions, as shown in Fig. 13.7, assuming one-slot periodicity for the monitoring occasions. It can be seen that the granularity is relatively coarse, especially on the frequency domain. This is to avoid too much signaling overhead for pre-emption indication.

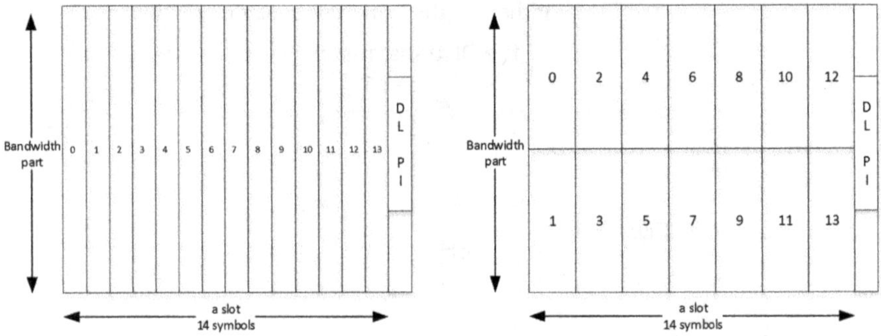

Fig. 13.7 Two patterns for the partitioning of the reference time-frequency region

4 URLLC Support in NR Rel-16

While basic support for URLLC has been specified in NR Rel-15, further enhancements for URLLC were pursued in Rel-16 to improve the URLLC performance from various aspects, with two study items [14, 15] followed by two work items [16, 17]. The enhancements were targeted to further improve the performance for Rel-15-enabled use cases such as AR/VR and effectively support new use cases such as factory automation, transport industry, and electrical power distribution. The outcome of the study items was documented in the technical reports [18, 19], which also include the performance evaluation for different use cases.

While the design targets in Rel-15 were 1 ms latency and 10^{-5} reliability, the design targets in Rel-16 were tightened to 0.5 ms latency and 10^{-6} reliability. At the same time, the enhancements cover both single-UE performance and inter-UE handling for improved system performance. All the channels, including PDCCH, PUCCH, PDSCH, and PUSCH, have been enhanced in certain ways to improve latency and/or reliability or system efficiency. Table 13.6 summarizes the enhancements with respect to the following:

- Which channels are addressed
- Whether the design was mainly targeted for URLLC-only traffic or mixed eMBB (enhanced Mobile Broadband) and URLLC traffic
  - In case of mixed traffic, whether it is mixed traffic within a UE (intra-UE) or across different UEs (inter-UE)
- Whether the design was mainly targeted to improve latency, reliability, and/or efficiency

Note that Table 13.6 shows only the main design target. A feature designed for one purpose does not prevent it from being used for another purpose when applicable. For example, features for URLLC traffic only can certainly be used in case of mixed eMBB and URLLC traffic. Moreover, a feature targeted at improving

Table 13.6 Summary of URLLC enhancements in NR Rel-16

Feature	PDCCH	PUCCH	PDSCH	PUSCH	URLLC only	Mixed eMBB/URLLC	Intra-UE	Inter-UE	Latency	Reliability	Efficiency
New DCI formats	√				√					√	
Enhanced PDCCH monitoring	√				√				√		
Sub-slot-based HARQ-ACK feedback		√			√				√		
PUSCH repetition type B				√	√				√	√	
Enhanced configured grant				√	√				√		
Enhanced SPS			√		√						√
Intra-UE prioritization/multiplexing		√		√		√	√			√	
Inter-UE UL cancellation				√		√		√	√		
Enhanced TPC				√		√	√	√		√	

one aspect can have impact on other aspect(s) as well. Latency, reliability, and efficiency are inter-connected performance metrics. Improving one performance metric typically has impact on the other performance metrics.

In the subsections, we will discuss each of the enhancement features in more detail.

4.1 New DCI Formats

Two new DCI formats, one DL DCI format (format 1_2) and one UL DCI format (format 0_2), are introduced in Rel-16. Compared to DCI formats 0_1/1_1, the main difference is that many fields in DCI format 0_2/1_2 now have configurable field sizes. This allows the gNB to configure smaller sizes for some fields that leads to smaller DCI size overall, which can improve the DCI reliability for a UE with poor RF condition. A smaller DCI can also improve the efficiency in PDCCH transmission by possibly allowing a lower aggregation level and reduce the PDCCH blocking probability.

Some examples of configurable DCI field sizes include:

- Frequency domain resource assignment field
- Redundancy version field
- HARQ process number field
- PUCCH resource indicator field
- Antenna port field

The new DCI formats provide a tool to significantly reduce the DCI payload size if needed (e.g., >16 bits reduction). The drawback is that with smaller field sizes, the scheduling flexibility is potentially reduced, which could impact the scheduling efficiency. Therefore, it would typically be used when either a UE is in poor RF condition or PDCCH is too congested to schedule the UEs in time.

The new DCI formats also differ from DCI format 0_1/1_1 in the sense that the new DCI formats do not support two transport blocks for DL or code block group-based transmission. The limitation was introduced because the new formats were designed targeting URLLC traffic, and these operation modes were not considered as important for URLLC.

4.2 Enhanced PDCCH Monitoring Capability

In Rel-15, to manage the UE implementation complexity and cost, the maximum number of control channel elements (CCEs) for channel estimation and the maximum number of PDCCH candidates are defined per slot for PDCCH monitoring, as shown in Table 13.7. These may be sufficient for non-URLLC traffic, where only one monitoring occasion at the beginning of a slot is needed. For URLLC

Table 13.7 Maximum number of CCEs and candidates for PDCCH monitoring in NR Rel-15

SCS (kHz)	15	30	60	120
Max # of CCEs per slot	56	56	48	32
Max # of PDCCH candidates per slot	44	36	22	20

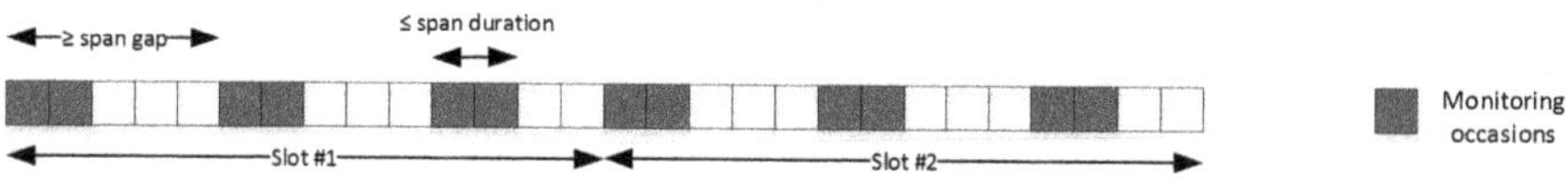

Fig. 13.8 Span pattern (span gap and span duration) in NR Rel-16

traffic with very stringent delay requirement, multiple monitoring occasions in a slot may be needed in order to minimize the waiting time for DCI scheduling. In this case, the maximum number of CCEs and PDCCH candidates needs to be distributed in these multiple monitoring occasions. For example, if we assume four monitoring occasions per slot, only up to 14 CCEs and 11 PDCCH candidates can be supported per monitoring occasion for 15 kHz. This becomes very limited and can cause significant PDCCH blocking. It is especially true for the maximum number of CCEs, considering that even a single blind decode candidate with aggregation level 16 would require 16 CCEs. Therefore, it was seen necessary to increase these numbers.

If we continue to use Rel-15 framework to define the maximum numbers on a per-slot basis, there is no restriction on how these numbers are distributed within a slot. This would require the UE to dimension the implementation based on the worst case assuming most of the candidates may occur within one monitoring occasion. This would certainly be very demanding for the UE implementation. On the other hand, in a practical configuration, it is typical that multiple monitoring occasions are configured to be more or less evenly spread out in a slot, and the PDCCH candidates are evenly distributed in these monitoring occasions. Therefore, it makes sense to define the limits on a so-called "per-span" basis, where the span is essentially the time duration from one monitoring occasion to the next. This could allow the UE to more effectively dimension the resource needed for PDCCH and allow efficient pipelining design to reuse resources for different spans.

With these considerations, it has been agreed to introduce a Rel-16 PDCCH monitoring capability that defines the maximum number of CCEs and PDCCH candidates per span for 15 kHz and 30 kHz SCSs because multiple monitoring occasions within a slot are more critical for smaller SCS with longer slot duration. Figure 13.8 illustrates how a span pattern is defined. A span pattern is defined as a pair of values (X, Y), where X is the span gap and Y is the span duration. Three pairs of values are supported: (2, 2), (4, 3), and (7, 3).

The exact values for the maximum number of CCEs and PDCCH candidates per span are still being discussed at the time of writing, and it is expected that these values will be finalized very soon.

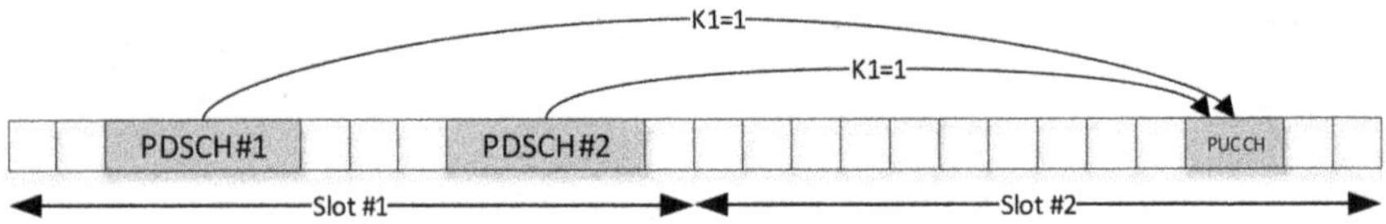

Fig. 13.9 Slot-based HARQ-ACK feedback in NR Rel-15

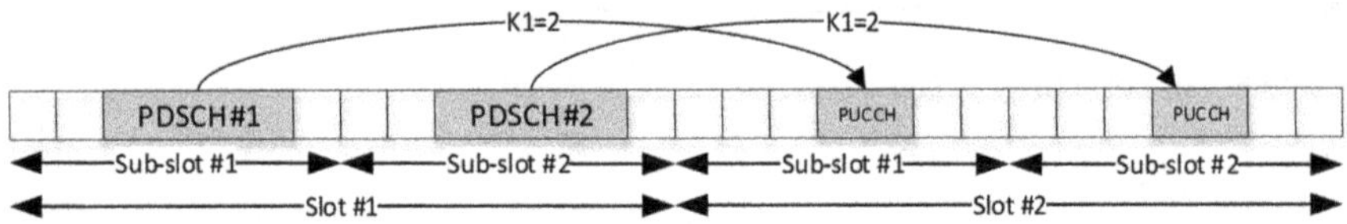

Fig. 13.10 Sub-slot-based HARQ-ACK feedback in NR Rel-16 with a sub-slot length of seven symbols

4.3 Sub-slot-Based HARQ-ACK Feedback

In Rel-15, there can be at most one PUCCH carrying HARQ-ACK in a slot. If there are multiple PDSCHs pointing to PUCCH resources in the same slot, the HARQ-ACK bits for these PDSCHs are multiplexed together and transmitted on a PUCCH resource (which is determined based on clearly specified rules). This is shown in Fig. 13.9. Since only one PUCCH carrying HARQ-ACK can be transmitted in a slot, the HARQ-ACK feedback for the earlier PDSCH may need to be delayed to the later part of a slot. This issue is addressed in Rel-16 by defining sub-slot-based HARQ-ACK feedback, which follows the same principle as Rel-15 but replacing slot with sub-slot. This is illustrated in Fig. 13.10, where the sub-slot length is seven symbols (i.e., there are two sub-slots in a slot). In this case, the PDSCH-to-HARQ timing *K1* is indicated in the unit of sub-slot, and there can be one PUCCH carrying HARQ-ACK within each sub-slot. The HARQ-ACK feedback for the first PDSCH can now be transmitted in the earlier part of a slot. This gives shorter RTT and can potentially allow more HARQ retransmissions within the tight latency bound. In Rel-16, two sub-slot lengths, 2-symbol and 7-symbol, are supported.

4.4 PUSCH Repetition Type B

Slot aggregation was specified in Rel-15, where one PDSCH or one PUSCH is transmitted over multiple slots to improve the reliability. In Rel-16, a new repetition Type B is introduced, which can be used for both dynamic PUSCH and configured grant PUSCH. It differs from slot aggregation in the following ways:

- The time domain resources for the multiple repetitions are back to back in time.

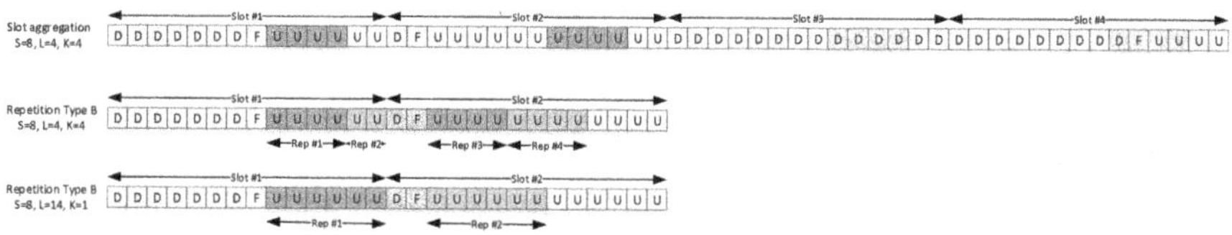

Fig. 13.11 Comparison between PUSCH repetition Type B in Rel-16 and slot aggregation in Rel-15

- The allocated time resources for each "nominal" repetition can span across DL/invalid UL symbols and/or slot boundary, in which case the nominal repetition is segmented into one or multiple actual repetitions around DL/invalid UL symbols and/or slot boundary.

The main advantage of PUSCH repetition Type B is that one UL grant can be used to schedule resources for PUSCH even if the scheduled resources span across slot boundary or DL symbols. This allows the maximum utilization of all the available UL symbols, and it is beneficial for both latency and reliability. For example, if the data comes late and can only be scheduled toward the end of a slot, without PUSCH repetition Type B, the gNB can choose to schedule the PUSCH immediately but with limited number of symbols for transmission because the transmission needs to stop at the end of the slot. Note that in this case, the gNB cannot schedule an HARQ retransmission until the HARQ-ACK feedback is received for the initial transmission. So both latency and reliability are affected. Alternatively, the gNB can wait until the next slot to schedule more symbols for transmission, which incurs more scheduling delay.

There were some other factors that motivated PUSCH repetition Type B at the early stage, but the benefit did not materialize eventually. For example, frequency diversity was considered as one of the potential advantages. But it was agreed that only a maximum of two hops is supported for PUSCH repetition Type B, so it does not provide any additional frequency diversity compared to Rel-15 schemes. Beam diversity (using different beams in different repetitions) was also one of the motivating factors. The supporting mechanism was not specified either, and there was concern on the overhead associated with beam switching between repetitions.

Both slot aggregation and PUSCH repetition Type B are illustrated in Fig. 13.11. With slot aggregation, the same time domain resources are used in a certain number of consecutive slots. If any of the symbols in a slot conflict with, for example, a DL symbol, the corresponding repetition is not transmitted. For repetition Type B, the nominal repetitions are back to back in time. If any nominal repetition conflicts with semi-static DL symbol(s), it is segmented around the semi-static DL symbol(s), resulting in one or more actual repetitions. There are additional rules that define whether an actual repetition is transmitted or not, depending on whether it is dynamic grant or configured grant, whether dynamic slot format indictor (SFI) is configured or not, and what is indicated by the dynamic SFI. These rules basically follow Rel-15 principles.

Invalid symbol pattern is also introduced for PUSCH repetition Type B, which indicates the symbols that shall not be used for PUSCH repetition Type B transmission. These symbols are treated in the same way as semi-static DL symbols, and a nominal repetition is segmented around these invalid symbols. The invalid symbol pattern can be used to avoid the collision with Sounding Reference Signal (SRS) or PUCCH (e.g., in a UL symbol at the end of a slot) or to reserve a DL to UL switching gap for the UE.

For simplicity, the transport block size of a PUSCH repetition Type B is determined based on the duration of a nominal repetition. Similarly, the transmit power control is also performed based on the duration of a nominal repetition.

4.5 Enhanced Configured Grant and Enhanced SPS

Semi-persistent scheduling (SPS) in DL allows the configuration of periodic PDSCH resources for a UE so that PDSCH can be transmitted periodically without requiring the associated DL assignment on PDCCH. This is very beneficial for the support of periodic traffic with relatively stable payload size, as it can greatly reduce PDCCH overhead. Configured grant (CG) is the counterpart in UL, which configures periodic PUSCH resources for a UE. Other than the PDCCH overhead saving, another important benefit of CG in the context of URLLC is that it allows the UE to skip the steps of transmitting SR and waiting for UL grant before transmitting PUSCH. This saving in delay can be critical for the most delay-sensitive applications. Therefore, DL SPS is typically used for periodic traffic only, while UL CG can be used either for periodic traffic or for aperiodic traffic for the purpose of latency reduction. Some enhancements have been introduced in Rel-16 in order to provide better support for different applications, including more periodicities to support a larger range of traffic characteristics and multiple configurations to support multiple data flows within a UE.

More Periodicities for CG and SPS To support different URLLC applications that may have various traffic characteristics, more periodicities have been introduced in Rel-16 for both CG and SPS. For SPS, the minimum periodicity is reduced from 10 ms in Rel-15 to 1 slot in Rel-16. Finer granularities are also supported for the periodicities for both CG and SPS.

Multiple CG Configurations and Multiple SPS Configurations In Rel-15, only one CG configuration is supported for UL per bandwidth part, and only one SPS configuration is supported for DL per cell group. Considering that some applications may have multiple service flows with different traffic characteristics (e.g., periodicity, latency requirements, payload size, etc.), a single CG configuration or a single SPS configuration would not be sufficient to support such use cases. Therefore, the support of multiple CG configurations and multiple SPS configurations is introduced in Rel-16, where up to 12 CG configurations and up to eight SPS configurations can be supported per bandwidth part.

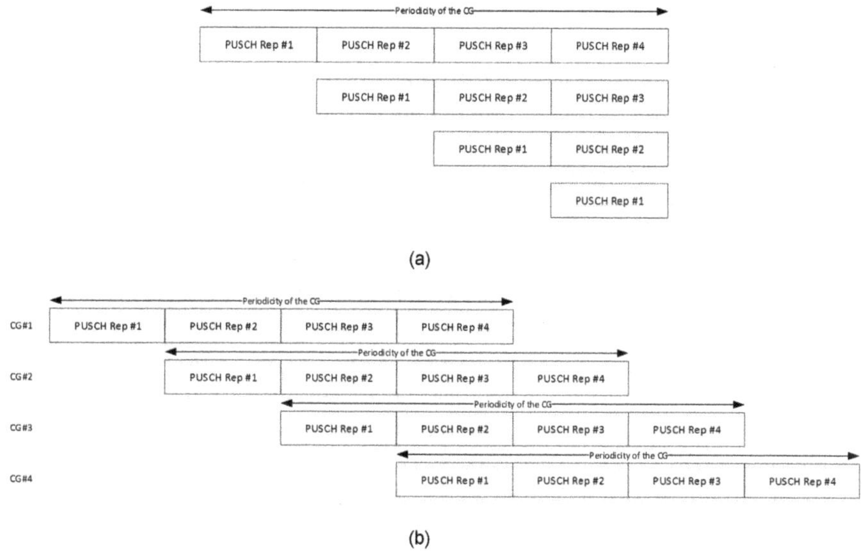

Fig. 13.12 Use of multiple CG configurations to reduce latency without affecting reliability in Rel-16. (**a**) Rel-15: a single CG configuration. (**b**) Rel-16: four CG configurations with staggered offsets

As mentioned earlier, CG is one critical feature for achieving low latency for UL. Rel-15 mechanism is limiting in the sense that a CG transmission cannot go across the periodicity boundary. If a CG needs to be configured with short periodicity (e.g., two symbols) to guarantee the latency performance, it also restricts how long the CG PUSCH transmission can be. As an example, Fig. 13.12 shows a case where four PUSCH repetitions are necessary to achieve the required reliability. In Rel-15, the minimum periodicity that can be used for the CG is the transmission duration of the four repetitions, as shown in Fig. 13.12a. Even though PUSCH transmission is still allowed to start at a later time within the periodicity (in case it corresponds to RV0) if the data comes late, it cannot be transmitted beyond the periodicity boundary. Therefore, if it starts at a later time, less repetitions can be transmitted, which means reduced reliability. Alternatively, if four repetitions are required, the UE has to wait until the next periodicity boundary to start the transmission, which introduces additional latency.

With the support of multiple CGs, this issue can be avoided by using multiple CGs with staggered offsets, as shown in Fig. 13.12b. In this example, four CGs are configured, and each starts at a different offset. Depending on when the data comes, the UE can choose one of the four CGs that has the closest starting position to transmit, and the full four repetitions can be transmitted. Both latency and reliability can be guaranteed.

To support multiple CGs and multiple SPSs, the activation and release DCIs for SPS and CG have been enhanced. In addition, for the support of multiple SPS configurations:

- The behavior in case of SPS conflict is defined, in which case the UE only decodes the one with lowest SPS configuration index. This clearly defined behavior allows the gNB and the UE to have common understanding on which one is supposed to be used for transmission and reception.
- The HARQ-ACK feedback is enhanced for both Type 1 and Type 2 HARQ-ACK codebooks to accommodate multiple SPS configurations.

4.6 Intra-UE Prioritization and Multiplexing

In Rel-15, for a UE with a mix of URLLC and non-URLLC traffic, there is no differentiation of traffic types in physical layer, and all the traffic is handled in the same manner transparent to what type of traffic is carried. With the stringent requirement of URLLC, it was seen necessary to provide some preferential handling for URLLC traffic so that it can be handled with higher priority compared to other traffic.

This requires physical layer to introduce the concept of priority. It is mainly for UL transmissions, including PUSCH data transmission, HARQ-ACK, SR, and CSI. The priority is determined as follows:

- PUSCH priority is indicated explicitly via a priority indicator field in UL grant for dynamic PUSCH. For CG PUSCH, the PUSCH priority is configured via an RRC parameter in the corresponding CG configuration.
  - There had been some debate on how to determine the PUSCH priority. An alternative approach is to use the priority level from MAC layer for the data mapped to a UL grant. Note that the priority level is configured for each logical channel in MAC. This approach provides accurate information on the priority of the data that is actually carried on PUSCH. The drawback, however, is that the gNB would not know the priority of the PUSCH and would not be able to predict the UE behavior in terms of prioritization and multiplexing in physical layer. Due to this, the gNB may need to perform more hypotheses to determine what has been transmitted by the UE. Explicit indication of PUSCH priority in UL grant removes this ambiguity and can simplify the UE implementation because the UE does not need to look into MAC priority level to decide the PHY behavior.
- For HARQ-ACK, the priority indication is indicated explicitly via a priority indicator field in DL assignment for dynamic PDSCH. For SPS PDSCH, the HARQ-ACK priority is configured via an RRC parameter in the corresponding SPS configuration.
- SR priority is configured via an RRC parameter in the corresponding SR resource configuration.
- CSI on PUCCH is always considered as low priority, given that CSI is not considered so critical in terms of either latency or reliability.

- CSI triggered on PUSCH is treated in the same way as other PUSCH in order to keep unified behavior. In addition, it allows the gNB to trigger high priority CSI report if necessary.

With the PHY priority determined, the next step is to define how a UE treats different priority levels. Only two PHY priority levels are defined in Rel-16 in order not to complicate UE behavior and system design too much. With the limited time in Rel-16, the UE behavior was also significantly simplified to only support prioritization between channels with different priorities, while further enhancements such as multiplexing between different priorities are left to Rel-17. To summarize, on the high level the UE prioritization and multiplexing behaviors in case of overlapping channels are defined as follows:

- Step 1: Within each priority level, the UE follows Rel-15 behavior for prioritization and multiplexing of the overlapping channels.
- Step 2: If a high priority channel conflicts with a low priority channel in time, high priority channel is transmitted and low priority channel is cancelled or stopped in order to transmit the high priority channel.

It should be noted that in Rel-16, overlapping dynamic PDSCHs or overlapping dynamic PUSCHs are not supported due to no consensus on the exact mechanisms to support them.

4.7 *Inter-UE UL Cancellation*

As discussed in Sect. 3.3, DL pre-emption allows better dynamic resource sharing between URLLC and non-URLLC traffic in DL. Inter-UE UL cancellation can be considered as the UL counterpart of the DL pre-emption. This is illustrated in Fig. 13.13. It allows the gNB to schedule a more urgent URLLC packet using the resources that had been allocated to other UE(s) earlier (e.g., eMBB). Different from DL pre-emption which could be purely gNB implementation behavior, for UL, the gNB needs to notify the other UE(s) that they need to cancel the transmission(s) so that it does not interfere with the URLLC transmission.

The UL cancellation indication, which indicates the time-frequency resources for which the PUSCH transmission shall be cancelled, follows many design principles of DL pre-emption indication in Rel-15. It is carried over a group common PDCCH, namely, DCI format 2_4, so that the same information can be received by multiple UEs. The time-frequency resources to be cancelled are indicated via a two-dimensional bitmap. Different from DL pre-emption indication, the monitoring of UL cancellation indicator is required to support the sub-slot level configuration, similar to URLLC traffic. Otherwise, the UE would not have sufficient time to cancel the intended PUSCH transmission.

The timeline for UL cancelation follows the PUSCH preparation time for capability 2. This is basically following the processing time for URLLC UEs, even

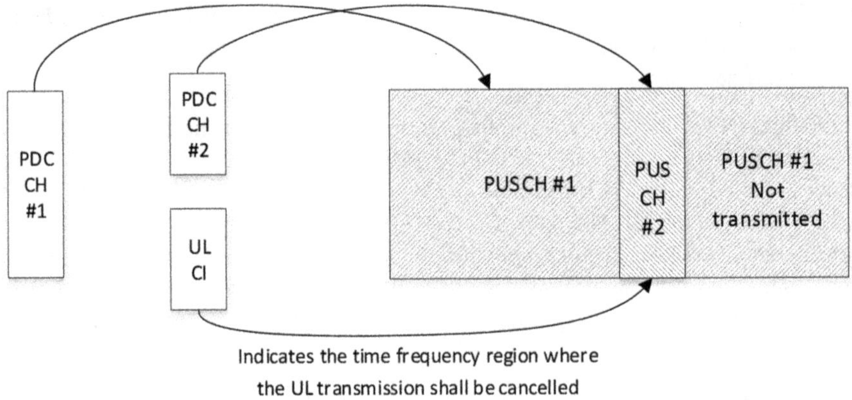

Fig. 13.13 Inter-UE UL cancellation

though the UL cancellation may be performed by UEs with PUSCH processing time capability 1. This is necessary because otherwise the scheduling delay for the URLLC traffic would be limited by this cancellation timeline. Certainly, this puts additional requirements on timeline for non-URLLC UEs. But it may not be extremely challenging because UL cancellation involves cancelling a transmission only. Cancelling a transmission should naturally take less time than the regular PUSCH preparation time, which includes the coding and modulation of PUSCH transmission.

4.8 PUSCH Transmit Power Control Enhancements

The PUSCH transmit power control (TPC) enhancements address the issue of mixed traffic for both intra-UE and inter-UE cases.

For a UE supporting both URLLC and non-URLLC traffic, URLLC traffic has different latency and reliability requirements from non-URLLC traffic. This naturally results in different operating points (e.g., in terms of BLER target) for different types of traffic. Different TPC parameters may be necessary to achieve these different operating points. For example, the UE may need to boost the transmit power for URLLC compared to non-URLLC for the same MCS in order to reach a lower BLER.

For inter-UE cases, if the gNB intends to support the overlay transmissions of URLLC PUSCH and non-URLLC PUSCH for improved system capacity (i.e., the URLLC PUSCH and non-URLLC PUSCH overlap in the allocated time-frequency resources), it becomes necessary to further boost the transmit power of URLLC PUSCH compared to the case without overlay transmission.

To support these different cases, multiple open loop power control parameter sets are introduced in Rel-16, and dynamic indication (up to 2 bits) of the power control parameter set can be included in the UL grant. This allows the gNB to dynamically indicate the parameters based on the scheduling decision.

5 Future Work and Conclusions

Even though there have been many new URLLC features introduced in Rel-16, there is still room for further enhancements in many aspects.

A new Rel-17 work item ("New WID on enhanced Industrial Internet of Things (IoT) and URLLC support") [20] has been agreed in December 2019, which includes the following enhancements for physical layer:

- Enhancements for intra-UE multiplexing and prioritization
  - Rel-16 only supports prioritization between channels with different priorities. This will further investigate the possibility of allowing multiplexing of channels with different priorities.
- HARQ-ACK enhancements
  - There is no clear scope defined for this, but it will most likely touch the aspects of HARQ-ACK handling when conflicting with DL symbols, Type-1 HARQ-ACK codebook size reduction, etc.
- CSI feedback enhancements
  - There is no clear scope defined for this either. Aperiodic CSI on PUCCH is a potential topic, and other CSI feedback enhancements targeted specifically for URLLC will be further discussed.

The Rel-17 work item is a relatively small one due to the short timeframe. A lot of other interesting areas have been pushed out of Rel-16 and Rel-17 and may be worthwhile to further investigate in further releases, such as the following:

- Out-of-order scheduling and HARQ
- TPC enhancements for configured grant
- Further reduced UE processing time
- URLLC enhancements for FR2
- URLLC enhancements for unlicensed spectrum
- Time-sensitive network specific enhancements

To summarize, the NR specifications already provide a large set of tools that can be used to satisfy the requirements of different URLLC applications. At the same time, it is clear that the standards will continue to evolve in order to better support all different types of URLLC applications, to turn URLLC on 5G into reality.

References

1. NGMN Alliance, 5G White Paper, Feb 2015
2. 3GPPP, White paper on eHealth Vertical Sector, Oct 2015
3. 3GPPP, White paper on Factories-of-the-Future Vertical Sector, Oct 2015
4. 3GPPP, White paper on Energy Vertical Sector, Oct 2015
5. 3GPPP, White paper on Automotive Vertical Sector, Oct 2015
6. 3GPPP, White paper on Media & Entertainment Vertical Sector, Jan 2016
7. NGMN Alliance, Verticals URLLC use cases and requirements, July 2019
8. 5G-ACIA, 5G for connected industries and automation, Feb 2019
9. 5GAA, White Paper on C-V2X use cases: Methodology, examples and service level requirements, June 2019
10. 3GPP TR 22.804, Technical Specification Group Services and System Aspects; Study on Communication for Automation in Vertical Domains
11. 3GPP TS 22.261, Technical Specification Group Services and System Aspects; Service requirements for the 5G system
12. 3GPP TS 22.186, Technical Specification Group Services and System Aspects; Enhancement of 3GPP support for V2X scenarios
13. 3GPP TS 22.104, Technical Specification Group Services and System Aspects; Service requirements for cyber-physical control applications in vertical domains
14. RP-182089, New SID on Physical layer enhancements for NR Ultra-Reliable and Low Latency Communication (URLLC), Huawei, HiSilicon, RAN#81, Sept 2018
15. RP-182090, Revised SID: Study on NR Industrial Internet of Things (IoT), Nokia, Nokia Shanghai Bell, RAN#81, Dec 2018
16. RP-191584, Revised WID: Physical layer enhancements for NR Ultra-Reliable and Low Latency Communication (URLLC), Huawei, HiSilicon, RAN#84, June 2019
17. RP-192324, Revised WID: Support of NR Industrial Internet of Things (IoT), Nokia, Nokia Shanghai Bell, RAN#85, Sept 2019
18. 3GPP TR 38.824, Technical Specification Group Radio Access Network; Study on physical layer enhancements for NR ultra-reliable and low latency case (URLLC)
19. 3GPP TR 38.825, Technical Specification Group Radio Access Network; Study on NR Industrial Internet of Things (IoT)
20. RP-193233, New WID on enhanced Industrial Internet of Things (IoT) and URLLC support, Nokia, Nokia Shanghai Bell, RAN#86, Dec 2019

Chapter 14
5G New Radio in Unlicensed Spectrum

Reem Karaki

1 Background: Motivation

It is crucial to obtain more bandwidth to support the rapid increase in traffic and use cases of various types, especially high-quality video traffic. Usually, getting access to new licensed spectrum is expensive and takes long time to be granted. On the other hand, unlicensed spectrum is open for use by any system as long as the regional regulations are not violated. Not only that but also worldwide availability in large amount makes it very attractive for cellular operators. Generally, it is more challenging to meet high reliability and coverage requirements when operating on unlicensed spectrum since exclusive access rights cannot be granted. However, it is still considered as an important tool to meet the increasing customer demands.

Operating in unlicensed spectrum, specifically 5GHz spectrum range, was introduced for the first time in Rel-13 licensed-assisted access using carrier aggregation framework. The primary cell is always operated on a licensed carrier to guarantee certain reliability. Additionally, the network capacity can be boosted by aggregating multiple secondary unlicensed carriers with the primary licensed carrier.

Similarly, New Radio (NR) [1] operation in unlicensed spectrum includes the support of a license-assisted mode based on carrier aggregation with another NR licensed carrier or dual connectivity between licensed band LTE/NR and NR unlicensed carrier. In addition, it is also possible to operate the entire system on unlicensed spectrum. This mode of operation is refered to a standalone deployment in unlicensed spectrum. In fact, the latter is the most challenging deployment. Several fundamental new enhancements were introduced in Rel-16 to enable

R. Karaki (✉)
Ericsson Research – Radio Network Communication Standards, Ericsson GmbH, Herzogenrath, Germany
e-mail: reem.karaki@ericsson.com

X. Lin, N. Lee (eds.), *5G and Beyond*, https://doi.org/10.1007/978-3-030-58197-8_14

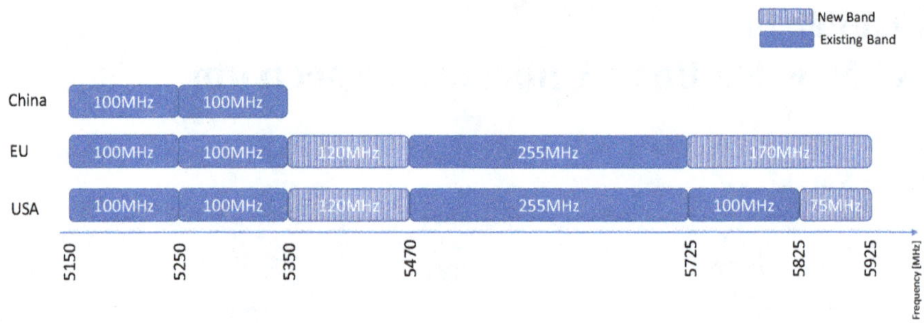

Fig. 14.1 Frequency allocation for 5 GHz in different regions

efficient standalone deployments in unlicensed spectrum. These are to be discussed in detail throughout this chapter.

Rel-16 NR-U focuses on the 5GHz unlicensed band and the 6GHz band under discussion for unlicensed use in both the USA and Europe (e.g., 5925–7125 MHz and 5925–6425 MHz, respectively) [2, 3]. Some bands are available worldwide and some bands are limited to certain regions as shown in Fig. 14.1. The European requirements on 5 GHz unlicensed deployment are specified in ETSI harmonized standards [4]. The use of unlicensed 5 GHz spectrum in the USA is governed by FCC part 15 regulations [5]. In principle, the European regulations are the most stringent.

NR-U is designed to support configurable functionalities that can fulfill the regulatory requirements of different regions and bands. Some of the functionalities are to be used worldwide, even though strictly speaking, they are not mandated by the regional regulations. Those requirements are specifically related to coexistence with other systems operating on the same band. At the moment, those technologies include Wi-Fi (equipment implementing the IEEE 802.11 Wireless Local Area Network standard) and LTE LAA. Among those regulations is listen-before-talk (LBT) mechanism (Sect. 2) which is mainly mandated in certain regions (e.g., Europe, Japan).

Last but not least, certain regulations required the support of new features in Rel-16 NR. Generally speaking, Rel-16 NR-U is Rel-15 NR with additional enhancements and functionality for operation in unlicensed spectrum. Those enhancements will be described in detail throughout this chapter.

2 Channel Access Mechanisms

The unlicensed spectrum can be used by any system if the regulatory requirements are fulfilled. These requirements may differ from one region to another. However, they are fundamental to govern the operation of different technologies operating on the same spectrum by mandating coexistence mechanisms that facilitate fair sharing of the unlicensed spectrum.

Listen before talk (LBT) is one of the most important requirements in some regions (e.g., Europe, Japan). Using this technique, the device can sense for presence of transmissions from other devices in the channel before performing the transmission. If the channel is found to be free, i.e., energy detected is below a certain threshold, the device can transmit for limited duration after which it has to sense the channel again. Otherwise, the device must restrain from transmitting. Such a mechanism facilitates fair sharing among different devices using the channel and more importantly limits the interference. NR-U supports two flavors of LBT, which are described in detail in Sects. 2.1 and 2.2.

2.1 Channel Access Procedures for Dynamic Channel Occupancy

The clear channel assessment in case of dynamic channel occupancy consists of a minimum sensing duration followed by a back-off phase. The parameters related to both phases depend on the type of traffic to be sent, which maps to a priority class, as shown in Table 14.1. The sensing duration during the back-off phase consists of N 9 μs observation slots (Tsl), where N is a random number between zero and contention window size (CW). This procedure, commonly known as exponential back-off, or Type 1 according to [6], is designed to randomize the start of transmissions from different nodes that want to access the channel at the same time.

If the device intends to transmit, it has first to sense the channel to be idle for a minimum sensing duration and then initiate the back-off phase. During the back-off phase, N is decremented after each idle Tsl. The slot is considered idle if the detected energy is lower than the energy detection (ED) threshold. If the detected energy within a Tsl is sensed to be above the ED threshold, the device suspends the back-off phase and continuously senses until the channel is free for a minimum sensing duration after which the back-off phase can be resumed again. Once the counter reaches zero, the node is allowed to initiate a transmit opportunity (txOP) and transmit for up to a maximum channel occupancy time (MCOT) (Fig. 14.2).

The sensing procedure considers the different latency and reliability requirements of different traffic types. Essentially, the sensing duration required for higher priority classes is smaller than that required for traffic with lower priority, e.g., the maximum N for VoIP traffic is 7, while for best effort, it can be as large as 1023 slots. Each traffic type maps to a priority class value, e.g., VoIP is highest priority class, while best effort is lowest priority class ($p = 4$). Each priority class value is characterized by (m_p, $CW_{\min,\ p}$, $CW_{\max,p}$, $T_{m\,\mathrm{cot},p}$), which is reflected in Table 14.1 for both downlink (DL) and uplink (UL) transmissions. The minimum sensing time is equal to 16+mp*Tsl slots. In the absence of other technologies, for certain

Table 14.1 Channel access priority class for both DL and UL [6]

Channel access priority class (p)	Downlink					Uplink				
	m_p	$CW_{min,p}$	$CW_{max,p}$	$T_{m\,cot,p}$	Allowed CW_p sizes	m_p	$CW_{min,p}$	$CW_{max,p}$	$T_{ulmcot,p}$	Allowed CW_p sizes
1	1	3	7	2 ms	{3, 7}	2	3	7	2 ms	{3, 7}
2	1	7	15	3 ms	{7, 15}	2	7	15	4 ms	{7, 15}
3	3	15	63	8 or 10 ms	{15, 31, 63}	3	15	1023	6 ms or 10 ms	{15, 31, 63, 127, 255, 511, 1023}
4	7	15	1023	8 or 10 ms	{15, 31, 63, 127, 255, 511, 1023}	7	15	1023	6 ms or 10 ms	{15, 31, 63, 127, 255, 511, 1023}

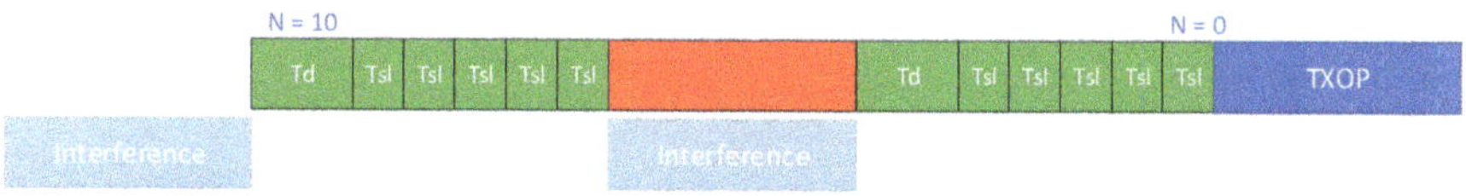

Fig. 14.2 Random back-off procedure for dynamic channel occupancy

priority class, MCOT can be extended to 10 ms. the minimum sensing time is equal to 16+mp*Tsl slots.

The default value of the contention window, CW, is CWmin. Before initiating a COT, the CW for every priority class should be updated. If the transmission corresponding to the latest COT is successful, the node will reset the CW to the minimum value. Otherwise, the node doubles the existing contention window size $(2 \times \mathrm{CW} + 1)$ for each occurrence of transmission problems until the maximum value CWmax is reached. In case the feedback for the latest COT is not yet available, the node can initiate a new COT without necessarily updating the CW, if the new COT starts within a certain small time interval from the start of the latest COT.

Typically, exponential back-off procedure is performed before initiating a COT. The only exception is discovery burst transmission, where the gNB is allowed to transmit Discovery Reference Signal (DRS) immediately after successfully sensing the channel for a fixed duration of 25 μs, as long as the DRS duty cycle is $\leq$1/20, and the total duration is up to 1 ms.

ED Threshold

ETSI EN 301 893 [4] regulates the maximum allowed ED threshold that can be used in Europe, thus controlling devices' aggressiveness in accessing the channel. Typically, for NR-U, the threshold used when preforming the channel assessment is -72 dBm but can be modified based on the transmit power of the node and/or the presence or absence of other technologies in the same deployment area. The network has more flexibility in selecting the energy detection threshold, while still fulfilling the regulatory requirements, if absence of other technologies with the deployment area can be guaranteed on a long term. If that cannot be guaranteed, the maximum energy detection threshold of $X_{\mathrm{Thresh\,max}}$ when gNB performs CCA is calculated as follows [6]:

$$X_{\mathrm{Thresh\,max}} = \max\left\{\begin{array}{c} -72 + 10\cdot \log 10\,(\mathrm{BW\ MHz}/20\ \mathrm{MHz})\ \mathrm{dBm}, \\ \min\left\{\begin{array}{c} T_{\max}, \\ T_{\max} - TA + (P_H + 10\cdot \log 10\,(\mathrm{BW\ MHz}/20\ \mathrm{MHz}) - P_{\mathrm{TX}}) \end{array}\right\} \end{array}\right\}$$

where:

Parameter	Value
T_A	10 dB *or* 5 dB for transmission(s) including PDSCH or discovery burst(s)
P_H	$23dBm$
P_{TX}	The maximum gNB output power in dBm for the channel
T_{max}(dBm)	$10 \cdot \log 10(3.16228 \cdot 10^{-8}(\text{mW/MHz}) \cdot \text{BW MHz (MHz)})$
BW MHz	Single channel bandwidth in MHz

RRC signaling has the ability to indicate to a UE the ED threshold that should be used. If no RRC signaling is received, the UE uses a default maximum energy detection threshold value that is derived using the same procedure to derive gNB's ED threshold.

COT Sharing

A channel occupancy, acquired by an initiating node (gNB/UE) performing clear channel assessment based on exponential random back-off, can be shared with other nodes, donated as responding nodes. Depending on the gap between initiating and responding nodes, the responding node might or might not be required to perform an LBT for a single observation duration, referred to as Type 2 LBT. NR-U supports two flavors of Type 2 LBT, 16 μs or 25 μs, as listed in Table 14.2.

In case the responding device proceeds without preforming any sensing, there is a possibility that the transmission of the responding device collides with an ongoing transmission from a device that is hidden from the initiating device. Therefore, an upper limit on the duration of the transmission of the responding device is introduced to limit the aggressiveness of nodes that transmit without preforming any sensing.

Table 14.2 LBT type and conditions for responding device transmissions in gNB-initiated COT

LBT type for responding device	Condition
Immediate transmission without sensing	Gap between end of the gNB's transmission and beginning of UE's transmission is not more than 16 μs, and UE's transmission duration does not exceed 0.584 ms
16 μs observation duration	Gap between end of the gNB's transmission and beginning of UE's transmission is exactly 16 μs
25 μs observation duration	Gap between end of the gNB's transmission and beginning of UE's transmission is equal to 25 μs or more than 100 μs

Table 14.3 Examples of gNB-initiated COT shared with one or more UEs

Example	LBT type for responding/initiating node within the gNB COT
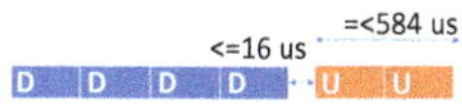	Immediate transmission without sensing
	Sensing for 16 μs immediately before the start of the UL transmission
	Sensing for 25 μs immediately before the start of the UL transmission
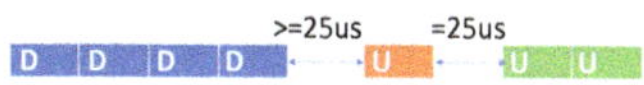	Sensing for 25 μs immediately before the start of the UL transmission by both first and second UEs
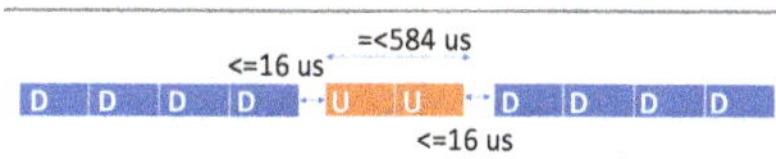	Immediate transmission without sensing
	Sending for 25 μs immediately before the start of the UL/DL transmission

gNB-Initiated COT

A gNB may share the channel with one or more UE according to the conditions listed in Table 14.2. Following the transmission of a UL burst from the UE, the gNB may continue to transmit within the COT initiated by the gNB. In this case, there should not be any gap larger than 25 μs between any two transmissions in the COT. If that is guaranteed, the gNB may proceed without performing any sensing when the gap from the end of the UL transmission to the beginning of the DL transmission is up to 16 μs. Otherwise, when the gap from the end of the UL transmission to the beginning of the DL transmission is larger than 16 μs but not more than 25 μs, sensing for 25 μs is applied by the gNB. Table 14.3 shows different examples of sharing gNB COT with one or more UEs.

UE-Initiated COT

The conditions for sharing a UE-initiated COT are more restrictive. In some cases, the gNB maximum transmit power might be higher than the one used by UEs. Even though this might not be a common case in unlicensed spectrum, it can be seen problematic. The ED threshold used by the UE depends on the maximum transmit power supported by the UE. If there is a power mismatch between the UE and gNB, the UE will initiate the channel using a higher ED threshold as compared to what the gNB might have used and give the remaining of the COT to a gNB that transmits using higher power and therefore interfere with a much wider coverage range than the one sensed by the UE. Nonetheless, this situation can be avoided by controlling

Table 14.4 LBT type and conditions for gNB's transmissions in UE-initiated COT

LBT type for responding node	Condition	ED threshold configured by the gNB	ED threshold not configured by the gNB
Immediate transmission without sensing	Gap between end of the UE's transmission and beginning of gNB's transmission is not more than 16 μs, and gNB's transmission duration does not exceed 0.584 ms	The gNB can transmit control/broadcast signals/channels for any UEs as long as the transmission contains transmissions for the UE that initiated the channel occupancy The transmission of the gNB shall not exceed 2/4/8 OFDM symbols in duration for 15/30/60 kHz subcarrier spacing	The gNB can transmit control/broadcast signals/channels for any UEs as long as the transmission contains transmissions for the UE that initiated the channel occupancy The gNB can also include unicast data to the UE that initiated the COT
16 μs observation duration	Gap between end of the UE's transmission and beginning of gNB's transmission is exactly 16 μs		
25 μs observation duration	Gap between end of the UE's transmission and beginning of gNB's transmission is exactly 25 μs		

the ED threshold used by the UE to match the one that would have been used by the gNB. NR-U supports configurability of UE's ED threshold. If the gNB does not configure the ED threshold, i.e., the ED threshold is derived by the UE depending on its maximum transmit power, the UE can still share its COT with the gNB; however, the DL part can only include control data that does not exceed few symbols. If the ED threshold is configured, the latter restriction is relaxed. In either case, the gap between any two transmissions in UE-initiated COT should not exceed 25 μs. A UE-initiated COT can be shared with gNB according to the conditions listed in Table 14.4.

Multichannel Operation

DL Multichannel Operation

There are two approaches for transmitting on multiple carriers in the DL. In the first approach, the gNB runs parallel independent back-off procedures on each of the channels. Multiple channels need to each have individually completed Type 1 LBT before transmitting simultaneously. This is equivalent to individually performing the single-carrier LBT procedure on each channel (Fig. 14.3).

In the second approach, the gNB runs a single random back-off procedure on one of the channels. The gNB chooses the carrier requiring full-fledged back-off procedure uniformly randomly before each transmission burst or fixes the carrier at least for 1 s. Upon the completion of the back-off procedure, a single 25 μs CCA check is required on the other channels immediately before the transmission starts. The gNB transmits only on the channel that is not busy (Fig. 14.4).

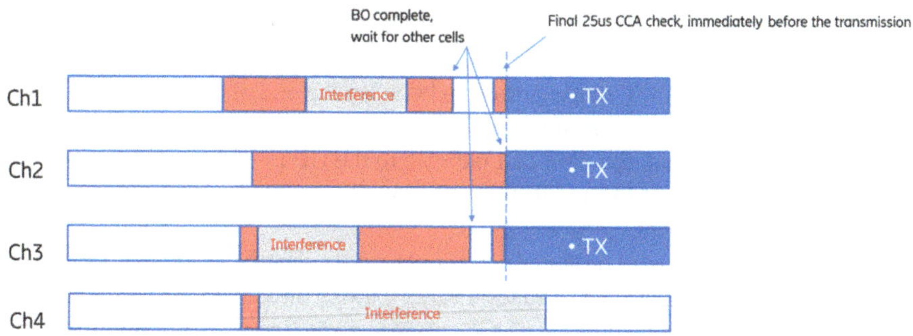

Fig. 14.3 Multichannel operation based on independent back-off

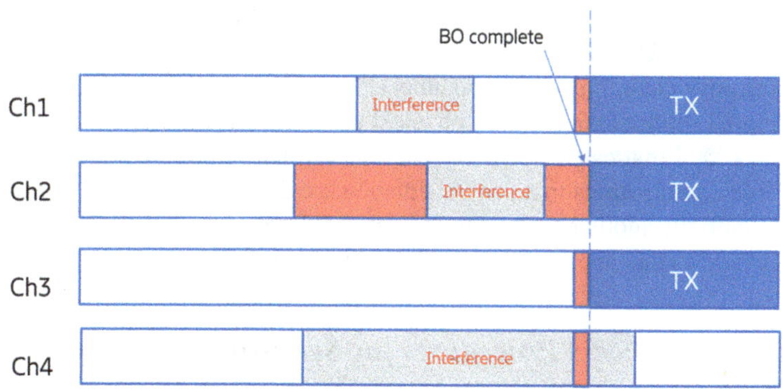

Fig. 14.4 Multichannel operation based on single back-off procedure

UL Multichannel Operation

The UL multichannel operation is the extension of the single channel operation when the gNB signals the same starting UL transmission time on multiple channels. A UE that has scheduled or configured UL resources on a set of channels in which random back-off procedure should be performed can switch to single 25 μs CCA check immediately before transmission on a channel in the set if back-off procedure has successfully completed on a designated channel in the set. The UE must select one channel uniformly randomly among the set of carriers which were scheduled/configured with random back-off procedure as the designated carrier prior to starting the back-off procedure on any of the carriers in the set.

Signaling LBT Information to the UEs

The channel access mechanism for the UL transmission depends on the COT initiation or COT sharing situation. In case of gNB sharing the COT with the UE, the UE transmission should follow the same or higher priority class as the one used when initiating the COT. This information, priority class and LBT category, is signaled to the UE.

For some cases for COT sharing, the gap between the end of the DL transmission and the start of the UL transmission should not exceed or be exactly equal to certain values. To guarantee that, the gNB can indicate to the UE to perform a longer cyclic prefix of the first symbol of the UL transmission. There is a limit on CP extension to not exceed one symbol.

To minimize the amount of signaling of parameters specific to unlicensed access within the downlink control information (DCI), those three parameters are jointly encoded. The value indicated in the DCI provides a row index to an radio resource control (RRC) configured table. The indexed row provides a value for each LBT category, channel access priority class (CAPC), and CP extension. The bit field in the non-fallback DCI depends on how many combinations the RRC signaling indicates for the UE. The indication is not only limited to the UL grant DCI scheduling physical uplink shared channel (PUSCH) but also to the DL assignments scheduling UL transmissions (e.g., physical uplink control channel (PUCCH)). For the fallback DCI that can be possibly used before UE specific RRC configuration, the possible combinations that can be signaled are defined in the specifications.

NR-U supports another method to indicate LBT category to the UE via DCI 2_0. Details related to this approach are explained in Sect. 6.1.

2.2 *Channel Access Procedures for Semi-static Channel Occupancy*

The semi-static channel occupancy allows gNB to perform a clear channel assessment (CCA) per fixed frame period for a duration of single 9 μs observation slot. If the channel is found to be busy after CCA procedure, the gNB must abstain from transmission during this fixed frame period. The fixed frame period (FFP) can be

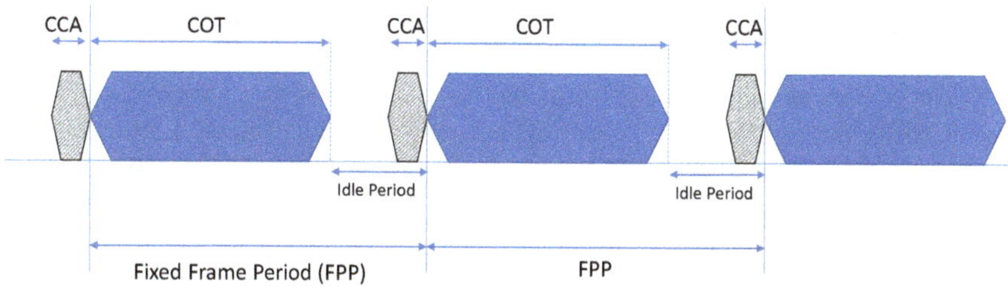

Fig. 14.5 Channel access procedure for semi-static channel occupancy

set to a value between 1 and 10 ms and can be adjusted once every 200 ms. If the channel is found to be idle, the device can transmit immediately up to a duration referred to as channel occupancy time, after which the device must remain silent for at least 5% of said channel occupancy time. At the end of the required idle period, the device can perform another CCA for channel access before the next FPP (Fig. 14.5).

The semi-static channel occupancy generally has difficulty competing with devices that use dynamic channel occupancy (such as LAA or NR-U) for channel access. Dynamic channel occupancy device has the flexibility to access the channel at any time after a successful LBT procedure, while the semi-static channel occupancy device has one chance for grabbing the channel every fixed frame period. The problems become more exacerbated with longer fixed frame period and higher traffic load. Secondly, the frame-based LBT can be rather inflexible for coordinating channel access between networks. If all the nodes are synchronized, then all nodes will find the channel available and transmit simultaneously and cause interference. If the nodes are not synchronized, then some nodes may have definitive advantages in getting access to the channel over some other nodes. Nonetheless, semi-static channel occupancy can be a good choice for controlled environments, where a network owner can guarantee the absence of dynamic channel occupancy devices and is in control of the behavior of all devices competing to access the channel. In fact, in such deployment, semi-static channel occupancy is an attractive solution because access latencies can be reduced to the minimum and lower complexity is required for channel access due to lack of necessity to perform random back-off.

COT Sharing

gNB COT sharing is also supported in case of semi-static channel access. UE transmissions within a fixed frame period can occur under the condition that a DL transmission for the serving gNB within the fixed frame period is detected. The detection of any DL transmission confirms that the gNB has initiated the COT. For this to work, the UE should be aware of the start and end of every FFP cycle. For this reason, the FFP configuration is signaled to the UEs in SIB1 and/or with UE-specific RRC signaling.

UE may transmit within a gNB-initiated COT without sensing the channel if the gap between the end of the DL transmission and the start of the UL transmission does not exceed 16 μs. Otherwise, the UE may transmit after sensing the channel to be idle for at least a 9 μs sensing slot duration immediately before the UL transmission.

The gNB can resume transmission after any previous transmission (UL or DL) within the gNB-initiated COT. If the gap between the DL and an earlier DL or UL transmission bursts is at most 16 μs, no sensing is required. Otherwise, if the gap between an earlier DL or UL and another DL transmission is more than 16 μs, the gNB must sense the channel for at least 9 μs slot immediately before the DL transmission within a gNB-initiated channel occupancy time (gNB-initiated COT).

3 Discovery Burst

For efficient operation in unlicensed spectrum, periodic transmissions should be kept at a minimum and be concentrated in time as much as possible. For instance, in Rel-13 LAA, LTE periodic signals included PSS, SSS, CRS, and CSI-RS. Sending those signals/channels separately would mean performing separately LBT procedure and also increasing the number of nodes contending to access the channel at a time. Instead, those signals are grouped together which would not only benefit the transmitting gNB but also the overall channel utilization. This led to the support of Discovery Reference Signal (DRS) which includes the existing PSS, SSS, CRS, and CSI-RS. For the same reason, discovery burst was introduced for NR-U.

A discovery burst can be seen as grouping of already existing set of signals in one downlink burst that repeats according to a configured duty cycle. The discovery burst includes as last one SS/Physical Broadcast Channel (PBCH) block and may also include CORESET for PDCCH scheduling Physical Downlink Shared Channel (PDSCH) with SIB1 and PDSCH carrying SIB1 and/or non-zero power CSI reference signals (CSI-RS). An example with one SS/PBCH and SIB1 is given in Fig. 14.6.

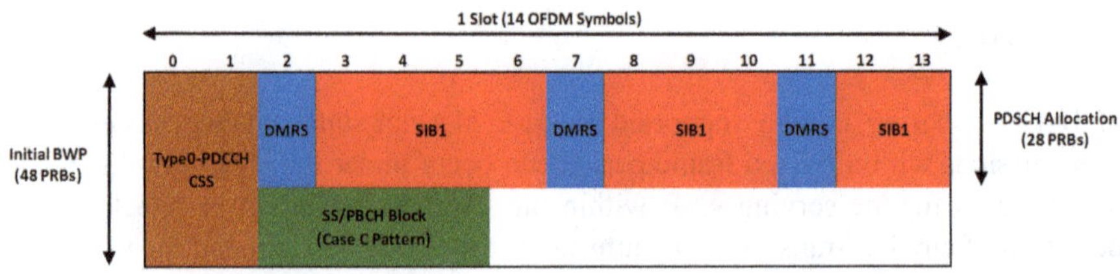

Fig. 14.6 Example of discovery burst for 30 kHz SCS, with one SS/PBCH block, two-symbol CORESET, and a large SIB1 allocation

Similar to LAA, the transmission of discovery burst can float within a transmission window, i.e., the SS/PBCH location shift in time domain. The support of a window in which the gNB can repeatedly attempt to send the discovery burst increases the probability to successfully access the medium. The duration of discovery burst transmission window is configurable. A maximum length of 5 ms is supported.

Shifting the SS/PBCH complicates the determination of the SS/PBCH block index. When SS/PBCH blocks are shifting, SS/PBCH block index is determined by the PBCH DMRS sequence index modulo Q, where Q is signaled to the UE. Q can take the values {1, 2, 4, 8}, thus supporting 1, 2, 4, and 8 beams.

4 Wideband Operation

Similar to NR, NR-U supports transmissions with wide bandwidth, e.g., up to several hundreds of MHz bandwidth. Two approaches for the device to use in wideband transmissions are the following:

- Carrier aggregation (CA)-based wideband operation
- Single wideband carrier operation based on a single active bandwidth part (BWP)

In CA mode, the UE is configured with multiple carriers (CCs) which are activated prior to reception/transmission. The gNB or UE is allowed to transmit on the carriers that succeed LBT as explained in Sect. 2.1.3. LBT is performed in units of 20 MHz, referred to here as LBT bandwidth. Hence, it is preferred to configure the individual CC's with a bandwidth of 20 MHz such that the CC bandwidth and the LBT bandwidth are one and the same. In terms of scheduling, different transport blocks (TB) are generated independently for each carrier. Failure to access one of the carriers has no impact on the transmission on another carrier that succeeded the LBT procedure.

For wideband operation, LBT is also performed in units of 20 MHz. ETSI-BRAN harmonized standard EN 301 893 [4] mandates certain RF requirements for spectral emission masks on an unlicensed channel and limits on the allowed leakage to the adjacent LBT channel. Accordingly, intra-carrier guard bands at both edges of the LBT bandwidth are introduced to avoid in-carrier leakage to the adjacent LBT channel. The intra-carrier guard bands on a carrier can be semi-statically adjusted with an RB level granularity. Besides, to limit the RF complexity, transmissions on only contiguous LBT bandwidth are supported.

In wideband operation the gNB transmits on the LBT bandwidth that succeed the LBT procedure described in Sect. 2.1.3. It makes sense to configure CORESET(s) and corresponding search space sets that are confined within an LBT bandwidth. Nonetheless, the wideband operation is not restricted to such a configuration. For uplink transmission, the UE may perform the transmission only if the LBT succeeds on all the scheduled LBT bandwidth. In that sense, wideband operation is more constrained as compared to CA mode where LBT and carrier bandwidth are aligned.

5 UL Interlacing

Two components that are commonly enforced by regulations to secure fair sharing of the channel and efficient channel utilization between different nodes are the maximum power spectral density and the requirements on the occupied channel bandwidth.

Both ETSI BRAN EN 301 893 and US regulations enforce a maximum power spectral density (PSD) limit. For instance, the ETSI BRAN EN 301 893 regulations require 10 dBm/MHz for 5150–5350 MHz. Such requirement has implication on both transmit power and coverage especially for small narrow bandwidth transmission. To overcome the transmit power limitation per 1 MHz, the resource allocation needed to be spread over a wider bandwidth by assigning non-contiguous resources.

The occupied bandwidth requirement is enforced by ETSI BRAN EN 301 893 only. The requirement mandates that bandwidth containing 99% of the power of the signal must be between 80% and 100% of the declared Nominal Channel Bandwidth [4]. Nonetheless, temporary exceptions are allowed. The frequency allocations for one UE must be assigned in such a way that the requirement is fulfilled.

Physical resource blocks (PRB)-based interlacing transmissions are supported for NR-U to give PUSCH and PUCCH transmission with small bandwidth higher transmission powers when needed and to satisfy the transmission BW requirement. For a 20 MHz carrier bandwidth, the bandwidth is divided into M interlaces, each interlace consisting of N equally spaced in frequency domain. M depends on the subcarrier spacing M = 10 for 15 kHz and M = 5 for 30 kHz. An interlace starts on a first CRB of a carrier and occupies every Mth PRB within the carrier. For a 20 MHz carrier, this yields 10 or 11 PRBs in each interlace depending on interlace index as shown in Fig. 14.7. For 30 kHz, the PUSCH frequency domain resource allocation consists of a bitmap that indicates which combination of M interlaces is allocated to the UE. For 15 kHz SCS, RIV indicates a start interlace index and number of contiguous indices.

For a carrier bandwidth larger than 20 MHz, the interlaces are defined over the whole carrier with the same definition as for 20 MHz. The only difference is that the number of PRBs per interlace scales with the bandwidth. For PUSCH, the carrier bandwidth is divided into 20 MHz RB sets that are separated by intra-carrier guard bands. The frequency domain resource allocation (FDRA) field includes, on top of the bitmap that indicates the allocated interlaces, an RB set indicator which indicates a first RB set and a number of contiguous RB sets allocated to the UE. Figure 14.8 illustrates an example where the UE is allocated 1st and 4th interlaces in the 2nd and 3rd RB sets of the carrier.

For PUCCH, a PUCCH resource is confined within one RB set. The PUCCH resource configuration includes an index of the allocated interlace(s) and an index of the allocated RB set.

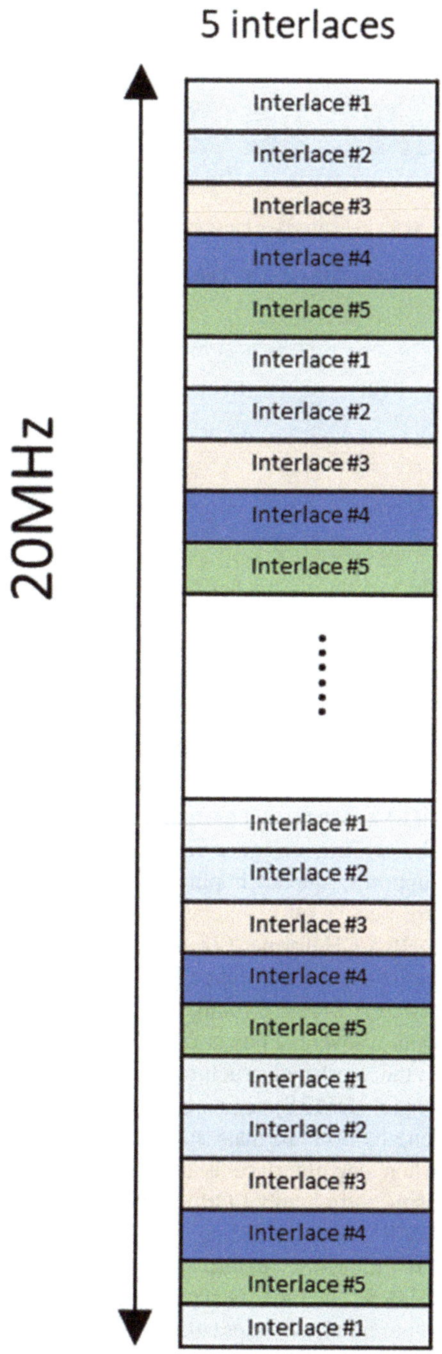

Fig. 14.7 Interlace design for 30 KHz subcarrier spaces in 20 MHz carrier bandwidth

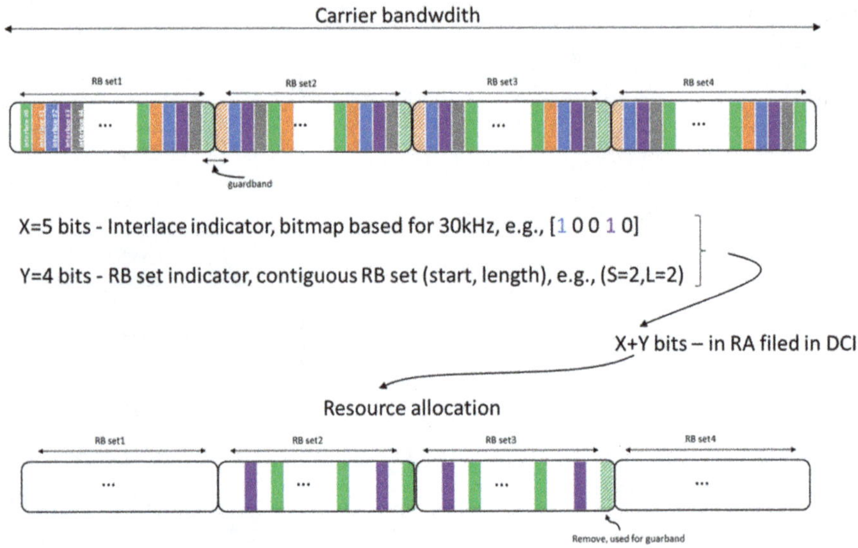

Fig. 14.8 Interlace design for 30 KHz subcarrier spaces in >20 MHz carrier bandwidth

5.1 PUCCH Formats

In principle, the Rel-15 PUCCH formats can still be used as long as they are used in a way that complies with the regulations. Additionally, PUCCH formats 0, 1, 2, and 3 are enhanced to support the interlaced structure. Since the interlaced structure already provides good frequency diversity, the enhancements of the formats do not support frequency hopping. PUCCH is always confined within one 20 MHz RB set irrespective of the operational bandwidth.

PUCCH format 0/1 occupies one full interlace by repeating the Rel-15 base sequence in each of the PRBs of the interlace and cycling the cyclic shifts across the different PRBs. Cycling the cyclic shifts across the PRBs is intended to reduce the peak-to-average power ratio (PAPR)/cubic metric (CM). PUCCH 2/3 is extended to fit the interlaced structure using one or two interlaces. The Demodulation Reference Signal (DMRS) sequence is adjusted by generating a sequence of the appropriate length. Both formats support multi-user multiplexing of one, two, or four UEs when one interlace is allocated. User multiplexing in case of PF2 is achieved using intra-symbol Orthogonal Cover Codes (OCCs) separately on both UCI and DMRS with spreading over the resource elements after modulation. In case of PF3, multi-user multiplexing is supported using OCCs on UCI and cyclic shifts on DMRS. The OCC is applied before the discrete Fourier transform (DFT) spreading using block-wise spreading over all resource elements available in each orthogonal frequency-division multiplexing (OFDM) symbol. This causes the different users to become orthogonal in frequency.

6 GC-PDCCH Enhancements

It is of interest to increase the channel access granularity on unlicensed carrier; otherwise, the competitiveness to acquire the channel is reduced. Rel-15 NR supports PDCCH monitoring on symbol level granularity, if needed. The gNB has the ability to access the channel at every symbol boundary. However, configuring such a frequent PDCCH occasion can be demanding in terms of UE power consumption. The problem becomes more severe if the UE is configured with a wideband or multiple carrier. To counterbalance that, NR-U introduced a mechanism that allows the UE to switch to a less frequent PDCCH monitoring when applicable.

6.1 Channel Occupancy Indication

The LBT parameters to be used by the UE for scheduled UL transmission are in principle indicated in the scheduling grant. Configured UL transmission (PUCCH, configured PUSCH, etc.) applies Type 1 LBT by default. Nonetheless, NR-U supports additional implicit indication via DCI format 2_0 to switch the UL transmission, scheduled or configured, LBT category from Type 1 to Type 2 sensing slot within a gNB acquired COT where Type 1 LBT is not needed for the UL. For this purpose, a COT duration field per serving cell is introduced in DCI 2_0. If the field is configured, the UE follows the indicated value for the purpose of LBT category switching. If the field is not present, the UE assumes that the COT duration is given by the duration of the SFI indicated by DCI 2_0.

6.2 SFI in Frequency Domain

Even though the UE might be configured with multiple channels, or a wide carrier consisting of multiple RB sets (corresponding to LBT bandwidths), the gNB might not simultaneously use all of them for many reasons including possibility of LBT failure on some of the carriers or RB sets. It is beneficial from UE power consumption perspective to indicate to the UE which carrier/RB set that will not be used for DL transmission by the gNB at a certain time. The indication of the availability of a carrier and/or RB set is done in the form of an explicit bitmap via DCI format 2_0 carried by GC-PDCCH. The indication is valid until the implicitly or explicitly indicated COT duration as explained in Sect. 6.1.

6.3 PDCCH Monitoring

In principle, frequent PDCCH occasions can be useful before the gNB initiates the COT to enable to access the channel at a finer granularity. The moment the gNB initiates its COT, it is enough to serve the UEs on slot basis. For NR operation in unlicensed spectrum, UEs can be configured with two groups of search space sets for PDCCH monitoring. The UE is indicated to switch between the two groups.

Two mechanisms are supported to indicate the dynamic switching, one based on explicit indication and the other based on implicit indication. If the UE is configured with the explicit method, DCI 2_0 includes an additional 1-bit flag. If the flag is set to 1, the UE starts a timer and switches to the PDCCH monitoring periodicity corresponding to the second group. Upon timer expiry, or reception on DCI 2_0 indicating flag equal to 0, or end of the gNB-initiated COT, the UE switches back to the PDCCH monitoring periodicity corresponding to group 1.

If the UE is configured with the implicit method, the default PDCCH monitoring corresponds to the first group. Upon the detection of gNB-initiated COT, the UE starts a timer and switches to the PDCCH monitoring corresponding to the second group until the timer expires.

7 Initial Access Enhancements

To fulfill the maximum power spectral density and 80% occupied channel bandwidth, interlacing is supported for PUCCH and PUSCH transmissions, as discussed in Sect. 5. Following the same reasoning, physical random access channel (PRACH) is extended to support larger bandwidth PRACH occasions. Spreading the transmission over a larger bandwidth is beneficial because it allows larger transmit power without exceeding the power spectral density (PSD) restrictions set by the regulations. In contrast to PUSCH and PUCCH, the larger bandwidth for PRACH is achieved with fully consecutive subcarriers without the support of interlacing resource allocation.

Rel-15 RACH formats are supported and can be used when applicable. Additionally, Rel-16 provides support for configuring larger-bandwidth PRACH formats obtained using longer sequence length. For PRACH with 15 kHz and 30 KHz subcarrier spacing, it is possible to configure sequence length $L = 1151$ and $L = 571$, respectively. In both cases, the resulting PRACH occupies almost 20 MHz. Signaling of PRACH bandwidth for initial access is in SIB1.

Larger-bandwidth PRACH occasions reduce maximum PRACH cell capacity by reducing the maximum possible degree of frequency multiplexing of PRACH occasions by a factor of 8 (15 kHz) or 4 (30 kHz). Additionally, new cyclic shift tables for the larger-bandwidth PRACH are defined. Small shift steps are used to allow many different preambles and hence high PRACH system capacity.

Another enhancement related to the initial access procedure is the support of two-step RACH. In two-step RACH procedure, messages of four-step RA in Rel-15 are compressed in two steps instead of four. PRACH preamble and Msg3/PUSCH are combined into a single uplink message (MsgA), while the two downlink messages (Msg2/RAR and Msg4) are combined into a single downlink message (MsgB). By reducing the number of steps in the RACH procedure, the number of required LBTs is reduced, which yields to improved latency in NR-U system.

8 HARQ Enhancements

In NR, the DL feedback timing (K1) between DL data transmission and acknowledgment can be flexibly signaled in PDSCH-to-HARQ-timing-indicator field in DCI. The field can be up to 3 bits. Eight values can be indicated. Those values correspond to a K1 value that ranges from 0 to 15. The mapping is RRC configured. If HARQ feedback transmission is subject to LBT, which is the case when operating on unlicensed spectrum, there is a risk that the UE fails to perform the transmission depending on the LBT outcome. Due to the one-to-one mapping between PDSCH and corresponding feedback in the time domain, if the UE fails to transmit the feedback on the predefined time location, the gNB will have to assume NACK and retransmit all the corresponding PDSCHs. The latter can be considered as an inefficient utilization of the band and causes unnecessary increase in the channel contention. Lack of feedback, or delayed feedback, significantly impacts the overall performance in terms of UE's throughput and in terms of inefficient use of the channel if the delayed feedback triggers unnecessary retransmissions [7]. NR-U supports variety of enhancement to overcome these issues.

8.1 LBT for Feedback on PUCCH

According to ETSI BRAN EN 301 893, the responding node may start transmission immediately within less than or equal 16 μs from the end of the initiating node transmission without performing clear channel assessment. Fortunately, NR-U takes advantage of this to send feedback without performing an LBT. It is possible to transmit PUCCH/UCI feedback within a gNB initiated COT without performing an LBT if the gap between DL and the immediately following UL transmission is less than or equal to 16 μs.

8.2 Non-numerical K1 Values

While transmissions without LBT protect the performance from the drawbacks of delayed feedback, it is not always sufficient or possible. NR supports small processing delays, but not as small as providing feedback within the same slot, at least for capability 1 UEs. For instance, with a subcarrier spacing of 30 KHz, L1 processing delay from end of PDSCH until beginning of PUCCH is minimum ten OFDM symbols assuming capability 1 UE. Therefore, there are cases in which the feedback of the last PDSCH(s) in the COT cannot be reported in the same COT. Instead, the UE is assigned resources outside of the gNB's COT, and transmissions on those resources are subject to LBT.

As an enhancement to Rel-15, the gNB can signal a non-numerical value in the PDSCH-to-HARQ-timing-indicator field in the DCI. When signaled, it indicates that the UE should hold on the HARQ-ACK feedback for the corresponding PDSCH until the timing and resource for the HARQ-ACK feedback are provided by the gNB in another DCI. Hence, the gNB can request the pending feedback as part of gNB's COT which increases the chances of successful reception of the feedback due to shorter or no LBT before the UL transmission. The HARQ-ACK timing for PDSCH scheduled with non-numerical value for K1 is derived by the next DCI scheduling a PDSCH and indicating a numerical value in the PDSCH-to-HARQ-timing-indicator field.

8.3 One-Shot Feedback

The dynamic HARQ codebook in Rel-15 was used as a starting point for the design of an enhanced mechanism for operation on unlicensed spectrum. In Rel-15, the timing control for the PUCCH/UCI feedback is quite limited. For instance, if the UE fails to provide the aggregated feedback on the specified resources due to unsuccessful LBT, all PDSCHs included in this feedback need to be retransmitted. gNB does not have the flexibility to ask for feedback retransmission. The gNB will have to assume NACK and retransmit all the corresponding PDSCHs. Another issue is the PUCCH misdetection at the gNB side. Even if the UE successfully transmits the HARQ feedback, there are chances that the gNB may not be able to detect it. From gNB perspective, failed LBT or missed UCI transmissions are indistinguishable. Due to the one-to-one mapping between PDSCH and corresponding feedback in the time domain, if the gNB fails to detect the feedback in the predefined time location, the gNB will have to assume NACK and retransmit all the corresponding PDSCHs. Generally, the likelihood of misdetecting a PUCCH transmission is more critical on an unlicensed band due to collisions with other concurrent transmissions [7].

In NR-U, to solve the above issues, the UE can be configured to monitor feedback request of an HARQ-ACK codebook containing all DL HARQ processes. The feedback can be requested in DL DCI 1_1. In response to the trigger, the UE reports the HARQ-ACK feedback for all DL HARQ processes. The format of the feedback, Code Block Group (CBG)-based HARQ-ACK or TB-based HARQ-ACK, can be configured to be part of the one-shot HARQ feedback for the CCs configured with CBG.

Additionally, to resolve any possible ambiguity between the gNB and the UE that might be caused by possible misdetection of PDCCH(s), the UE can be configured to report the corresponding latest new data indicator (NDI) value for a latest received PDSCH for that HARQ process along with the corresponding HARQ-ACK for the received PDSCH. From gNB perspective, if the NDI value matches the last transmitted value, it indicates that the reported HARQ-ACK feedback correctly corresponds to the HARQ process with pending feedback. Otherwise, the mismatch suggests that the UE is reporting an outdated feedback.

8.4 Enhanced Dynamic Codebook Enhancements

As described in Sect. 8.3, one-shot feedback is a method to enable retransmission of feedback in case the UE fails to provide the feedback due to failed LBT. In principle, this option will cause unnecessary overhead on the UCI especially if the HARQ based on code block group is activated. Nonetheless, this option can be beneficial as a fallback scheme, e.g., when the eNB fails to receive any feedback for some time. However, it should not be the main mode of operation.

NR-U Rel-16 supports another mechanism that enables retransmission of the HARQ feedback corresponding to the used HARQ processes instead of all HARQ processes when enhanced codebook is configured. If, for any reason, the scheduled codebook was not received, the retransmission of the feedback can be requested by the gNB. A toggle bit, new feedback indicator (NFI), is added in the DCI to indicate whether the HARQ-ACK feedback from the UE was received by the gNB or not. If toggled (as in Fig. 14.9), the UE assumes that the reported feedback was correctly received. Otherwise, if the gNB fails to receive the scheduled PUCCH (Fig. 14.10), the UE is expected to retransmit the feedback. In the latter case, the DAI (C/T-DAI) is not reset; instead, the DAI is accumulated within a PDSCH group until NFI for the PDSCH group is toggled.

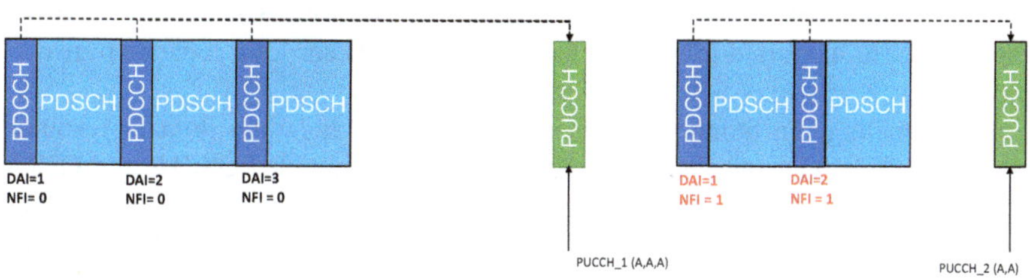

Fig. 14.9 PUCCH successfully received by the gNB

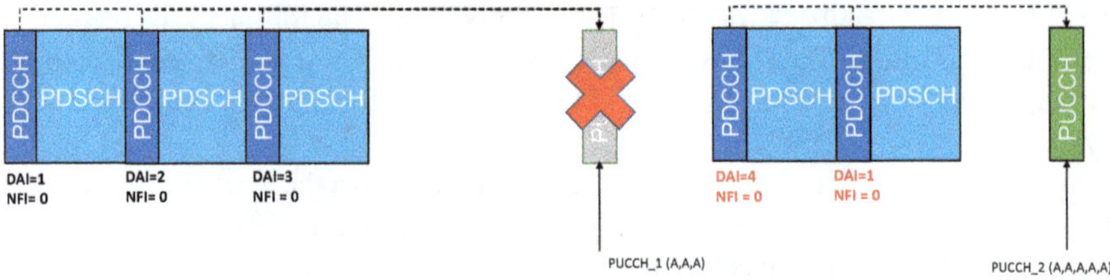

Fig. 14.10 PUCCH misdetection case; gNB requests retransmission of earlier feedback

As the triggering of additional HARQ feedback reporting occurs with ambiguous timing relation to the associated PDSCHs, PDSCH grouping is introduced. PDSCH group is defined as the PDSCH(s) for which the HARQ-ACK information is originally indicated to be carried in a same PUCCH. PDSCH grouping allows the gNB to explicitly indicate which exact codebook is missing. The group index is explicitly signaled in the scheduling DCI. If enhanced dynamic codebook is configured, two PDSCH groups are supported. Together with the group ID, the gNB signals a request group ID which is a 1-bit field. If set to 0, the gNB is requesting feedback for the scheduled group, otherwise, for both the group scheduling using the DCI and the other one. By referring to the group ID (ID), request ID (RI), and the value of the NFI field in the DCI, the UE can figure out if the next feedback occasion should include only initial transmission or also retransmission of feedback corresponding to PDSCH(s) associated with the indicated group.

Similar to NR, the DAI value is also included in the UL grant scheduling PUSCH. As an additional functionality, the gNB can indicate the DAI value for each group separately in the UL grant to resolve any possible ambiguity at the UE side.

9 Scheduling Enhancements

9.1 Scheduling Multiple PUSCHs Using Single Grant

Generally, grant transmission consumes channel access time and introduces high signaling overhead especially when it is not multiplexed with other transmissions, e.g., when downlink traffic is low. For this reason, multi-PUSCH scheduling using single grant is supported in NR-U, where each PUSCH carries a separate transport block.

For signaling the number of scheduled PUSCHs and time domain resource allocation (TDRA) in one DCI format 0_1 scheduling multiple PUSCHs, the TDRA table is extended such that each row indicates multiple PUSCHs that are contiguous in time domain. Each PUSCH has a separate start and length indicator value (SLIV) and mapping type, but same frequency resource allocation. The number of

TDRA index	K2	SLIV	Mapping Type
1	{2,3,4,5,6}	S {0,0,0,0,0} L {14,14,14,14,14}	{A,A,A,A,A}
2	{2,3,3,4,4}	S {7,0,7,0,7} L {7,7,7,7,7}	{B,A,B,A,B}
2	{2,3,3,4,5,6}	S {7,0,7,0,0,0} L {7,7,7,14,14}	{B,A,B,A,A,A}

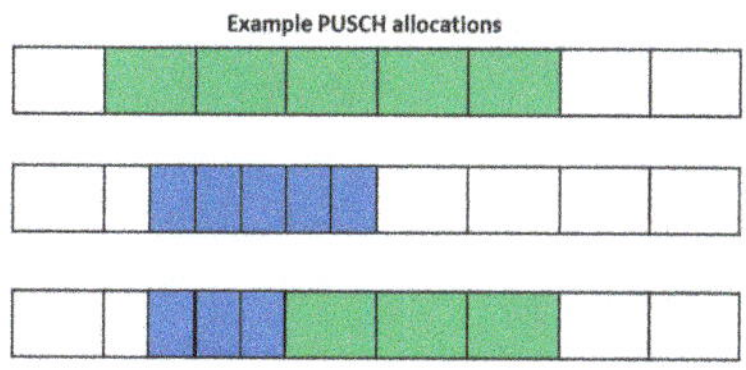

Fig. 14.11 Examples for multi-PUSCH scheduling [8]

scheduled PUSCHs is signaled by the number of indicated valid SLIVs in the row of the TDRA table signaled in DCI (Fig. 14.11).

To maintain the same blind decoding complexity at the UE side as in Rel-15, the same DCI format is used to schedule one or more PUSCHs. The payload of the two cases is matched. Almost all the bit fields are kept the same as compared to single PUSCH scheduling; the only fields that are extended are redundancy version (RV) and NDI. In case more than one PUSCH is scheduled, the DCI format includes multi-bit RV and NDI. The number of NDI bits and RV bits in DCI format 0_1 is determined based on the maximum number of PUSCHs that can be scheduled using single UL grant.

Some fields, even though not changed in terms of bit field length, are repurposed or interpreted differently. For instance, in case more than one PUSCH is scheduled, the HARQ process ID signaled in the DCI applies to the first scheduled PUSCH. HARQ process ID is then incremented by 1 for subsequent PUSCHs in the scheduled order, with modulo operation as needed. Additionally, when the DCI is scheduling only a single PUSCH, the multi-bit RV and NDI are repurposed to indicate CBGTI in case of single PUSCH scheduling, if CBG-based transmission is configured. This way, the benefits of enabling CBG-based retransmission are retained but limited to the case of scheduling single PUSCH [8]. Table 14.5 clarifies the interpretation of DCI 1_0 depending on the number of scheduled PUSCHs.

10 Configured UL Enhancements

One or even multiple message exchange has to occur before a PUSCH is scheduled for transmission. If UL data arrives at the UE, the UE informs the gNB about it by sending a scheduling request (SR) on predefined resources. On unlicensed spectrum, the transmission of the SR is subject to LBT success. If UE manages to successfully access the channel immediately before the predefined SR resources, it can transmit. If the gNB successfully decode the SR, it sends a UL grant, which is also subject to LBT. It is only then that the UE is allowed to transmit PUSCH on the granted resources if it successfully grabbed the channel by means of LBT. As you can see, a PUSCH transmission is subject to success of multiple LBT procedures at

Table 14.5 DCI 1_0 scheduling single or multiple PUSCHs

Field	DCI format 1_0	
	Single PUSCH scheduling	Multi-PUSCH scheduling
HARQ	Same as Rel-15	Applies to the first scheduled PUSCH. HARQ process ID is then incremented by 1 for subsequent PUSCHs in the scheduled order, with modulo operation as needed
NDI	1 bit	1 bit per PUSCH
RV	2bits	1 bit per PUSCH
CSI request field	Same as Rel-15	Applies to single PUSCH. The PUSCH that carries the aperiodic CSI feedback is: When M <= 2: The (M)-th scheduled PUSCH When M > 2: The (M-1)-th scheduled PUSCH
UL-SCH indicator	Same as Rel-15	Not present
CBGTI, if configured	Same as Rel-15	Not present
DAI fields	Same as Rel-15	Same as multi-slot PUSCH scheduling in Rel-15
Remaining fields	Common functionality	

different times and nodes. This procedure may degrade uplink performance due to unnecessarily high overhead and delays caused by the multiple LBT procedures.

Fortunately, NR also supported UL transmissions on preconfigured resources. Using this mode, latency can be reduced. Unlike scheduled PUSCH, configured PUSCH transmission is subject to single LBT procedure at the UE side. NR-U supports both Rel-15 NR schemes (types 1 and 2) with some extensions.

10.1 Time Resource Assignment

For unlicensed operation, it is of high interest to enable configured UL transmissions on consecutive slots without gaps in between, to avoid multiple LBTs. However, with Rel-15 NR, the only way to do so is to configure SLIV in a way that the UE is provided with CG resources on all symbols, e.g., Configuration 1: Periodicity 2, $S = 0$, $L = 2$ or Configuration 2: Periodicity 14, $S = 0$, $L = 14$. However, it would be very restrictive if the only way to efficiently use configured UL on unlicensed channel is by allowing the UE to transmit in every slot. The gNB should have some flexibility to assign or exclude certain slots for configured UL. For this reason, enhancements for the configured UL time resources assignment were adopted for NR-U (Fig. 14.12).

For NR-U, the time domain resource assignment in configured grant repeats over the multiple slots within the CG-allocated slots. The same symbol allocation, SLIV, and mapping type are used for the first PUSCH in every slot of the CG-allocated

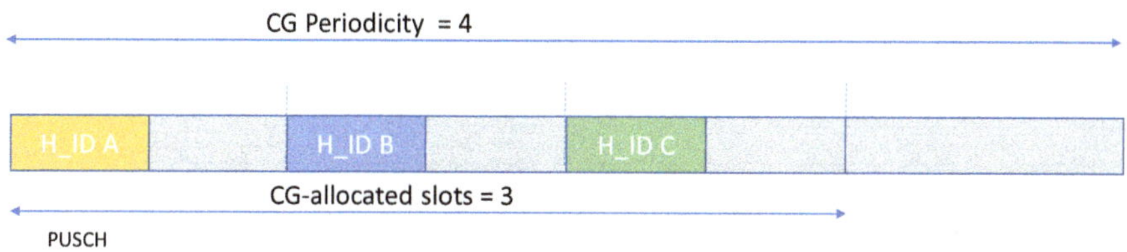

Fig. 14.12 CG time domain resource allocation

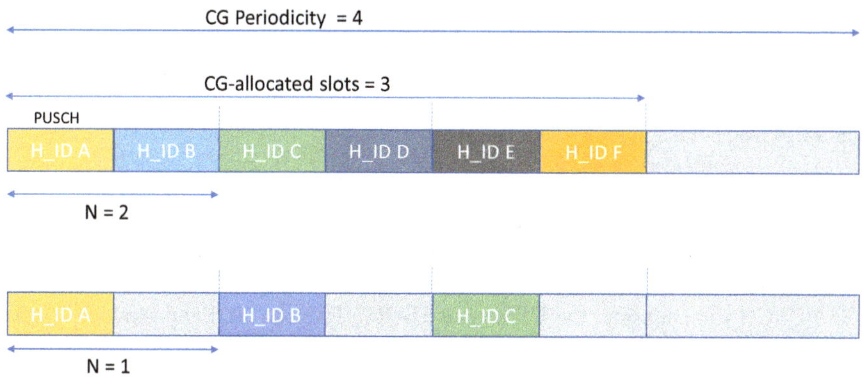

Fig. 14.13 TDRA for configured UL with N = 1 and N = 2

slots. In Fig. 14.13, the CG-slot window is configured as three slots. SLIV is set so that the PUSCH transmission starts at S = 0 and L = 7. In every slot of the CG slots, a 7 symbol PUSCH is configured, each starting at symbol 0.

Additionally, a parameter N indicates how many consecutive PUSCHs are allocated within a slot, as shown in Fig. 14.13. For configured PUSCHs within a single configuration, length of all PUSCHs is the same.

If repetition is configured for a CG configuration, the UE repeats the transport block in the earliest consecutive transmission occasion candidates within the same configuration instead of consecutive slots. Figure 14.14 shows an example. The UE may drop repetition transmissions that fall into a subsequent configured period.

10.2 HARQ Enhancements

In NR, HARQ ID corresponding to a transmission on a preconfigured resource is derived from a formula similar to LTE SPS, i.e., the HARQ process ID is determined based on the resources that are used for transmission. This synchronous behavior imposes large delays due to the uncertainty of channel availability on unlicensed bands and therefore is inefficient for unlicensed channel operation. Asynchronous HARQ, where the timing relationship between configured UL transmission and

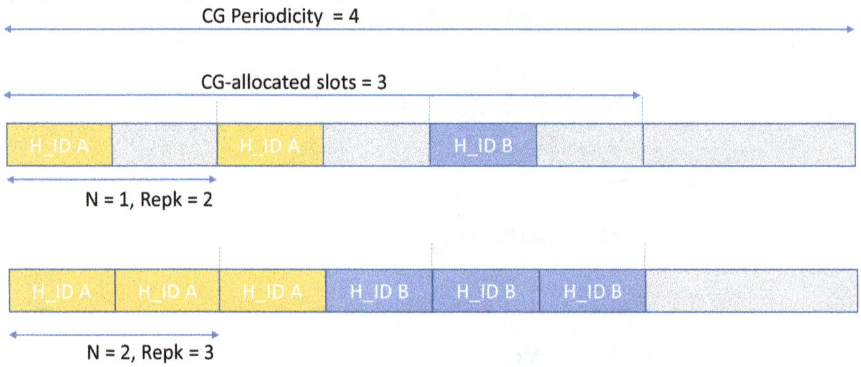

Fig. 14.14 NR-U configured UL with repetition

corresponding UL HARQ feedback is not fixed, was found to be more beneficial for operation in unlicensed spectrum. Accordingly, a new uplink control information (UCI) was introduced to enable fully asynchronous HARQ for NR-U CG. For every CG-PUSCH transmission, the UE selects HARQ, RV, and NDI and report it on the new CG-UCI.

NR does not support nonadaptive HARQ operation. ACK feedback is implicit and NACK is explicit. A timer starts when a TB is transmitted, and if no explicit NACK (dynamic grant) is received before the timer expires, the UE assumes ACK. This approach does not work well on the unlicensed carrier since the absence of feedback might be due to failed LBT. The UE may misinterpret a delayed retransmission grant as an ACK. Since the channel availability is not guaranteed on the unlicensed channel, the UE might run into this situation often. For this reason, CG NR-U follows the opposite behavior, where ACK feedback is explicit and NACK is implicit. A timer starts when a TB is transmitted, and if no implicit ACK is received before the timer expires, the UE assumes NACK and performs nonadaptive retransmission. To reduce the signaling overhead corresponding to explicit feedback transmission, NR-U supports a new DCI format, downlink feedback information ("CG-DFI"), that carries HARQ-ACK bitmap for all UL HARQ processes from the same UE. Additionally, the gNB may trigger an adaptive retransmission using a dynamic grant.

10.3 Configured UL UCI

As described in the earlier section, CG-UCI is included in every CG-PUSCH transmission and includes the information listed in Table 14.6. CG-UCI is mapped as per Rel-15 rules with CG-UCI having the highest priority. It is mapped on the symbols starting after first DMRS symbol. To determine the number of REs used for CG-UCI, the mechanism of beta-offset in Rel-15 NR for HARQ-ACK on CG-

Table 14.6 CG-UCI content

UCI content
HARQ
RV
NDI
COT sharing information
CRC

PUSCH is reused. Nonetheless, a new RRC configured beta-offset for CG-UCI is defined.

If CG-PUSCH resources overlap with PUCCH carrying CSI part 1 and/or CSI part 2, the latter can be sent on CG-PUSCH. RRC configuration can be provided to the UE indicating whether to multiplex CG-UCI and HARQ-ACK. If configured, in the case of PUCCH overlapping with CG-PUSCH(s) within a PUCCH group, the CG-UCI and HARQ-ACK are jointly encoded as one UCI type. Otherwise, configured grant PUSCH is skipped if CG-PUSCH overlaps with PUCCH that carries HARQ-ACK feedback.

10.4 Intra-cell Collision Reduction

At low load situations, where latency matters the most and the contention is minimum, it might be beneficial for UL efficiency reasons to allocate the same UL resources to multiple UEs. UEs assigned the same CG resources in time domain, and same SLIV may start simultaneous transmission and collide. To overcome this problem and reduce the chances that different UEs start their transmission at the same time, the gNB can spread out the earliest possible transmission time for CG UEs by assigning different starting time offset. The starting time offset applied by a UE at the beginning of a transmitted burst with a CG resource at the start of the transmission burst is RRC configured and defined as the length of a CP extension of the first symbol that is located before the configured resource. The CP extension is up to 72 microseconds with a granularity of 9 microseconds. The UE is assigned different offset values for full or partial bandwidth transmissions inside or outside the gNB's COT.

References

1. RP-151045, New work item on licensed-assisted access to unlicensed spectrum, 2015
2. 3GPP, 3GPP TR 38.889 3rd generation partnership project; Technical specification group radio access network; Study on NR-based access to unlicensed spectrum; (Release 16), 2019
3. QUALCOMM, RP-RP-182878, New WID on NR-based access to unlicensed spectrum, 3GPP, 2019

4. ETSI BRAN EN 301 893, Broadband Radio Access Networks (BRAN); 5 GHz high performance RLAN; Harmonized EN covering the essential requirements of article 3.2 of the R&TTE Directive, ETSI BRAN EN 301 893
5. FCC Part 15 ruling, [Online]. Available: http://www.ecfr.gov/cgi-bin/text-idx?SID=3c5e2d1533490603e0131fcdc041030d&node=pt47.1.15&rgn=div5
6. 3GPP, ETSI TS 137 213, Physical layer procedures for shared spectrum channel access, 2019
7. R1-1811303, Potential HARQ enhancements, Ericsson, RAN1#94 Meeting, Chendgu
8. R1-1912711, HARQ and scheduling enhancements for NR-U, Ericsson, RAN1#99, Reno

Chapter 15
5G NR Positioning

Sven Fischer

1 Introduction

Historically, the main driver for location-based services has been requirements from regulatory authorities (e.g. emergency caller location requirements). However, today, many public and private entities demand delivery of location information to enable commercially motivated location-based services, which often require higher location accuracy and precision. The range of use cases that benefit from improved accuracy positioning services is broad, including industry, asset tracking, automotive, traffic management, hospitals (e.g. person and medical equipment location), shared bikes, aerial vehicles (drones), augmented reality, wearables, and so on. Consequently, 5G New Radio (NR) will offer a variety of positioning technologies delivering position information of UEs depending on the needs of specific use cases.

The first phase of 5G as specified in 3GPP Release-15 includes support for location services (LCS) but restricted to regulatory use cases (emergency calls and lawful intercept). This support has been enabled by the addition of a 5G Core Network (5GCN) location server, referred to as Location Management Function (LMF). Similar to previous generations (2G, 3G, 4G), a Gateway Mobile Location Centre (GMLC) is used for external clients to access a UE's location. Roaming and commercial LCS capabilities are not supported in 5G Phase 1. With the exception of NR Cell-ID positioning, no new NR positioning methods have been added in Release-15, and the same positioning methods as specified for LTE are reused.

The second phase of 5G as specified in 3GPP Release-16 will enhance the location services capabilities to support roaming and commercial use cases. This includes (among others) support for mobile-terminated location requests (MT-LR),

S. Fischer (✉)
Qualcomm, Nuremberg, Germany
e-mail: sfischer@qti.qualcomm.com

X. Lin, N. Lee (eds.), *5G and Beyond*, https://doi.org/10.1007/978-3-030-58197-8_15

mobile-originated location requests (MO-LR) and deferred location requests for periodic, triggered and UE available events; support for UE privacy notification and verification and update of privacy preference; support for bulk (multi-UE) location request to a GMLC; and support for efficient periodic/triggered location reporting.

To address the diverse location requirements resulting from new applications and industry verticals, 5G Phase 2 will support multiple NR native positioning technologies, including methods based on time difference of arrival (TDOA) measurements, round-trip time (RTT) measurements with multiple base stations (multi-RTT) and angle of arrival (AoA) or angle of departure (AoD) measurements. New NR positioning reference signals (PRS) are defined to improve the performance of the NR positioning technologies.

Additional positioning enhancements enabled in 5G Phase 2 include support for broadcast of location assistance data, support for UE-based NR native positioning, enhancements to the location services architecture and additional GNSS assistance data to support high accuracy location. With the needs of diverse industry verticals in mind, future releases will aim for very high accuracies and very low latencies.

2 Location Services in the 5G System

2.1 General Concepts

Position location of a UE can be performed by cellular positioning technologies, assisted-GNSS technologies and other external methods.

The *cellular positioning technologies* make use of signal measurements from cellular base stations and devices and, therefore, typically rely on existing cellular infrastructure. The most fundamental technique for providing mobile location information is the Cell-ID technique where the mobile device location is determined based on the coverage area of the cell the mobile is using for communication and data services (geographical location of a serving base station). The basic Cell-ID method can be improved by using additional measurements available or required for the communication system, such as measurements for radio resource management (RRM). The technique is then referred to as an Enhanced Cell-ID (E-CID) method. More advanced cellular positioning technologies are based on time of arrival (TOA) measurements, which can be performed at the mobile device or at the base stations. The cellular positioning techniques for 5G NR are described in Sect. 3.3.

Assisted-GNSS (A-GNSS) technologies require a Global Navigation Satellite System (GNSS) receiver integrated in the UE. GNSS refers collectively to multiple satellite systems including GPS, GLONASS, Galileo, BeiDou, etc. GNSS operates by measuring the propagation times of radio signals sent by satellites. The GNSS satellites continuously broadcast signals whose transmission times are synchronized across all satellites. The user can then calculate his position using the TOA method (see also section "Time of Arrival (TOA)"). The time stability and synchronization of the transmitted signals are excellent due to the use of multiple atomic clocks onboard the satellites and monitoring by ground-based stations that upload time corrections to the satellites. With stand-alone (conventional) GNSS, the GNSS receiver integrated in the mobile device is receiving satellite signals and computes its location without assistance from the cellular network. The receiver needs to acquire satellite signals through a multi-dimensional search process. The GNSS receiver must acquire signals from at least four satellites in order to compute a three-dimensional position. This acquisition or search process can be demanding in terms of battery consumption and processing power, and the time-to-first-fix (TTFF) can be long (e.g. 30 seconds or more). The performance of stand-alone GNSS can be improved if the cellular network provides assistance data to the device. The technique is then referred to as assisted-GNSS. The assistance data generally consists of the approximate time and position of the receiver, as well as the ephemeris data of the GNSS constellation. By sending the assistance data from the location server to the UE, the device knows which satellites are visible at the approximate device location and where in time and frequency to search for the satellite signals. This allows the receiver to perform a so-called hot start, and determining its position within a few seconds rather than the long time-to-first-fix typically required without assistance data under normal situations. Moreover, A-GNSS also increases the sensitivity of the receiver which also allows acquiring the satellite signals within, e.g. urban canyons or light buildings. It also helps to save battery power of the UE. A detailed description of A-GNSS receivers and technologies can be found in [17].

Other external positioning techniques supported in 5G similar to previous generations include positioning based on Wireless Local Area Network (WLAN) and Bluetooth signals, as well as sensors and Terrestrial Beacon Systems (TBS). WLAN positioning makes use of the WLAN measurements (Access Point (AP) identifiers and, e.g. received signal strength measurements) and typically uses a reference database to determine the location of the UE. Similar to WLAN positioning, Bluetooth positioning uses Bluetooth measurements (beacon identifiers and received signal strength measurements) to determine the location of the UE. TBS positioning makes use of a network of ground-based transmitters, broadcasting signals only for positioning purposes. The sensor measurements can include barometric pressure measurements to aid altitude determination or inertial measurement sensors.

2.2 Positioning Modes

Determining the location of a UE generally involves two main steps: radio signal measurements (e.g. NR native signal measurements or external signal measurements such as measurements from GNSS satellite signals) and position computation/estimation based on these measurements.

For performing the signal measurements and/or to enable position calculation, the network may provide assistance data to the UE. The various positioning methods may be supported in one or more of the following positioning modes.

In *UE-assisted mode*, the UE provides position measurements to a location server for computation of a location estimate by the location server. The network typically provides assistance data to the UE to enable position measurements or improve measurement performance.

In *UE-based mode*, the UE performs both position measurements and computation of a location estimate, and assistance data for one or both of these functions is provided to the UE by a location server.

In *stand-alone mode*, the UE obtains position measurements and computes a location estimate without making use of assistance data provided by the serving PLMN.

In *network-based mode*, a serving PLMN obtains position measurements of signals transmitted by a UE and computes a location estimate. The transmission of the UE's signals for a network-based mode may or may not be transparent to the UE. For example, the device-transmitted signals may be any of the normal communication and reference signals or may be a positioning-specific signal (e.g. a positioning reference signal (PRS); see also Sect. 3.4).

2.3 Location Services Architecture

Overall System Architecture

The overall system architecture can be divided into the *Radio Access Network* (RAN) and the 5G *Core Network* (5GCN). The 5G Core Network is an evolution of the 4G Evolved Packet Core (EPC). In particular, a service-based architecture (SBA) is the basis for the 5G Core Network. In a service-based architecture, the network functionality is delivered using a set of interconnected *network functions* (NFs), each with authorization to access each other's services. Each NF exposes its functionality through a *Service-Based Interface* (SBI) using HTTP/2 application layer protocol.

The location services (LCS) architecture for the non-roaming case using SBI representation is shown in Fig. 15.1 [2]. As in previous generations (e.g. LTE, UMTS, etc.), the Gateway Mobile Location Centre (GMLC) is used by external LCS clients (e.g. an emergency call centre) to access location services using the Le reference point, which is typically based on the Mobile Location Protocol (MLP) defined by the Open Mobile Alliance (OMA) [11]. A Location Retrieval Function (LRF) could be collocated with a GMLC or separate and is responsible for retrieving or validating location information and providing routing information for a UE which has initiated an IMS emergency session. An LRF may also provide the external Le interface to an emergency services LCS client (e.g. a Public Safety Answering Point

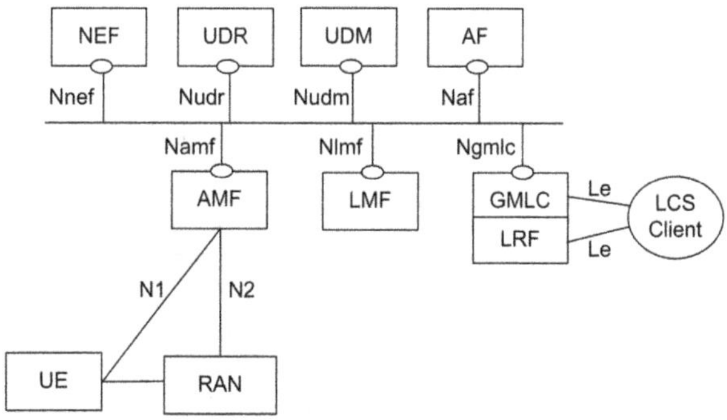

Fig. 15.1 Non-roaming architecture for location services in SBI representation

(PSAP)) in some countries like the USA. In this case, the GMLC can still be present to support location requests but may not have an external Le interface.

Application functions (AFs) and network functions (NFs) may access location services from a GMLC using the Ngmlc interface or event exposure with location information from an AMF using the Namf interface. External AFs may access location services from a Network Exposure Function (NEF) using an API. The NEF may then obtain requested location information either from an AMF using the Namf interface or from a GMLC using the Ngmlc interface.

The key network function added to support positioning of a target UE is the *Location Management Function* (LMF). The LMF is similar to a Serving Mobile Location Centre (SMLC) in previous generations (e.g. LTE, UMTS, etc.) and generally manages the overall coordination and scheduling of resources required for the location of a target UE. This includes positioning of UEs and delivery of assistance data to UEs. The LMF receives location requests for a target UE from the serving AMF using the Nlmf interface. The LMF is also responsible for the provision of broadcast assistance data to UEs via the RAN in ciphered or unciphered form. If the broadcast assistance data were ciphered by the LMF, the LMF forwards any ciphering keys to subscribed UEs via the AMF.

The other NFs shown in Fig. 15.1 perform the following main functions to support location services:

AMF The Access and Mobility Management Function (AMF) is responsible for managing positioning for a target UE. The AMF is accessible to the GMLC and NEF via the Namf interface, to the RAN via the N2 reference point and to the UE via the N1 reference point. AMF functions to support location services include initiating a location request for a UE with an IMS emergency call, receiving and managing location requests from a GMLC or UE, receiving and managing event exposure requests for location information from an NEF and selecting an LMF for positioning a target UE. When assistance data is broadcast by the RAN in a ciphered form, the

AMF receives the ciphering keys from an LMF and forwards the ciphering keys to suitably subscribed UEs using mobility management procedures.

UDM The Unified Data Management (UDM) contains the subscriber LCS privacy profiles and routing information. The UDM is accessible from an AMF, GMLC or NEF via the Nudm interface.

UDR The Unified Data Repository (UDR) contains privacy data information for target UEs and may be updated by a serving AMF via the UDM with new privacy information received from a UE.

NEF The Network Exposure Function (NEF) provides means for accessing location services by an external AF or internal AF. AFs access location services from an NEF using an API. A NEF can forward a location request to a GMLC or request an event exposure for location information from a serving AMF (via a UDM, if needed). An NEF may request routing information and/or target UE privacy information from the UDM via the Nudm interface.

AF The application function (AF) interacts with the 3GPP Core Network via the NEF in order to access network capabilities.

RAN The Radio Access Network (RAN) is involved in the handling of various positioning procedures including position location of a UE, provision of location-related information not associated with a particular UE and transfer of positioning messages between an AMF or LMF and a UE. The RAN may also broadcast assistance data in Position System Information (posSI) messages in ciphered or unciphered form.

The location services (LCS) architecture for the roaming case using SBI representation is shown in Fig. 15.2 [2]. The Home-GMLC (HGMLC) is the GMLC residing in the target UE's home PLMN. The HGMLC performs the authorization of an external LCS client or AF and verifies the target UE's subscription and privacy requirements. After successful authorization of an external LCS client or AF, the HGMLC forwards location requests for a roaming UE either to a visited GMLC (VGMLC) using the Ngmlc interface or to the serving AMF in the VPLMN using the Namf interface. The VGMLC is the GMLC which is associated with the serving node of the target UE.

NG-RAN Positioning Architecture

The *Radio Access Network* (RAN) in Figs. 15.1 and 15.2 consists of both ng-eNBs for LTE access and gNBs for NR access and is referred to as NG-RAN in Fig. 15.3 [3].

The Xn interface is connecting gNBs to each other, as well as gNBs and ng-eNBs to support, for example, mobility and dual connectivity. The Xn interface is currently not used for location services and positioning signalling. The gNBs and ng-eNBs are connected to the 5G Core Network using the NG interface which

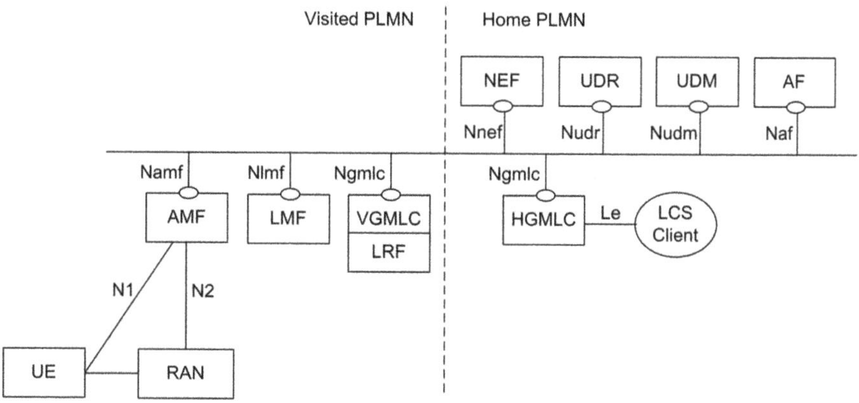

Fig. 15.2 Roaming architecture for location services in SBI representation

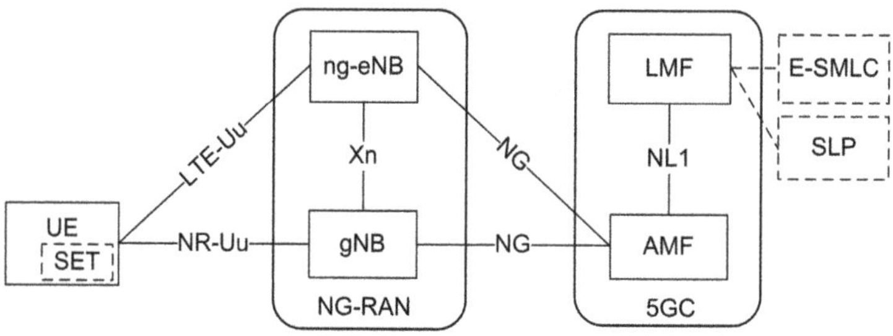

Fig. 15.3 UE positioning architecture applicable to NG-RAN

supports transfer of positioning messages between an LMF and a RAN node via an AMF.

Figure 15.3 also shows the LMF in reference point representation which is connected to the AMF via the NL1 interface. In a reference point representation, the interactions between the services in the network functions are described by point-to-point reference points (e.g. NL1) between any two network functions (e.g. AMF and LMF).

An LMF may have a (non-standardized) signalling connection to an Evolved-Serving Mobile Location Centre (E-SMLC) which may enable an LMF to access information from E-UTRA (e.g. to support the "OTDOA for LTE" positioning method using downlink measurements obtained by a target UE of signals from eNBs in LTE). An LMF may also have a (non-standardized) signalling connection to a SUPL Location Platform (SLP). The SLP is the Secure User Plane Location (SUPL) entity responsible for positioning over the user plane. A mobile terminal supporting SUPL is referred to as SUPL-enabled terminal (SET). User plane positioning using SUPL is standardized by the Open Mobile Alliance (OMA). Further details of user plane positioning can be found in [12].

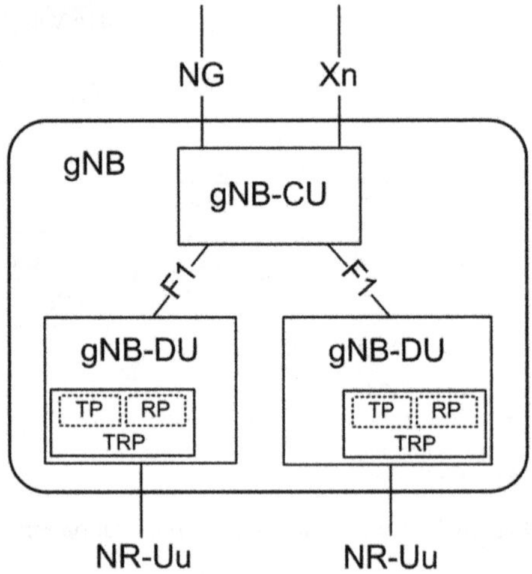

Fig. 15.4 gNB positioning architecture

The 4G RAN architecture was based on a "monolithic" building block, the eNB (the LTE logical node (aka base station)). In 5G the gNB can be split up between central units (gNB-CUs) and distributed units (gNB-DUs). This allows a more flexible hardware implementation and coordination of performance features, load management, etc. The gNB-CU and gNB-DU are connected using a standardized F1 interface. In the case of a split gNB, the RRC, PDCP and SDAP protocol entities reside in the gNB-CU and the lower-layer protocol entities (RLC, MAC, PHY) in the gNB-DU.

Both the gNB-CU and gNB-DU are also part of the positioning architecture. For transmitting DL positioning signals, and measuring UL positioning signals, a "Transmission Measurement Function" is defined which is realized by a "Transmission-Reception Point" (TRP). The TRPs are part of the gNB-DU, as shown in Fig. 15.4, but may also be implemented as a separate entity. A TRP may support TP (transmission point) or RP (reception point) functionality, or both TP and RP functionality. A positioning-only gNB-DU with TRP is also supported which does not need to offer cell services to the gNB-CU.

In certain deployment scenarios where sufficient communication coverage exists, but additional positioning coverage is desired (e.g. factory IIoT scenarios, etc.), a physically separate positioning gNB-DU/TRP may be more economic, since only a small subset of gNB-DU functionality is required for positioning purposes. In addition, positioning gNB-DUs/TRPs may be able to broadcast positioning signal configurations which cannot be supported by a normal gNB-DU (e.g. in case of a Terrestrial Beacon System (TBS)).

The general TRP capabilities comprise the transmission of DL positioning reference signals (PRS) according to a selected configuration, and performing UL-PRS signal measurements and reporting the measurement result in an LMF.

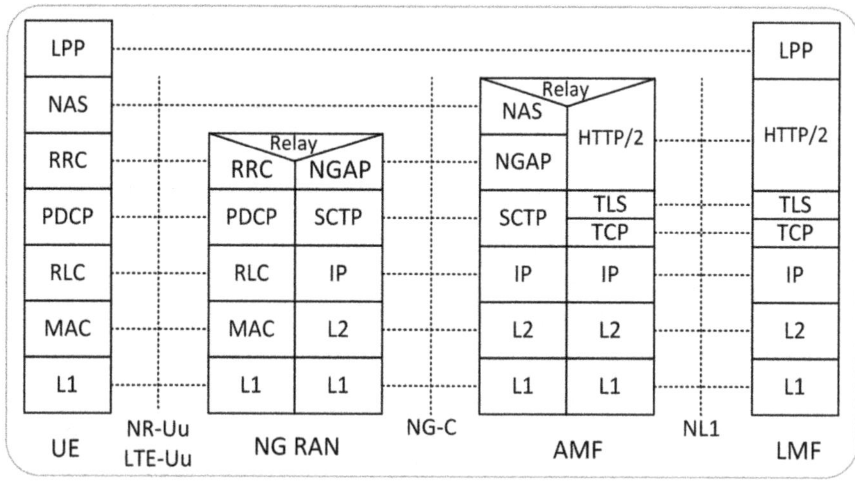

Fig. 15.5 Protocol stack for LMF to UE signalling

Positioning Protocol Architecture

The protocol architecture for the positioning signalling between an LMF and target UE is shown in Fig. 15.5 [3]. The Long-Term Evolution (LTE) Positioning Protocol (LPP) specified in [4] is used for the positioning signalling between an LMF and target UE also for 5G NR positioning (despite its name). LPP has been defined initially for LTE access (during 3GPP Release-9), but in a forward-compatible way, which avoids creating new positioning protocols for future access types developed by 3GPP. The LPP enables position location using a variety of different positioning methods while separating the details of any particular positioning method and the specifics of the underlying transport from each another. Therefore, also NR native positioning methods can be supported with LPP.

In addition, LPP supports the inclusion of External Protocol Data Units (EPDUs). This allows support for additional positioning methods not standardized in 3GPP. An example of external positioning methods which can be supported by LPP is the Open Mobile Alliance LPP Extension (LPPe) specified in [13].

The LTE Positioning Protocol (LPP) is terminated between a UE and a positioning server, which in case of 5G is the LMF. LPP messages are carried as transparent Protocol Data Units (PDUs) across intermediate network interfaces using the appropriate protocols (e.g. NGAP over the NG-C interface, NAS/RRC over the LTE-Uu and NR-Uu interfaces).

LPP operates on a transaction basis between a target UE and an LMF. Each LPP transaction is realized as an independent procedure with each procedure performing a single operation. An LPP procedure may include a request-response pairing of messages or one or more unsolicited messages. Each LPP procedure has a single objective, including the transfer of assistance data, the exchange of positioning related capabilities or the positioning of a target UE using one or more positioning method. More than one LPP procedure may be in progress at any time during a location session to achieve more complex objectives (e.g. positioning of a target UE together with the transfer of assistance data and exchange of positioning related capabilities). The main functions of LPP are:

- To provide the location server (LMF) with the positioning capabilities of the target UE
- To transfer assistance data from the LMF to the UE
- To provide the LMF with coordinate position information (UE-based or stand-alone) or UE measured signals (UE-assisted)
- To report errors during the positioning session

LPP also supports RRC broadcast of location assistance data information using data types defined in relation to LPP which are embedded in positioning System Information Blocks (SIBs). This enables an LMF and a UE to support broadcast location assistance data using the same data structures which are used for point-to-point location.

In the case of NR positioning techniques (see Sect. 3.3), the LMF may require information from an NG-RAN node (gNB or ng-eNB), such as UL positioning measurements from RPs or gNB (TRP) configuration information, for example. A protocol called the NRPP-Annex (NRPPa) is used to transport this information. NRPPa carries information between an NG-RAN node and the LMF as specified in [5]. It is used to support the following main functions:

- Network-based positioning where assistance data or measurements are exchanged between an NG-RAN node and LMF
- Data collection from NG-RAN nodes (e.g. gNB (TRP) coordinates and radio configuration) for support of various positioning methods
- Exchange of information between LMF and NG-RAN for the purpose of assistance data broadcasting

The protocol architecture for the signalling between an LMF and NG-RAN node is shown in Fig. 15.6 [3]. The NRPPa protocol is transparent to the AMF.

2.4 *Location Procedures*

Types of Location Requests

Location information for UEs may be requested by an authorized LCS client or application function (AF) (e.g. emergency centre, friend-finder application, etc.). A location request may be an *immediate location request*, where the LCS client or AF expects to receive a response containing location information for the target UE within a short time period which may be specified using a quality of service (QoS) parameter (referred to as response time). An immediate location request may be

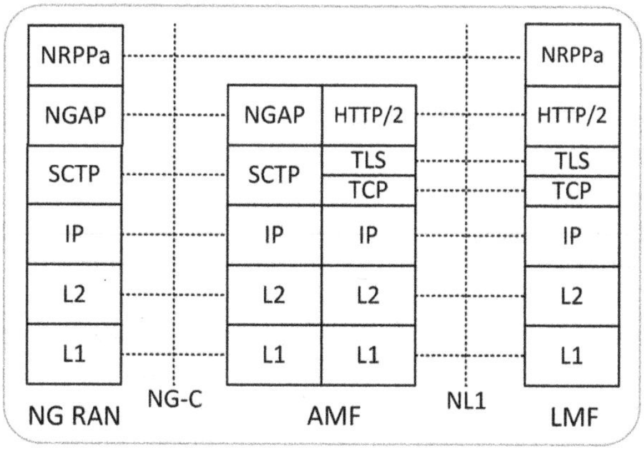

Fig. 15.6 Protocol stack for LMF to NG-RAN signalling

used for a mobile-terminated location request (MT-LR), mobile-originated location request (MO-LR) or network-induced location request (NI-LR). With a *mobile-terminated location request*, an LCS client or application function sends a location request to the PLMN (which may be the home PLMN or visited PLMN) for the location of a UE. With a *mobile-originated location request*, a UE sends a request to a serving PLMN for location-related information. This may include a request for a location estimate of the UE (e.g. in case of UE-assisted or network-based mode) or a request for location assistance data (e.g. for UE-based mode). A location estimate may be returned to the UE, or the UE may request transfer of the location estimate to an external LCS client or external AF via a GMLC and/or NEF. With a *network-induced location request*, a serving AMF for a UE initiates location of the UE for some regulatory service (e.g. emergency call service or lawful intercept).

A location request may also be a *deferred location request*. With a deferred location request, an LCS client or AF sends a location request to a PLMN for a target UE and expects to receive a response containing the indication of event occurrence and location information at some future time (or times). Deferred location requests are defined for MT-LR and in 5G support the following event reports.

A *UE availability event* is triggered when the network has established a contact with the UE. This event is, for example, applicable when the UE is temporarily unavailable due to an enforced idle period (e.g. to conserve UE power), or for temporary loss of radio connectivity and so on. The UE available event only requires one response to an LCS client or AF, and after this response, the UE available event is concluded.

An *area event* is triggered when the UE enters, leaves or remains within a predefined geographical area. The area event report contains the indication of the event occurrence and possibly a location estimate if that was requested by the LCS client or AF.

For *periodic location events*, the event report is triggered when a defined periodic timer expires in the UE.

For a *motion event*, the event is triggered when the UE moves by more than some predefined straight-line distance from a previous location. For successive motion event reports, motion is determined relative to the UE location corresponding to the immediately preceding event report.

An application may have the same location service requirements for multiple UEs, e.g. a group of UEs or any UE in a certain area (e.g. certain IoT applications). In this case, it is not necessary to initiate multiple individual location requests with the same requirements for each UE. The 5G system supports location requests for multiple UEs simultaneously. A mobile-terminated immediate or deferred location request can be used by an LCS client or AF to request the location of a group of UEs.

Privacy for Location Services

Support of location privacy for a UE is a 5G Core Network function. An LCS client or AF may or may not be authorized to retrieve the location of a UE, e.g. for commercial use. The 5G system allows a UE and AF to control which LCS clients are and are not allowed to access UE location information. UE LCS privacy can be supported via subscription, and a UE can update certain privacy preferences.

The Unified Data Repository (UDR) is an entity in the 5G Core Network which stores the subscriber information. Related to the UDR is the Unified Data Management (UDM) function which interfaces with NFs such as AMF, GMLC or NEF so that relevant subscription and privacy data becomes available to the AMF, GMLC or NEF. These subscription and privacy data may include the UE LCS privacy profile which can include an indication whether location requests from LCS clients are allowed or disallowed (referred to as a Location Privacy Indication parameter). The Location Privacy Indication can be provided and updated by the UE. Additional information for each identified LCS client may be stored in the UE LCS privacy profile, such as a time period or geographical area where positioning is allowed.

If indicated by the UE privacy profile, the UE can be notified about any location request, and can additionally deny or grant the request.

The UE privacy settings are evaluated before location information has been determined and may additionally be verified after a location estimate has been obtained but before sending the UE location information to the requesting LCS client or AF. For example, for a deferred location request, the UE privacy settings may have changed between initiation and event reporting, or the privacy profile may include restrictions for a geographical area or time.

Example Location Services Procedure

Figure 15.7 illustrates the general network positioning procedure for an 5GC MT-LR requested by an LCS client or AF external to the PLMN for commercial location services [2]. An external LCS client or the AF (via the NEF) sends a request to the (H)GMLC for the location of a UE in Step 1. The (H)GMLC then invokes a Nudm service operation towards the UDM of the target UE to get the privacy settings of the UE at Step 2 (Nudm Subscriber Data Management Service). The (H)GMLC checks the privacy settings and performs all the following steps only if the target UE is

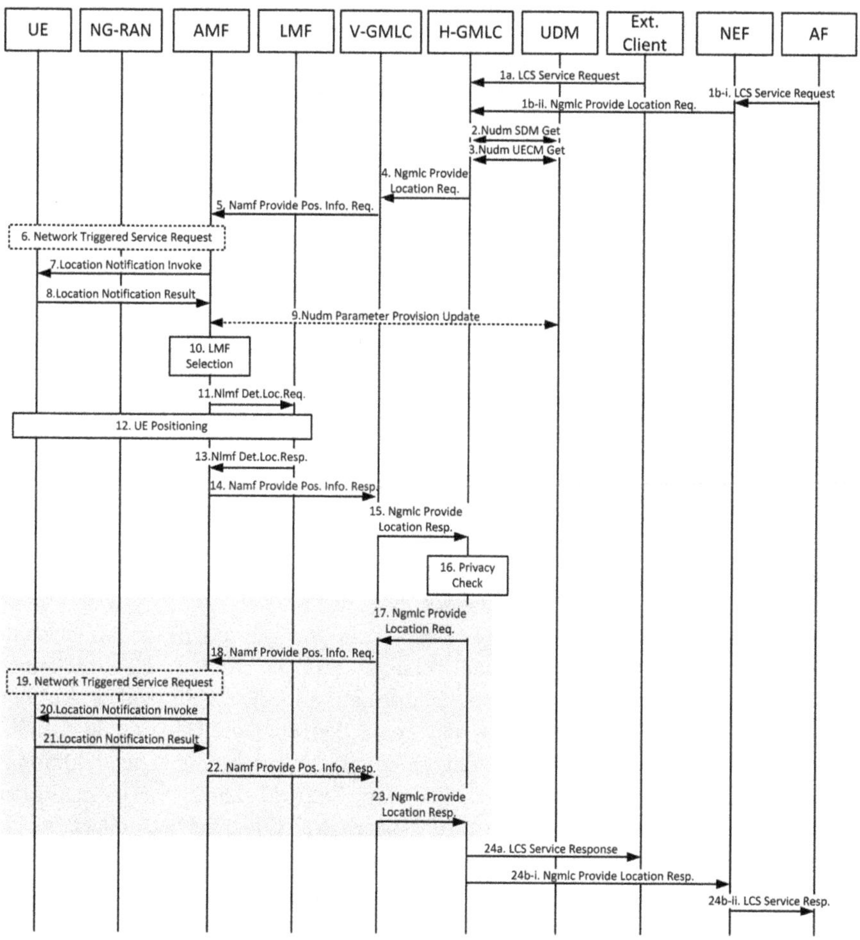

Fig. 15.7 5GC-MT-LR procedure for the commercial location services

allowed to be located. At Step 3, the (H)GMLC may invoke the Nudm_UECM_Get service operation (UE Context Management Service) towards the UDM of the target UE to obtain the network address of the current serving AMF and additionally the address of a V-GMLC (for roaming case). In the case of roaming, the HGMLC then sends the location request to the VGMLC by invoking a Ngmlc service operation towards the VGMLC in Step 4 (Ngmlc Provide Location Request). The VGMLC first authenticates that the location request is allowed from this HGMLC and, if allowed, invokes a Namf service operation towards the AMF to request the current location of the UE at Step 5 (Namf Provide Positioning Info Request). If the UE is in an idle state, the AMF initiates a network-triggered service request procedure to establish a signalling connection with the UE at Step 6.

If LCS privacy settings indicate that the UE must be notified, a notification invoke message is sent to the UE at Step 7, including an identity of the LCS client. The UE then notifies the user and sends any result to the AMF (e.g. location allowed or denied) at Step 8. This notification result may also include an update to the UE LCS privacy settings, e.g. for subsequent location requests. In that case, the AMF invokes an Nudm service operation at Step 9 to store in the UDM the new location privacy information received from the UE. The AMF then selects a suitable LMF at Step 10 and invokes an Nlmf service operation towards the LMF to request the current location of the UE at Step 11 (Nlmf Determine Location Request).

The LMF then performs the positioning procedures at Step 12 required to obtain the UE location. This may include providing assistance data to the UE and requesting location measurements or a location estimate for any of the supported positioning methods. If the UE provides location measurements (UE-assisted mode), the LMF calculates the location of the device. For UE-based mode, the UE provides the computed location to the LMF. The LMF then invokes an Nlmf service operation towards the AMF to provide the current location of the UE at Step 13, and the AMF invokes a Namf service operation towards the VGMLC to forward the UE location to the VGMLC at Step 14. In case of roaming, the VGMLC forwards the location estimate of the UE to the HGMLC at Step 15.

If the privacy check in Step 2 indicates that further privacy checks are needed, the (H)GMLC performs an additional privacy check in order to decide whether the (H)GMLC can forward the location information to the LCS client or AF. For example, an additional privacy check may be needed when the UE user has defined different privacy settings for different geographical areas. When an additional privacy check is not needed, the (H)GMLC skips Steps 17–23. The (H)GMLC then sends the location service response to the external location services client or AF (via the NEF) at Step 24. The (H)GMLC may record charging information both for the external LCS client or AF and inter-network revenue charges from the AMF's network.

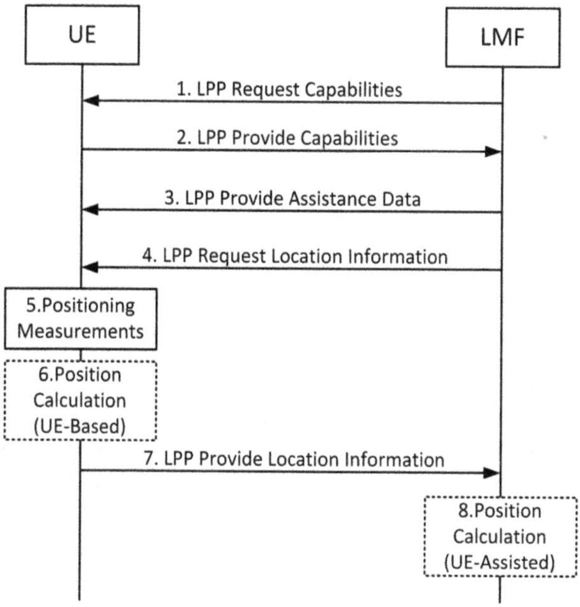

Fig. 15.8 Example UE positioning procedure

Example Positioning Procedure

The positioning of a UE (e.g. at Step 12 in Fig. 15.7) may involve multiple LPP transactions as illustrated in Fig. 15.8. The LMF may first request the positioning capabilities of the UE, which the UE provides then at Step 2 to the LMF. These capabilities include indications of which positioning methods are supported by the UE and any additional details for each method, such as positioning mode (UE-based and/or UE-assisted), specific assistance data supported or any details of the positioning measurements supported. The LMF uses this information to decide on the positioning method(s) to use in order to fulfil the location service request (e.g. requirements on accuracy and response time). The LMF then provides any necessary assistance data to the UE at Step 3. The specific assistance data depends on the selected positioning method, e.g. may include GNSS assistance data for A-GNSS or may include a list of TRPs for DL-TDOA positioning together with positioning reference signal configurations, etc. The LMF then sends a location request to the UE at Step 4 to request the positioning measurements (e.g. DL-TDOA measurements or a location estimate). The UE then performs the requested location measurements at Step 5 and may use these measurements to calculate the position at Step 6, if that was requested at Step 4 (UE-based mode). At Step 7, the UE provides the position estimate to the LMF (if Step 6 was performed); otherwise, the UE provides the position measurements (UE-assisted mode) and the LMF calculates the UE location at Step 8.

3 NR Positioning

3.1 Fundamentals of Position Location

The primary function of a position location system is to locate the coordinates of a UE with respect to a set of objects (e.g. cellular base stations or satellites) with known positions.

Lines of Position

Use of position location systems in a plane, with reference to known transmitter positions (e.g. base stations), always gives so-called lines of position (LOPs), and surfaces of position in three-dimensional position location systems [15]. These lines and surfaces differ depending on the position measurement principles that are used in the specific positioning system (see section "Position Measurements").

If the distance between a receiver and a point of reference is measured, the information implies that the receiver is on a circle or sphere with the point of reference (e.g. base station) at the centre and the measured distance as the radius. If the direction from the receiver to the reference point is measured, the receiver is on a straight line in the measured direction from this point of reference. If the difference in distance to two given reference points is measured, the receiver is on a hyperbola, the foci of which are the two given reference points. Several lines of position have to intersect in order to position the UE, as illustrated in Fig. 15.9.

Any measurement contains uncertainty and has statistical errors following a certain (but usually unknown) probability distribution. In addition, there are systematic errors that are often dominating and might be reduced or eliminated by proper calibration. Examples of systematic errors include signal group delays in transmitter or receiver or time and phase synchronization offsets of the transmitters. Systematic errors may vary too, e.g. as a function of time and position. Therefore, to

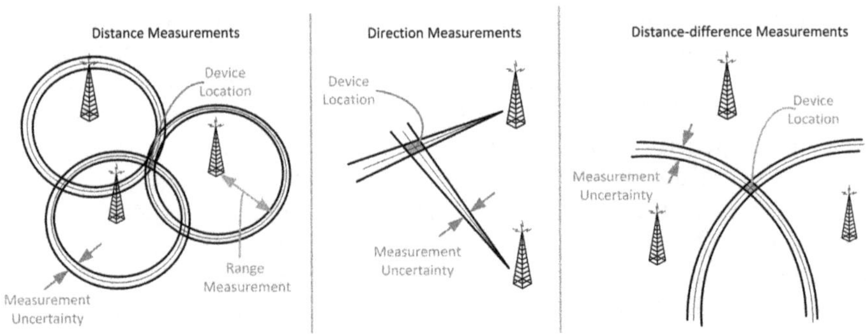

Fig. 15.9 Positioning with intersecting lines of position

some extent it is a matter of definition whether an error type is termed statistical or systematic. In particular, non-line-of-sight signal propagation conditions can involve considerable errors. All these errors correspond to a change of the lines of position, as illustrated in Fig. 15.9.

An angular error changes the direction of the corresponding line of position at the angular measurement, while a distance error gives a parallel displacement of the corresponding line of position. There is also a parallel displacement in case of the hyperbola if the transmitters are at great distance.

The measurement errors have many different causes, and any of these causes can have different non-Gaussian probability distributions. Because of the central limit theorem, the total error distribution is usually approximated by a Gaussian distribution in practice.

Position Measurements

To obtain lines of position, the fundamental quantities needed are distances or angles which can be measured using radio signals. The measurement model can be generalized as

$$\mathbf{r} = \mathbf{f}(\mathbf{x}) + \mathbf{n}, \tag{15.1}$$

where $\mathbf{r}$ is the measurement (column) vector, $\mathbf{x}$ is the (unknown) UE location to be determined, $\mathbf{f}(\mathbf{x})$ is a known non-linear function in $\mathbf{x}$ and $\mathbf{n}$ is the measurement noise/error (column) vector.

Time of Arrival (TOA)

Time of arrival (TOA) of radio signals can be used to measure distance based on an estimate of signal propagation time between a transmitter and a receiver since radio waves propagate at the constant speed of light. The distance between the transmitter and receiver is directly proportional to the propagation time. The usual method for measuring the time of arrival is to construct a signal replica (template) representing the known transmitted signal and to cross-correlate this replica with the actual received signal. Therefore, TOA is usually measured on a known portion of the transmitted signal.

The TOA measurements can be modelled as follows: assuming (without loss of generality) a two-dimensional Cartesian coordinate system with $\mathbf{x} = [x, y]^{\mathrm{T}}$ being the unknown UE location and $\mathbf{x}_i = [x_i, y_i]^{\mathrm{T}}$ for $i = 1, 2, \ldots, N$ being the (known) coordinates of the N base stations (TRP coordinates). The distance between the UE and the base station i denoted as d_i is then

$$d_i = \sqrt{(x - x_i)^2 + (y - y_i)^2} \quad i = 1, 2, \ldots, N. \tag{15.2}$$

Assuming the base stations i transmit a signal at time T_i and the UE receives this signal at time t_i later, there is a simple relationship between TOA t_i and distance d_i:

$$(t_i - T_i) = \frac{d_i}{c} \quad , \quad i = 1, 2, \ldots, N, \tag{15.3}$$

where c is the speed of the radio waves (speed of light). For TOA-based location, the transmitting signal must be tagged with a time stamp in order for the receiver to determine the distance the signal has travelled (i.e. T_i can be determined at the receiver by some means). The TOA measurements denoted as r_{TOA_i} are subject to measurement errors and can be modelled as

$$r_{\mathrm{TOA}_i} = d_i/c + n_i = \frac{1}{c} \cdot \sqrt{(x - x_i)^2 + (y - y_i)^2} + n_i \quad i = 1, 2, \ldots, N, \tag{15.4}$$

where n_i is the range error in r_{TOA_i}. Eq. (15.4) can then be written in form of Eq. (15.1):

$$\mathbf{r}_{\mathrm{TOA}} = \mathbf{f}_{\mathrm{TOA}}(\mathbf{x}) + \mathbf{n}_{\mathrm{TOA}} \tag{15.5}$$

with

$$\mathbf{r}_{\mathrm{TOA}} = \left[r_{\mathrm{TOA}_1} \; r_{\mathrm{TOA}_2} \; \cdots \; r_{\mathrm{TOA}_N} \right]^T \tag{15.6}$$

$$\mathbf{n}_{\mathrm{TOA}} = \left[n_1 \; n_2 \; \cdots \; n_N \right]^T \tag{15.7}$$

and

$$\mathbf{f}_{\mathrm{TOA}}(\mathbf{x}) = \frac{1}{c} \cdot \begin{bmatrix} \sqrt{(x - x_1)^2 + (y - y_1)^2} \\ \sqrt{(x - x_2)^2 + (y - y_2)^2} \\ \vdots \\ \sqrt{(x - x_N)^2 + (y - y_N)^2} \end{bmatrix} \tag{15.8}$$

The position location problem based on TOA measurements is then to estimate $\mathbf{x}$ given the TOA measurements $\{r_{TOA_i}\}$.

Direct TOA measurements for distance estimation result in two main problems. First, the TOA method requires that all transmitters and receivers in the positioning system have precisely synchronized clocks (e.g. 1 μs of timing error can result in 300 m distance error). Second, the transmitting signal must be labelled with a time stamp in order for the receiver to determine the distance the signal has travelled. For this reason, time difference of arrival (TDOA) measurements are a more practical means for position location in cellular systems.

Time Difference of Arrival (TDOA)

As the name implies, TDOA estimation requires the measurement of the difference in time of the received signals (either at the UE which receives signals from multiple base stations (DL-TDOA) or at multiple base stations which all receive the signal transmitted by the UE (UL-TDOA)). Unlike TOA measurements, the transmitted signals need not contain a time stamp, and all the transmitters and receivers in the system do not need to be precisely synchronized. Instead, positioning based on TDOA measurements requires that the base stations have precisely synchronized clocks (either for transmitting the ranging signal (DL-TDOA) or for measuring the UE uplink signal (UL-TDOA)). This corresponds to the timing standard typically available at base stations, making TDOA-based positioning more practical than requiring each UE to have a precise clock.

TDOA measurements could be made via cross-correlation of two received signals. For example, assume the transmitted signal from a base station i is $s_i(t)$ which is received at a target device at propagation delay d_i later $s_i(t-d_i)$. Similarly, the signal from a base station j is received d_j later $s_j(t-d_j)$. The cross-correlation function between the two signals $s_i(t-d_i)$ and $s_j(t-d_j)$ can provide the receive time difference of the two signals. However, this requires receiving and processing the signals from multiple transmitters simultaneously. Therefore, TDOA is typically not directly measured but calculated as the difference of two TOA measurements made relative to the same receive time base. Since this differencing operation cancels any receiver clock offset, for positioning the result is the same as direct cross-correlation: the difference of two TOAs provides a hyperbolic line of position.

The measurement model for TDOA is then the same as in Eq. (15.3). However, compared to TOA location, the absolute transmit time T_i is assumed to be unknown:

$$t_i = T_i + \frac{d_i}{c} \quad , \quad i = 1, 2, \ldots, N. \tag{15.9}$$

There are $N(N-1)/2$ distinct TDOAs from all possible pairs of TOA measurements; however, there are only $(N-1)$ nonredundant TDOAs. Typically, one TOA measurement is used as a reference (e.g. $i = 1$), and the nonredundant TDOAs are calculated as

$$t_j - t_1 = (T_j - T_1) + \left(\frac{d_j}{c} - \frac{d_1}{c}\right) \quad , \quad j = 2, 3, \ldots, N. \tag{15.10}$$

Similar to Eq. (15.4), the TDOA measurement r_{TDOA_j} can then be written as

$$r_{\text{TDOA}_j} = (T_j - T_1) + \left(\frac{d_j}{c} - \frac{d_1}{c}\right) + (n_j - n_1) \quad , \quad j = 2, 3, \ldots, N. \tag{15.11}$$

For TDOA location, the base stations need to be synchronized, and the relative synchronization offset $(T_j - T_1)$ must be known. The function $\mathbf{f}(\mathrm{x})$ in Eq. (15.1) for TDOA location is then

$$\mathbf{f}_{\mathrm{TDOA}}(\mathbf{x}) = \frac{1}{c} \cdot \begin{bmatrix} \sqrt{(x-x_2)^2 + (y-y_2)^2} - \sqrt{(x-x_1)^2 + (y-y_1)^2} \\ \sqrt{(x-x_3)^2 + (y-y_3)^2} - \sqrt{(x-x_1)^2 + (y-y_1)^2} \\ \vdots \\ \sqrt{(x-x_N)^2 + (y-y_N)^2} - \sqrt{(x-x_1)^2 + (y-y_1)^2} \end{bmatrix} \tag{15.12}$$

The position location problem based on TDOA measurements is then to estimate $\mathbf{x}$ given the TDOA measurements $\{r_{\mathrm{TDOA}_i}\}$.

The NR positioning methods which make use of TDOA measurements are the DL-TDOA and UL-TDOA positioning techniques described in sections "Downlink Time Difference of Arrival (DL-TDOA) Positioning" and "Uplink Time Difference of Arrival (UL-TDOA) Positioning", respectively.

Round-Trip Time (RTT)

If a pair of UE and base station are measuring receive and transmit time, a round-trip time (RTT) can be determined, which corresponds to the time of flight (and hence, distance) separating them. Figure 15.10 illustrates how time of flight is typically measured. It involves both uplink and downlink measurements.

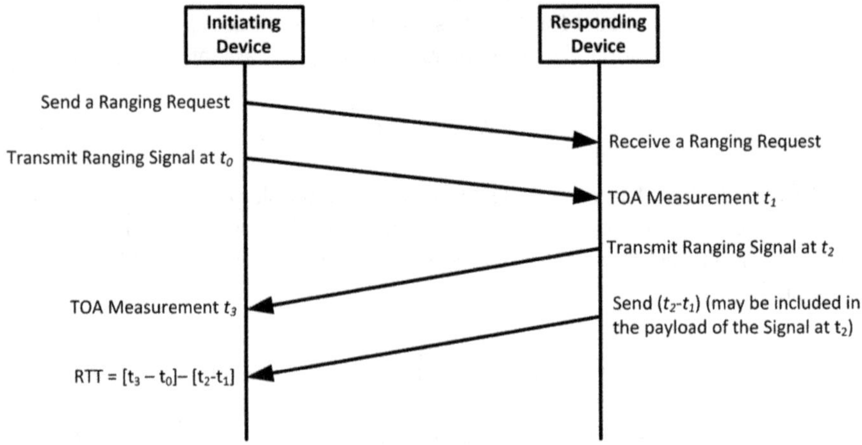

Fig. 15.10 Illustration of time-of-flight principle

The initiating device in Fig. 15.10 may be a base station or a UE and the responding device a UE or base station. The initiating device transmits a ranging signal and records the transmit t_0. This ranging signal is received at the propagation distance later at the responding device which records the receive time t_1. After some internal processing time, the responding device then transmits a ranging signal at t_2, which the initiating device receives at time t_3. The round-trip time (RTT) is then $(t_3\text{-}t_0) - (t_2\text{-}t_1)$. This RTT corresponds to the two-way time of flight if there are no internal processing delays between, e.g. receiving and processing the signals. However, there are usually additional time delays incurred as the signals propagate through various processing stages at the devices. Knowledge of these processing delays is required for accurate distance estimation using RTT measurements. However, since the time differences $(t_3\text{-}t_0)$ and $(t_2\text{-}t_1)$ involve only the local clock of the initiating and responding device, respectively, positioning based on multiple RTT measurements (e.g. between a UE and multiple base stations) does not require synchronized base stations, which can be a significant advantage in practice.

Similar as for TDOA positioning in cellular networks, also for RTT-based positioning, the fundamental measurement is a TOA measurement from which transmit-receive time differences are determined at the individual devices. The RTT can then be calculated as the sum (or difference) of the transmit-receive time differences at the initiating device and responding device. The measurement model is then the same as for TOA-based location, denoting $r_{\mathrm{RTT}_i} = [(t_3 - t_0) - (t_2 - t_1)]_i$ for a pair i of devices with

$$\mathbf{f}_{\mathrm{RTT}}(\mathbf{x}) = \frac{2}{c} \cdot \begin{bmatrix} \sqrt{(x - x_1)^2 + (y - y_1)^2} \\ \sqrt{(x - x_2)^2 + (y - y_2)^2} \\ \vdots \\ \sqrt{(x - x_N)^2 + (y - y_N)^2} \end{bmatrix}. \tag{15.13}$$

The position location problem based on RTT measurements is then to estimate $\mathbf{x}$ given the RTT measurements $\{r_{\mathrm{RTT},\,i}\}$.

The NR positioning method which makes use of RTT measurements is the multi-RTT technique described in section "Multi-Round-Trip-Time (Multi-RTT) Positioning".

Received Signal Strength (RSS)

Another way for obtaining distance information is based on received signal strength (RSS) measurements. The strength of DL radio signals from nearby base stations is usually measured at each UE for radio resource management purposes. Since signal strength reduces the further the signal travels, signal strength can represent an estimate of distance when comparing the received signal strength to the transmission power. Signal strength varies non-linearly with distance, so RSS is typically a better indicator of distance when devices are close to each another.

Signal strength in cellular systems is highly variable since radio signals are highly susceptible to environmental conditions, and thus can only be characterized statistically. However, the positioning scheme is simpler compared to TOA- or TDOA-based location.

Assuming that the base station i transmitted power is $P_{t,\,i}$ and in the absence of disturbance, the average receive power at a UE $P_{r,\,i}$ can be modelled as

$$P_{r,i} = K_i P_{t,i} d_i^{-\alpha} \quad , \quad i = 1, 2, \ldots, N \tag{15.14}$$

where K_i accounts for all other factors that may affect the receive power (e.g. antenna height, antenna gains, etc.) and α is the pathloss exponent. Depending on the propagation environment, α can vary from 2 to about 5 ($\alpha = 2$ corresponds to free space wave propagation). It is assumed that $P_{t,\,i}$, K_i and α are known a priori. The disturbance in the received signal strength is typically lognormal distributed. Accordingly, the lognormal measurement (pathloss) model from Eq. (15.14) can be expressed as

$$\ln\left(P_{r,i}\right) = \ln\left(K_i\right) + \ln\left(P_{t,i}\right) - \alpha \ln\left(d_i\right) + n_{\mathrm{RSS},i} \quad , \quad i = 1, 2, \ldots, N, \tag{15.15}$$

where the disturbance $n_{RSS,\,i}$ is now Gaussian distributed. Denoting

$$r_{\mathrm{RSS},i} = \ln\left(P_{r,i}\right) - \ln\left(K_i\right) - \ln\left(P_{t,i}\right), \tag{15.16}$$

the RSS signal model can be simplified as

$$r_{\mathrm{RSS},i} = -\alpha \ln\left(d_i\right) + n_{\mathrm{RSS},i} \quad , \quad i = 1, 2, \ldots, N, \tag{15.17}$$

which can be written in matrix notation according to Eq. (15.1):

$$\mathbf{r}_{\mathrm{RSS}} = \mathbf{f}_{\mathrm{RSS}}\left(\mathbf{x}\right) + \mathbf{n}_{\mathrm{RSS}} \tag{15.18}$$

with

$$\mathbf{f}_{\mathrm{RSS}}\left(\mathbf{x}\right) = -\alpha \cdot \begin{bmatrix} \ln\left(\sqrt{(x - x_1)^2 + (y - y_1)^2}\right) \\ \ln\left(\sqrt{(x - x_2)^2 + (y - y_2)^2}\right) \\ \vdots \\ \ln\left(\sqrt{(x - x_N)^2 + (y - y_N)^2}\right) \end{bmatrix} . \tag{15.19}$$

The position location problem based on RSS measurements is then to estimate $\mathbf{x}$ given the RSS measurements $\{r_{\text{RSS},\,i}\}$.

RSS-based ranging techniques require knowledge of $P_{t,\,i}$, K_i and in particular α, which depends on the propagation environment. RSS-based ranging techniques are seldom accurate, being more suited to proximity determination than continuous positioning. RSS measurements are typically used for "RF Pattern Matching" which uses a survey of the area with signal strength measurements. The method is called "RF Pattern Matching" or "RF Fingerprinting" because it takes a map of predicted or surveyed received signal strength at grid points of the network coverage area and compares the UE RSS measurements with this map. Each "fingerprint" can be associated with a specific location. The best match between the UE measured RSS or RF pattern (usually from multiple base stations) with entries in the database determines the device location. The main issue in practice with this approach is the construction of an accurate map of predicted "RF Fingerprints", which also requires regular updates of this map since cellular networks are inherently time varying.

The NR positioning methods which may make use of received signal strength measurements are the general E-CID technique described in section "Enhanced Cell-ID (E-CID) Positioning" and the DL-AoD technique described in section "Downlink Angle-of-Departure (DL-AoD) Positioning". Both methods may make use of a database of radio signal properties ("RF Fingerprint"), such as a received signal strength map from multiple TRPs at a particular location, or received signal strength at different angles received from the same TRP.

Angle of Arrival (AoA)

An angle of arrival (AoA) estimate is made from base stations using directional antennas such as phased arrays to determine which direction a signal was transmitted from. This permits a heading to be determined from the base station (line of position). When combined with a range estimate (e.g. using RTT), this can provide a location estimate from a single base station.

Directional antennas can be used to steer radio signals in a particular direction. The signal produced by some directional antennas can be controlled such that narrow beams can be transmitted. If the signal beam is rotated in a sweeping pattern, then the location of a UE can be detected by determining which beams generate responses. The shaped beam isn't perfectly narrow, so several responses are likely to be generated. In addition, accuracy can be limited by multipath reflections arriving from misleading directions. However, AoA positioning does not require synchronization among the base stations.

Denoting ϕ_i the angle of arrival between a base station i and a UE, we obtain:

$$\tan(\phi_i) = \frac{(y - y_i)}{(x - x_i)} \quad , \quad i = 1, 2, \ldots, N \ . \tag{15.20}$$

Geometrically, ϕ_i is the angle between the line of bearing from the UE to the i-th base station and the x-axis. The AoA measurements in the presence of measurement errors, denoted by $\{r_{AOA,\,i}\}$, can then modelled as

$$r_{\text{AOA},i} = \phi_i + n_{\text{AOA},i} = \tan^{-1}\left(\frac{y - y_i}{x - x_i}\right) + n_{\text{AOA},i} \ , \quad i = 1, 2, \ldots N. \qquad (15.21)$$

and the function **f**(x) in Eq. (15.1) for AoA location is then

$$\mathbf{f}_{\text{AOA}}(\mathbf{x}) = \begin{bmatrix} \tan^{-1}\left(\frac{y-y_1}{x-x_1}\right) \\ \tan^{-1}\left(\frac{y-y_2}{x-x_2}\right) \\ \vdots \\ \tan^{-1}\left(\frac{y-y_N}{x-x_N}\right) \end{bmatrix}. \qquad (15.22)$$

The position location problem based on AoA measurements is then to estimate **x** given the AoA measurements $\{r_{\text{AOA},\,i}\}$.

The NR positioning method which makes use of AoA measurements is the general UL-AoA technique described in section "Uplink Time Difference of Arrival (UL-TDOA) Positioning".

Measurement Errors

The fundamental position location measurements and principles described in the sections above can work well if the radio signals are not affected by noise, interference or multipath. In cellular communication systems, however, position location errors occur due to imperfections of the propagation channel. While positioning techniques typically require three or more TRPs to determine a unique position, the cellular communication system is typically designed to ensure only one high signal-to-noise ratio (SNR) link between a transmitter (e.g. base station) and receiver (e.g. UE). The ability of the UE to receive ("hear") signals from multiple TRPs or the ability of multiple TRPs to receive the UE signal is essential to the design of position location systems. This problem is referred to as *hearability*, and it is where the design principles of position location systems are typically different from those of cellular communication systems. Hearability is more of a problem in rural or suburban environments, where coverage requirements determine the communication system design and deployment rather than capacity demands, which leads to extra base stations with redundant coverage in urban areas.

Both time and angle measurement techniques rely on a direct line-of-sight propagation path from the transmitter to the receiver. However, most propagation environments induce significant propagation path blockage and multipath due to reflections and diffractions from and around buildings and terrains. Multipath components may appear as a signal arriving from a completely different direction which can lead to large errors in an angle-based position location system. Although diffraction may have less severe consequences for the relative TOA of a signal in a TDOA-based system, signal reflections from distant objects can lead to time distortions and multipath excess delays of several microseconds.

The propagation channel impairments typically require special signal processing techniques to improve resistance of both angle and time delay methods, to noise, interference and multipath. It is often beneficial to use more than the minimum number of TRPs for a unique solution, in order to average out errors induced by the propagation channel.

Geometrical Influence on Position Errors

In any ranging- or angle-based positioning system, receiver-transmitter geometry influences position precision. Figure 15.11 illustrates the influence of receiver-transmitter geometry to position precision for a ranging (distance)-based system. In Fig. 15.11, the uncertainty in the UE position is indicated by the shaded areas. In Fig. 15.11a, this position uncertainty is relatively small. In Fig. 15.11b, the two transmitters are moved closer together, and, although the measurement uncertainty is the same, the position uncertainty is considerably larger. This effect is referred to as dilution of precision (DOP).

A formal derivation of the DOP relation requires the linearization of the measurement equations in section "Position Measurements" (Eq. (15.1)):

$$\mathbf{r} = \mathbf{f}(\mathbf{x}) + \mathbf{n}. \tag{15.23}$$

The measurement error $\mathbf{n}$ is assumed to be a multivariate random vector with a $N \times N$ covariance matrix:

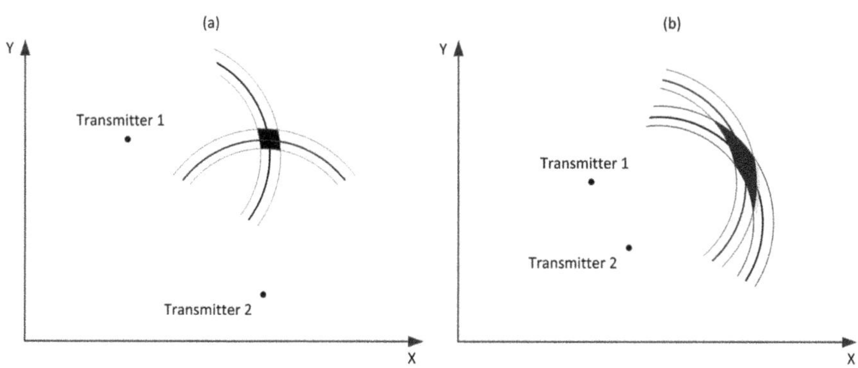

Fig. 15.11 Illustration of dilution of precision (DOP)

$$\mathbf{N} = E\left\{(\mathbf{n} - E\{\mathbf{n}\})(\mathbf{n} - E\{\mathbf{n}\})^T\right\}, \tag{15.24}$$

where $E\{\cdot\}$ denotes the expected value and the subscript T denotes the transpose. If $\mathbf{x}$ is regarded as an unknown but non-random vector and $\mathbf{n}$ is assumed to have a zero-mean Gaussian distribution, then the conditional probability density function of $\mathbf{r}$ given $\mathbf{x}$ is

$$p(\mathbf{r}|\mathbf{x}) = \frac{1}{(2\pi)^{N/2}|\mathbf{N}|^{1/2}} \exp\left\{-(1/2)[\mathbf{r} - \mathbf{f}(\mathbf{x})]^T \mathbf{N}^{-1}[\mathbf{r} - \mathbf{f}(\mathbf{x})]\right\} \tag{15.25}$$

where $|\mathbf{N}|$denotes the determinant of $\mathbf{N}$ and subscript -1 denotes the inverse. The maximum likelihood estimator is therefore the value $\mathbf{x}$ which minimize the following cost function:

$$Q(\mathbf{x}) = [\mathbf{r} - \mathbf{f}(\mathbf{x})]^T \mathbf{N}^{-1}[\mathbf{r} - \mathbf{f}(\mathbf{x})], \tag{15.26}$$

and therefore

$$\hat{\mathbf{x}} = \text{argmin}\left\{[\mathbf{r} - \mathbf{f}(\mathbf{x})]^T \mathbf{N}^{-1}[\mathbf{r} - \mathbf{f}(\mathbf{x})]\right\}. \tag{15.27}$$

The minimization of the cost function Q is a reasonable criterion for determination of $\mathbf{x}$ even when the additive error cannot be assumed to be Gaussian. In this case, the resulting estimator is the least-squares estimator and $\mathbf{N}^{-1}$ is the matrix of weighting coefficients.

The function $\mathbf{f}(\mathbf{x})$ in Eq. (15.23) is a non-linear vector function (e.g. see section "Position Measurements"). A common approach to minimize the cost function (15.26) is to linearize $\mathbf{f}(\mathbf{x})$. $\mathbf{f}(\mathbf{x})$ can be expanded in a Taylor series about a reference point $\mathbf{x}_0$ and the second and higher terms can be neglected:

$$\mathbf{f}(\mathbf{x}) \cong \mathbf{f}(\mathbf{x}_0) + \mathbf{G} \cdot (\mathbf{x} - \mathbf{x}_0) \tag{15.28}$$

where $\mathbf{x}$ and $\mathbf{x}_0$ are $N \times 1$ column vectors and $\mathbf{G}$ is the $N \times 2$ (in case of 2D $\mathbf{x} = [x, y]^{\mathrm{T}}$) matrix of derivatives evaluated at $\mathbf{x}_0$:

$$\mathbf{G} = \begin{bmatrix} \left.\frac{\partial f_1}{\partial x}\right|_{\mathbf{x}=\mathbf{x}_0} & \left.\frac{\partial f_1}{\partial y}\right|_{\mathbf{x}=\mathbf{x}_0} \\ \left.\frac{\partial f_2}{\partial x}\right|_{\mathbf{x}=\mathbf{x}_0} & \left.\frac{\partial f_2}{\partial y}\right|_{\mathbf{x}=\mathbf{x}_0} \\ \vdots & \vdots \\ \left.\frac{\partial f_N}{\partial x}\right|_{\mathbf{x}=\mathbf{x}_0} & \left.\frac{\partial f_N}{\partial y}\right|_{\mathbf{x}=\mathbf{x}_0} \end{bmatrix} \tag{15.29}$$

The solution of (27) is then given by (e.g. [14]):

$$\hat{\mathbf{x}} = \mathbf{x}_0 + \left(\mathbf{G}^{\mathrm{T}}\mathbf{N}^{-1}\mathbf{G}\right)^{-1}\mathbf{G}^{\mathrm{T}}\mathbf{N}^{-1}\left(\mathbf{r} - \mathbf{f}\left(\boldsymbol{x}_0\right)\right). \tag{15.30}$$

The covariance matrix of the position estimate $\hat{\mathbf{x}}$ is

$$\mathbf{R} = E\left\{\left(\hat{x} - E\left\{\hat{x}\right\}\right)\left(\hat{x} - E\left\{\hat{x}\right\}\right)^T\right\} \tag{15.31}$$

which gives (e.g. [14])

$$\mathbf{R} = \left(\mathbf{G}^{\mathrm{T}}\mathbf{N}^{-1}\mathbf{G}\right)^{-1} = \begin{bmatrix} \sigma_x^2 & \rho_{xy} \\ \rho_{xy} & \sigma_y^2 \end{bmatrix} \tag{15.32}$$

For the definition of DOP, the assumption is that the components of the measurement error $\mathbf{n}$ are identically distributed and independent with the same variance σ_{meas}^2 :

$$\mathbf{N} = \sigma_{\text{meas}}^2\mathbf{I} \tag{15.33}$$

where $\mathbf{I}$ is the $N \times N$ identity matrix. Substitution into (15.32) yields

$$\mathbf{R} = \left(\mathbf{G}^{\mathrm{T}}\mathbf{G}\right)^{-1}\sigma_{\text{meas}}^2. \tag{15.34}$$

Under the stated assumptions, the covariance of the errors in the computed position is just a scalar multiple of the matrix $(\mathbf{G}^{\mathrm{T}}\mathbf{G})^{-1}$. The geometric dilution of precision (GDOP) is defined as the ratio of the standard deviation of errors in the least-squares solution to the standard deviation of the measurement errors. For this two-dimensional example, GDOP is given by

$$\text{GDOP} = \frac{\sqrt{\sigma_x^2 + \sigma_y^2}}{\sigma_{\text{meas}}} = \frac{\sqrt{\text{trace}\left\{\left(\mathbf{G}^{\mathrm{T}}\mathbf{G}\right)^{-1}\right\}}}{\sigma_{\text{meas}}} \tag{15.35}$$

or

$$\sqrt{\sigma_x^2 + \sigma_y^2} = \text{GDOP} \times \sigma_{\text{meas}} \tag{15.36}$$

The square root term on the left side characterizes the location error. GDOP depends only on transmitter-receiver geometry. It represents the amplification of the standard deviation of the measurement errors onto the solution. For example, considering a TOA-based system according to section "Time of Arrival (TOA)", the matrix $\mathbf{G}$ can be determined with (15.8) and (15.29), which only depends on mobile and base station coordinates/locations:

$$\mathbf{G} = \begin{bmatrix} \frac{(x-x_1)}{\sqrt{(x-x_1)^2+(y-y_1)^2}} & \frac{(y-y_1)}{\sqrt{(x-x_1)^2+(y-y_1)^2}} \\ \frac{(x-x_2)}{\sqrt{(x-x_2)^2+(y-y_2)^2}} & \frac{(y-y_2)}{\sqrt{(x-x_2)^2+(y-y_2)^2}} \\ \vdots & \vdots \\ \frac{(x-x_N)}{\sqrt{(x-x_N)^2+(y-y_N)^2}} & \frac{(y-y_N)}{\sqrt{(x-x_N)^2+(y-y_N)^2}} \end{bmatrix} \quad (15.37)$$

3.2 Enhanced Location Capabilities with 5G NR

5G NR provides a number of significant improvements when compared to 4G in areas like flexibility, scalability and efficiency, both in terms of power usage and spectrum [18]. There are several cornerstones to the new radio used for 5G that can be exploited to expand the range of cellular positioning techniques available, including larger bandwidths at higher frequencies, denser cellular networks and more antennas combined into complex antenna arrays.

Spectrum for 5G NR

The demanding mobile broadband usage scenario and related new services require higher data rates and capacity in dense deployments. Initial 5G deployments will typically be in the frequency bands which are already used for previous communication systems generations. However, frequency bands above 24 GHz are supported in 5G NR from the beginning to complement to the frequency bands below 6 GHz. With the 5G requirements for extremely high data rates, even higher frequencies above 60 GHz will be considered. These bands are often called mm-wave bands, referring to the wavelength in these bands.

The advantage of the higher-frequency bands is that they are much wider which allows much higher signal bandwidths and hence support much higher data rates. Large signal bandwidth is also one of the key enablers for more accurate cellular positioning.

Larger signal bandwidth means that signal time can be more accurately resolved (there is an inverse relationship between time resolution and signal bandwidth). Therefore, larger signal bandwidths offer an improved ability to resolve multipath effects, which is one of the main error sources in cellular positioning, in particular in urban environments and indoor scenarios. The signals that travel different paths arrive at different times, and therefore, it is more challenging to obtain accurate TOA estimates of the desired line-of-sight component.

5G NR supports channel bandwidth of up to 100 MHz in frequency bands below 6 GHz (e.g. LTE supports up to 20 MHz). The frequency bands below 6 GHz are referred to as frequency range 1 (FR1) in 3GPP specifications, which includes all existing and new bands below 6 GHz. In frequency bands above 24 GHz, channel bandwidth of up to 400 MHz is supported. The frequency bands in the range of 24.25–52.6 GHz are referred to as frequency range 2 (FR2) in 3GPP specifications. And even higher bandwidths are possible through carrier aggregation.

Cellular Base Station Deployment

The disadvantage of the higher-frequency bands in some respects is that they will have a much shorter range, but this is also an advantage because it will also allow much greater frequency reuse – and it further benefits cellular positioning. Due to smaller cell sizes, positioning with basic Cell-ID may already provide accuracies sufficient for some user cases and applications. Cell-ID positioning allows very low complexity and very low power consuming "always-on" positioning in the background (e.g. when supported in UE-based mode), which can then be used to speed up positioning measurement time for more advanced location technologies. Having reasonably good a priori UE location information available allows selecting proper base station candidates for positioning measurements, or predicting shorter signal search windows which reduces the signal search and measurement time not only for cellular positioning techniques but also for, e.g. A-GNSS-based methods.

The higher density of base stations also leads to a higher probability of line-of-sight signal propagation conditions between base station and UE and a better geometry (DOP) of the positioning solution.

Multi-antenna Transmission and Reception in NR

The use of multiple antennas for transmission and/or reception is a fundamental component of NR, in particular at higher frequencies [18]. By adjusting the phase, and possibly also the amplitude, of each antenna element, multiple transmit antenna elements can provide directivity; i.e. focus the overall transmitted power in a certain direction (including azimuth and elevation). Similar, using multiple receive antennas can provide receiver-side directivity by focusing the reception in the direction of a desired signal. This directional signal transmission or reception is referred to as beamforming or spatial filtering..

Beamforming in NR can be divided into two categories: analog and digital beamforming. In digital beamforming different signals are generated for each antenna in the digital domain (Fig. 15.12a). The signal is precoded (amplitude and phase modifications) in baseband processing before radio transmission. This allows forming multiple beams simultaneously from the same set of antenna elements. Digital beamforming can improve the network capacity since the same time and frequency resources can be used to transmit data simultaneously for multiple users (spatial multiplexing). Digital beamforming (where the antenna weights can be flexibly controlled) is often referred to as multi-antenna precoding.

Digital beamforming requires one digital-to-analog/analog-to-digital converter per antenna element. In particular at higher frequencies with a large number

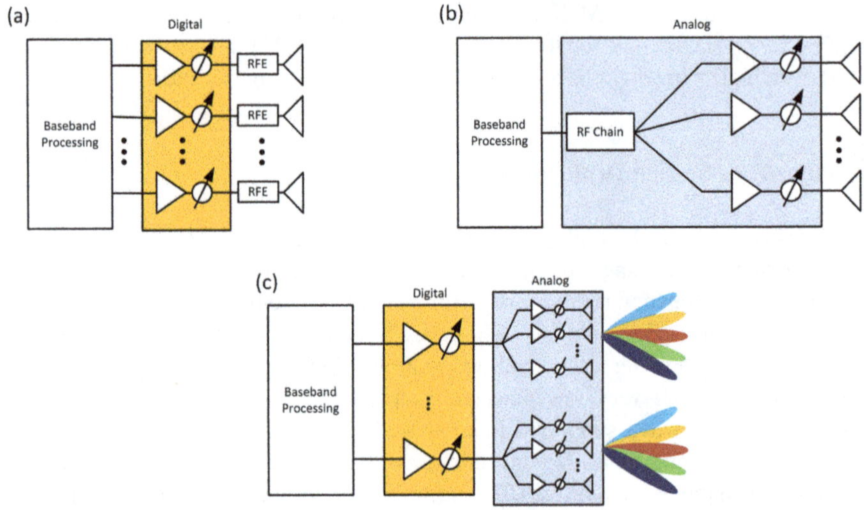

Fig. 15.12 (**a**) Digital beamforming; (**b**) analog beamforming; (**c**) hybrid digital-analog beamforming

of antenna elements, analog beamforming is used due to the circuit complexity and power consumption of digital-to-analog/analog-to-digital converters at high sampling frequencies.

In analog beamforming the same signal is fed to each antenna, and then analog phase shifters are used to steer the signal emitted by the array (phased antenna array; Fig. 15.12b). Since the signal phases of individual antenna signals are adjusted in RF domain, only one beam per set of antenna elements can be formed. Analog beamforming requires beam sweeping where the same signal is repeated but in different transmit beams. With beam sweeping, it is ensured that the signals can be transmitted with a gain sufficiently high to reach the entire intended coverage area. Hybrid beamforming combines analog beamforming and digital beamforming (Fig. 15.12c). As an example, analog beamforming may be used for coarse beamforming, and inside the analog beam a digital beamforming scheme may be used as appropriate.

At higher-frequency bands, multiple antennas are primarily used for beamforming to extend coverage (i.e. to cope with the large pathloss at higher frequencies), while at lower-frequency bands, they enable full-dimensional MIMO, also referred to as massive MIMO. In any case, multiple antennas (antenna arrays) can be used to provide directional information which forms a line of position. Beamforming can also minimize the multipath propagation effects for TOA-based ranging methods and reduce interference. Additionally, it may become possible to localize devices using a single base station (e.g. together with RTT measurements).

NR positioning supports directional positioning using both uplink and downlink signals (see Sect. 3.3). With uplink signals, the method is referred to as angle-of-arrival positioning; with downlink signals the method is referred to as angle-of-departure positioning. In both cases, the directional information is based on the gNB antenna beam orientation.

3.3 *NR Positioning Methods*

The NR native positioning technologies supported in 5G NR include DL-only, UL-only and DL + UL positioning methods:

- Downlink (DL)-based positioning:
  - DL time difference of arrival (DL-TDOA)
  - DL angle of departure (DL-AoD)
- Uplink (UL)-based positioning:
  - UL time difference of arrival (UL-TDOA)
  - UL angle of arrival (UL-AoA)
- Combined DL- and UL-based positioning:
  - Round-trip time (RTT) with one or more neighbouring base station (multi-RTT)

In addition, Enhanced Cell-ID (E-CID) based on radio resource management (RRM) measurements is supported in 5G NR Rel-16.

Downlink Time Difference of Arrival (DL-TDOA) Positioning

DL time difference of arrival (DL-TDOA) positioning is similarly supported in previous generations of mobile communication systems, where the method is referred to as observed time difference of arrival (OTDOA) in LTE and UMTS, advanced forward link trilateration (AFLT) in cdma or enhanced observed time difference (E-OTD) in GSM. The method is based on TOA measurements of DL signals received from multiple TRPs at the UE. The DL signal measured to determine the TOA is the DL-PRS, as described in section "Downlink Positioning Reference Signal (DL-PRS)". From the individual UE TOA measurements, the TDOAs are calculated (see also section "Time Difference of Arrival (TDOA)"). The calculated TDOAs are referred to as *DL reference signal time difference* (DL-RSTD) measurements [10].

There are three basic quantities associated with the DL-TDOA or OTDOA positioning method:

- *Observed time difference of arrival* (OTDOA). This defines the time interval that is observed by a UE between the reception of DL reference signals from two

different TRPs. If a signal from TRP_1 is received at the moment t_1, and a signal from TRP_2 is received at the moment t_2, the OTDOA value is OTDOA $= t_2 - t_1$.
- *Real-time difference* (RTD). This defines the relative synchronization difference between two TRPs. If the TRP_1 transmits a signal at the moment t_3, and the TRP_2 at the moment t_4, the RTD between them is RTD $= t_4 - t_3$. If the TRPs transmit exactly at the same time, that means that the network is perfectly synchronized and hence RTDs $= 0$.
- *Geometric time difference* (GTD). This defines the time difference between the reception of DL signals from two different TRPs due to geometry. If the length of the propagation path between TRP_1 and the UE is d_1, and the length of the propagation path between the TRP_2 and the UE is d_2, then GTD $= (d_2 - d_1)/c$, where c is the speed of the radio waves (speed of light).

There is a simple relationship between these three quantities: OTDOA = RTD + GTD.

With the measurement model from Eq. (15.11)

$$r_{\mathrm{TDOA}_j} = (T_j - T_1) + \left(\frac{d_j}{c} - \frac{d_1}{c}\right) + (n_j - n_1) \quad , \quad j = 2, 3, \ldots, N. \tag{15.38}$$

we have:

$$\mathrm{OTDOA} = r_{\mathrm{TDOA}_j} \tag{15.39}$$

$$\mathrm{RTD} = (T_j - T_1) \tag{15.40}$$

$$\mathrm{GTD} = \left(\frac{d_j}{c} - \frac{d_1}{c}\right) \tag{15.41}$$

OTDOA is the quantity being measured by the UE to be located (DL-RSTD measurements). RTD is a quantity related to the network (TRP) synchronization. GTD is a quantity related to the geometry of the location situation (positions of the mobile and TRPs). GTD is the actual quantity that is useful for location purposes, since it contains information about the position of the device and defines a hyperbolic line of position as described in section "Lines of Position". To obtain GTDs, both OTDOAs and RTDs must be known.

To assist the UE in performing the DL-RSTD measurements, the UE receives assistance data from an LMF. These assistance data may be delivered in an LPP Provide Assistance Data message, as described in section "Example Positioning Procedure", or may be available from broadcast. Broadcast of location assistance data can reduce network signalling and load on network elements (e.g. LMF, AMF, gNB), reduce latency in obtaining assistance data at a UE and reduce UE signalling and power usage. Ciphering of the broadcast assistance data is supported in NR, which enables controlled access to location assistance data by a network operator.

The location assistance data contains a list of candidate TRPs for TOA measurements together with the DL-PRS signal configuration (see section "Downlink Positioning Reference Signal (DL-PRS)"). The list of candidate TRPs is selected by an LMF based on an a priori location estimate of the UE (which is usually the UE serving cell, e.g. obtained via basic Cell-ID positioning method (see section "Enhanced Cell-ID (E-CID) Positioning")). The candidate TRPs for measurements may be selected by an LMF such that a good measurement geometry would be obtained (i.e. low GDOP; as described in section "Geometrical Influence on Position Errors"). The UE then tries to detect the DL-PRS from each candidate TRP and measures the TOAs from which the TDOAs are calculated. In case of UE-assisted mode, the UE provides the measurements to the LMF in an LPP Provide Location Information message, as described in section "Example Positioning Procedure". For UE-based mode, the location assistance data include, in addition, the geographical locations of the candidate TRP antenna reference points and the RTDs. With this information, the UE is able to calculate the location (possibly using other location measurements available at the UE in case of "hybrid location") and may provide the location estimate to the LMF in an LPP Provide Location Information message, as also summarized in section "Example Positioning Procedure".

Downlink Angle-of-Departure (DL-AoD) Positioning

Downlink angle-of-departure (AoD) positioning can be based on per-beam RSRP measurements of DL-PRS (see section "Downlink Positioning Reference Signal (DL-PRS)") performed at the UE (DL-PRS Reference Signal Received Power (DL-PRS RSRP) measurement [10]). The TRPs may transmit beam-formed DL-PRS in a beam sweeping manner that may be measured by the UE as illustrated in Fig. 15.13. Figure 15.13 illustrates two UEs at different locations which use a fixed RX beam to receive multiple TRP TX beams. In the example of Fig. 15.13, UE_1 receives the strongest signal from DL-PRS beam #2 resulting in the highest RSRP. With the same UE_1 RX beam, the DL-PRS TX beams 1,3,4 may also be received, but with lower signal strength. Similarly, UE_2 may measure multiple TRP DL-PRS TX beams with the same RX beam, resulting in a different RSRP measurement pattern at the UE. The UE RSRP measurement vector can be considered as a "RF fingerprint", and AoD calculation may be performed using a pattern-matching approach (see also section "Received Signal Strength (RSS)"). By comparing the measured DL-PRS RSRP vector to the fingerprints of all pre-stored angles, a e.g. maximum likelihood algorithm can be used to estimate the azimuth angle of departure (AoD) and zenith angle of departure (ZOD). As with any pattern-matching approach, a database of reference "fingerprints" must be available, which, in this case, requires the beam spatial information.

Similar to DL-TDOA, the UE requires assistance data for performing the measurements, including a list of candidate TRPs together with the DL-PRS signal

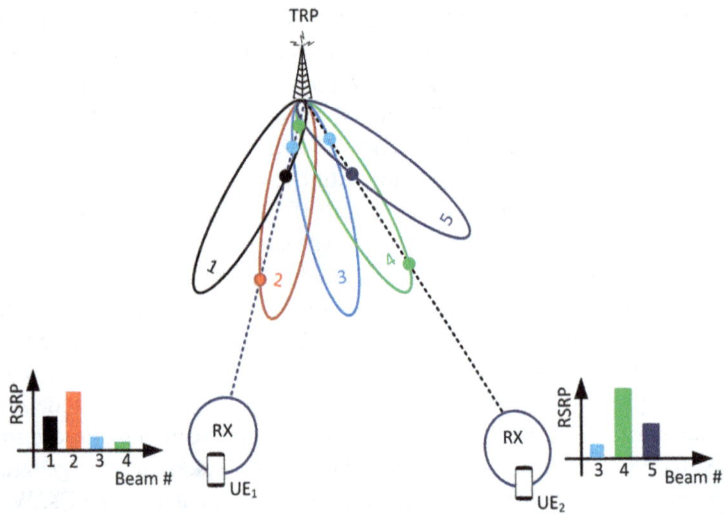

Fig. 15.13 Illustration of DL-AoD method using multiple RSRP measurements with a single UE RX beam

configuration. These assistance data may be delivered in an LPP Provide Assistance Data message or may be available from broadcast. Since the UE should use a fixed RX beam for receiving multiple DL-PRS Resources, the *DL-PRS Quasi-Colocation Information* as described in section "Downlink Positioning Reference Signal (DL-PRS)" may be selected for DL-AoD positioning in such a way that a UE uses the same RX beam for all DL-PRS RSRP measurements. Since the UE may have to perform other operations for normal communication in parallel to positioning measurements, it may not always be possible to use a fixed/same RX beam for the DL-PRS RSRP measurements. The UE may inform the LMF in the LPP Provide Location Information message whether a fixed RX beam was used for the RSRP measurements or not.

The measured RSRPs per DL-PRS TX beam may then be reported in an LPP Provide Location Information message to the LMF for each measured TRP where the corresponding AoDs are estimated and position calculation is performed (in case of UE-assisted mode). For UE-based mode, the UE requires additional assistance data including the TRP geographical locations (similar to DL-TDOA) and the DL-PRS beam information (e.g. beam azimuth, elevation, beamwidth, etc.). However, TRP synchronization information (e.g. RTDs) are not required for DL-AoD positioning; DL-AoD positioning can also be deployed in unsynchronized networks. With the additional assistance data, the UE is able to calculate the location (possibly using additional location measurements available at the UE) and may provide the location estimate to the LMF in an LPP Provide Location Information message.

Uplink Time Difference of Arrival (UL-TDOA) Positioning

In UL time difference of arrival (UL-TDOA) positioning, the UL signal transmitted by the UE is received by multiple TRPs which determine the TOA of the received signal. The UL-TDOA and DL-TDOA methods are in principle uplink/downlink duals of one another. The UL TOA is measured at the TRPs relative to a common time scale (e.g. GPS time or any other time base which can be shared among all the TRPs). The measurement is referred to as UL Relative Time of Arrival (RTOA) [10], since the TOA is measured relative to a common reference time. Therefore, similar to DL-TDOA where the DL transmission of different TRPs must be synchronized (i.e. the RTDs must be known), the UL reception points for UL-TDOA must be synchronized (i.e. must use the same common time base for the RTOA measurements). The UL signal is based on Sounding Reference Signals (SRS) (see also section “Uplink Positioning Reference Signal (UL-PRS)”). For obtaining uplink measurements, the TRPs need to know the characteristics of the SRS transmitted by the UE for the time period required to determine uplink RTOA measurements. These characteristics need to be static over the transmission of SRS during the uplink measurements. The TRPs receive this information from an LMF in an NRPPa message (see also section “Multi-Round-Trip-Time (Multi-RTT) Positioning”) such that the TRPs are able to generate a signal replica for RTOA measurements.

The RTOA measurements of all participating TRPs are sent to the LMF, which calculates TDOAs. The differencing operation cancels the unknown UE transmit time, and position calculation can be based on hyperbolic trilateration, similar to DL-TDOA.

With the measurement model of Eq. (15.3), the measured RTOA at TRP i can be written as

$$t_i = T + \frac{\mathrm{d}_i}{c} + n_i \quad , \quad i = 1, 2, \ldots, N. \tag{15.42}$$

where t_i is the RTOA measured at the TRP_i relative to a common time base (which is the same among all TRPs), T is the (unknown) UE signal transmission time, and n_i is a measurement error. By calculating the difference between two TRP measurements, the unknown (common) transmit time cancels, and one obtains hyperbolic equations similar to DL-TDOA:

$$r_{\mathrm{TDOA}_j} = \left(t_j - t_1\right) = \left(\frac{d_j}{c} - \frac{d_1}{c}\right) + \left(n_j - n_1\right) \quad , \quad j = 2, 3, \ldots, N. \tag{15.43}$$

UL-TDOA is a network-based positioning method (see also Sect. 2.2), where the positioning operation can be transparent to the UE; i.e. in case of a Rel-15 uplink signal is used for the RTOA measurements. The UE is only required to transmit the UL signal, but does not perform any positioning measurements.

Uplink Angle-of-Arrival (UL-AoA) Positioning

Uplink angle-of-arrival (AoA) positioning is a network-based positioning method (similar to UL-TDOA). The gNB uses the received signal transmitted by the UE to derive the angle of arrival (AoA) in azimuth and zenith. The UL signal may be a signal defined for communication purposes (e.g. Sounding Reference Signals (SRS); in which case UL-AoA positioning can be transparent to the UE), or a UL positioning reference signal (see section "Uplink Positioning Reference Signal (UL-PRS)"). To estimate the angle of arrival of incident signals, a directional antenna is required, which is available at the gNB due to the inherent support of multi-antenna transmission and reception in 5G NR (see section "Multi-antenna Transmission and Reception in NR"). The AoA can be determined in multiple ways. The straightforward way is to use the gNB received beam information, which is steered to the direction of arrival of the UE UL signal. In classical AoA estimation, this technique is typically referred to as "conventional beamforming". These techniques do not exploit any assumption on the model of the received signal and noise. These conventional AoA estimation techniques electronically steer beams in all possible directions (or in a set of fixed directions) and look for peaks in the output power. Conventional beamforming is an application of Fourier-based spectral analysis applied to spatiotemporal samples. However, in such approaches, the angular resolution is limited by the beamwidth of the array. Therefore, a large number of antenna elements are required to achieve high accuracy.

Subspace-based methods are high-resolution signal processing techniques, which exploit the eigenstructure of the incident signal when assembled in a space-time matrix. One of the most often used subspace-based techniques is the MUltiple SIgnal Classification (MUSIC) algorithm. The geometric concepts of MUSIC define the basis of a broader class of subspace-based algorithms. Another popular method of this family is the Estimation of Signal Parameters through Rotational Invariance Technique (ESPRIT). A summary of all these techniques can be found in, e.g. reference [16].

The non-conventional AoA estimation techniques require digital samples of each antenna element output, e.g. to determine spatial covariance matrices for eigenstructure processing, and, therefore, are better suited for lower frequencies (FR1), where digital beamformers are typically available. Since the array aperture is relatively small at lower frequencies, the spatial resolution is limited (i.e. conventional beams are relatively broad). Therefore, high-resolution techniques are beneficial in particular at lower frequencies since they are able to increase angular resolution to values smaller than the beamwidth of the array without the need of increasing the array aperture.

The procedures for UL-AoA positioning are similar to UL-TDOA (see also section "Multi-Round-Trip-Time (Multi-RTT) Positioning"). The UE is triggered by the network to transmit an UL signal, and selected TRPs in the neighbourhood of the UE are configured by an LMF via NRPPa to listen to the UE transmission and measure the UL AoA. However, compared to UL-TDOA, no common time reference is needed at the TRPs, and therefore, UL-AoA positioning is also suitable for non-synchronized networks.

Multi-Round-Trip-Time (Multi-RTT) Positioning

DL-TDOA and UL-TDOA positioning methods are based on time-of-arrival (TOA) measurements performed on downlink signals or uplink signals, respectively. Although these methods have been shown to be effective, they require installing and maintaining hardware for very precise base station time synchronization.

Round-trip-time (RTT) positioning uses two-way time-of-arrival measurements and requires in principle no time synchronization between TRPs as also described in section "Round-Trip-Time (RTT)". However, a coarse base station time synchronization is desired in order to reduce interference and increase hearability from multiple transmission points. This time synchronization requirement is similar to the TDD synchronization requirements (e.g. microsecond level synchronization instead of nano-seconds as in the case of TDOA positioning).

Figure 15.14 illustrates the principle of obtaining distance information from two-way time-of-arrival (time-of-flight) measurements (UL and DL measurements).

The measurements to support multi-RTT are the *UE Rx-Tx Time Difference* and *gNB Rx-Tx Time Difference* Measurement [10]. In the example of Fig. 15.14, the *UE Rx-Tx Time Difference* corresponds to $(t_3\text{-}t_0)$, and the *gNB Rx-Tx Time Difference* corresponds to $(t_1\text{-}t_2)$, and therefore, the RTT would be *UE Rx-Tx Time Difference* + *gNB Rx-Tx Time Difference* (note that the $(t_1\text{-}t_2)$ difference would be negative in this example, since t_2 occurs after t_1). Therefore, similar to TDOA location, the basic UE and gNB measurements are TOA measurements. Those measurements can be related to the appropriate TX time so that Rx-minus-Tx time differences can be reported to determine RTT.

Multi-RTT positioning requires both DL and UL positioning procedures as illustrated in Fig. 15.15. Figure 15.15 shows a typical procedure, although a different

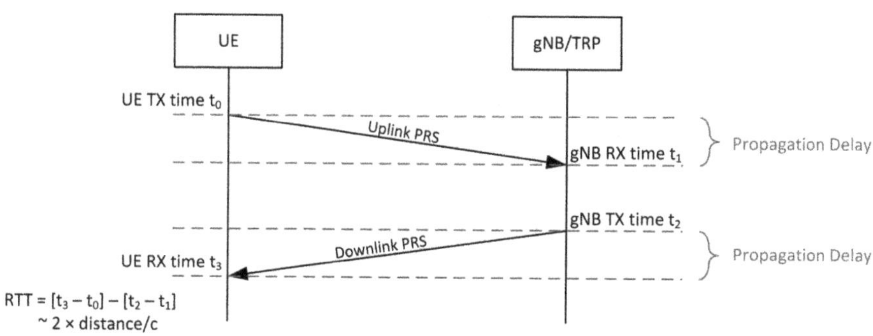

Fig. 15.14 Principle of determining distance between two devices using UL and DL measurements

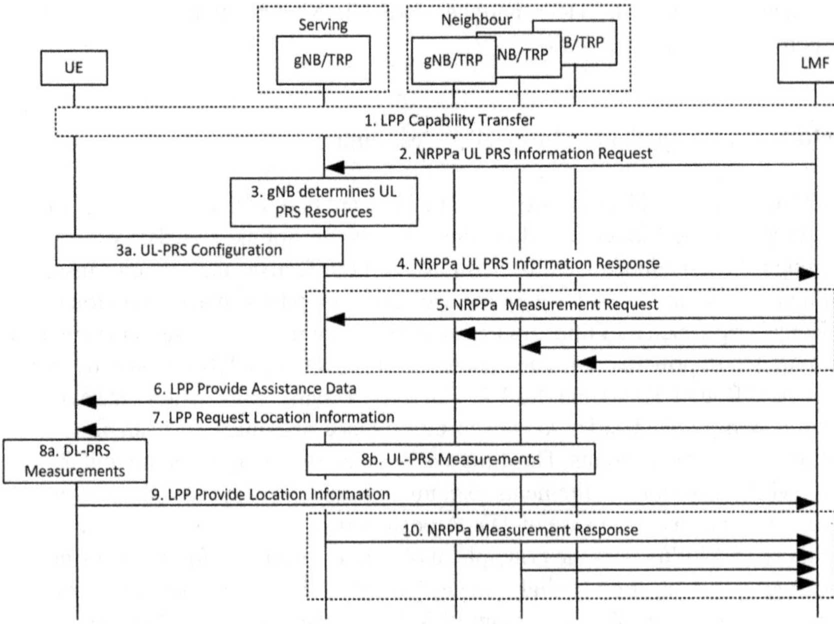

Fig. 15.15 Exemplary multi-RTT positioning procedure

arrangement of the various steps may be possible. The figure also summarizes the application of LPP and NRPPa procedures described previously in sections "Positioning Protocol Architecture" and "Example Positioning Procedure". The procedure in Fig. 15.15 may be performed at Step 12 in Fig. 15.7.

At Step 1 of Fig. 15.15, the LMF may request the positioning capabilities of the UE using the LPP Capability Transfer procedure described in section "Example Positioning Procedure". In order to obtain UE UL signal information, the LMF may send a NRPPa message to the serving gNB of the UE to request UL-PRS configuration information for the UE at Step 2. The serving gNB determines the resources available for UL-PRS at Step 3 and configures the UE with the UL-PRS resources at Step 3a. At Step 4, the serving gNB provides the UL-PRS configuration information to the LMF in a NRPPa message. The LMF then provides the UL-PRS configuration to the selected gNBs (TRPs) in a NRPPa message at Step 5. The message includes all information required to enable the gNBs/TRPs to perform the UL measurements. At Step 6, the LMF sends an LPP Provide Assistance Data message to the UE which includes any required assistance data for the UE to perform the UE Rx-Tx Time Difference measurements (i.e. TRP list and DL-PRS signal configuration). An LPP Request Location Information message is sent at Step 7 to request the UE measurements. The UE transmits the UL-PRS according to the time domain behaviour of UL-PRS resource configuration from Step 3a. At Step 8, the UE performs the UE Rx-Tx Time Difference measurements from all gNBs provided in the assistance data at Step 6, and each gNB configured at Step

5 measures the gNB Rx-Tx Time Difference based on UL transmissions from the UE. The UE reports the UE Rx-Tx Time Difference measurements to the LMF in an LPP Provide Location Information message at Step 9, and each gNB reports the gNB Rx-Tx Time Difference measurements from the TRPs to the LMF in an NRPPa message at Step 10. The LMF determines the RTTs from the UE and gNB Rx-Tx Time Difference Measurements for each gNB for which corresponding UL and DL measurements were provided at Steps 9 and 10 and calculates the position of the UE.

In general, all measurements that are used to generate a location estimate should be made concurrently. If measurements from different points in time are used for generating a location estimate, UE motion as well as changes to UE clock and TRP clocks would result in measurement errors that ultimately produce position errors. When it comes to multi-RTT, it should be possible to conduct and associate as close in time as possible the RxTx Time Difference Measurements produced by the UE with the RxTx Time Difference Measurements produced by the corresponding TRPs. The UE and TRP measurements contain time stamps, and the multi-RTT positioning procedures allow a coordination of the DL-PRS and UL-PRS measurement times accordingly.

The procedure shown in Fig. 15.15 includes also DL-only (DL-TDOA, DL-AoD) and UL-only (UL-TDOA, UL-AoA) positioning as special cases. For DL-only positioning, Steps 2, 3, 4, 5, 8b and 10 in Fig. 15.15 are not needed. For UL-only positioning, Steps 6, 7, 8a and 9 in Fig. 15.15 are not needed.

Enhanced Cell-ID (E-CID) Positioning

Cell-ID (CID) positioning is a network-based method that can be used to estimate the position of the UE immediately, but typically with relative low accuracy. The position of the UE can be estimated to be the position of the base station the UE is camped on. Cell-ID positioning performance can be enhanced by measuring additional network attributes; the technique is then referred to as Enhanced Cell-ID (E-CID). In 5G NR, the radio resource management (RRM) measurements can be used for position location of a UE. For operations such as handover to a neighbour cell, the UE typically measures cell quality, such as Reference Signal Received Power (RSRP) or Reference Signal Received Quality (RSRQ). With 5G NR, the cell quality can be measured by using the SS/PBCH Blocks (SSBs) or Channel State Information Reference Signals (CSI-RS).

For ECID positioning, the UE typically reports the RRM measurements already available (i.e. not positioning specific) to the LMF. The LMF may use various techniques to determine the location of the UE. For example, it may use some ranging technique as described in section "Received Signal Strength (RSS)" and/or a "RF Pattern Matching" approach, etc.

3.4 NR Positioning Reference Signals (PRS)

Positioning reference signals (PRS) are defined for NR positioning to enable UEs to detect and measure more neighbour TRPs (for DL positioning), or to enable multiple TRPs to detect and measure the UE signal (for UL positioning), i.e. to increase hearability.

As summarized in section "Multi-antenna Transmission and Reception in NR", the use of multiple antennas for transmission and/or reception is a fundamental component of NR, and therefore, the NR PRS design supports multi-beam operation. The ultimate task of beam operation is to find suitable beam pairs for a multitude of TRPs (e.g. serving and a list of neighbour TRPs). That is, to find a transmitter-side PRS beam direction and a corresponding receiver-side beam direction that allows positioning measurements (e.g. DL and/or UL TOAs). For communication, the best beam pair may not necessarily be the transmitter and receiver beams that are pointing directly towards each other. However, for positioning, beam pair pointing directly to each other is a fundamental requirement to reduce, e.g. TOA or AoA measurement errors, as illustrated in Fig. 15.16. A direct path between transmitter and receiver may be blocked (e.g. beam pair 1-1 in Fig. 15.16) due to obstacles in the propagation environment, and a reflected path (beam pair 3–3 in Fig. 15.16) may provide a stronger signal. However, for positioning measurements, the beam pair 1-1 in Fig. 15.16 would be desired, since beam pair 3-3 would result in additional excess delay for TOA measurements or in wrong direction measurements (e.g. AoA). This typically requires that a measurement entity (TRP or UE) measures multiple transmit-receive beam pairs to determine which pair results in the earliest arrival path (which may require finding a weak signal in the presence of a strong signal). To support PRS beam operation, beam sweeping and beam alignment are supported for PRS.

Beamforming is also relevant for the uplink direction with beam-formed transmission by the UE and corresponding beam-formed reception by the TRP. Typically, a suitable positioning beam pair for the downlink direction is also a suitable positioning beam pair for the uplink direction. This is referred to as beam correspondence [18]. In case of positioning beam correspondence, it is sufficient to determine a suitable positioning beam pair in one of the transmission directions. The same positioning beam pair can then also be used in the opposite transmission direction. This positioning beam correspondence can in particular be exploited for multi-RTT positioning described in section "Multi-Round-Trip-Time (Multi-RTT) Positioning", which requires both DL- and UL-PRS transmissions.

Downlink Positioning Reference Signal (DL-PRS)

Similar to LTE, the DL-PRS corresponds to a set of resource elements (REs) used for DL positioning measurements but does not carry information originating from higher layers. The DL-PRS sequence is generated for a specific OFDM symbol

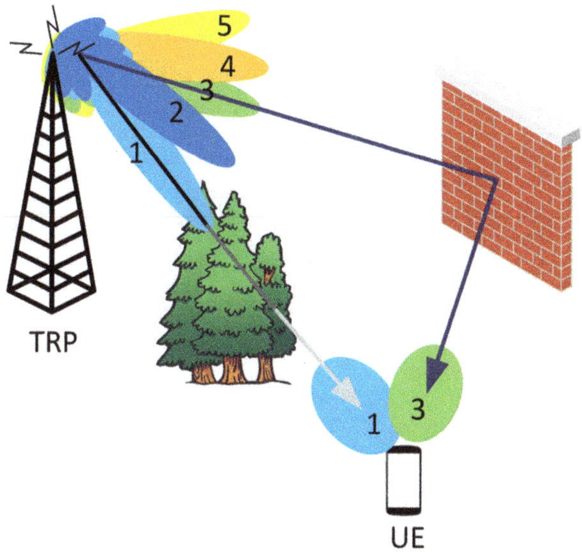

Fig. 15.16 Illustration of positioning beam pair

within a slot in a radio frame using a length-31 Gold sequence, which is mapped to complex-valued QPSK modulation symbols.

The DL-PRS resource elements are arranged in a particular time/frequency pattern. The DL-PRS may span 2, 4, 6 or 12 consecutive OFDM symbols and can be transmitted in any downlink or flexible symbol configured anywhere in a slot. In the frequency domain, the DL-PRS has a comb-like pattern, as illustrated in Fig. 15.17. A comb-N structure implies that DL-PRS is transmitted on every N^{th} subcarrier, where N can take the values 2, 4, 6 or 12.

For the comb-2 pattern, every second subcarrier contains a DL-PRS RE with an RE-offset of 1-symbol between consecutive OFDM-symbols, resulting in a staggered pattern. From basic Fourier transform properties, a comb-N frequency domain signal produces time domain equivalents with N repetitions ("aliasing"). OFDM communication systems are robust against such ambiguities as long as the delay spread of the channel is less than the ambiguity distance. However, such ambiguities are detrimental to positioning systems that depend on measurements of time of arrivals, because the ambiguities may produce false peaks in the correlation function which can lead to wrong position estimates. After coherent combining (integration) of the two staggered comb-2 DL-PRS symbols, an equivalent comb-1 pattern is obtained without alias peaks in the correlation function ("de-staggering"). The comb size and the number of symbols are trade-offs between measurement latency, avoiding ambiguity/aliasing in the correlation function, power boosting and frequency reuse. Since not all subcarriers are occupied with a comb-N signal with $N > 1$, more transmit power is available per RE. With a comb-N pattern, DL-PRS from N TRPs can be frequency multiplexed within the same frequency range by

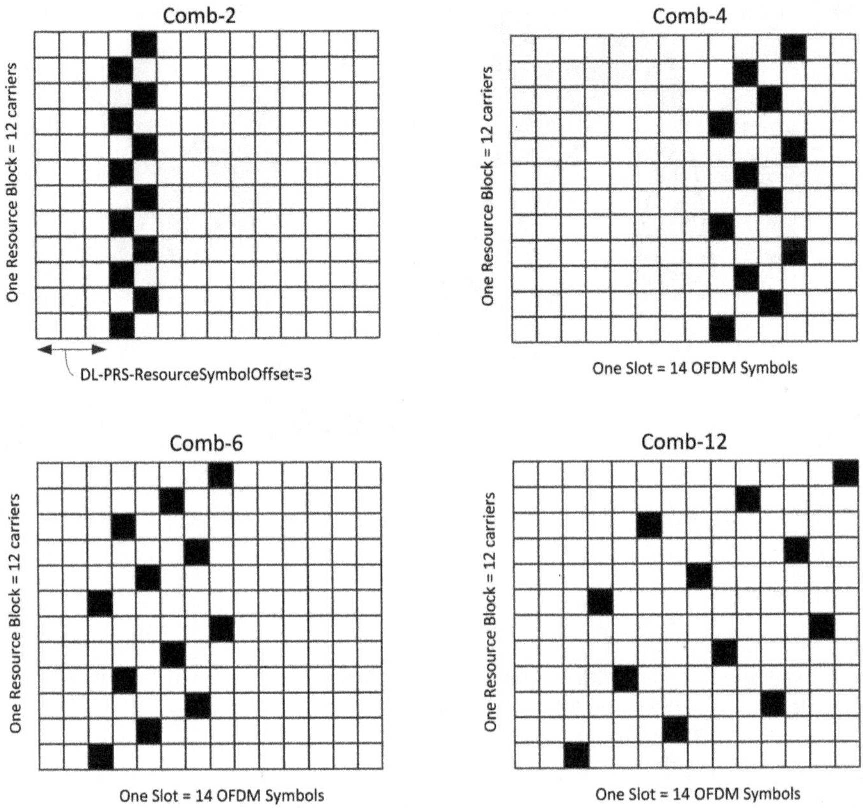

Fig. 15.17 Example of DL-PRS time/frequency pattern

assigning different frequency offsets (frequency reuse of N). For example, for a comb-2 DL-PRS, one TRP can use RE-offsets {0, 1}, and a second TRP could use RE offsets {1, 0}. In case of comb-6, 6 TRPs can be frequency multiplexed (similar to LTE PRS).

The comb-4, comb-6 and comb-12 DL-PRS are not arranged in a staircase pattern. For the comb-4 pattern for example, the comb offsets of each of the four symbols are {0, 2, 1, 3}. This has the advantage that already the first few symbols have a better effective comb size. In a high mobility scenario, for example, the UE speed may limit the coherent integration time, and therefore, not all DL-PRS symbols may be coherently combined. If only the first two symbols of the comb-4 pattern are used for the TOA measurement, the effective pattern would be a comb-2 pattern. In addition, if some DL-PRS symbols are punctured by other high priority signalling and data, the remaining DL-PRS symbols may still generate a rather uniform comb pattern. For the comb-6 pattern, the DL-PRS symbols have the comb offsets {0, 3, 1, 4, 2, 5}, and for the comb-12 pattern, the DL-PRS symbols have the comb offsets {0, 6, 3, 9, 1, 7, 4, 10, 2, 8, 5, 11} as illustrated in Fig. 15.17.

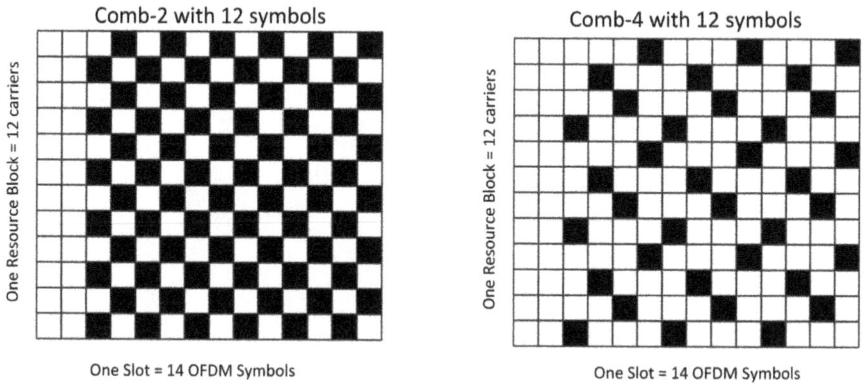

Fig. 15.18 Example of DL-PRS time/frequency pattern

Table 15.1 Resource element offsets for pairs of DL-PRS comb size and number of symbols

Number of symbols	2	4	6	12
Comb size				
2	{0, 1}	{0, 1, 0, 1}	{0, 1, 0, 1, 0,1}	{0, 1, 0, 1, 0,1, 0, 1, 0, 1, 0,1}
4	–	{0, 2, 1, 3}	–	{0, 2, 1, 3, 0, 2, 1, 3, 0, 2, 1, 3}
6	–	–	{0, 3, 1, 4, 2, 5}	{0, 3, 1, 4, 2, 5, 0, 3, 1, 4, 2, 5}
12	–	–	–	{0, 6, 3, 9, 1, 7, 4, 10, 2, 8, 5, 11}

The comb-N time/frequency pattern of DL-PRS can be repeated over several OFDM symbols. The comb-2 pattern may span 2, 4, 6 or 12 consecutive OFDM symbols. In this case, the pattern is repeated which allows longer coherent integration in low mobility scenarios. Similar, the comb-4 pattern may span 4 or 12 consecutive OFDM symbols, and the comb-6 pattern may span 6 or 12 symbols. For example, Fig. 15.18 shows a comb-2 DL-PRS with 12 symbols and a comb-4 DL-PRS with 12 symbols.

Table 15.1 summarizes the RE offsets for the combinations of number of symbols {2, 4, 6, 12} and the comb sizes {2, 4, 6, 12}. The resource element pattern of DL-PRS is configured with a comb offset for the first symbol (this comb offset for the first symbol is 0 in the examples of Figs. 15.17 and 15.18), and the relative RE offsets of the following symbols are defined relative to the comb offset of the first symbol.

A "DL-PRS Resource" with 2, 4, 6 or 12 consecutive OFDM symbols has a minimum transmission bandwidth of 24 contiguous Physical Resource Blocks (PRBs), and a maximum transmission bandwidth of 272 PRBs. The granularity of the DL-PRS bandwidth configuration is 4 PRBs. The actual frequency bandwidth for the DL-PRS depends then on the subcarrier spacing. For 15 kHz subcarrier spacing, the minimum DL-PRS bandwidth is about 5 MHz and the maximum about 50 MHz. With a 120 kHz subcarrier spacing, the DL-PRS bandwidth can be up to about 400 MHz.

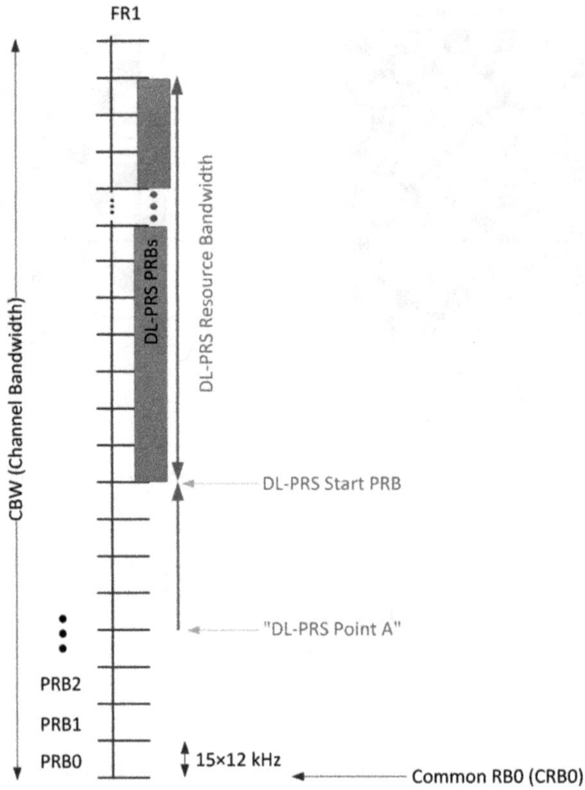

Fig. 15.19 DL-PRS frequency occupancy

A DL-PRS Resource can be located anywhere in the frequency grid. A common reference point for the DL-PRS is defined which is named "DL-PRS Point A". The "DL-PRS Point A" serves as a common reference point for the DL-PRS resource block grid and is represented by an Absolute Radio Frequency Channel Number (ARFCN). The DL-PRS Start PRB is then defined as a frequency offset between DL-PRS Point A and the lowest subcarrier of the lowest DL-PRS resource block expressed in units of resource blocks, as illustrated in Fig. 15.19. (for FR1 in this example. Note that the "DL-PRS Point A" is different from the "Point A" used for communication purposes.)

The DL-PRS configuration is provided to a UE in the assistance data (i.e. in an LPP Provide Assistance Data message or via broadcast messages) and is typically cell (TRP) specific. A UE in RRC connected state and without measurement gaps configured is required to measure the DL-PRS only within the active DL bandwidth part (BWP) and with the same numerology as the active DL BWP. If DL-PRS is available outside the active BWP of the UE, the UE may request measurement gaps from the serving gNB. With measurement gaps configured, a UE is able to measure the DL-PRS Resource outside the active DL BWP (or with a numerology different from the numerology of the active DL BWP).

In summary, a DL-PRS Resource description includes the following parameters [9]:

- A *DL-PRS Resource Identity*, defining the particular DL-PRS Resource. A *DL-PRS Resource ID* in a DL-PRS Resource Set is associated with a single spatial transmission filter (beam) and is transmitted from a single TRP (see also below).
- A *DL-PRS Sequence Identity*, defining the initialization seed for the pseudo-random Gold sequence generator for the DL-PRS Resource.
- A *DL-PRS Comb Size N*, defining the resource element spacing in frequency domain for each symbol of the DL-PRS Resource. The value N can take the values {2, 4, 6, 12}.
- A *DL-PRS RE Offset*, defining the resource element offset in frequency domain for the first symbol in the DL-PRS Resource. The relative RE offsets of the following symbols are defined relative to the RE offset in frequency domain of the first symbol in the DL-PRS Resource.
- A *DL-PRS-Resource Slot Offset*, defining the starting slot of the DL-PRS Resource with respect to the corresponding *DL-PRS-Resource Set Slot Offset* (see below).
- A *DL-PRS Resource Symbol Offset* with values {0, 1, 2, . . . , 12} defining the starting symbol of the DL-PRS Resource within a slot determined by *DL-PRS-Resource Slot Offset*.
- A *DL-PRS Number of Symbols*, defining the number of symbols per DL-PRS Resource within a slot. Values of {2, 4, 6, 12} are defined.
- A *DL-PRS Subcarrier Spacing*, defining the Subcarrier Spacing for the DL-PRS Resource (15, 30, 60 kHz for FR1; and 60, 120 kHz for FR2).
- A *DL-PRS Cyclic Prefix*, defining the cyclic prefix length of the DL-PRS Resource (normal or extended).
- A *DL-PRS Point A*, defining the absolute frequency of the reference resource block for the DL-PRS. Its lowest subcarrier is named “DL-PRS Point A”.
- A *DL-PRS-Start PRB*, defining the start PRB index as an offset from *DL-PRS Point A*, in multiples of 1 PRB.
- A *DL-PRS Resource Bandwidth*, defining the number of PRBs allocated for the DL-PRS Resource (allocated DL-PRS bandwidth), in multiples of 4 PRBs.
- A *DL-PRS Quasi-Colocation Information*, providing quasi-colocation (QCL) information between DL-PRS and other reference signals.

Generally speaking, if a signal A is said to be quasi-colocated with a signal B, it means that the two signals have propagated through very similar channel conditions. In order for signals A and B to propagate through similar channels, they must come from the same location (and same antenna). The DL-PRS Resource can be quasi-colocated in terms of average delay and Doppler shift (referred to as “QCL Type C” in 3GPP specifications) with a Synchronization Signal Block (SSB), meaning that if the receiver can detect the SSB and determine the channel properties, it can assume that the DL-PRS has the same properties in terms of average delay and Doppler

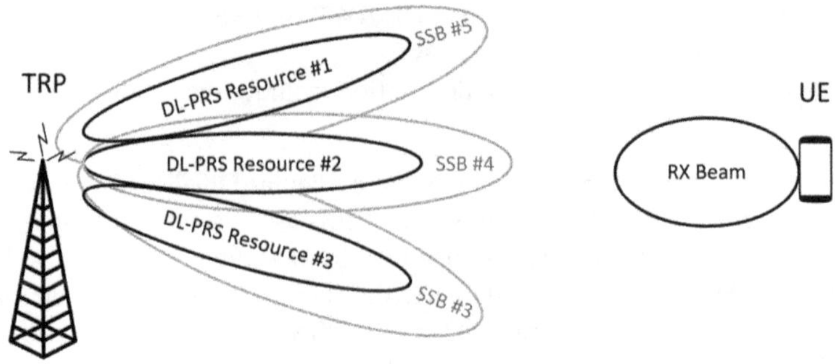

Fig. 15.20 Illustration of quasi-colocation (QCL) indication for a "fixed" UE RX beam

shift. Another colocation type is defined for spatial receive parameter (referred to as "QCL Type D" in 3GPP specifications), which in practice means that the two signals are transmitted from the same place and same beam. If the device knows that a certain receiver beam direction is appropriate for the reception of one of the signals, it can assume that the same beam direction is suitable also for the reception of the other signal. The DL-PRS can be configured to be QCL Type D with a DL Reference Signal from a serving or neighbouring TRP where an SSB or DL-PRS can be the QCL Type D source for the DL-PRS. Therefore, if it happens that the UE has received an SSB from a TRP, and if the DL-PRS from the TRP is quasi-colocated with the SSB, the UE can use the same RX beam as found for the SSB also for the DL-PRS.

The QCL Type D between two DL-PRS Resources from the same TRP can be used to indicate to the UE to use the same RX beam on multiple DL-PRS Resources. As described in section "Downlink Angle-of-Departure (DL-AoD) Positioning", for the DL-AoD method, the UE should use a fixed RX beam to receive multiple DL-PRS Resources from a TRP. Figure 15.20 illustrates a TRP with three DL-PRS Resources (beams). DL-PRS Resource #2 has a spatial relation with the SSB#4 ("DL-PRS Resource #2 is QCLed Type D with SSB#4"). The DL-PRS Resources #1 and #3 are QCLed Type D with DL-PRS Resource #2 in this example. Therefore, the UE would use the same RX beam for all three DL-PRS Resources as it uses for receiving the SSB#4.

A "DL-PRS Resource Set" is introduced which is defined as a set of DL-PRS Resources. The DL-PRS Resources in a DL-PRS Resource Set are associated with the same TRP. Multiple DL-PRS Resource Sets may be transmitted from the same TRP, for example, with multiple (different) periodicities, bandwidths or beamwidths.

Each DL-PRS Resource in a DL-PRS Resource Set can have multiple repetitions within one transmission periodicity. The parameter *DL-PRS Resource Repetition Factor* is introduced which defines how many times each DL-PRS Resource is repeated across multiple slots in a single instance of the DL-PRS Resource Set. The *DL-PRS Resource Repetition Factor* can have values {1, 2, 4, 6, 8, 16, 32}

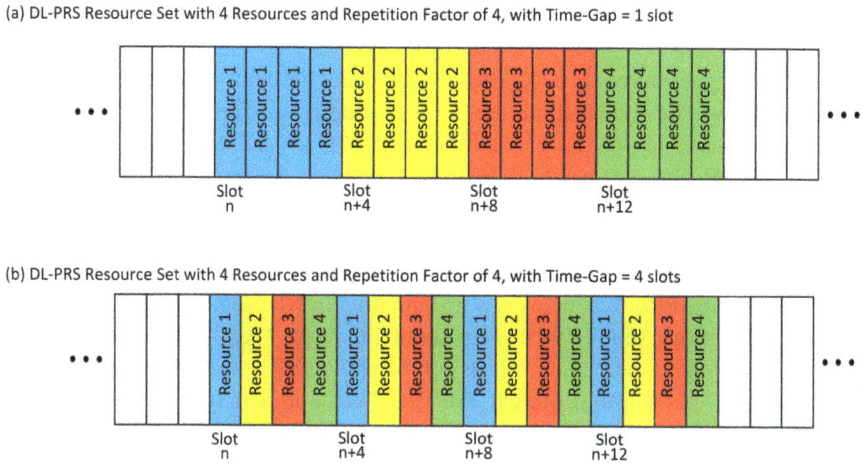

Fig. 15.21 Illustration of DL-PRS resource repetition factor and DL-PRS resource time gap for a single instance of the PRS resource set

repetitions. The repetitions of one DL-PRS Resource are transmitted with the same downlink spatial domain transmission filter (i.e. the same beam). The repetitions allow for the UE receiver-side beam to sweep over the repetitions of DL-PRS Resources. A *DL-PRS Resource Time Gap* indicates the offset in units of slots between two repeated instances of a DL-PRS Resource with the same DL-PRS Resource ID. The *DL-PRS Resource Time Gap* can take the values {1, 2, 4, 8, 16, 32} slots. These two parameters, *DL-PRS Resource Repetition Factor* and *DL-PRS Resource Time Gap*, essentially control combined beam sweeping and repetition for the two possibilities, (a) "repeat and sweep" and (b) "sweep and repeat" as illustrated in Fig. 15.21. Each DL-PRS Resource #n (with n = 1,2,3,4 in the example of Fig. 15.21) is transmitted using the same beam. The "repeat and sweep" in Fig. 15.21a enables beam sweeping of repeated DL-PRS Resources and has the advantage of allowing combining over the repeated DL-PRS Resources at the UE. The "sweep and repeat" illustrated in Fig. 15.21b allows faster beam sweeping.

Figure 15.21 illustrates one DL-PRS Resource per slot. However, multiple DL-PRS Resources can be present in a slot of 14 OFDM symbols. For example, with a comb-2 resource element pattern with two symbols, seven DL-PRS Resources may fit into one slot (assuming all symbols of the slot can be available for DL-PRS). The parameter *DL-PRS Resource Symbol Offset* controls the starting symbol of each DL-PRS Resource in a slot.

Therefore, there are two levels of repetition defined for DL-PRS: (a) "symbol level repetition", where the DL-PRS symbols are repeated multiple times, as described above; e.g., a comb-2 DL-PRS Resource may span more than two symbols; and (b) "DL-PRS Resource repetition", where each DL-PRS Resource is repeated and transmitted with the same downlink spatial domain transmission filter across multiple slots. Note that each repetition instance of a DL-PRS Resource can only occur in one slot; i.e., a slot cannot carry the repetition instances of the same DL-PRS Resource.

All DL-PRS Resources in a DL-PRS Resource Set have the same periodicity; i.e., DL-PRS Periodicity is defined per DL-PRS Resource Set. The *DL-PRS Periodicity* can take values of $2^{\mu}\cdot${4, 5, 8, 10, 16, 20, 32, 40, 64, 80, 160, 320, 640, 1280, 2560, 5120, 10,240} slots, with $\mu = 0, 1, 2, 3$ for SCS 15, 30, 60 and 120 kHz, respectively. Also, the *DL-PRS Number of Symbols*, which defines the number of symbols per DL-PRS Resource within a slot as mentioned above, is the same for all Resources in a DL-PRS Resource Set.

Muting of DL-PRS Resources is also supported in NR, similar to LTE. DL-PRS Resources from different TRPs can be isolated in space (i.e. different beams), frequency domain (different comb offsets for different TRPs), time domain (different symbol offsets in a slot for different TRPs) and code domain (different scrambling sequences for the DL-PRS Resources of different TRPs). If the DL-PRS Resources from different TRPs collide in time at the UE receiver with the same frequency pattern (comb offset), they would only be isolated by the scrambling code, which usually does not provide sufficient isolation between near and far TRPs. Muting can turn off DL-PRS Resources to reduce the interference in case of colliding DL-PRS Resources.

Muting in NR is signalled using a bit map to indicate which configured DL-PRS Resources are transmitted with zero power, similar to LTE. Two muting options are supported for NR as illustrated in Fig. 15.22. The muting bit map can have a length of {2, 4, 6, 8, 16, 32} bits. For Option 1, muting is applied on each transmission instance of a DL-PRS Resource Set (assuming a periodic transmission of DL-PRS Resource Sets). Each bit in the bit map corresponds to a configurable number of consecutive instances of a DL-PRS Resource Set. All DL-PRS Resources within a DL-PRS Resource Set instance are muted (transmitted with zero power) if the corresponding bit in the bit map indicates a "0". The number of consecutive instances is controlled by the parameter *DL-PRS Muting-Bit Repetition Factor*, which can have the values {1, 2, 4, 8}.

For Option 2, muting is applied on each repetition of each of the DL-PRS Resources. Each bit in the bit map corresponds to a single repetition of the DL-PRS Resource within an instance of a DL-PRS Resource Set. For Option 2, the length of the bit map is equal to the *DL-PRS Resource Repetition Factor*.

Option 1 and Option 2 muting can also be used together. If the *DL-PRS Muting Pattern* is provided in the assistance data for both Option 1 and Option 2 muting, the logical AND operation is applied, and a DL-PRS Resource is transmitted when both bits in Option 1 and Option 2 bit strings have the value 1.

In summary, a DL-PRS Resource Set description includes the following parameterss [9]:

- A *DL-PRS Resource Set Identity*, which defines an identity of the DL-PRS resource set configuration.
- A *DL-PRS Periodicity*, which defines the DL-PRS Resource periodicity in the number of slots. The periodicity depends on the subcarrier spacing (SCS) and

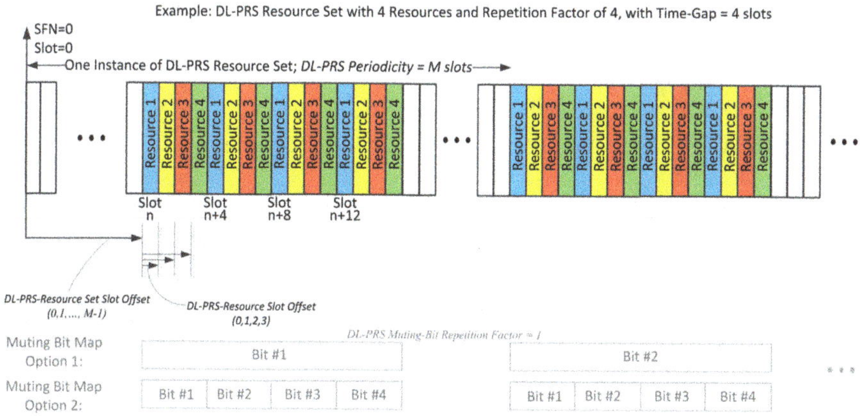

Fig. 15.22 Illustration of DL-PRS muting options

can take the values of $2^{\mu}\cdot\{4, 5, 8, 10, 16, 20, 32, 40, 64, 80, 160, 320, 640, 1280, 2560, 5120, 10{,}240\}$ slots, with $\mu = 0, 1, 2, 3$ for SCS 15, 30, 60 and 120 kHz, respectively. All the DL-PRS Resources within one resource set have the same periodicity.

- A *DL-PRS Resource Repetition Factor*, which defines how many times each DL-PRS Resource is repeated for a single instance of the DL-PRS Resource Set. Values of {1, 2, 4, 6, 8, 16, 32} are supported. All the DL-PRS resources within one resource set have the same Resource Repetition Factor.
- A *DL-PRS Resource Time Gap*, which defines the offset in the number of slots between two repeated instances of a DL-PRS Resource with the same DL-PRS Resource ID within a single instance of the DL-PRS Resource Set. Values of {1, 2, 4, 8, 16, 32} are supported.
- A *DL-PRS Muting Pattern*, which defines a bit map of the time locations where the DL-PRS resource is transmitted or not for a DL-PRS Resource Set. The bit map size can be {2, 4, 8, 16, 32} bits long.
- A *DL-PRS Muting-Bit Repetition Factor*, which defines the number of consecutive instances of a DL-PRS Resource Set corresponding to single bit of the *DL-PRS Muting Pattern* for Option 1 muting.
- A *DL-PRS Resource Set Slot Offset*, which defines the slot offset with respect to SFN#0/slot#0 of the TRP (i.e. defines the slot where the first DL-PRS Resource of the DL-PRS Resource Set occurs).
- The DL-PRS Resource list, defining the configuration for each resource in the set, as described above.

A *DL-PRS Positioning Frequency Layer* is defined as a collection of DL-PRS Resource Sets which have the following common parameters:

- *DL-PRS Subcarrier Spacing*: All DL-PRS Resources and DL-PRS Resource Sets in the same DL-PRS-Positioning Frequency Layer have the same subcarrier spacing for the DL-PRS.
- *DL-PRS Cyclic Prefix*: All DL-PRS Resources and DL-PRS Resource Sets in the same DL-PRS-Positioning Frequency Layer have the same cyclic prefix length for the DL-PRS.
- *DL-PRS Point A*: All DL-PRS Resources belonging to the same DL-PRS Resource Set of the same DL-PRS-Positioning Frequency Layer have a common "DL-PRS Point A".

All DL-PRS Resource Sets belonging to the same Positioning Frequency Layer have the same value of *DL-PRS-Start PRB*, *DL-PRS Resource Bandwidth* and *DL-PRS Comb Size N*.

A DL-PRS measurement (e.g. RSTD, UE Rx-Tx Time Difference Measurement, RSRP) is made per DL-PRS Resource. The measurement report (in case of UE-assisted mode) can include the DL-PRS Resource ID of the DL-PRS Resource Set which is used by the UE for the measurement. Measurements may be made from multiple DL-PRS Resource IDs (in practice, multiple beams) from the same TRP (or pair of TRPs in case of RSTD). For example, a UE may be instructed to measure up to four DL-PRS RSTD measurements for the same pair of TRPs with each measurement between a different pair of DL-PRS Resources or DL-PRS Resource Sets.

Uplink Positioning Reference Signal (UL-PRS)

The UL positioning reference signal is based on the Rel-15 Sounding Reference Signals (SRS) with enhancements for positioning purposes. In the 3GPP specifications, the UL-PRS is referred to as "SRS for positioning". In some respects, the UL-PRS can be seen as the uplink equivalence to the DL-PRS. Both DL-PRS and UL-PRS can also serve as spatial QCL references to establish positioning beam pairs. That is, given the knowledge of a suitable receiver beam for DL-PRS, the receiver knows that the same receiver beam should be suitable for UL-PRS.

The UL-PRS sequence is based on the Zadoff-Chu sequences also used for the Rel-15 SRS [7].

The UL-PRS resource elements are arranged in a particular time/frequency pattern. In contrast to the Rel-15 SRS, the UL-PRS may span 1, 2, 4, 8 or 12 consecutive OFDM symbols which can be located anywhere in a slot. Note that for the Rel-15 SRS, the possible number of symbols per resource is {1, 2, 4} and is located within the last six symbols of a slot only. In the frequency domain, the UL-PRS has also a comb-N pattern. For the Rel-15 SRS, N can take the values 2 or 4, but for the UL-PRS N is extended to the set of {2, 4, 8}. For the UL-PRS, the number of symbols can be larger or smaller than the comb size. For example, a comb-2 UL-PRS with one symbol is supported, or a comb-4 with eight symbols. The defined options are summarized in Table 15.2. For each pair of comb size and

Table 15.2 Resource element offsets for pairs of UL-PRS comb size and number of symbols

Number of symbols Comb-size	1	2	4	8	12
2	{0}	{0, 1}	{0, 1, 0, 1}	–	–
4	–	{0, 2}	{0, 2, 1, 3}	{0, 2, 1, 3, 0, 2, 1, 3}	{0, 2, 1, 3, 0, 2, 1, 3, 0, 2, 1, 3}
8	–	–	{0, 4, 2, 6}	{0, 4, 2, 6, 1, 5, 3, 7}	{0, 4, 2, 6, 1, 5, 3, 7, 0, 4, 2, 6}

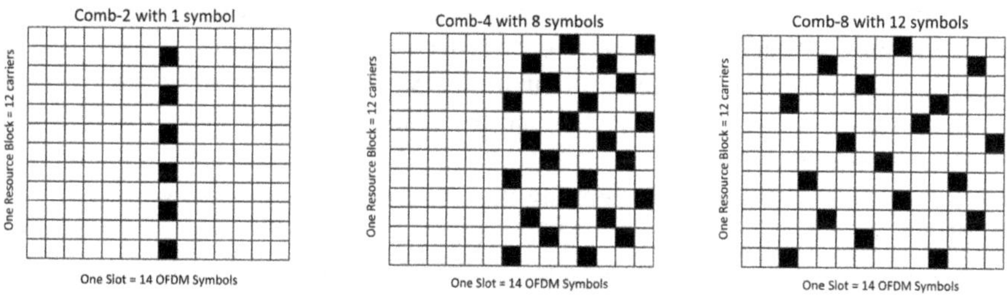

Fig. 15.23 Example of UL-PRS time/frequency pattern

number of symbols, there is one RE pattern. Table 15.2 shows the RE offsets for the combinations of the number of symbols {1, 2, 4, 8, 12} and the comb sizes {2, 4, 8}. The resource element pattern of an UL-PRS resource is configured with a comb offset for the first symbol in the UL-PRS. The relative RE offsets of the following symbols are defined relative to the comb offset of the first symbol.

Examples of UL-PRS resource element pattern are shown in Fig. 15.23. Similar to the DL-PRS, the RE mapping is not arranged in a staircase pattern. This has the advantage that already the first few symbols have a better effective comb size. For example, if only the first few symbols are considered for the TOA measurement, the effect of the alias correlation peaks is better supressed. For example, the first two symbols of the comb-4 UL-PRS provide an effective comb-2 resource element pattern.

UL-PRS transmissions from different UEs can be frequency multiplexed within the same frequency range by using different comb offsets. For comb-2, for example, two UL-PRS can be frequency multiplexed. In the case of comb-12, up to 12 UL-PRS can be frequency multiplexed.

An "UL-PRS Resource" with 1, 2, 4, 8 or 12 consecutive OFDM symbols is transmitted in the active UL bandwidth part (BWP) of the UE. Frequency hopping for the UL-PRS is not supported in Rel-16 (one single transmission of the UL-PRS typically covers the whole configured bandwidth).

Similar to Rel-15 SRS, an UL-PRS can be configured for periodic, semi-persistent or aperiodic transmission. A periodic UL-PRS is transmitted with a certain periodicity and slot offset within that periodicity. A semi-persistent UL-PRS has a periodicity and slot offset in the same way as a periodic UL-PRS. However, actual UL-PRS transmission according to the periodicity and slot offset is activated and deactivated by means of MAC Control Element (CE) signalling. An aperiodic UL-PRS is only transmitted when explicitly triggered by means of Downlink Control Information (DCI).

In summary, an UL-PRS Resource description includes the following parameters, which are similar or extensions of the Rel-15 SRS parameter [7]:

- An *UL-PRS Resource Identity* (*SRS-PosResourceId* in 3GPP specifications), defining the particular UL-PRS Resource.
- A *Transmission Comb*, defining the comb size N of the UL-PRS ($N = 2$, 4 or 8), the comb offset of the first symbol of the UL-PRS Resource ($0 \ldots N$-1) and the cyclic shift for generating the reference sequence [7].
- A *Resource Mapping*, defining the first OFDM symbol location of the UL-PRS Resource in a slot (0,1,2, ...,13) and the number of symbols of the UL-PRS Resource (1, 2, 4, 8 or 12).
- A *Frequency Domain Shift*, defining the frequency domain position of the UL-PRS Resource (same as for Rel-15 SRS [7]).
- A *Frequency Hopping*, defining the bandwidth of the UL-PRS Resource. Note that the name is reused from the Rel-15 SRS [7], although frequency hopping for UL-PRS is not supported. However, part of the frequency hopping parameter is the bandwidth indication, which is the only parameter applicable for UL-PRS.
- A *Group Or Sequence Hopping*, defining whether group or sequence hopping is used (same as for Rel-15 SRS [7]). The hopping modes are used to randomize the reuse of a sequence in the system.
- A *Resource Type*, defining the UL-PRS Resource type (periodic, semi-persistent, aperiodic) and the periodicity for semi-persistent and periodic UL-PRS. In addition to the Rel-15 SRS periodicities of {1, 2, 4, 5, 8, 10, 16, 20, 32, 40, 64, 80, 160, 320, 640, 1280, 2560} slots, the periodicities of {5120, 10,240, 20,480, 40,960, 81,920} slots are supported for UL-PRS. The periodicity of 20,480 slots is applicable for 30, 60 and 120 kHz SCS only; the periodicity of 40,960 slots is applicable for 60 and 120 kHz SCS only; and the periodicity of 81,920 slots is applicable for 120 kHz SCS only.
- A *Sequence ID*, defining the sequence ID used to initialize the pseudo-random group and sequence hopping [7]. For the UL-PRS, the number of different sequence group hopping pattern is increased from 1024 (Rel-15 SRS) to 65,536. Since UL-PRS should be received by neighbouring TRPs, increasing the available number of UL-PRS sequences can be beneficial for reducing UL-PRS collision and further mitigating the UL interference.
- A *Spatial Relation Info*, defining the spatial relation between a reference RS and the target UL-PRS. The reference RS can be a SSB, CSI-RS (for serving cell only), DL-PRS, SRS or UL-PRS.

Spatial relation indication for UL-PRS Resources is supported, where the spatial relation can be either to a downlink reference signal (SSB, CSI-RS (for serving cell only) or DL-PRS) or by the UE previously transmitted SRS or UL-PRS.

In order to provide connectivity, NR UEs supporting millimetre wave operation typically include multiple antenna panels pointing in different directions. The spatial

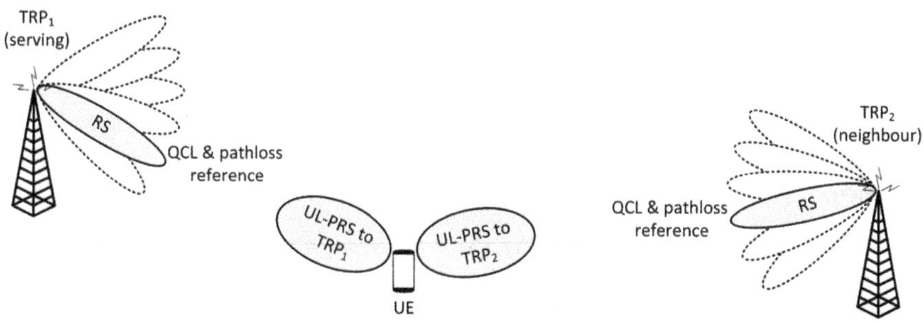

Fig. 15.24 Illustration of spatial relation and pathloss reference for UL-PRS

relation for both serving and neighbouring TRPs is primarily used to indicate which uplink transmission beam the UE may use for the UL-PRS. The UL-PRS beam may be derived from the spatial relation to an indicated downlink reference signal, whereupon the UE may transmit the UL-PRS in the reciprocal direction to how it set its receive beam when receiving the downlink reference signal, as illustrated in Fig. 15.24. An additional procedure may be used by the network, where the UE transmits an UL-PRS or SRS beam sweep and the gNB refers back to one of the swept beams in a previously transmitted UL-PRS or SRS resource to indicate the spatial relation to the UL-PRS resource.

Compared to the Rel-15 SRS, the UL-PRS can have a spatial relation to a neighbour TRP. For positioning, the UL-PRS generally needs to be received also by neighbouring TRPs. To determine an appropriate UL-PRS spatial domain transmission filter in FR2 (i.e. beam) towards neighbouring TRPs, the UE may receive the reference signal for the UL-PRS beam from the same direction as that of the UL-PRS desired direction. The reference signal from neighbour TRPs may be an SSB or DL-PRS.

Similar to Rel-15 SRS, a device can be configured with one or more UL-PRS Resource Sets, where each UL-PRS Resource Set can include one or more UL-PRS Resources. As described above, an UL-PRS can be configured for periodic, semi-persistent or aperiodic transmission. All UL-PRS Resources included within an UL-PRS Resource Set are of the same type; i.e. periodic, semi-persistent or aperiodic UL-PRS transmission is a property of an UL-PRS Resource Set.

Another property of the UL-PRS Resource Set is the transmission power control. For the UL-PRS, open-loop power control is supported, including support for (fractional) pathloss compensation to serving and neighbouring TRPs, where the UE estimates the uplink pathloss for serving and neighbouring TRPs based on downlink measurements and sets the UL-PRS transmit power accordingly.

The UE may estimate the pathloss from a downlink reference signal, which may be an SSB or DL-PRS not only from the serving TRP but also from neighbour TRPs. As the serving TRP is likely to be closer to the UE than a neighbouring TRP, the downlink pathloss estimate based on the serving TRP may result in too small transmit power for the UL-PRS to be detectable at the neighbour TRPs (UL hearability). The pathloss estimate based on a reference signal from neighbour TRPs can be used to transmit the UL-PRS with an appropriate power towards the intended TRPs. That is, a smaller pathloss estimate results in higher UL-PRS transmit power towards the intended TRP.

In summary, an UL-PRS Resource Set description includes the following parameters, which are similar or extensions of the Rel-15 SRS parameter [7]:

- An *UL-PRS Resource Set Identity* (*SRS-PosResourceSetId* in 3GPP specifications), defining the particular UL-PRS Resource Set. It is unique in the context of the BWP in which the UL-PRS is defined.
- A *Resource Type*, defining the time domain behaviour of the UL-PRS resource configuration. The network configures the UL-PRS Resources in the same Resource Set with the same time domain behaviour on periodic, aperiodic and semi-persistent.
- An *alpha* value for the UL-PRS power control [8], defining the fractional pathloss compensation. The alpha value is multiplied by the UE with the pathloss estimate. For full pathloss compensation, alpha is equal to 1.
- A *p0* value for the UL-PRS power control [8], which can be described as the "desired receive power" at the TRP. That is, the UL-PRS transmit power determination is based on $p0 + alpha \cdot PL$, where *PL* is the pathloss estimate [8].
- A *Pathloss Reference RS*, defining the reference DL signal to be used for pathloss estimation. The DL reference signal can be an SSB or DL-PRS from the serving or neighbouring TRP.
- The UL-PRS Resource list, defining the configuration for each resource in the set, as described above.

4 Outlook

The previous sections in this chapter summarized the main location services and positioning features supported in 5G NR as specified in 3GPP Release-16. Five NR native positioning techniques were specified in Rel-16, based on time and angle measurements. The positioning features have been evaluated in various simulations [6] and were designed for accuracy targets suitable for regulatory and some commercial use cases. The performance targets for commercial use cases were set as follows [6]:

- Horizontal and vertical positioning error < 3 m for 80% of UEs in indoor deployment scenarios (i.e. indoor base stations and indoor UEs)
- Horizontal positioning error < 10 m and vertical positioning error < 3 m for 80% of UEs in outdoor deployment scenarios (i.e. outdoor base stations and outdoor UEs)
- Latency <1 s

The 5G service requirements specified in [1] also include "High Accuracy Positioning" requirements, which are characterized by ambitious system requirements for positioning accuracy in many verticals. For example, on the factory floor, it is important to locate assets and moving objects such as forklifts, or parts to be assembled. Similar needs exist in transportation and logistics, for example. In some road user cases, UE's supporting Vehicle-to-Everything (V2X) communication applications are also applicable to such needs. With the needs of diverse industry verticals in mind, future 3GPP Releases will aim for very high accuracies (e.g. < 1 metre) and very low latencies (e.g. 10's of milliseconds).

References

1. 3GPP TS 22.261: Service requirements for the 5G system; Stage 1
2. 3GPP TS 23.273: 5G System (5GS) Location Services (LCS); Stage 2
3. 3GPP TS 38.305: Stage 2 functional specification of User Equipment (UE) positioning in NG-RAN
4. 3GPP TS 37.355: LTE Positioning Protocol (LPP)
5. 3GPP TS 38.455: NG-RAN; NR Positioning Protocol A (NRPPa)
6. 3GPP TR 38.855: Study on NR positioning support
7. 3GPP TS 38.211: Physical channels and modulation
8. 3GPP TS 38.213: Physical layer procedures for control
9. 3GPP TS 38.214: Physical layer procedures for data
10. 3GPP TS 38.215: Physical layer measurements
11. OMA MLP TS: Mobile Location Protocol. Open Mobile Alliance
12. OMA-AD-SUPL: Secure User Plane Location Architecture. Open Mobile Alliance
13. OMA-TS-LPPe: LPP Extensions Specification. Open Mobile Alliance
14. D.J. Torrieri, Statistical theory of passive location systems. IEEE Transactions on Aerospace and Electronic Systems **AES-20**(2), 183 (1984)
15. B. Forssell, *Radionavigation Systems* (Prentice Hall, New York, 1991)
16. S.A. Zekavat, R.M. Buehrer, *Handbook of position location* (IEEE Press, New Jersey, 2012)
17. F. van Diggelen, *Assisted GPS, GNSS, and SBAS, Boston, London* (Artech House, 2009)
18. E. Dahlman, S. Parkwall, J. Sköld, *5G NR – The next generation wireless access technology* (Academic Press, London, 2018)

Chapter 16
NR Integrated Access and Backhaul

Qian (Clara) Li, Thomas Novlan, and Erik Dahlman

1 Introduction

Integrated access and backhaul (IAB) is specified in 3GPP Rel-16 with the goal to enable NR self-backhauling for flexible network deployment without being constrained by backhaul transport network [1]. Network deployment flexibility is very much needed to achieve goals of 5G system, e.g.:

- Most of the frequency bands available for 5G systems are at high frequency and mmWave bands, e.g., 4 GHz, 28 GHz, and 39 GHz. As the carrier frequency goes higher, coverage of each base station (BS) becomes lower due to high propagation loss. In addition, at mmWave bands, radio propagation is subjected to blockage, which further affects coverage. To address the high propagation loss and blockage and ensure network coverage, dense network deployment is needed. Network densification imposes challenges on network deployment, especially on deployment of backhaul transport networks.
- One of the main objectives of 5G system is to address vertical use cases, e.g., industrial automation, intelligent transportation, etc. As verticals have diverse requirements on communication network, allowing flexible network deployment could make NR technology better suit for vertical use cases.

Q. Li (✉)
Intel Corporation, Hillsboro, OR, USA
e-mail: clara.q.li@intel.com

T. Novlan
AT&T, Austin, TX, USA

E. Dahlman
Ericsson, Stockholm, Sweden

X. Lin, N. Lee (eds.), *5G and Beyond*, https://doi.org/10.1007/978-3-030-58197-8_16

Main features of 3GPP Rel-16 IAB design include:

- Support operation on both sub-6GHz and mmWave bands. As NR air interface is a unified air interface supporting both sub-6GHz and mmWave band, the self-backhauling design can be applied to both carrier frequencies.
- Support both in-band and out-of-band backhauling. For in-band backhauling, the backhaul and access links operate in a same carrier band. For out-band backhauling, the backhaul and access links operate in different carrier bands. In-band backhauling requires dynamic multiplexing between access and backhaul links and, therefore, higher design complexity.
- Support multi-hop backhauling with directed-acyclic-graph (DAG) topology and topology adaptation. Multi-hop backhauling is needed to extend network coverage. Topology adaptation could enable the network to adapt its backhaul topology to tackle cases such as backhaul link failure, backhaul overload, addition of new IAB nodes/hops, etc.
- Support dynamic resource multiplexing between access and backhaul links. Resource multiplexing between backhaul and access links can be done in time domain, spatial domain, and frequency domain. Granularity of time domain multiplexing can be in symbol level.
- Support both NSA and SA deployment in both access and backhaul links. For access link in NSA mode, UE's MCG path is in LTE, and UE's SCG path can be over NR IAB backhauling. For backhaul link in NSA mode, MCG path of IAB node's MT (mobile terminal) function is in LTE, and SCG path of IAB node's MT function is in NR. For both cases, only NR Uu-based backhaul is supported.

In Rel-16, the design of IAB is confined within the RAN and is transparent to UE. Impact on core network is minimized from both specification and signaling perspectives. No specification and implementation impacts on the UE side. This allows a NR UE of an earlier release (i.e., Rel-15) to access to an IAB node without noticing the difference in network. Rel-16 IAB design only considered fixed relay deployment. Mobile relay is a potential item to be studied in Rel-17. Half-duplex constraint is assumed at an IAB node, i.e., an IAB node cannot transmit and receive simultaneously on one frequency band using a same RF chain. The half-duplex constraint imposes design challenges especially on resource multiplexing between access and backhaul links. Also, half-duplex constraint degrades performance in terms of throughput and latency. As the number of hops increase, the performance degradation will be more severe.

In the following sections, we will provide an overview on NR IAB system architecture, the key issues and designs. We will conclude this chapter by discussing future directions on IAB technology.

2 System Description

A NR IAB RAN contains donor nodes and IAB nodes. A donor node connects to core network (CN) via wired transport link. An IAB node could connect to a donor node or a parent IAB node via wireless backhaul. For a NR RAN with CU-DU split architecture, the donor node could contain a CU and multiple DUs. Each IAB node contains a DU function (IAB-DU) and a MT (mobile terminal) function (IAB-MT). Figure 16.1 illustrates the NR IAB RAN architecture with CU-DU split.

The IAB-DU function allows an IAB node to communicate with child IAB nodes or access UEs. The IAB-MT function allows an IAB node to communicate with donor node or parent IAB node. Interface between donor node and 5G core network is based on NG interface. Interface between donor CU and IAB-DU is based on F1 interface. Interface between donor DU (or a parent IAB-DU) and IAB-MT is NR Uu interface. Interface between IAB-DU and access UE is NR Uu interface. From network perspective, an IAB-MT appears as an UE. Donor CU generates RRC message for both IAB-MT and access UEs. For IAB nodes in NSA mode connecting with EPC, IAB donor connects with EPC using master eNB as control plane anchor via X2 interface and connect with S-GW in user plane via S1 interface. IAB-MT connects with MeNB via LTE Uu interface. MeNB servers as control plane anchor for IAB-MTs. Figure 16.2 shows the IAB network reference architecture.

Figures 16.3 and 16.4 show the user plane (UP) and the control plane (CP) protocol stacks among donor CU, donor DU, IAB nodes, and access UE. An adaptation layer is added at above RLC layer in IAB nodes and donor DU. Functions of the backhaul adaptation layer protocol (BAP) include:

- Bearer mapping and de-mapping. This includes functions such as mapping of UE UP PDUs to backhaul RLC channels, differentiating traffics to be delivered to upper layers from traffics to be forwarded to egress RLC channels, and mapping of ingress backhaul RLC channel to egress backhaul RLC channel.
- Routing. This includes functions such as encode routing ID to the BAP header, maintaining a routing table for route selection, handling routing priority, choosing routing strategy based on routing table, and considering network load balancing.

Some basic IAB operations are listed in the following. Those operations often involve both MT function and DU function and have design impact on physical layer, L2 radio protocol, and backhaul network.

- IAB node initial access and integration to network. This includes IAB MT initial access with parent DU and CU, IAB MT function registration and connection establishment with core network, RLC channel establishment at IAB MT for backhauling between parent DU and CU, IAB DU initial connection setup with CU, and IAB network synchronization.
- IAB node configuration update. This includes IAB MT function configuration update via RRC and IAB DU function configuration update via F1-AP.

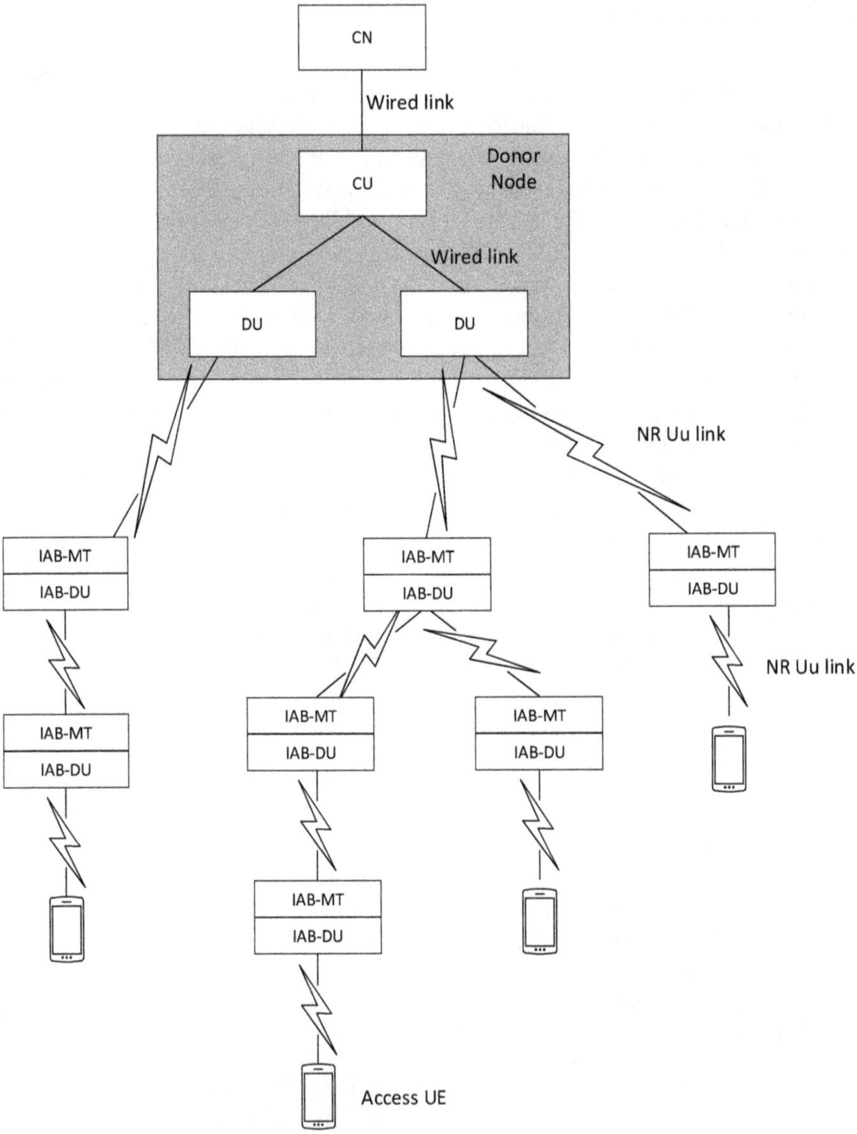

Fig. 16.1 Illustration on NR IAB RAN architecture with CU-DU split

- CP and UP data transportation. This includes radio resource scheduling, flow control and routing, and QoS management.
- IAB network topology adaptation. This includes inter-IAB node measurement and discovery, backhaul link failure management, parent node reselection, and multiple backhaul connectivity.

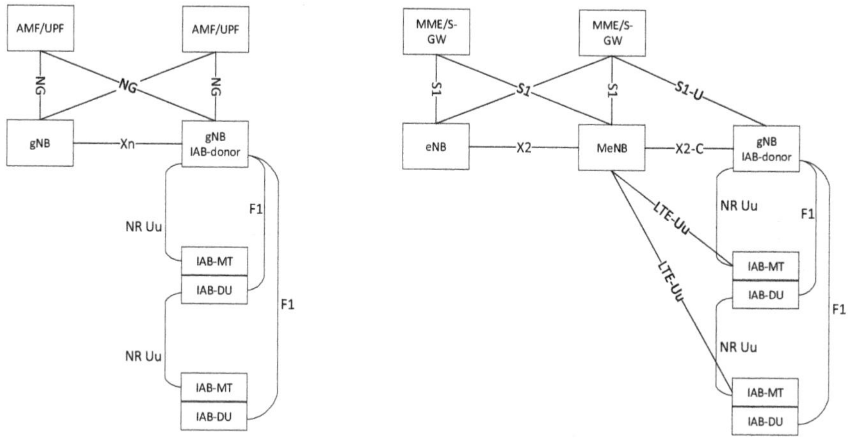

(a) IAB node in SA mode connecting with 5GC

(b) IAB node in EN-DC mode connecting with EPC

Fig. 16.2 IAB network reference architecture

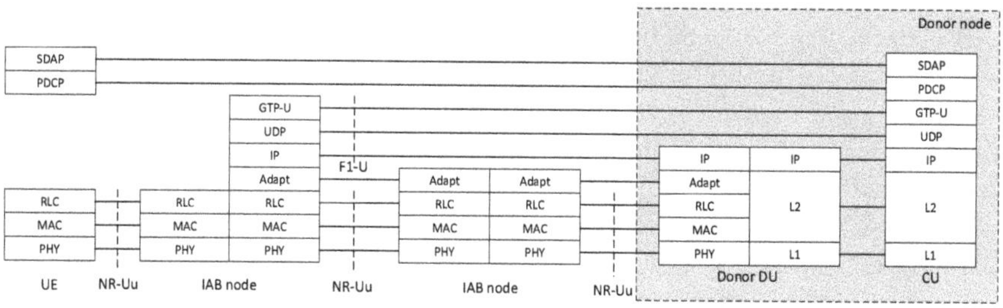

Fig. 16.3 User plane protocol stack

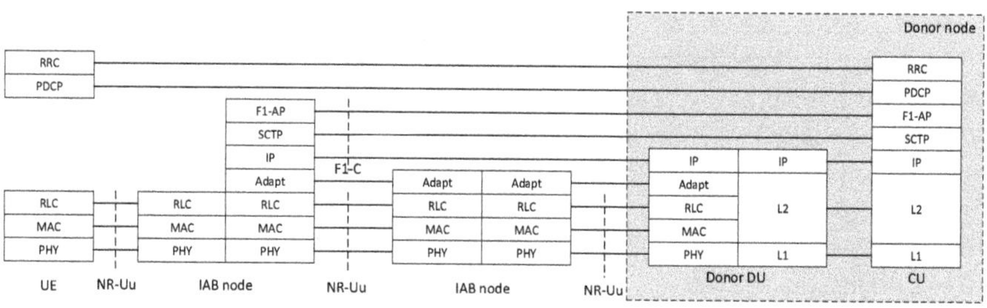

Fig. 16.4 Control plane protocol stack

In the following sections, we will look into details on IAB operations from physical layer, radio protocol, and backhaul aspects.

3 Physical Layer Aspects

3.1 Backhaul Link Discovery and Measurement

Backhaul link discovery and measurement has two stages: initial backhaul link selection and backhaul link reselection. Initial backhaul link selection is conducted when MT performs initial access to parent IAB node. A same measurement procedure as used by a Rel-15 UE is applied, i.e., MT measures parent nodes' SSB for initial access. The assumed SSB transmission periodicity in MT initial access is 160 ms which is different from the 20 ms SSB transmission periodicity assumed in access UE's initial access.

Backhaul link reselection can be performed based on SSB measurement or CSI-RS measurement. Due to the half-duplex constraint at an IAB node and the need to support multi-hop backhauling, additional SSB resources and SSB measurement windows need to be defined at DU and MT besides the ones defined for access UE's measurement. In general, when the IAB nodes are configured with M SSB resources and N SSB measurement windows, the number of mutually discoverable IAB nodes is C_{M+N}^{N}. It can be seen that C_{M+N}^{N} could achieve its maximum value when $M = N$. In 3GPP Rel-16, it was agreed to support $M = N = 4$, i.e., up to 4 SSB transmission resources and 4 SSB measurement resources can be defined for inter-IAB node discovery and measurement. This allows maximum 70 IAB nodes to be mutually discovered. Considering the SSB transmission resource defined for UE/MT initial access, the total number of SSB transmission resources at an IAB DU will be maximum 5. SSB used for inter-IAB node discovery and measurement can be transmitted in the same sync raster or in different sync rasters as SSBs used for UE initial access. To enable IAB node mutual discovery, two IAB nodes need to have at least one non-overlapping SSB transmission occasion and SSB measurement window. Figure 16.5 shows one example on mutual discovery among three IAB nodes based on SSB. In the example, 2 SSB transmission occasions and 1 SSB measurement window are defined for each IAB node. This allows maximum 3 IAB nodes to be mutually discovered and measured. To reduce overhead, longer SSB transmission periodicity and SSB measurement periodicity can be used. In 3GPP Rel-16, the SSB transmission periodicity can be 5 ms, 10 ms, 20 ms, 40 ms, 80 ms, 160 ms, 320 ms, and 640 ms. The SSB measurement periodicity can be 5 ms, 10 ms, 20 ms, 40 ms, 80 ms, 160 ms, 320 ms, 640 ms, and 1280 ms.

3.2 IAB PRACH

Considering half-duplex constraint and multi-hop relaying, additional time orthogonal PRACH resources need to be introduced for IAB operation so that PRACH occasions in adjacent hops can be time-domain multiplexed. Based on the PRACH occasions defined in 3GPP Rel-15 for access UEs, additional PRACH occasions for

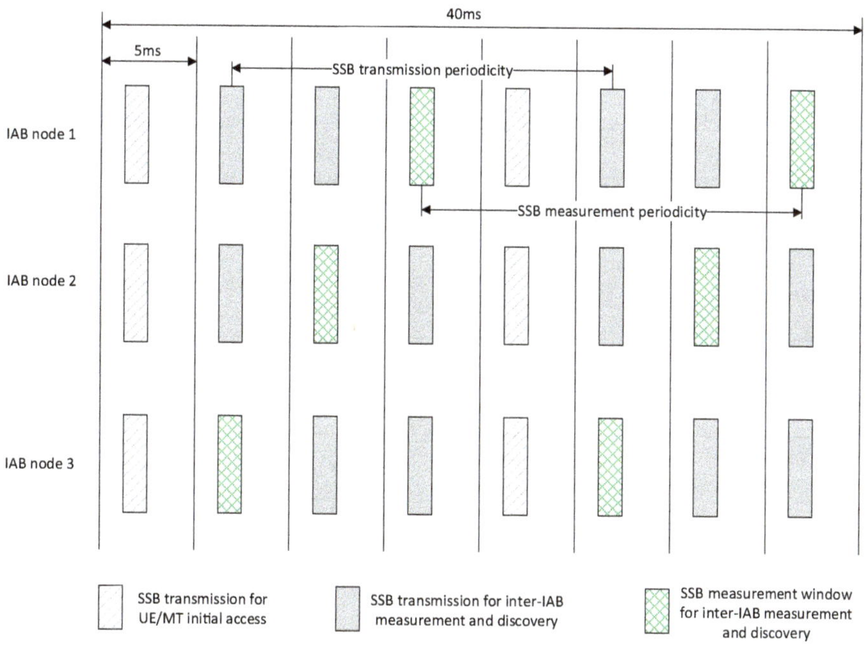

Fig. 16.5 Example on SSB based inter-IAB node discovery and measurement

IAB MT are introduced by applying scaling factors on the periodicity of Rel-15 PRACH occasions, and/or applying subframe offsets on Rel-15 PRACH occasions, and/or applying slot offsets on Rel-15 PRACH occasions. With the introduction of the scaling factor on PRACH periodicity, the maximum PRACH periodicity is extended to 640 ms. The scaling on PRACH periodicity also leads to scaling on the association between SSB and PRACH. The association pattern between PRACH occasions and SSB repeats at most every 640 ms.

3.3 Resource Multiplexing Among Backhaul and Access Links

From an IAB node's perspective, resource multiplexing between parent backhaul link and child parent link or access link can be done by means of TDM, FDM, or SDM. By TDM, communications in parent backhaul link and in child backhaul link or access link are conducted in orthogonal time resources. By FDM, communications in parent backhaul link and in child backhaul link or access link are conducted in orthogonal frequency resources. By SDM, an IAB node can simultaneously transmit or receive in parent backhaul link and in child backhaul link or access link using different beams/antenna panels in a same time/frequency resource. Examples on TDM, FDM, and SDM are illustrated in Fig. 16.6. As

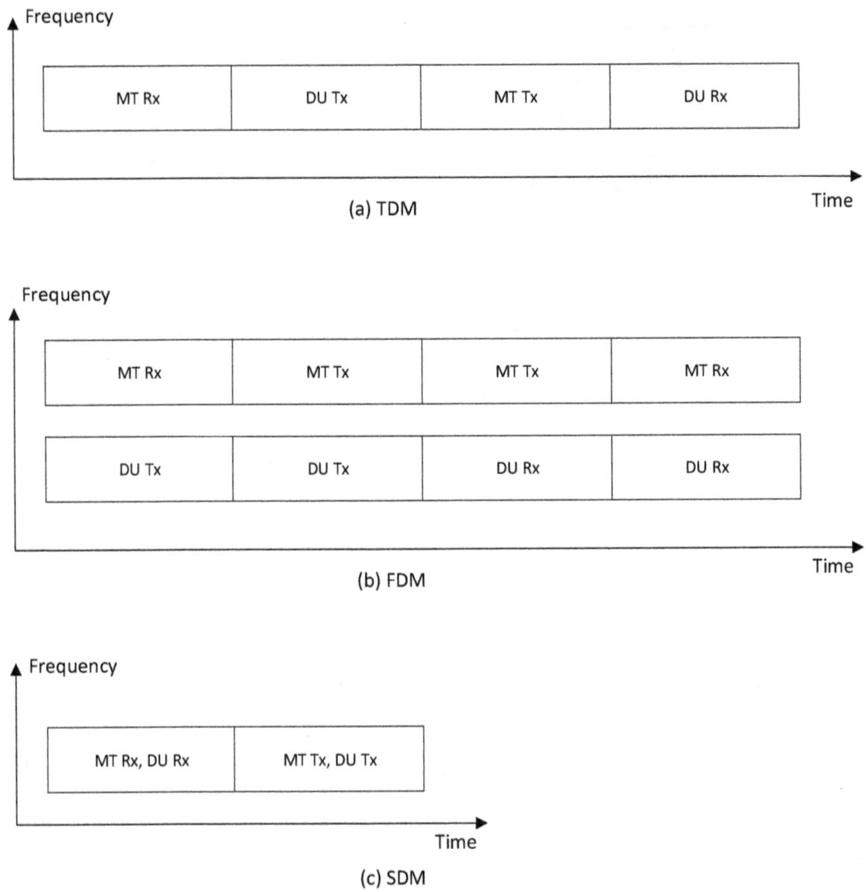

Fig. 16.6 Examples of TDM, FDM, and SDM

SDM requires simultaneous transmission or reception at MT and DU over a same frequency resource, symbol-level and slot-level time alignment are needed between MT *R*x and DU *R*x and between MT *T*x and DU *T*x.

TDM and SDM both impose constraints on how resource is used at an IAB node. For TDM, MT *T*x/*R*x and DU *T*x/*R*x need to be in different time occasions. For SDM, MT *T*x (or *R*x) and DU *R*x (or *T*x) need to be in different time occasions. From a DU perspective, there are resources that it cannot use or can only conditionally use (e.g., conditioned on if its co-locate MT uses the resource or not). Three new resource types, Hard, Soft, and Not Available (NA), are therefore introduced to IAB DU in addition to the downlink/uplink/flexible resource types. A Hard resource means the DU can always use the resource. A Soft resource means the DU can only use the resource if the resource is indicated as available. A NA resource means the DU cannot use the resource. The H/S/NA resources of an IAB

DU are semi-statically configured by CU via F1-AP signaling. To enable more flexible use of the resources and to support SDM between backhaul links, NR IAB allows UL/F/DL resource type configuration, i.e., resource configurations that start from UL resource.

A Soft resource at an IAB node is typically associated with a Hard resource assigned to its parent node. When the parent node is not using the resource or when the parent node is using the resource to serve other MTs or UEs, the Soft resource could be made available to the IAB node. The availability of the Soft resource depends on parent node's scheduling and therefore needs to be dynamically indicated to the IAB node. The dynamic indication can be implicit or explicit. For implicit indication, the IAB DU can perceive the Soft resource availability based on scheduling of its parent backhaul link, i.e., if its co-located MT is scheduled with *T*x or *R*x in the parent backhaul link, the IAB DU can derive whether it can use the resource. For explicit indication, parent node will provide an explicit L1 signal to IAB node's MT regarding the Soft resource availability on the IAB node's DU. The explicit availability indication is provided per DL/UL/F resource type of a slot. Considering the similarity between the availability indication and the slot format indication, DCI Format 4_0 is defined for the availability indication which follows similar structure as DCI Format 2_0. Each field value in the DCI format 4_0 is used as the index in a RRC-configured Availability Combination table similar to the SFI *SlotFormatCombination* table. Each entry in the Availability Combination table indicates the resource availability for a set of consecutive slots. Each element in an entry of the Availability Combination table indicates the resource availability in a slot. The resource availability can take eight values. The eight values correspond to the eight combination of DL/UL/F resource availability. As the DU Soft resource availability indication and the MT's slot format indication can be signaled to MT at a same time using DCI Format 4_0 over different PDCCH resources, to differentiate the two information, a new RANI (AI-RNTI) is introduced to scramble the PDCCH carrying the availability indication.

When switching between MT *T*x/*R*x and DU *T*x/*R*x in one IAB node, there could be time conflicts at the slot boundaries of the switch, due to propagation delay and timing advanced in parent backhaul link and child backhaul link or access link. Figure 16.7 illustrates the slot boundary time conflict scenarios. To resolve the conflicts, one approach is to let parent node to schedule conservatively, i.e., parent node can avoid scheduling the last few symbols or the first few symbols in the parent backhaul link to avoid conflict with child backhaul link. Another approach is to let IAB node to schedule conservatively, i.e., if IAB node anticipates conflict, it will avoid scheduling the symbols in slot boundaries. A more balanced approach is to let IAB node and parent node coordinate on conflict resolution, i.e., IAB node can report to parent node on the number of symbols it would like parent node to yield, and parent node can inform IAB node on the number of symbols it plans to yield.

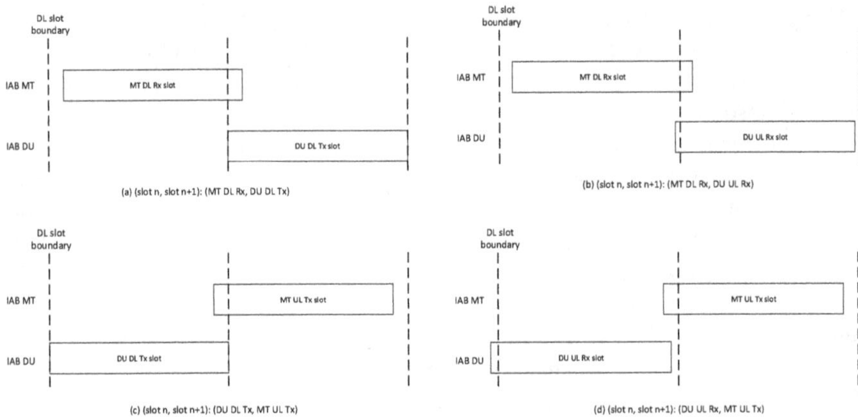

Fig. 16.7 Illustration on time conflicts at slot boundaries for switching between MT *T*x/*R*x and DU *T*x/*R*x

3.4 OTA-Based DL Timing Alignment

For IAB nodes to acquire DL timing, the synchronization source can be GNSS or synchronization signals from parent nodes. If synchronization signal in parent node is used, the IAB node needs to be able to derive the propagation time (T_p) between parent DU's DL *T*x ($T_{\mathrm{pDU}-Tx}$) and MT's DL *R*x ($T_{\mathrm{MT}-Rx}$), so that IAB node DU's DL *T*x timing can be calculated as $T_{\mathrm{MT}-Rx} - T_\mathrm{p}$. Figure 16.8 shows the timing relationships between parent DU *T*x/*R*x and MT *T*x/*R*x. It can be seen that T_p can be derived using the following formula:

$$T_\mathrm{p} = \left(\mathrm{TA} - T_\mathrm{g}\right)/2 = \frac{\left(N_{\mathrm{TA}} + N_{\mathrm{TA-offset}}\right) \times T_\mathrm{C} - N_{\mathrm{TA-offset}} \times T_\mathrm{C} - T_\Delta}{2}$$

$$= \frac{N_{\mathrm{TA}} \times T_\mathrm{C} - T_\Delta}{2},$$

where TA is the timing advanced value applied at MT for UL transmission; N_{TA} is the value obtained at MT based on the TA command from parent DU to MT; $N_{\mathrm{TA-offset}}$ is the TA offset value configured to MT based on the operation carrier frequency; T_g is the time gap between TA and $2 \times T_\mathrm{p}$ which includes parent DU's RF switch time, other implementation related time gaps, and quantization introduced time gaps; $T_\Delta = T_\mathrm{g} - N_{\mathrm{TA-offset}} \times T_\mathrm{C}$, which captures RF and implementation time gaps other than that one counted in $N_{\mathrm{TA-offset}} \times T_\mathrm{C}$ and quantization-introduced time gaps.

From the above formula, we can see that T_Δ needs to be additionally signaled to MT to derive T_p. Similar to TA command, signaling of T_Δ can be done via MAC CE signaling.

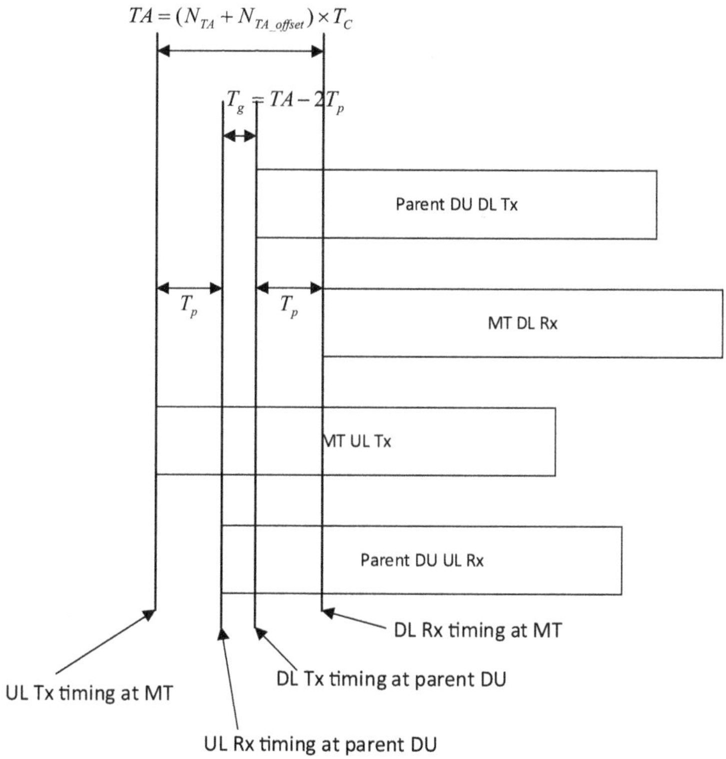

Fig. 16.8 Illustration on time relationship among parent DU *Tx*/*Rx* and MT *Tx*/*Rx* for OTA-based timing synchronization

4 Radio Protocol Aspects

4.1 Backhaul Adaptation Protocol (BAP)

As illustrated in Figs. 16.3 and 16.4, backhaul adaptation protocol is introduced in the CP and UP protocol stacks to enable routing and bearer mapping. Main functions of routing are to determine the egress link for the packet. Main function of bearer mapping is to determine the RLC channel in the egress backhaul link to which the packet should be mapped on. Detailed functions of BAP include [2]:

- F1: Retrieve packets from ingress RLC layer
- F2: Deliver packets to egress RLC layer
- F3: Retrieve packets from upper layer
- F4: Deliver packets to upper layer
- F5: Differentiate traffic to be delivered to upper layers from traffic to be delivered to egress RLC layer

- F6: Perform bearer mapping and routing for packets delivered to egress RLC layer
- F7: Selection/addition of BAP identifiers for packets received from upper layer

To enable the BAP functions, the following information need to be carried in the BAP header includes:

- Routing ID (20 bits), which includes BAP address (10 bits for both DL and UL) and BAP path ID (10 bits for both DL and UL)
- C/D bit (1 bit), which is used to indicate CP and UP packet
- Reserved bits (3 bits)

The BAP address is used to identify the destination node. Path ID is used to identify the path to destination (i.e., there could be multiple paths to a destination). The BAP header is added at the donor DU (for DL) or at the access IAB node (for UL) based on information received from higher layers. An intermediate IAB node will check the BAP header and decide the egress link based on its routing table.

4.2 *Bearer Mapping*

For packet forwarding in an IAB node, UE radio bearers need to be multiplexed on to backhaul RLC channels. Two bearer mapping options are considered: 1:1 mapping and N:1 mapping. Figure 16.9 illustrates the two mapping options for UP. In an 1:1 mapping, each UE DRB is mapped to separate RLC channel along the path between an IAB access node and an IAB donor node. One ingress RLC channel

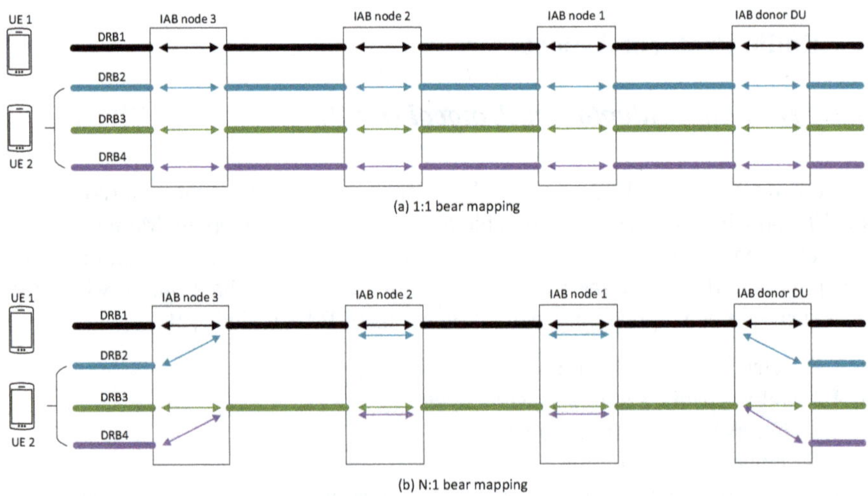

Fig. 16.9 Illustration on 1:1 bear mapping and N:1 bear mapping

identified by ingress LCID is mapped to only one egress RLC channel identified by egress LCID. The number of RLC channels supported by an intermediate IAB node will be the same as the number of UE DRBs served by the intermediate IAB node. In a N:1 mapping, several UE DRBs are multiplexed onto a single BH RLC-channel based on specific parameters such as bearer QoS profile, i.e., if DRBs for different UEs require same or similar QoS requirement, those DRBs can be multiplexed to one backhaul RLC channel at the access IAB node or IAB donor DU. In the access IAB node, e.g., IAB node 3 in Fig. 16.9, the received packets from the ingress BH RLC channel is forwarded to the associated access RLC channel using based on the GTP-U TEID in the GTP-U header. In the intermediate IAB nodes, e.g., IAB nodes 1 and 2 in Fig. 16.9, LCIDs of the egress and ingress backhaul RLC channels are used to perform mapping between egress and ingress links. The same bearer mapping mechanism is applied for CP and UP packets in the DL and UL.

4.3 Flow Control

Congestion and packet discard may happen in the intermediate IAB nodes. Flow control is therefore supported in both UL and DL directions in order to avoid congestion-related packet drops on IAB-nodes and IAB-donor DU. In the uplink, flow control can be achieved by hop-by-hop UL scheduling, i.e., when a MT experiences congestion in the parent backhaul link, the MT can provide its buffer status information to the co-locate DU, which can then reduce resource allocation in its UL grants to the child IAB MTs or UE. In the downlink, end-to-end flow control is needed to manage congestion, as a congestion in an IAB node may not be known by its parent IAB node, the parent IAB node may schedule a high ingress data rate beyond the IAB node's capability. E2E flow control (e.g., via F1-U) allows the intermediate IAB nodes to report the congestion condition to the CU which can then control the influx data rate at the source. In addition to E2E flow control, hop-by-hop flow control can also be applied in the DL to manage the local traffic congestion. The congested IAB node can feedback flow control information via BAP layer to its parent IAB node.

4.4 Scheduling and QoS

Multi-hop transmission in an IAB network would lead to increased delay. Figure 16.10 shows one example on UL transmission delay. In each hop along the link between UE and donor, a packet originated from UE would experience delays due to scheduling request, buffer status report and UL grant. Low latency scheduling is therefore needed in IAB networks. One way to reduce latency is to allow MT in an intermediate IAB node to send SR or BSR before receiving UL packet from access link or child backhaul link, i.e., since the co-locate DU has information on

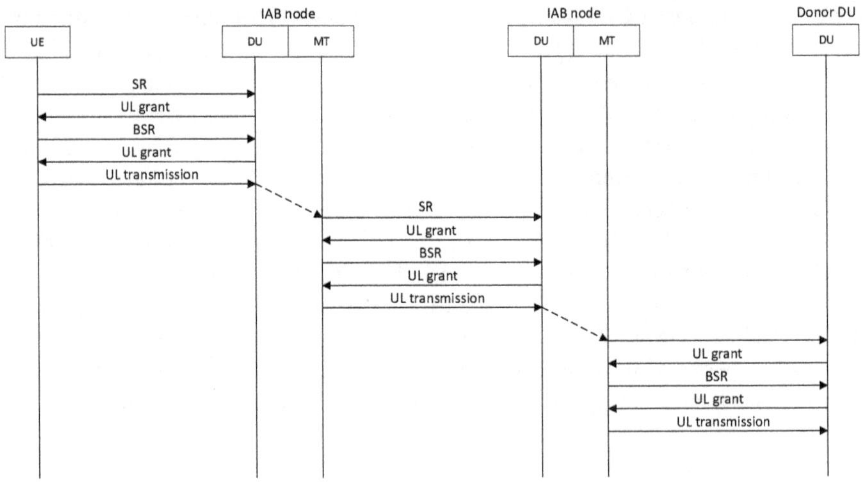

Fig. 16.10 Illustration on UL delays

when an UL packet is expected to arrive at the DU, the DU can inform the co-locate MT on this information so that the MT can send SR or BSR in advance to request for UL grant before the actual data arrival. It would enable MT to obtain UL resource prior to the actual packet arrival and therefore save UL scheduling time in the backhaul links. Considering the complexity of enhancing SR to support this feature, in 3GPP Rel-16, only enhanced BSR is supported. A new MCE CE is designed for the enhanced BSR.

4.5 Radio Link Failure Handling

Radio link failure (RLF) in the backhaul link can happen due to various reasons such as blockage, weather condition, etc. IAB MT performs RLF detection following the same approach as access UEs. The following RLF detection criteria as defined in [3] for access UEs also apply to backhaul link RLF detection.

- Expiry of a timer started after indication of radio problems from the physical layer (if radio problems are recovered before the timer is expired, the UE stops the timer)
- Random access procedure failure
- RLC failure

After RLF is declared, recovery from RLF in the backhaul link also follows the same approach as access UEs defined in [3], i.e., the MT will:

- Stay in RRC_CONNECTED.
- Select a suitable cell and then initiate RRC reestablishment.

- Enter RRC_IDLE if a suitable cell was not found within a certain time after RLF was declared.

As RLF in a backhaul link would impact all the downstream IAB nodes, when RLF is detected by an IAB node MT, the IAB node DU would inform the downstream IAB nodes on the RLF so that the downstream IAB nodes can take action (e.g., start to search for a new route) accordingly.

RLF in a backhaul will trigger topology adaptation and routing table update as described in Sect. 5.2.

5 Backhaul Network Aspect

5.1 IAB Node Initial Integration

IAB node integration into network would go through three phases as showed in Fig. 16.11 [4].

- Phase 1: IAB MT setup. In this phase, the new IAB node's (e.g., IAB node 2 in Fig. 16.11) MT connects to network following same procedure as UE, i.e., MT will perform initial access to parent IAB node, RRC connection setup with donor CU, registration and authentication with 5GC, MT-related configuration and context establishment, MT-related radio bearer establishment, etc.
- Phase 2: Backhaul link setup. In this phase, backhaul RLC links are established between parent IAB DU and the new IAB node's MT (Phase 2-1) for at least CP traffic, so that CP message (e.g., F1-AP) can be transported from CU to the IAB node for DU setup in phase 3. Also, BAP layer is updated to support routing between the new IAB node and donor DU (Phase 2-2). This includes BAP routing

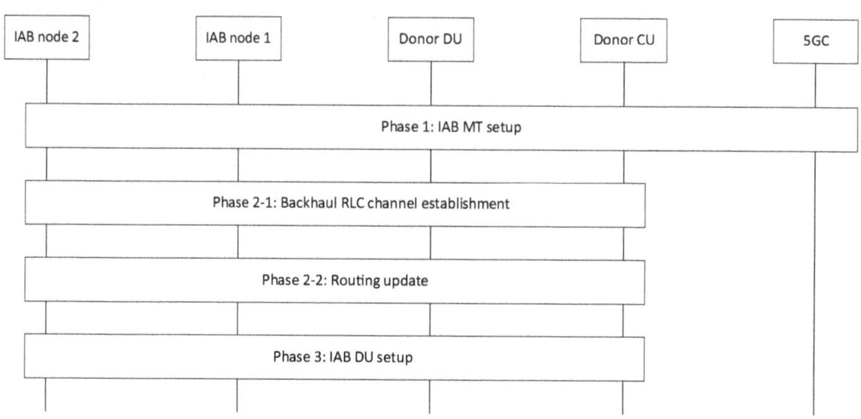

Fig. 16.11 High-level procedure on IAB node initial integration

ID configuration for the downstream direction at the donor DU and BAP routing ID configuration for the upstream direction at the new IAB node's MT. Routing tables at the donor DU and all ancestor DUs are updated with the BAP routing ID.

- Phase 3: IAB DU setup. In this phase, the DU function of the new IAB node is configured. The configuration is done using F1-AP messages via F1 procedures. The F1-AP messages are transported via the RLC channel set up in phase 2. After IAB DU setup, the new IAB node can start to serve access UEs.

5.2 *Routing and Topology Adaptation*

The IAB network topology follows a directed acyclic graph (DAG) topology, where an IAB node can have multiple parent nodes and could have multiple routes to another ancestor node or donor node. Figure 16.12 shows one example of the DAG topology. The IAB network topology can change due to events such as backhaul RLF, addition or removal of redundant backhaul links, etc. From an IAB node perspective, scenarios of topology adaption could include switching to a new parent IAB node under a same donor DU, switching to a new parent IAB node under different donor DUs of a same donor CU, switching to a new parent IAB node under

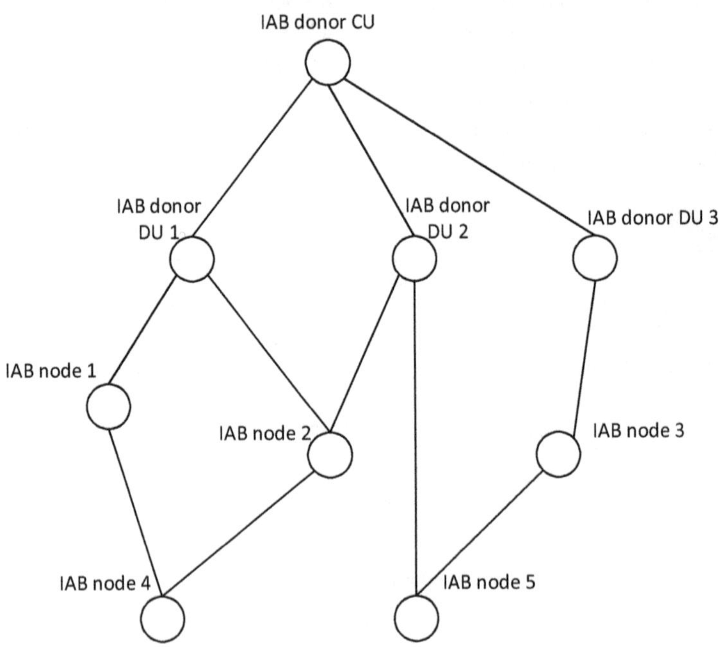

Fig. 16.12 Example on DAG topology

different donor CUs, and adding a redundant route with a new parent node. Detailed procedures for different topology adaption scenarios could differ. But generally speaking, topology adaption procedure would include the following steps:

- Step 1: Measurement and information collection. This step could involve IAB node measurement on backhaul link quality and IAB node and donor node collect information on backhaul link condition and network load over a sufficient large area of the IAB topology.
- Step 2: Topology determination. Based on the information collected, a topology is determined following performance objectives.
- Step 3: Topology reconfiguration. This setup could include routing table update, BAP layer update, IAB DU context update, IAB MT reconfiguration, and donor CU/DU context update.

6 Conclusion and Future Work

This chapter introduced basic IAB features as specified in 3GPP Rel-16. This sets the technical foundation for enabling IAB capabilities in 5G network. Moving forward, further enhancement on IAB is under discussion around aspects including mobile relaying, routing and topology adaptation for mesh network, supporting of advanced multiplexing modes such as SDM and FDM, and supporting on using unlicensed spectrum in IAB network.

References

1. 3GPP RP-172290, Study on integrated access and backhaul for NR
2. RAN 2 106 chairman notes
3. 3GPP TS 38.331 v15.7.0, NR; Radio Resource Control (RRC); Protocol specification
4. 3GPP TS 38.401 v15.6.0, NR-RAN; Architecture description

Chapter 17
Sky High 5G: New Radio for Air-to-Ground Communications

Xingqin Lin, Anders Furuskär, Olof Liberg, and Sebastian Euler

1 Introduction

The proliferation of smartphones and their use to access the Internet have grown to the point where lack of connectivity may result in discomfort and frustration. As current mobile trends develop, there is a clear global demand for ubiquitous wireless connectivity. While mobile broadband service is becoming prevalent on land, in-flight connectivity (IFC) remains limited, and its quality of service is often perceived by consumers as poor [1]. Provision of home-quality broadband IFC is an attractive market considering 4.3 billion passengers being carried by airlines in 2018 [2].

IFC can be provided by satellite communication systems and cellular-based direct air-to-ground (A2G) communications. In the case of satellite communications, the connectivity between aircraft and ground stations is established by utilizing satellites. The currently available satellite-based IFC solutions mostly operate over the Ku and Ka frequency bands [1, 3]. The satellite-based IFC solutions are particularly suitable for intercontinental flights over the ocean, but they usually suffer from limited system capacity and long transmission latencies. Direct A2G communications utilize cellular technology to establish direct connectivity between aircraft and ground stations. The ground stations play a role similar to cellular towers, but their antennas are up-tilted toward the sky. The inter-site distances (ISD) of the ground stations for direct A2G communications are also much greater than their counterparts deployed for terrestrial communications. Compared to the

X. Lin (✉)
Ericsson, Santa Clara, CA, USA
e-mail: xingqin.lin@ericsson.com

A. Furuskär · O. Liberg · S. Euler
Ericsson, Stockholm, Sweden
e-mail: anders.furuskar@ericsson.com; olof.liberg@ericsson.com; sebastian.euler@ericsson.com

X. Lin, N. Lee (eds.), *5G and Beyond*, https://doi.org/10.1007/978-3-030-58197-8_17

satellite-based solutions, the cellular-based direct A2G solutions have the potential of offering larger system capacity and shorter latencies for IFC and are particularly attractive for short- and medium-haul continental flights and long-haul flights over or near land. But the direct A2G solutions have difficulties in providing connectivity for intercontinental flights over the oceans. Therefore, the satellite-based and cellular-based solutions complement each other, and both are needed to achieve full-scale IFC in the skies.

In this chapter, we focus on direct A2G communications for IFC. The existing A2G systems for public mobile communications utilize cellular technologies [4]. For example, in the USA, the Gogo Biz network uses a modified version of the Third-Generation (3G) Code Division Multiple Access (CDMA) 2000 technology to provide IFC. Another example is the European Aviation Network that utilizes Fourth-Generation (4G) Long-Term Evolution (LTE) ground network (in combination with satellite coverage) [5]. As described in more detail in Sect. 2, the existing A2G systems are limited in capacity (typically up to tens of Mbps). They cannot fulfill the vision of providing home-quality broadband to every seat of every aircraft [6].

To provide significantly improved IFC experience for the passengers, the A2G systems will need to be evolved to exploit the Fifth-Generation (5G) wireless access technology, known as New Radio (NR). 5G NR will become a dominant access technology in the next several years, addressing a wide range of use cases from enhanced mobile broadband (eMBB) to ultra-reliable low-latency communications (URLLC) to massive machine type communications (mMTC) [7]. NR features ultra-lean transmission, support for low latency, advanced antenna technologies, and spectrum flexibility including operation in high frequency bands, interworking between high and low frequency bands, and dynamic time-division duplex (TDD) [8]. Embracing NR in A2G systems is expected to provide enhanced performance and vastly improved user experience across a range of flight paths, use cases, and aircraft types. It is worth noting that the 3rd Generation Partnership Project (3GPP) work on NR non-terrestrial networks (NTN) in Release 17 also includes the support of A2G communications [9].

The recent research on A2G communications has mostly focused on low altitude (e.g., a few hundreds of meters) unmanned aerial vehicles [10, 11]. For IFC in commercial aircraft typically flying at an altitude between 9.5 km and 12 km, the work of [12] discussed the technical possibilities of enhancing the existing LTE for A2G communications. The work of [13] presented a simulation study for the compatibility of an in-cabin LTE femto-cellular system with the current terrestrial LTE systems. In [14], the authors conducted a performance comparison of a 4G A2G network, a 5G A2G network, and a satellite network for IFC. The studied networks were mainly based on the LTE standards, though one of the networks is called "5G A2G network." Preliminary link and system level evaluations of NR A2G systems were carried out in [15]. In contrast, this chapter provides a more in-depth performance evaluation of NR A2G systems in a range of bands (low, mid, and high). Further, we identify the major challenges associated with NR-based direct A2G communications and delve into the detailed NR technical specifications to discuss enhancements to address the challenges. Additionally, we provide an overview of the existing A2G systems and point out some fruitful avenues for future research.

2 Overview of A2G Systems

In this section, we provide an overview of the existing exemplary A2G systems for public communications [4]. Figure 17.1 gives an illustration of the system architecture for such systems, which consists of (•) cabin access network providing, for example, WiFi connectivity to end users, (•) A2G network equipment onboard aircraft for communicating with ground stations, (•) ground radio access network for establishing direct A2G radio links to aircraft, and (•) core network for connection management and connectivity to external packet data networks.

2.1 A2G Systems in North America

Gogo Biz's A2G network has more than 200 towers in the continental USA, Alaska, and Canada for providing in-flight WiFi connectivity. It operates in the frequency bands 849–851 MHz (downlink) and 894–896 MHz (uplink). The connection between aircraft and ground stations uses modified Evolution-Data Optimized (EVDO), which is part of the CDMA2000 standard. The maximum total download data rate is 9.8 Mbps. Enhancements were made to handle extended cell size (up to 400 km) and aircraft speed (resulting in an extended range of Doppler shifts and complexities of the airborne handover procedure).

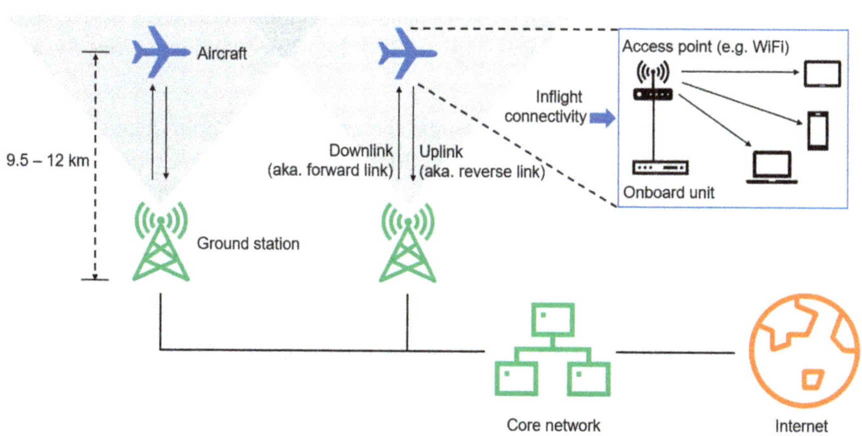

Fig. 17.1 An illustration of the system architecture for cellular-based direct A2G communications

Back in 2011, Qualcomm submitted a petition to the Federal Communications Commission (FCC) on deploying an A2G system (dubbed as the *Next-Gen AG* system) operating in the Ku band (14–14.5 GHz) sharing with fixed-satellite service. The proposed system would use between 150 and 250 ground stations scattered around the USA to provide an aggregated data rate up to 300 Gbps. The proposed air interface was based on orthogonal frequency-division multiplexing (OFDM), with TDD being the communication mode. The proposed system, however, has not been deployed as of today.

2.2 A2G Systems in Europe

Extensive studies have been carried out in Europe to identify suitable frequency band(s) for A2G systems, including 1900–1920 MHz, 2010–2025 MHz, 2400–2483.5 MHz, 3400–3600 MHz, and 5855–5875 MHz.

- A2G system identified in ETSI TR 103 054 [16]: This A2G system was based on the LTE specifications, using paired spectrum of 2 × 10 MHz for frequency division duplex (FDD) operation. Trials were conducted in Germany within the 2.6 GHz FDD bands. The trial results demonstrated peak data rates of up to 30 Mbps in the downlink and 17 Mbps in the uplink.
- A2G system identified in ETSI TR 101 599 [17]: This A2G system was optimized to operate within the bands 2400–2483.5 MHz and 5855–5875 MHz, utilizing 20 MHz TDD spectrum or 2 × 10 MHz FDD spectrum. The air interface was based on OFDM. This system featured adaptive beamforming antennas and used four separate phased array antennas at each ground station. Each phased array antenna could generate multiple spatially separated beams to serve the aircraft.
- A2G system identified in ETSI TR 103 108 [18]: The system was designed to operate in the 5855–5875 MHz TDD band and could use 5 MHz or 10 MHz bandwidth. The air interface was Universal Mobile Telecommunications System (UMTS) based on CDMA.

Despite the extensive studies and trials, the commercial deployments of these systems have not yet emerged. In July 2018, the Electronic Communications Committee (ECC) withdrew the previous decision on the harmonized use of A2G systems in the 1900–1920 MHz band. That said, the European Aviation Network, with integrated S-band satellite connection and complementary LTE-based terrestrial network, was launched in 2018. The ground network uses LTE band 65 (2100 MHz) and includes 300 ground stations (up to 75 km cell radius) spread across Europe. The network can provide data rates of up to 75 Mbps in the downlink and 20 Mbps in the uplink.

2.3 A2G Systems in Asia

In October 2012, the Civil Aviation Administration of China (CAAC) started China's A2G system project. The initial system was based on the synchronous CDMA (SCDMA) specifications and employed TDD as the communication mode [19]. Trials were conducted in the 1785–1805 MHz band. Later, the focus changed to LTE-based A2G. Extensive experimental verifications for LTE-based A2G in civil aviation applications have been conducted in China in the last few years. It is expected that the A2G system will be commercially available in China in the next few years.

In Japan, several trials were conducted in 2012 to test the performance of a prototype A2G system operating in the 40 GHz frequency range. The system used FDD and employed antenna tracking for proper operation in the millimeter wave frequency range. The trial results demonstrated 141.7 Mbps data rate for quadrature phase shift keying (QPSK) modulation and 106.3 Mbps data rate for eight phase shift keying (8PSK) modulation. In 2017, a trial A2G system based on TDD LTE was tested in the very high frequency (VHF) band in Japan. The trial results showed a maximum downlink data rate of 27 Mbps at a flight speed of 430 km/h.

3 Performance Study of NR A2G Systems

The different A2G systems, as described in Sect. 2, use different cellular technologies and operate in different frequency ranges, from below 1 GHz to millimeter wave frequencies. It will be desirable to adopt a unified standard globally to reap the benefits of economies of scale to provide enhanced IFC performance and improved user experience. To this end, 5G NR, the latest wireless access technology, will be the natural technology choice for future A2G systems. In this section, we present performance evaluation of NR-based A2G systems for a range of bands (low, mid, and high) to shed light on the potential of NR for IFC in the 5G era.

3.1 NR A2G at Low Band

We first consider an NR A2G system operating at low-band spectrum (below 1 GHz). In the simulation, there are 19 ground stations placed on a hexagonal grid. The NR A2G system uses 2×10 MHz FDD spectrum at 700 MHz carrier frequency. Each ground station uses an antenna array with parameters (M, N, P) $=$ (2, 2, 2) to produce a beam, where M denotes the number of rows in the array, N denotes the number of columns in the array, P denotes the number of polarizations, and the pattern of each antenna element follows the 3GPP TR 38.901 [20]. These antenna arrays are laid flat facing the sky at a height of 35 m. The

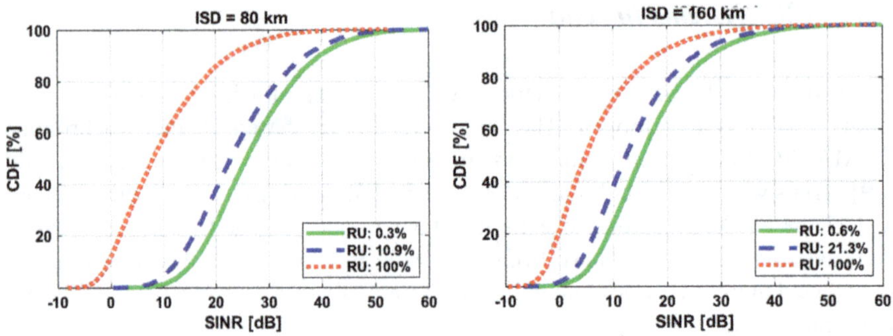

Fig. 17.2 Downlink SINR distributions under different traffic loads in an NR A2G system at the low band

aircraft are placed at a height of 12 km, and each has two cross-polarized isotropic antennas. The transmit powers of the ground stations and the aircraft are 80 W and 0.2 W, respectively. Figure 17.2 shows the downlink signal-to-interference-plus-noise ratio (SINR) distributions at different resource utilization (RU) levels for two different ISD values at the low band. RU level can indicate the interference level in the network: the higher the RU level, the more the co-channel interference. So, as expected, the SINR becomes worse as the RU level increases. It is also observed that the SINR distributions with 80 km ISD are better than their counterparts with 160 km ISD. For example, the 5-percentile SINR with 80 km ISD at 0.3% RU level is 13.2 dB, which is much higher than the 4.5 dB 5-percentile SINR with 160 km ISD at 0.6% RU level. This is because the received signal powers are lower in the larger cells due to larger path loss and smaller antenna gains experienced by the UEs in the network with 160 km ISD that uses a same antenna array as used in the network with 80 km ISD. The SINR difference becomes smaller at the high RU levels where co-channel interference becomes more pronounced. For example, at the 100% RU level, the 5-percentile SINR with 80 km ISD is −2 dB, which is slightly higher than the −3.3 dB 5-percentile SINR with 160 km ISD.

Next, we turn to the throughput performance with 80 km ISD in a single beam setting. Figure 17.3 shows the downlink and uplink throughput distributions at different RU levels at the low band. Since in-flight traffic is typically downlink heavy, we focus on examining the throughput performance at high load in the downlink and at low load in the uplink. At the RU level of 79%, the 5-, 50-, and 99-percentile downlink throughput values are 1.6 Mbps, 8.5 Mbps, and 32.2 Mbps, respectively. At the RU level of 1.7%, the 5-, 50-, and 90-percentile uplink throughput values are 0.35 Mbps, 7.6 Mbps, and 19.2 Mbps, respectively. These throughput values appear to be on par with the data rates offered by the LTE-based European Aviation Network (up to 75 Mbps in the downlink and 20 Mbps in the uplink).

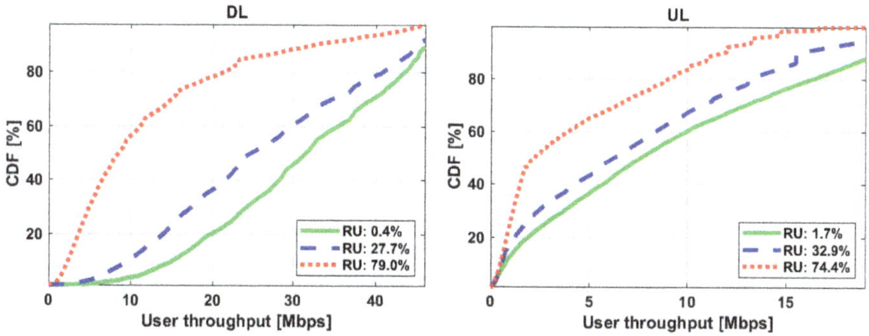

Fig. 17.3 Throughput distributions under different traffic loads in an NR A2G system at the low band

3.2 *NR A2G at Mid- and High Band*

NR features spectrum flexibility and supports operation in the spectrum ranging from sub-1 GHz to millimeter wave bands. To further explore the potential of NR-based A2G systems, we next turn to NR A2G systems operating in mid-band (1–7 GHz) and high band (millimeter wave frequencies).

The NR A2G system at the mid-band spectrum uses 2 × 100 MHz FDD spectrum at 3.5 GHz carrier frequency and an antenna array with parameters (M, N, P) = (4, 4, 2) at the ground station. One mid-band beam covers one-fourth of the area covered by one beam at the low band. So, if 80 km ISD is kept in the mid-band, each ground station should produce four beams to cover a cell. Figure 17.4 shows the downlink and uplink throughput distributions at different RU levels at the mid-band. At the RU level of 82.3% in the downlink, the 5-, 50-, and 99-percentile throughput values are 5.9 Mbps, 40.6 Mbps, and 175.6 Mbps, respectively. At the RU level of 2.3% in the uplink, the 5-, 50-, and 99-percentile uplink throughput values are 0.72 Mbps, 19.3 Mbps, and 112.0 Mbps, respectively. The highest downlink and uplink throughput values are 454.9 Mbps and 197.5 Mbps, respectively. We can see that by exploiting the large bandwidth in the mid-band, this NR A2G system offers much higher throughput values than its counterpart in the low band.

The NR A2G system at the high band uses 2 × 400 MHz FDD spectrum at 28 GHz carrier frequency and an antenna array with parameters (M, N, P) = (8, 8, 2) at the ground station. One mid-band beam covers 1/64 of the area covered by one beam at the low band. So, if 80 km ISD is kept in the high band, each ground station should produce 64 beams to cover a cell. Figure 17.5 shows the downlink and uplink throughput distributions at different RU levels at the high band. The highest downlink and uplink throughput values are 1.5 Gbps and 563.9 Mbps, respectively. This NR A2G system is capable of providing Gbps links to the aircraft.

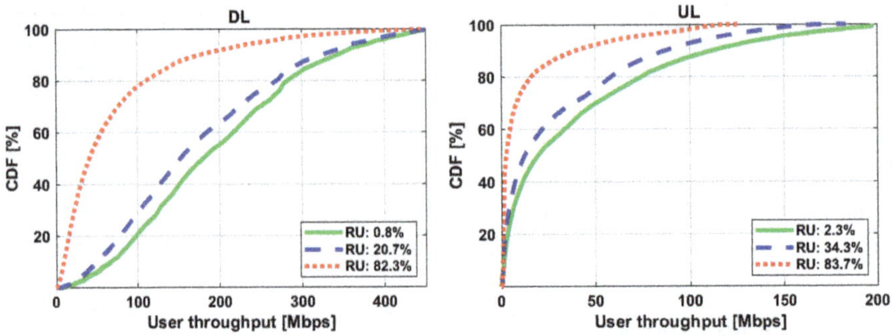

Fig. 17.4 Throughput distributions under different traffic loads in an NR A2G system at the mid-band

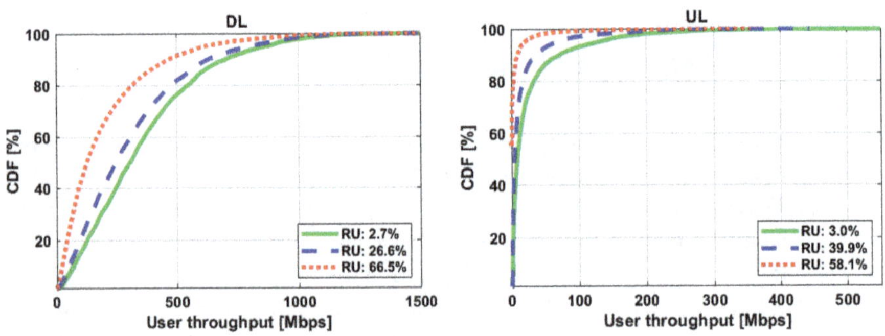

Fig. 17.5 Throughput distributions under different traffic loads in an NR A2G system at the high band

4 Potential Enhancements for NR A2G Systems

In the previous section, we have presented a performance study of NR-based A2G systems for a range of frequency bands, illustrating the potential of NR for IFC. Though the inherent flexibility of NR allows it to be used to support A2G communications, NR has been designed mainly targeting terrestrial mobile communications. In this section, we discuss performance enhancing solutions to optimize NR connectivity to provide further improved performance for IFC.

4.1 Large Cells

A2G systems feature large cells to limit network deployment cost for serving sparsely scattered aircraft in the sky. Typical ISD in A2G systems is expected to range from 80 to 200 km. In some cases, a larger cell size may be needed, for example, to enable offshore aircraft flying close to coast to communicate with nearest ground stations. To accommodate these cases, we consider a maximum cell radius of 300 km as an appropriate design target for NR A2G systems.

Supporting NR A2G systems with up to 300 km cell radius across a range of bands requires revisiting the many timing relationships defined in NR specifications. For example, timing advance is used at the User Equipment (UE) to adjust uplink frame timing relative to downlink frame timing. The required timing advance for a UE is roughly equal to the round-trip delay between the UE and the serving 5G NodeB (gNB), e.g., up to ~2 ms for an A2G system with cell radii up to 300 km. During a random-access procedure, the gNB estimates the required timing advance value by processing the received random-access preamble and sends the value to the UE in a random-access response message. The maximum timing advance applied during initial access in NR is equal to $2^{-\mu} \times 2$ ms for subcarrier spacing values of $2^{\mu} \times 15$ kHz, where $\mu = 0, 1, 2, 3, 4$. So, except for the 15 kHz subcarrier spacing, the timing advance value range is not sufficient to support NR A2G systems across a range of bands and, thus, would need to be extended.

In case of TDD-based NR A2G systems, a guard period is required to isolate downlink slots from uplink slots. The maximum guard period would be 2 ms to support 300 km cell radius. To limit the overhead of guard period to be no more than 10%, the TDD period would need to be at least 20 ms. Consider, for example, the 30 kHz subcarrier spacing with 0.5 ms slot duration, which is a popular design choice for NR deployments in the mid-band spectrum. In this case, the 2 ms guard period translates into four guard period slots. Assuming 20 ms TDD period, one example TDD frame structure could include 24 downlink slots, four guard period slots, and 12 uplink slots in one period, where the number of downlink slots is twice the number of uplink slots to adapt to the downlink heavy in-flight traffic. For a physical downlink shared channel (PDSCH) reception ending in the slot n, a UE may need to transmit hybrid automatic repeat request (HARQ) feedback in the slot $n + k_1$, where k_1 indicates the slot offset between the PDSCH ending slot and the slot for HARQ transmission. The maximum possible value of k_1 in NR is 15, while the aforementioned example TDD structure with 30 kHz subcarrier spacing would require the value of k_1 to be configurable up to 39.

4.2 High Mobility

Commercial aircraft cruise at about 740–930 km/h. We consider a maximum UE speed of 1200 km/h to be the design target for NR A2G systems.

The high UE speeds in A2G systems result in pronounced Doppler effects. At a speed of 1200 km/h, a UE would experience Doppler shifts of up to about ± 1.11 ppm in the downlink, i.e., about ± 0.78 kHz, ± 3.89 kHz, and ± 31.08 kHz at the carrier frequencies of 700 MHz, 3.5 GHz, and 28 GHz, respectively. Handling such high Doppler shifts may need new UE performance requirements for NR A2G systems. The Doppler shifts in the uplink would be about twice the Doppler shifts in the downlink. The severe Doppler effects may cause inter-carrier interference in the uplink. To mitigate the inter-carrier interference, a gNB may schedule different UEs in different frequencies with sufficient guard spectrum in between. However, this is not a spectrally efficient solution. An enhancement which can help the transmissions from different UEs in a cell to be frequency aligned at the gNB would be more desirable. This can be achieved by applying different frequency adjustment values at different UEs in the uplink to compensate for their different Doppler shifts.

The high aircraft speeds in A2G systems also bring in challenges for antenna beam tracking, though the NR channels, signals, and procedures have been designed to support beamforming. The volumes of cone-shaped beams become larger due to the larger ISD values in A2G systems, increasing the probability of intersecting beams and potentially larger inter-beam interference. Nonetheless, the A2G channels are line-of-sight (LOS) dominated. The beamforming can exploit the location information of aircraft which usually have fixed flight routes and stable mobility patterns. The ground stations may obtain the location information of the aircraft by listening to, for example, the Automatic Dependent Surveillance Broadcasting (ADS-B) signal that includes the position, speed, and altitude of aircraft. The beamforming at the UE may be further facilitated if the location information of the ground stations is made available at the UE side. The location information-assisted beamforming can help improve signal strength and reduce interference. The information transmitted in ADS-B may also be utilized for Doppler estimation and compensation.

Despite the high UE speeds in A2G systems, handover events are not expected to be frequent due to the large cell sizes. For example, it would take a few minutes for an aircraft cruising at 1200 km/h to traverse through a cell with 50 km radius. This handover rate is lower than the handover rate that a high-speed train would experience in a terrestrial network where cells are of much smaller sizes. As NR is capable of serving high-speed trains, mobility management is not expected to be challenging for NR A2G systems. Certain enhancements may be considered to further improve the mobility performance in NR A2G systems. For example, the triggering of measurement reporting and/or conditional handover may be made dependent on aircraft UE location.

4.3 Coexistence with Terrestrial and Satellite Systems

As indicated in the performance study presented in Sect. 3, large bandwidths are crucial for providing high data rates in NR A2G systems. Securing harmonized large bandwidths dedicated to the A2G systems might be challenging. An alternative could be that mobile operators reuse their terrestrial spectrum for A2G services if permitted by regulation.

Using the same spectrum for both terrestrial 5G network and A2G network requires careful deployment planning to ensure that mutual interference between the terrestrial 5G and A2G networks is acceptable. Such deployment coordination may be easier for the upper portion of mid-band spectrum and high-band spectrum, which are typically used for local deployment for capacity enhancement in terrestrial networks. Thus, the terrestrial gNBs may be geographically separated from the A2G gNBs which can be placed in the remote areas such as on remote mountains. Nonetheless, there would likely be many complications obstructing co-channel deployment of terrestrial 5G and A2G networks, for which radio frequency (RF) requirements of base stations for A2G and aircraft UEs would need to be studied.

Note that the A2G ground stations are deployed to serve aircraft UEs, and they do not aim to directly serve the passengers' UEs. The use of mobile phones, tablets, and laptops during the flight typically requires the devices to be switched to airplane mode. The devices without airplane mode switched on may continuously try to access the A2G cells, draining the batteries of the devices and causing vast access loads to the A2G system. To prevent this, NR A2G would need an access control mechanism to give access exclusively to aircraft.

One might also consider reusing the satellite spectrum for A2G services, similar to Qualcomm's petition on deploying an A2G system operating in the Ku band used for fixed-satellite service. The incumbent satellite services should be protected from interference from such A2G systems. Regulation and interference coordination issues in this case appear to be even more challenging than the co-channel deployment of terrestrial 5G and A2G networks.

5 Conclusions and Research Directions

Wireless broadband connectivity is becoming ubiquitous, and the sky should not impose a limit on that. The quality of the current in-flight connectivity service is unsatisfactory. It is of importance to develop solutions to provide true broadband connectivity in the cabin. 5G NR will become the new normal in the next several years. The existing A2G systems are based on earlier generations of mobile technologies. This chapter has presented a performance study of NR A2G systems in a range of bands. The results show that embracing 5G NR in the A2G systems has the potential of providing enhanced performance and vastly improved user experience. This chapter has also identified the major challenges and discussed enhancements for NR A2G systems.

Making sky high 5G broadband connectivity a reality requires breaking down many barriers along the road. We conclude by pointing out some fruitful avenues for future research.

Aircraft UE Beamforming The performance study of NR A2G systems in this chapter has assumed beamforming at the ground stations. Beamforming at the aircraft UEs has the potential of further improving the system performance. It is important to make the antenna design and operation compatible with aircraft engineering and operations.

Interference Management for A2G Systems Beamforming and beam-steering techniques deliver a directional signal to the aircraft. The beams may intersect in the skies and cause mutual interference. Coordinating resource allocation and beam management for interference mitigation in A2G systems is an interesting research problem. Besides, coexistence studies between A2G systems and other terrestrial/satellite systems are of high interest to ensure that the systems do not cause harmful interference to each other.

Prototyping of NR A2G Systems For further understanding the potential and challenges of NR A2G systems, it is important to develop early prototypes and collect feedback. The prototypes may help identify potential shortcomings in the NR specifications for A2G systems. The prompt feedback would facilitate the adoption of appropriate enhancements in the NR standards.

Integrated Terrestrial 5G, A2G, and Satellite Networks The future of connectivity will be seamless, regardless of where you are. True seamless connectivity will need a network of networks that integrate terrestrial 5G, A2G, and satellite networks, among others. Designing and managing the network of networks to provide transparent service continuity to users is challenging, but important.

Spectrum, Regulation, and Business Models for A2G Systems Harmonized spectrum allocations and unified regulatory frameworks across national borders are key to a significant uptake of A2G systems. The right business models should also be in place to help achieve sufficient market scale for A2G systems. It is vital to develop an agreed set of international standards to build a successful A2G ecosystem to achieve seamless in-flight broadband connectivity.

References

1. J.P. Rula, J. Newman, F.E. Bustamante, A.M. Kakhki, D. Choffnes, Mile high wifi: A first look at in-flight internet connectivity, in *Proceedings of the 2018 World Wide Web Conference*, 2018, pp. 1449–1458
2. International Civil Aviation Organization, The world of air transport in 2018, 2018. Available at https://www.icao.int/annual-report-2018/Pages/the-world-of-air-transport-in-2018.aspx. Accessed on 4 Nov 2020
3. E. Dinc, M. Vondra, S. Hofmann, D. Schupke, M. Prytz, S. Bovelli, M. Frodigh, J. Zander, C. Cavdar, In-flight broadband connectivity: Architectures and business models for high capacity air-to-ground communications. IEEE Commun. Mag. **55**(9), 142–149 (2017)
4. ITU-R, Systems for public mobile communications with aircraft, Report ITU-R M.2282-0, Dec 2013
5. European Aviation Network, The fastest connectivity service made for European Skies. Available at https://www.europeanaviationnetwork.com/content/dam/inmarsat/aviation/services/InmarsatAviation_EuropeanAviationNetwork.pdf. Accessed on 4 Nov 2020
6. Seamless Air Alliance, How the seamless air alliance will collaborate to make in-flight broadband access out of this world. Available at https://www.seamlessalliance.com/wp-content/uploads/Seamless-Whitepaper-FINAL.pdf. Accessed on 4 Nov 2020

7. X. Lin, Debunking seven myths about 5G new radio, Aug 2019. Available at https://arxiv.org/ftp/arxiv/papers/1908/1908.06152.pdf. Accessed on 4 Nov 2020
8. X. Lin, J. Li, R. Baldemair, T. Cheng, S. Parkvall, D. Larsson, H. Koorapaty, M. Frenne, S. Falahati, A. Grövlen, K. Werner, 5G new radio: Unveiling the essentials of the next generation wireless access technology. IEEE Commun. Stand. Mag. **3**(3), 30–37 (2019)
9. RP-193234, Solutions for NR to support non-terrestrial networks (NTN), 3GPP TSG RAN meeting #86, Sitges, Spain, Dec 2019. Available at http://www.3gpp.org/ftp/tsg_ran/TSG_RAN/TSGR_86/Docs/RP-193234.zip. Accessed on 4 Nov 2020
10. M. Mozaffari, W. Saad, M. Bennis, Y. Nam, M. Debbah, A tutorial on UAVs for wireless networks: applications, challenges, and open problems. IEEE Commun. Surv. Tuts. **21**(3), 2334–2360, third quarter (2019)
11. X. Lin, V. Yajnanarayana, S. Muruganathan, S. Gao, H. Asplund, H.-L. Maattanen, M. Bergstrom, S. Euler, E. Wang, The sky is not the limit: LTE for unmanned aerial vehicles. IEEE Commun. Mag. **56**(4), 204–210 (2018)
12. N. Tadayon, G. Kaddoum, R. Noumeir, Inflight broadband connectivity using cellular networks. IEEE Access **4**, 1595–1606 (2016)
13. T. Cogalan, S. Videv, H. Haas, Operating an in-cabin femto-cellular system within a given LTE cellular network. IEEE Trans. Veh. Technol. **67**(8), 7677–7689 (2018)
14. M. Vondra, E. Dinc, M. Prytz, M. Frodigh, D. Schupke, M. Nilson, S. Hofmann, C. Cavdar, Performance study on seamless DA2GC for aircraft passengers toward 5G. IEEE Commun. Mag. **55**(11), 194–201 (2017)
15. L. Liu, Performance evaluation of direct air-to-ground communication using new radio (5G), M.S. thesis, School of Electrical Engineering, KTH, Aug 2017. Available at http://www.diva-portal.org/smash/get/diva2:1129315/FULLTEXT01.pdf. Accessed on 4 Nov 2020
16. ETSI TR 103 054, Broadband direct-air-to-ground communications operating in part of the frequency range from 790 MHz to 5 150 MHz, V1.1.1, July 2010. Available at https://www.etsi.org/deliver/etsi_tr/103000_103099/103054/01.01.01_60/tr_103054v010101p.pdf. Accessed on 4 Nov 2020
17. ETSI TR 101 099, Broadband direct-air-to-ground communications system employing beamforming antennas, Operating in the 2,4 GHz and 5,8 GHz bands, V1.1.3, Sept 2012. Available at https://www.etsi.org/deliver/etsi_tr/101500_101599/101599/01.01.03_60/tr_101599v010103p.pdf. Accessed on 4 Nov 2020
18. ETSI TR 103 108, Broadband direct-air-to-ground communications system operating in the 5,855 GHz to 5,875 GHz band using 3G technology, V1.1.1, July 2013. Available at https://www.etsi.org/deliver/etsi_tr/103100_103199/103108/01.01.01_60/tr_103108v010101p.pdf. Accessed on 4 Nov 2020
19. ECC Report 214, Broadband direct-air-to-ground communications (DA2GC), May 2014. Available at https://docdb.cept.org/download/27d4b5f0-025c/ECCREP214.PDF. Accessed on 4 Nov 2020
20. TR 38.901, Study on channel model for frequencies from 0.5 to 100 GHz (Release 16), V16.0.0, Sept 2019. Available at http://www.3gpp.org/ftp//Specs/archive/38_series/38.901/38901-g00.zip. Accessed on 4 Nov 2020

Chapter 18
5G New Radio Evolution Meets Satellite Communications: Opportunities, Challenges, and Solutions

Xingqin Lin, Björn Hofström, Y.-P. Eric Wang, Gino Masini, Helka-Liina Maattanen, Henrik Rydén, Jonas Sedin, Magnus Stattin, Olof Liberg, Sebastian Euler, Siva Muruganathan, Stefan Eriksson Löwenmark, and Talha Khan

1 Introduction

The 3rd Generation Partnership Project (3GPP) has in its Release 15 developed the fifth-generation (5G) wireless access technology, known as the 5G system (5GS). 5GS defines the 5G core (5GC) and the new radio (NR) air interface, which features spectrum flexibility, ultra-lean design, forward compatibility, low latency support, and advanced antenna technologies [1]. Built on the first release of NR, the evolutions of NR including 3GPP Release 16 and the ongoing Release 17 are bringing additional capabilities to improve performance and address new use cases. Meanwhile, we are witnessing a resurgent interest in providing connectivity from space. In the past few years, there has been a surge of proposals about using large constellations of low earth orbit (LEO) satellites, such as *OneWeb* and *SpaceX*, to provide broadband access. There are also several European H2020 research projects dedicated to integration of satellite and 5G networks such as Shared Access

X. Lin (✉) · Y.-P. E. Wang · T. Khan
Ericsson, Santa Clara, CA, USA
e-mail: xingqin.lin@ericsson.com

B. Hofström
Ericsson, Linköping, Sweden

G. Masini
DU Radio - Systems and Technologies, Ericsson AB, Stockholm, Sweden; 3GPP RAN3 Chairman

H. Rydén, · J. Sedin, · M. Stattin, · O. Liberg, · S. Euler, · S. E. Löwenmark
Ericsson, Stockholm, Sweden

H.-L. Maattanen
Ericsson, Helsinki, Finland

S. Muruganathan
Ericsson, Ottawa, Canada

X. Lin, N. Lee (eds.), *5G and Beyond*, https://doi.org/10.1007/978-3-030-58197-8_18

Terrestrial-Satellite Backhaul Network Enabled by Smart Antennas (SANSA) and Virtualized Hybrid Satellite-Terrestrial Systems for Resilient and Flexible Future Networks (VITAL) [2].

The ambition of providing connectivity from space is not new. A series of satellite communications projects (e.g., *Iridium* and *Globalstar*) were launched in the 1990s, but the services were limited to voice and low-data-rate communications. A resurgence of interest in providing connectivity from space started around 2014, stimulated by technology advancement and demand for ubiquitous connectivity services. The advancement of microelectronics following Moore's law has paved the way for using advanced technologies in satellite communications such as multi-spot beam technologies, onboard digital processing, and advanced modulation and coding schemes. Meanwhile, the development cycle and the costs of satellite manufacturing and launching processes have been dramatically reduced.

A major driver of the success of terrestrial mobile networks over the past few decades has been the international standardization effort, which yields the benefits of significant economies of scale. 3GPP has been the dominating standardization development body for several generations of mobile technology. The international standardization effort helps ensure compatibility among vendors and reduce network operation and device costs. In contrast, the interoperability between different satellite solution vendors has been difficult to achieve, and the availability of devices is limited [3].

Already in 2014 it was pointed out that the satellite community must work closely with the mobile 5G community to realize the integration of satellite and 5G networks [4]. Indeed, the satellite industry has realized the need to embrace standardization and furthermore to join forces with the mobile industry in 3GPP. The ongoing evolution of 5G standards provides a unique opportunity to revisit satellite communications. The satellite work in 3GPP is commonly known as non-terrestrial networks (NTN). The objective is to achieve full integration of NTN in 5G by evolving the 5GC and the next-generation radio access network (NG-RAN) protocols and functions including the NR radio interface to support NTN [3].

3GPP Technical Specification Group (TSG) Radio Access Network (RAN) has completed a first NTN study in its Release 15, focusing on channel models, deployment scenarios, and identifying potential key impacts on NR [5]. 3GPP TSG RAN has also conducted a Release 16 NTN study to define and evaluate solutions for the identified key impacts [6]. 3GPP TSG Service and System Aspects (SA) working groups have also completed a study that identifies use cases and requirements when using satellite access in 5G [7, 8]. 3GPP TSG SA is currently conducting further studies on integrating satellite access in 5G including architecture aspects [9] and management and orchestration aspects [10].

In this chapter, we focus on the NR radio access network and study how to adapt the NR air interface for satellite links. The overview provided in this chapter covers the 3GPP state-of-the-art findings (including the recently completed 3GPP Release 16) on NTN in 3GPP TSG RAN. Integrating satellite with 5G networks can also occur in the core network based on recent developments in network softwarization using the tools such as software-defined networking (SDN), network functions virtualization (NFV), and network slicing. We refer interested readers to the recent *IEEE Network* special issue [11] for more comprehensive treatments on the integration of satellite and 5G networks.

This chapter is an accessible reference for researchers interested in learning the latest 3GPP findings on satellite access in 5G. There are several other works addressing the same topic [12–14]. The work [12] focused on using a LEO satellite system to provide backhaul connectivity to terrestrial 5G relay nodes. The work [13] discussed and assessed the impact of the satellite channel characteristics on the physical and medium access control (MAC) layers. The work [14] provided details on higher-layer standardization aspects for both connected mode and idle mode mobility as well as network architecture aspects in both GEO and non-GEO NTN systems. The discussions in this chapter are along similar lines and provide further insights by presenting link budget analysis and new system simulation results on path gain distribution, geometry signal-to-interference ratio (SIR) distribution, and packet delay distribution. In addition, this chapter provides solutions to the identified challenges for adapting NR air interface for satellite links.

2 Use Cases of Satellite Communications

2.1 Introduction to Satellite Communications Use Cases

Satellite access networks have been playing a complementary role in the mobile communications ecosystem. Satellite links can provide direct connectivity to user equipment (UE) or indirectly serve a UE by providing backhaul connectivity to terrestrial base stations or via relay nodes.

Despite the wide deployment of terrestrial mobile networks, there are unserved or underserved areas around the globe. Satellite access networks can augment the terrestrial networks to provide connectivity in rural and remote areas. Satellite connectivity can also be used for backhauling, fostering the rollout of 5G services.

Satellite access networks can benefit communication scenarios with airborne and maritime platforms (onboard aircrafts or vessels) while being attractive in certain machine-to-machine and telemetry applications. In case of natural disasters which disrupt terrestrial communications systems and services in some areas, satellites can help quickly restore the communications network in the affected areas by leveraging their wide coverage to enable rapid response in emergency situations.

Satellite communication is well positioned for broadcasting/multicasting data and media to a broad audience spread over a large geographical area. While television broadcasting has undoubtedly been the main satellite service in this area, there are other use cases as well. For instance, mobile operators and Internet service providers can utilize satellite communications to multicast content to the network edge to facilitate content caching for local distribution.

2.2 Understanding the Use Cases by Link Budget Analysis

To get a more concrete understanding of the use cases of satellite communications, we carry out a link budget analysis based on the assumptions in 3GPP Release 16 NTN study [6]:

- A LEO satellite operating in the S-band, with both the nominal downlink and uplink carrier frequencies of 2 GHz. The system bandwidth is 30 MHz. For the satellite, the effective isotropic radiated power (EIRP) is 34 dBW/MHz, and the antenna gain-to-noise-temperature (G/T) is 1.1 dB/K. The UE is assumed to be a handheld terminal with 23 dBm EIRP. The UE has two cross-polarized antenna elements and the G/T equals -31.6 dB/K.
- A LEO satellite operating in the Ka-band, with the nominal downlink and uplink carrier frequencies of 20 GHz and 30 GHz, respectively. The system bandwidth is 400 MHz. For the satellite, the EIRP is 4 dBW/MHz and the antenna G/T is 13 dB/K. The UE is assumed to be a very small aperture terminal (VSAT) with 76.2 dBm EIRP and G/T of 15.9 dB/K.

Table 18.1 presents the link budget calculation results assuming an orbit altitude of 600 km and an elevation angle of 30°. In S-band downlink with 30 MHz bandwidth, the signal-to-noise ratio (SNR) is 8.9 dB, which according to the Shannon formula can yield a spectral efficiency of 3.1 bps/Hz and a total throughput of 93.9 Mbps. In Ka-band downlink with 400 MHz bandwidth, the SNR is 9.4 dB, which can yield a spectral efficiency of 3.3 bps/Hz and a total throughput of 1.32 Gbps.

In S-band uplink, the handheld UE uses 180 kHz bandwidth to obtain the 8.1 dB SNR. The corresponding spectral efficiency is 2.9 bps/Hz and the achieved data rate is 0.52 Mbps. In Ka-band uplink, the VSAT with high transmit power and high gain antenna can use the whole 400 MHz bandwidth and achieves 19.3 dB SNR. The

Table 18.1 Link budget calculation for LEO with 600 km orbital height based on Set-1 satellite parameters in 3GPP TR 38.821. Note that the bandwidth values in this table are units for normalizing EIRP and are not system bandwidths

Link budget	S-band		Ka-band	
System	Downlink	Uplink	Downlink	Uplink
TX: EIRP/bandwidth	56.6	23.0	26.6	76.2
RX: G/T [dB/T]	-31.6	1.1	15.9	13.1
Bandwidth [Hz]	180000	180000	180000	400e6
Free space path loss (PL) [dB]	159.1	159.1	179.1	182.6
Atmospheric loss (LA)	0	0	0	0
Shadow fading margin (SF) [dB]	3	3	0	0
Polarization loss [dB]	0	0	0	0
Additional losses (AD) [dB]	0	0	0	0
SNR [dB]	8.9	8.1	9.4	19.3

corresponding spectral efficiency is 6.4 bps/Hz and the achieved data rate is 2.57 Gbps.

The above results show that LEO satellite can support use cases with medium-high data rate requirements.

3 A Primer on Satellite Communications

In this section, we provide a primer on satellite communications. We refer interested readers to [15] for a more in-depth introduction to satellite communications.

3.1 Satellite Communications System Architecture

Besides LEO satellite, there are other types of NTN platforms including geosynchronous earth orbit (GEO) satellite, medium earth orbit (MEO) satellite, high elliptical orbit (HEO) satellite, and high-altitude platform station (HAPS). A satellite communications system may consist of the following components [5]: satellite, terminal, gateway, feeder link, service link, and inter-satellite link. An illustration is given in Fig. 18.1. Depending on the implemented functionality of the communication payload of the satellite in the system, we can consider two payload options: *bent-pipe* transponder and *regenerative* transponder. With a bent-pipe transponder, the satellite receives signals from the earth, amplifies the received signals, and retransmits the signals to the earth after frequency conversion. With a regenerative transponder, the satellite performs onboard processing to demodulate and decode the received signals and regenerates the signals for further transmission.

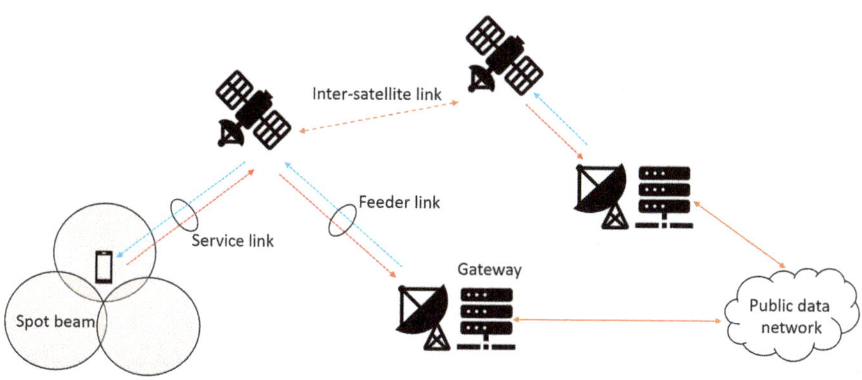

Fig. 18.1 An illustration of satellite communications system architecture

A modern satellite typically uses multi-spot beam technology to generate multiple high-power beams to cover a geographical area. For a non-geostationary satellite, the footprint may sweep over the earth's surface with the satellite movement or may be earth fixed with some beam pointing mechanism used by the satellite to compensate for its motion. The radii of the spot beams depend on the satellite communications system design and may range from tens of kilometers to a few thousands of kilometers.

Low latency is a key 5G requirement. While satellite communications systems by nature cannot provide ultra-low latency (e.g., 1 ms), a mega LEO constellation communication system with a sufficient number of gateways properly distributed over large geographic areas can offer low latency (on the order of tens of milliseconds) across long distances. Leveraging inter-satellite links can further reduce the latency.

3.2 *Example System-Level Simulation Results*

To get some intuition on path gain distribution and SIR distribution in satellite access networks, we present example system-level simulation results for a LEO communications system with 600 km orbital height and 2 GHz carrier frequency at an elevation angle of 90°. The satellite antenna pattern is generated based on a typical parabolic reflector antenna with a circular aperture, as described in the 3GPP TR 38.811 [5]. The diameter of the satellite antenna aperture is 2 m. The UE has two cross-polarized antenna elements, and each antenna element has 0 dBi antenna gain. The satellite creates a hexagonal pattern of spot beams on the ground. The maps in Fig. 18.2 show the center area of this pattern.

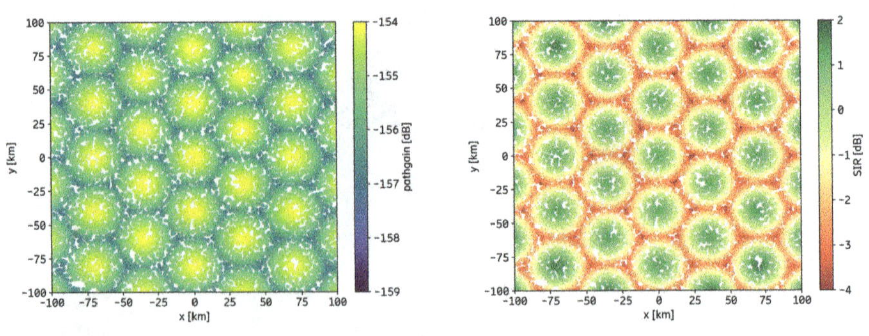

(a) (b)

Fig. 18.2 Example system-level simulation results for a LEO communications system with 600 km orbital height and 2 GHz carrier frequency at 90° elevation angle: subfigure (**a**) shows path gain and subfigure (**b**) shows geometry SIR

Figure 18.2a shows the path gain distribution over the simulation area. Path gain here is the sum of free-space path loss and normalized antenna gains. It can be seen that the path gain range is less than 5 dB, which is much smaller when compared to a macro terrestrial network where the path gain range may span over 100 dB.

Figure 18.2b shows the geometry SIR distribution over the same simulation area. Geometry SIR is a measure of the satellite to UE signal quality in a fully loaded network. It can be seen that the geometry SIR range is comparable to that of a macro terrestrial network.

3.3 Varying Coverage in Time and Space

The coverage of a GEO satellite is quite static, with infrequent updates of spot beam pointing directions to compensate for the GEO satellite movement. In contrast, the movements of non-GEO satellites, especially LEO satellites, lead to a varying coverage in time and space [5]. A typical LEO satellite is visible to a ground UE for a few minutes only. This implies that even in a LEO satellite communications system with earth-fixed beams, the serving spot beam associated with the serving satellite for a fixed position on the ground changes every few minutes. In a LEO satellite communications system with moving beams, from the perspective of a fixed position on the ground, a typical spot beam with a radius of tens of kilometers may serve the position for less than 1 minute before another spot beam starts to cover the position. The serving satellite stays the same if the consecutive spot beams covering the position are generated by the same satellite.

Figure 18.3 illustrates the varying coverage in LEO satellite communications with polar orbiting satellites for three different heights. Figure 18.3a shows satellite elevation angle trajectories observed by a static reference UE #0 as a function of

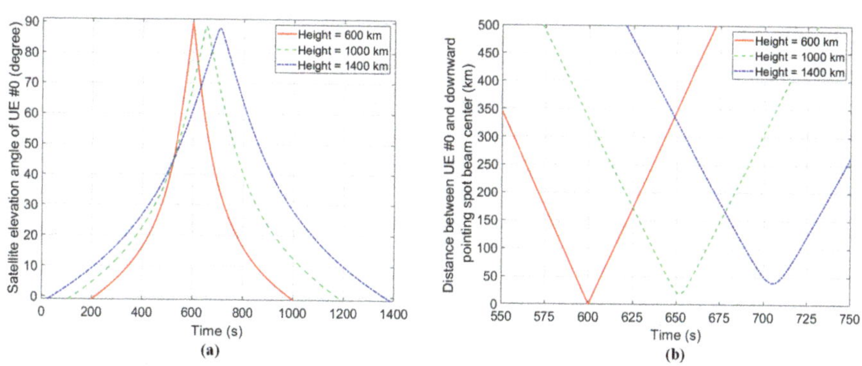

Fig. 18.3 Varying coverage in satellite communications with polar orbiting satellites at three different orbital heights: subfigure (**a**) shows satellite elevation angle trajectories of a static reference UE #0 as a function of time; subfigure (**b**) shows the trajectories of the distance between the reference UE #0 and the center of the downward pointing spot beam as a function of time

time. Assuming a typical 10° minimum satellite elevation angle for service link connection, the UE can stay connected to the satellite passing at 600 km height above for only about 450 s. Figure 18.3b shows the trajectories of the distance between the reference UE #0 and the center of downward pointing spot beam as a function of time. If the spot beam radius is 50 km, the spot beam from the satellite at the height of 600 km covers the UE for only about 15 s.

3.4 Propagation Delays

Rapid interactions between a UE and its serving base station in a terrestrial mobile communications system are possible since the propagation delay is usually within 1 ms. In contrast, the propagation delay in a satellite link is much longer [5]. The one-way propagation time between a GEO satellite and a ground UE is 119.3 ms, assuming that the radio signal propagates at the speed of light in a vacuum and that the UE is immediately underneath the GEO satellite. With 600 km LEO satellite height, the minimum service link propagation delay is 2 ms attained at 90° satellite elevation angle. The propagation delay may increase to ~6.5 ms at 10° satellite elevation angle.

Differential delay, which refers to the propagation delay difference of two selected points in the same spot beam, is of interest as it impacts the multi-access scheme. Since the feeder link is shared by the devices in the same spot beam, the differential delay mainly depends on the size of the spot beam, which results in different path lengths of the service links.

3.5 Doppler Effects

In terrestrial mobile communications systems, Doppler effects are typically caused by the movements of the UE and surrounding objects, while in satellite systems the satellite movement induces additional Doppler effects [5]. Doppler effect is pronounced in LEO systems. At the height of 600 km, a LEO satellite moves at the speed of 7.56 km/s, which can result in a Doppler shift value as large as about 48 kHz at the carrier frequency of 2 GHz. In addition, the Doppler shift value varies rapidly over time, and the rate of such variation is referred to as the Doppler variation rate.

The Doppler effects due to satellite movements in GEO systems in most cases can be negligible. Note that when a satellite is in near GEO orbit with inclination up to 6° relative to the equatorial plane, the Doppler shift can reach around 300 Hz at the carrier frequency of 2 GHz [5]. Terrestrial mobile technologies such as NR have been designed to handle this order of magnitude of Doppler shift values. For a satellite communications system operating at a higher frequency, the Doppler shift increases proportionally. NR supports a flexible numerology, which can handle the increased Doppler shift with increased carrier frequency. For example, 15 kHz subcarrier spacing in NR can be used to handle the ~300 Hz Doppler shift in the GEO satellite communications system operating at 2 GHz. The Doppler shift would increase to ~3000 Hz if the GEO satellite communications system would operate at 20 GHz. In this case, 120 kHz subcarrier spacing in NR can be used to handle the increased Doppler shift.

4 Design Aspects of NR over Satellite Links

In this section, we describe several key areas that require adaptation to evolve NR for satellite communications.

4.1 Uplink Timing Control

Problems

NR utilizes orthogonal frequency-division multiple access (OFDMA) as its multi-access scheme in the uplink. The transmissions from different UEs in a cell are time-aligned at the 5G NodeB (gNB) to maintain uplink orthogonality. Time alignment is achieved by using different timing advance values at different UEs to compensate for their different propagation delays. The gNB can estimate the timing advance value based on the UE's uplink signals such as the physical random-access channel (PRACH) preamble. The existing NR uplink timing control scheme, however, has not been designed to handle the large propagation delays incurred in satellite communications.

Potential Solutions

One promising approach is to rely on global navigation satellite system (GNSS)-based techniques. Each UE equipped with a GNSS chipset determines its position, calculates its propagation delay with respect to the serving satellite using ephemeris data of the satellite constellation, and derives the initial timing advance value. The UE then uses its initial timing advance value to initiate the random-access procedure, which can help to further refine the timing advance to cope with a residual timing error associated with the initial timing advance estimate.

Some low-cost, reduced complexity UEs may not be equipped with GNSS chipsets. Thus, non-GNSS-based techniques are also needed. One possible technique may work as follows. For each spot beam, the gNB may choose a reference point such as the center of the spot beam and adjust its uplink receiver timing with respect to the reference point. With this approach, the uplink timing control only needs to handle the delay difference between each UE and the reference point instead of the much larger absolute propagation delays. The existing uplink timing control can be directly used for spot beams with radii up to about 200–300 km. For spot beams with radii larger than 300 km, further adaptation of uplink timing control design may be needed.

4.2 *Frequency Synchronization*

Problems

NR uses orthogonal frequency-division multiplexing (OFDM) for both downlink and uplink transmissions and additionally supports the use of discrete Fourier transform (DFT) spread OFDM in the uplink. Maintaining the orthogonality of OFDM requires tight frequency synchronization between transmitter and receiver. The downlink synchronization can be treated as a point-to-point OFDM synchronization problem since each receiver in a cell tunes its downlink reference frequency based on the received synchronization signals. The uplink synchronization is more challenging since it is a multipoint-to-point synchronization problem in OFDMA-based NR. The transmissions from different UEs in a cell need to be frequency-aligned at the gNB to maintain uplink orthogonality. Therefore, different frequency adjustment values at different UEs are needed in the uplink to compensate for their different Doppler shifts.

Potential Solutions

GNSS-based techniques can be used for uplink frequency adjustment: Each UE equipped with a GNSS chipset determines its position and calculates its frequency adjustment value based on the estimated Doppler shift using its position information, satellite ephemeris data, and carrier frequencies. To mitigate the effects of large Doppler shifts due to satellite movements in non-GEO satellite communications systems, pre-compensation can be applied to forward link signals [5]: A time-varying frequency offset tracking the Doppler shift is applied to the forward link reference frequency such that the forward link signals for a spot beam received at a reference point in the spot beam appear to have zero Doppler shift. With pre-compensation, the Doppler shift of the forward link signals received at a given location in the spot beam becomes equal to the difference between the original Doppler shifts of the given location and the reference point.

The Doppler shift differences at different locations in the spot beam however are different and time-varying. Consider a spot beam with 100 km radius in a LEO satellite system with 600 km orbital height. Even if frequency pre-compensation is applied to the downlink of the service link with respect to the center of the spot beam, the Doppler shift difference of a point at the edge of the spot beam and a reference point in the center of the spot beam can still be as large as 8 kHz at 2 GHz carrier frequency.

For non-GNSS-based frequency adjustment techniques, the gNB may estimate the return link frequency shift of each UE and transmit is a corresponding frequency adjustment command to the UE. To establish the uplink orthogonality as early as possible, it is desirable that the gNB estimates the uplink frequency shift from the random-access preamble transmitted by the UE and includes the frequency adjustment command in the random-access response message. The challenge is that the gNB has to estimate both timing advance and frequency adjustment values based on the PRACH preamble.

The existing NR PRACH preambles are based on Zadoff-Chu sequences. In case of large timing and frequency uncertainties, there are several peaks in the ambiguity function of Zadoff-Chu sequences in the delay-Doppler plane, leading to timing and frequency ambiguities. In other words, due to the nature of Zadoff-Chu sequences, both timing delay and frequency shift cause cyclic shift in the observation window of the received Zadoff-Chu sequence at the gNB. One potential solution is to transmit not only the Zadoff-Chu sequence but also its complex conjugate R1-1912725, "On NTN synchronization, random access, and timing advance," Ericsson, 3GPP TSG-RAN WG1 Meeting #99, Reno, November 2019. The gNB can then detect a composite cyclic shift from the first transmission and another composite cyclic shift from the second transmission. Based on the two composite cyclic shifts, the effects of delay and frequency shifts can be separated.

4.3 Hybrid Automatic Repeat Request

Problems

To combat transmission errors, NR uses a combination of forward error correction and automatic repeat request (ARQ), which is known as hybrid ARQ (HARQ). NR supports 16 HARQ processes with stop-and-wait protocols per component carrier in both uplink and downlink. In a stop-and-wait protocol, the transmitter stops and waits for acknowledgement after each (re)transmission. Using 16 HARQ processes with stop-and-wait protocols would lead to significant throughput reduction especially in GEO communications systems [5].

Potential Solutions

One straightforward approach is to increase the number of HARQ processes to cope with the increased round-trip delays in satellite communications systems. This, however, comes at the cost of UE implementation complexity due to the increased UE HARQ soft buffer size. Another approach is to introduce a mechanism in NR to support the possibility of turning off retransmissions in HARQ processes.

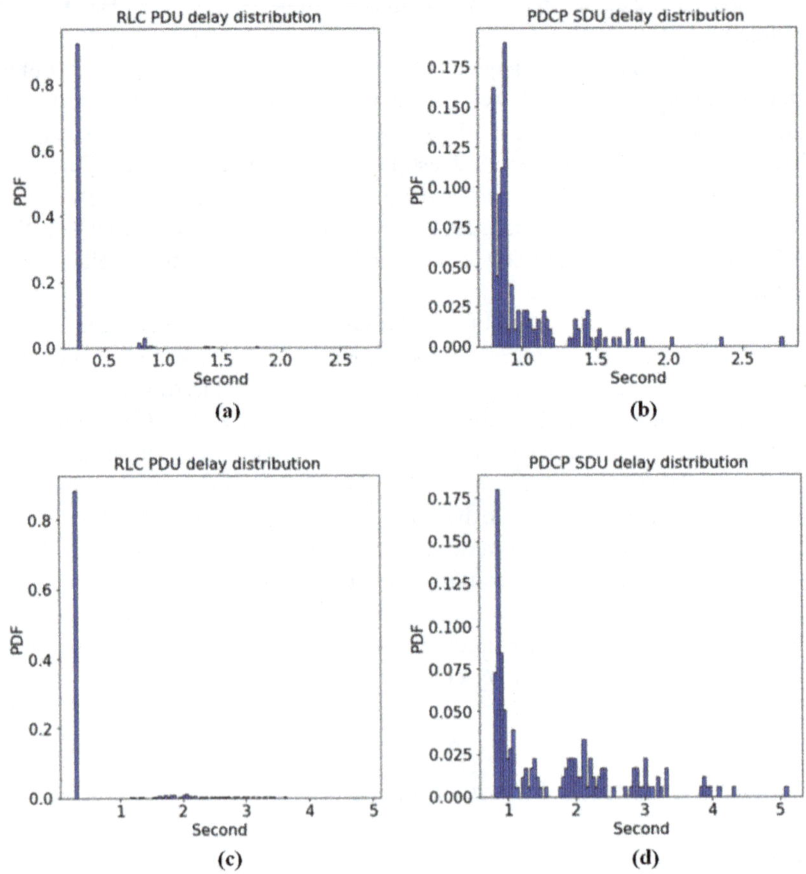

Fig. 18.4 Simulated delay distributions for a traffic model characterized by periodic packet arrival (1 kilobyte per second), 256 ms one-way propagation delay, RLC acknowledged mode, and good link quality: subfigures (**a**) and (**b**), respectively, show the delay distributions of RLC PDU and PDCP SDU when HARQ is used; subfigures (**c**) and (**d**), respectively, show the delay distributions of RLC PDU and PDCP SDU when HARQ is not used

Instead, the retransmissions are handled by the layers above MAC if error-free data units are required at the receiver. For example, the radio link control (RLC) layer supports an acknowledged mode that may be used for the retransmission of erroneous data. Retransmissions at the layers above MAC may lead to increased latency due to the slower feedback. Additionally, the communications system may need to operate with more conservative coding rate in the physical layer to avoid excessive retransmissions in the layers above MAC.

Figure 18.4 shows an example of simulated delay distributions for a traffic model defined by periodic packet arrival. The delay of RLC protocol data units (PDU) includes the delays in physical layer and MAC layer. The delay of packet data convergence protocol (PDCP) service data units (SDU) includes the delays in physical layer, MAC layer, RLC layer, and PDCP layer. Figure 18.4a, b, show the delay distributions of RLC PDU and PDCP SDU when HARQ is used and the transport block size is 1000 bits. Due to the good link quality, most of the RLC PDUs are successfully received without HARQ retransmission and thus have a ~256 ms delay, while a small fraction of the RLC PDUs are successfully received with one HARQ retransmission and thus have a ~800 ms delay. Accordingly, most of the PDCP SDUs have delays in the range of 0.8–2 s. In contrast, when HARQ is not used, Figure 18.4c, d, show the delay distributions of RLC PDU and PDCP SDU. While most of the RLC PDUs have ~256 ms delay due to the good link quality, small fraction of the RLC PDUs has a delay greater than ~1.25 s. Accordingly, compared to the case with HARQ, more PDCP SDUs have delays greater than ~2 s when HARQ is not used. This is because HARQ the feedback is faster than feedback in higher-layer protocols.

4.4 Idle Mode UE Tracking and Paging

Problems

When a UE in idle mode selects or reselects a cell, it reads the broadcast system information to learn which tracking area the cell belongs to. If the cell does not belong to any of the tracking areas to which the UE is registered, the UE performs a tracking area update to notify the network of the tracking area of the cell it is currently camping on. For a GEO satellite communications system where the cell's coverage area is usually fixed on the ground, the existing UE tracking and paging procedures in NR can be largely reused. However, for a non-GEO satellite communications system with moving beams, the cell's coverage area moves on the ground. Under the existing UE tracking and paging procedures in NR, the tracking area sweeps over the ground as well. As a result, a stationary UE would have to keep performing location registration in idle mode [6].

Potential Solutions

One potential solution to tracking idle mode UEs in non-GEO satellite communications system with moving beams would be to decouple the tracking area management from the moving cell pattern. While the beams and cells are moving, the registration and tracking areas are fixed on the ground. This implies that while the cells sweep over the ground, the broadcasted tracking area is changed when the cell enters the area of the next earth-fixed tracking area location [14].

5 Conclusions

It is an interesting time to witness the increasing interest in satellite communications. The ongoing evolution of 5G standards provides a unique opportunity to revisit satellite communications. Though NR has been designed mainly targeting terrestrial mobile communications, the inherent flexibility of NR allows it to be evolved to support NTN. As this chapter has highlighted, when adapting NR to support satellite communications, there are challenges including long propagation delays, large Doppler shifts, and moving cells. Addressing such challenges requires a rethinking of many of the working assumptions and models used to date for designing NR. Throughout this chapter, we have attempted to highlight ideas on how to overcome the key technical challenges faced by NR evolution for satellite communications in the 3GPP Release 16 NTN study.

References

1. X. Lin et al., 5G New Radio: Unveiling the Essentials of the Next Generation Wireless Access Technology, in IEEE Communications Standards Magazine, **3**(3), 30–37, September (2019)
2. L. Boero, R. Bruschi, F. Davoli, M. Marchese, F. Patrone, Satellite networking integration in the 5G ecosystem: Research trends and open challenges. IEEE Netw. **32**(5), 9–15 (2018)
3. EMEA Satellite Operators Association, ESOA satellite action plan for 5g standards, *white paper*. Available at https://esoa.net/cms-data/positions/1771%20ESOA%205G%20standards.pdf. Last Accessed on 09-09-2019
4. B. G. Evans, The role of satellites in 5G, 7th advanced satellite multimedia systems conference and the 13th signal processing for space communications workshop (ASMS/SPSC), Livorno, (2014), pp. 197–202
5. 3GPP TR 38.811, Study on new radio (NR) to support non-terrestrial networks, V15.0.0, August 2018. Available at http://www.3gpp.org/ftp//Specs/archive/38_series/38.811/38811-f00.zip. Last Accessed on 09-09-2019
6. 3GPP TR 38.821, Solutions for NR to support non-terrestrial networks (NTN) (Release 16), V0.6.0, May 2019. Available at http://ftp.3gpp.org//Specs/archive/38_series/38.821/38821-060.zip. Last Accessed on 09-09-2019
7. 3GPP TR 22.822, Study on using satellite access in 5G, V16.0.0, August 2018. Available at http://www.3gpp.org/ftp//Specs/archive/22_series/22.822/22822-g00.zip. Last Accessed on 03-25-2019. Last Accessed on 09-09-2019
8. 3GPP TS 26.261, Service requirements for the 5G system; Stage 1 (Release 16)," V0.0.1, April 2019. Available at http://www.3gpp.org/ftp//Specs/archive/26_series/26.261/26261-001.zip. Last Accessed on 09-09-2019
9. 3GPP TR 23.737, Study on architecture aspects for using satellite access in 5G (Release 16), V0.8.0, June 2019. Available at http://www.3gpp.org/ftp//Specs/archive/23_series/23.737/23737-080.zip. Last Accessed on 09-09-2019
10. 3GPP TR 28.808, Study on management and orchestration aspects of integrated satellite components in a 5G network, V0.1.0, April 2019. Available at http://www.3gpp.org/ftp//Specs/archive/28_series/28.808/28808-010.zip. Last Accessed on 09-09-2019
11. T. de Cola, A. Perez-Neira, R. Channasandra, S. Covaci, Guest editorial – integration of satellite and 5G networks. IEEE Netw. **32**(5), 6–8 (2018)
12. O. Kodheli, A. Guidotti and A. Vanelli-Coralli, Integration of satellites in 5G through LEO constellations, *IEEE Global Communications Conference*, Singapore, (2017), pp. 1–6

13. A. Guidotti et al., Architectures and key technical challenges for 5G systems incorporating satellites. IEEE Trans. Veh. Technol. **68**(3), 2624–2639 (2019)
14. H.-L. Määttänen et al. 5G NR communication over GEO or LEO satellite systems: 3GPP RAN higher layer standardization aspects, *2019 IEEE Global Communications Conference (GLOBECOM)*, Hawaii, USA, (2019)
15. G. Maral, M. Bousquet, *Satellite communications systems: Systems, techniques and technology*. (John Wiley & Sons, 2011)

Index

X. Lin, N. Lee (eds.), *5G and Beyond*, https://doi.org/10.1007/978-3-030-58197-8

O

P

GPSR Compliance
The European Union's (EU) General Product Safety Regulation (GPSR) is a set of rules that requires consumer products to be safe and our obligations to ensure this.

If you have any concerns about our products, you can contact us on

ProductSafety@springernature.com

In case Publisher is established outside the EU, the EU authorized representative is:

Springer Nature Customer Service Center GmbH
Europaplatz 3
69115 Heidelberg, Germany

www.ingramcontent.com/pod-product-compliance
Lightning Source LLC
Chambersburg PA
CBHW072005011225
36163CB00003B/50
9783030581961